Introduction to Statistical Signal Processing with Applications

PRENTICE HALL INFORMATION AND SYSTEM SCIENCES SERIES

Thomas Kailath, Editor

ANDERSON & MOORE	*Optimal Control: Linear Quadratic Methods*
ANDERSON & MOORE	*Optimal Filtering*
ASTROM & WITTENMARK	*Computer-Controlled Systems: Theory and Design, 2/E*
BASSEVILLE & NIKIROV	*Detection of Abrupt Changes: Theory & Application*
BOYD & BARRATT	*Linear Controller Design: Limits of Performance*
DICKINSON	*Systems: Analysis, Design and Computation*
FRIEDLAND	*Advanced Control System Design*
GARDNER	*Statistical Spectral Analysis: A Nonprobabilistic Theory*
GRAY & DAVISSON	*Random Processes: A Mathematical Approach for Engineers*
GREEN & LIMEBEER	*Linear Robust Control*
HAYKIN	*Adaptive Filter Theory*
HAYKIN	*Blind Deconvolution*
JAIN	*Fundamentals of Digital Image Processing*
JOHANSSON	*Modeling and System Identification*
JOHNSON	*Lectures on Adaptive Parameter Estimation*
KAILATH	*Linear Systems*
KUNG	*VLSI Array Processors*
KUNG, WHITEHOUSE, & KAILATH, EDS.	*VLSI and Modern Signal Processing*
KWAKERNAAK & SIVAN	*Signals and Systems*
LANDAU	*System Identification and Control Design Using P.I.M. + Software*
LJUNG	*System Identification: Theory for the User*
LJUNG & GLAD	*Modeling of Dynamic Systems*
MACOVSKI	*Medical Imaging Systems*
MOSCA	*Stochastic and Predictive Adaptive Control*
NARENDRA & ANNASWAMY	*Stable Adaptive Systems*
RUGH	*Linear System Theory*
RUGH	*Linear System Theory, Second Edition*
SASTRY & BODSON	*Adaptive Control: Stability, Convergence, and Robustness*
SOLIMAN & SRINATH	*Continuous and Discrete-Time Signals and Systems*
SOLO & KONG	*Adaptive Signal Processing Algorithms: Stability & Performance*
SRINATH, RAJASEKARAN, & VISWANATHAN	*Introduction to Statistical Signal Processing with Applications*
VISWANADHAM & NARAHARI	*Performance Modeling of Automated Manufacturing Systems*
WILLIAMS	*Designing Digital Filters*

Introduction to Statistical Signal Processing with Applications

M.D. Srinath

Southern Methodist University
Dallas, Texas

P.K. Rajasekaran

Texas Instruments, Inc.
Dallas, Texas

R. Viswanathan

Southern Illinois University
Carbondale, Illinois

Prentice Hall
Englewood Cliffs, New Jersey 07632

Library of Congress Cataloging-in-Publication Data

Srinath, Mandyam D. (Mandyam Dhati),
An introduction to statistical signal processing with applications / M.D. Srinath, P.K. Rajasekaran, R. Viswanathan.
p. cm. — (Prentice Hall information and system sciences series)
Includes bibliographical references and index.
ISBN 0-13-125295-X
1. Signal processing—Statistical methods. I. Rajasekaran, P. K., 1942– . II. Viswanathan, R. (Ramanarayanan), 1953– III. Title. IV. Series.
TK5102.9.S69 1996
621.382′2—dc20 94-45580
CIP

Acquisitions editor: Tom Robbins
Production editor: Irwin Zucker
Buyer: Julia Meehan
Cover design: Bruce Kenselaar
Editorial assistant: Phyllis Morgan

Printed in the United States of America

10 9 8 7 6 5 4 3 2 1

ISBN 0-13-125295-X

Prentice-Hall International (UK) Limited, *London*
Prentice-Hall of Australia Pty. Limited, *Sydney*
Prentice-Hall Canada Inc., *Toronto*
Prentice-Hall Hispanoamericana, S.A., *Mexico*
Prentice-Hall of India Private Limited, *New Delhi*
Prentice-Hall of Japan, Inc., *Tokyo*
Simon & Schuster Asia Pte. Ltd., *Singapore*
Editora Prentice-Hall do Brasil, Ltda., *Rio de Janeiro*

Contents

Preface

Signal processing may broadly be considered to involve the recovery of information from physical observations. In the presence of random disturbances, we must resort to statistical techniques for the recovery of the information. Applications of statistical signal processing arise in a wide variety of areas, such as communications, radio location of objects, seismic signal processing, and computer-aided medical diagnosis.

This book is intended to serve as a first-year-graduate-level text in statistical signal processing and aims at covering certain basic techniques in the processing of stochastic signals and illustrating their use in a few specific applications. The book is an expanded and revised version of the text by M.D.S. and P.K.R. published by Wiley-Interscience in 1979, and seeks to take into account new developments and shifting emphases in the field over the last few years. As noted in the preface to the original edition, the experience of the authors in teaching graduate-level courses in communication theory, radar systems, system identification, pattern recognition, spectral analysis, adaptive filters, and related areas led to the realization that all these areas require an essentially common background in certain techniques of statistical signal processing. In particular, the topics of detection and estimation theory constitute a common foundation in many of these courses. Most available texts provide a treatment of these two topics in varying detail but usually cover only one of the application areas in depth. The need for a text that covers both detection and estimation in a clear and concise fashion, and illustrates their application to specific problems, is as evident today as it was when the book was originally published. As in the original edition, the particular applications chosen for illustrating the basic techniques are communications, radar, pattern recognition, and system identification and are by no means exhaustive. Many of the techniques in the book can be either applied directly or extended to other areas, such as speech or image processing. The basic techniques are covered in sufficient depth as to enable first-year graduate students and practicing engineers in different disciplines to acquire a background suitable for advanced study and a better understanding of the fields in which they are involved.

The background required to use this book is a course on systems theory and a course on stochastic processes. The mathematical rigor takes a middle path so as not to obscure intuitive reasoning. The first chapter presents an introduction and overview of the book. Chapter 2 is intended to serve as a quick review of certain results in systems theory and stochastic processes that are useful in later developments. Chapters 3 through 7 provide the background in detection and estimation theory that is the essence of the book. These chapters are required for the applications chapters. The applications considered in Chapters 8 and 9 can be covered selectively and in any order, depending on the interests of the reader. For a class with a strong interest in communications, a good suggestion is to review Chapter 2 quickly, followed by Chapters 3 through 7 on theory, and cover Chapter 8. On the other hand, for a class with strong interests in pattern recognition, or control, Chapter 9 may be emphasized. The exercises at the ends of chapters serve the dual purpose of obtaining a better understanding of the text material and of presenting results or applications not covered in the text.

We wish to express our gratitude to the many graduate students whom it has been our pleasure to have taught over the last several years. Interacting with these students has given us a deeper understanding of the material. In particular, R.V. thanks his students Arif Ansari and Chandrakanth for their help in preparing some of the illustrations. R. V. also thanks Dr. Glafkos Galanos, chair of Electric Engineering, SIUC, for providing necessary resources during the writing of this book.

M. D. SRINATH
P. K. RAJASEKARAN
R. VISWANATHAN

Dallas, Texas
Carbondale, Illinois

Introduction to Statistical Signal Processing with Applications

1

Introduction

Statistical signal processing has application in a wide variety of human activity, ranging from processing of seismic signals to computer-assisted medical diagnosis and treatment, tracking of objects in space, and traffic control. Signal processing generally involves the recovery of information from physical observations. The processing required is relatively simple if the observation contains the information explicitly and any interference present is exactly described. Often, the physical characteristics and limitations of the devices used for observation, and/or the media through which the information is observed or communicated, make this impossible. In fact, the interference is usually random in nature and can only be described in terms of its average properties or statistics. The processing of such an observation to recover information can be termed *statistical signal processing*. This book is devoted to this topic and some applications. The applications considered are by no means exhaustive and are meant only to introduce the reader to the utility of statistical signal processing techniques and provide a flavor for the wide variety of problems which can be solved using these techniques. Thus, only basic approaches are considered, and no discussion of the issues involved in the practical implementation of these techniques is given. A reader who is interested in any particular application is invited to refer to the several recent books dealing with specific areas (see for example, [1-10]).

To illustrate the types of problems that arise in statistical signal processing, let us consider the radio location of a flying object (target) in space. We can do this by beaming a packet of electromagnetic energy in the direction of the target and by observing the reflected electromagnetic wave. We have two problems to consider at this point. The primary problem is to decide whether there is an object present at all (the detection problem). If we decide that the object is present, we may desire to determine certain parameters associated with the object, say its range or velocity (the estimation problem). If there is no interference and if the reflected wave is not distorted through the transmission medium, the solution is straightforward. We monitor the reflected signal and observe the time delay τ between the transmitted and reflected waves, by noting the time at which a peak occurs. If the target is not present, there will be no peak. If a target is present, we can estimate its range as $R = \tau c/2$, where c is the velocity of propagation of the electromagnetic wave.

In the presence of interference (noise), the solution is no longer simple. The interference could be due to distortions through the transmission media or thermal noise in the measurement device. The effect of the interference is to mask the peak that we are monitoring. We may get a spurious peak when no target is present or we may not be able to discern a peak when a target is present. In either case, the presence of noise can cause erroneous decisions. Our problem is to monitor the signal for a certain length of time and decide about the presence of the object. This is the detection problem and falls into the general topic of statistical decision making.

If we decide that a target is present and seek to determine its range by observing the delay, we may still have difficulties because of the interference causing the peak to appear at the wrong time. We then have the problem of recovering the information (target range) from the "noisy" observation—which is the estimation problem referred to earlier.

These two problems of detection and estimation arise in many other areas besides the radio location problem and are basic to all statistical signal processing techniques. Similar problems are encountered in areas such as communication, pattern recognition, and system identification.

In a digital communication system, the message is encoded into a sequence of binary digits (more generally into a sequence of several symbols). Typically, these digits (bits), represented by a 1 or a 0, are transmitted by sending suitably chosen pulses, which again are distorted during transmission. The effect of this distortion is that at the receiver it is no longer possible to determine which waveform was transmitted. We can again model the distortion during transmission as random noise in the receiver. The problem is one of deciding, on the basis of noisy observations, whether the transmitted waveform corresponds to a 1 or a 0.

Recent work in efficient speech transmission involves characterizing the speech waveform by means of certain parameters that describe its spectrum. These parameters are then transmitted to the receiver, which synthesizes the speech waveform from a knowledge of these parameters. The problem of extracting the parameters is an identification problem and is essentially one of estimating the parameters in a suitably chosen model for the speech waveform. These parameters undergo distortion during transmission so that at the receiver we have the problem of recovering (estimating) the parameters from noisy observations.

Pattern recognition systems also provide examples of systems involving signal processing. Consider that we want to design an automatic machine to distinguish between two handwritten characters, say "a" and "b." The characteristics of these letters will vary with the person writing them. Typically, a single specimen can be characterized as a sample from a population with a known statistical description. The machine is to be designed to distinguish between the two classes or categories—the letters "a" and "b"—when a sample is presented to it. Often the statistics of the population will be unknown and will have to be determined from samples from each of the two classes. Another pattern classification example arises in electroencephalograph (EEG) analysis. The EEG is a recording of electrical brain signals and may be used in a variety of clinical purposes. A typical application is in the determination of the sleep state of a patient. The EEG signals are corrupted by noise introduced by the measuring device so that the characteristic features of the EEG will have to be estimated to determine the sleep state. Other applications involve the study of evoked potentials or response to stimuli and can be used for determining sensory perception. The evoked response is superimposed on the normal EEG, which in this case may be treated as interference. The useful information from the evoked response is obtained in terms of certain characteristic lines in the power spectrum which can be used for diagnosis and study of any abnormalities.

The preceding has served to point out that the problems of detection and estimation arise either separately or jointly in a wide variety of applications. While there is a seeming dichotomy between the two, in fact, there is an underlying similarity in the structure of the two problems, which facilitates the solution of many signal-processing problems. The aim of this book is to develop the fundamentals of detection and estimation theory as the basis of statistical signal processing and illustrate the application of the concepts and techniques in various areas. The application areas chosen are communications, radar systems, pattern recognition, and system identification. These areas are by no means exhaustive, but have been chosen because they are general enough to be of interest to a fairly wide audience.

1.1 ORGANIZATION

The book is addressed to the first-year graduate student in the disciplines of electrical and systems engineering. It will also serve the purpose of a graduate engineer in industry who is exposed to a wide variety of signal-processing problems but has not had an opportunity to obtain a cohesive background. Typically, the student of this book may be expected to have had a prior course in systems and a course in probability and stochastic processes. The exposure to systems should preferably have included state-variable and transform techniques for both continuous- and discrete-time systems.

The book consists essentially of two parts. The first part, Chapters 2 through 7, provides the basic framework of detection and estimation theory, while the second part, Chapters 8 through 9, covers applications in the four areas mentioned earlier. Specifically, our coverage is as follows. In Chapter 2 we provide a quick review of basic concepts from systems theory and stochastic processes which is useful in the development in later chapters. In particular, we discuss Gauss–Markov models for signal generation, the complex representation of bandlimited signals, and the concepts of likelihood and sufficiency.

Chapter 3 is concerned with the first class of statistical signal-processing problems: namely, decision problems. In this chapter we discuss classical decision theory using statistical hypothesis-testing methods. Various decision criteria are introduced and decision rules obtained. The performance of the decision rules is also investigated. The sequential decision test of Wald is included for the sake of completeness. These concepts are extended in Chapter 4 to the detection of waveforms observed in noise. The results are derived primarily in terms of a binary hypothesis-testing problem in which we are required to detect which of two known waveforms is present in the observations. The correlation receiver and the matched filter are obtained for optimum processing when the interference is additive white noise. Receiver structures for signals observed in colored noise are derived. The chapter concludes with a brief section on the detection of signals with unknown parameters.

Chapter 5 deals with the problem of estimating parameters from observations that are corrupted by noise. The Bayes estimators for parameters are obtained. Bounds on the performance of these estimators are presented. The concept of a reproducing density is introduced and applied to the estimation of parameters of probability density functions.

In Chapter 6 we consider the minimum-mean-square estimation of signals observed in noise. For stationary processes, the classical Wiener filter is derived for both continuous- and discrete-time processes. The Kalman filter, which provides an attractive recursive solution to the estimation problem in nonstationary processes, is discussed. Estimation techniques for certain nonlinear signal models are also considered.

In Chapter 7 we present further topics in detection and estimation. Nonparametric techniques that do not assume a functional form for the underlying distributions are introduced, and nonparametric tests for deciding between two hypotheses are presented. The asymptotic relative efficiency, which is a measure of the performance of two tests, is discussed as are locally optimal tests. The chapter concludes with a discussion of robust detection and estimation techniques.

After developing the powerful tools of detection and estimation theory in these five chapters, in Chapter 8 we consider applications to digital communication systems and radar systems. We first consider applications to communication systems and develop optimum receiver structures for a variety of digital transmission schemes. We consider the performance of these receivers in various situations, including fading. Techniques of synchronization are presented and the problem of intersymbol interference discussed. Spread spectrum systems are introduced and the application of statistical techniques in evaluating their performance illustrated. In the second part of the chapter we consider radar systems. The models for radar targets are derived using the complex representation of signals developed in Chapter 2. In addition to studying the detection of fluctuating and nonfluctuating targets, the problem of estimating such radar parameters as delay and Doppler frequency is investigated. The chapter concludes with an application of the Kalman filter for dynamic target tracking and maneuver detection.

In Chapter 9 we consider applications to pattern classification and system identification. The problem of pattern classification is posed as a hypothesis-testing problem by associating the various pattern classes with hypotheses. The problem of learning the parameters of the associated density functions under both supervised and unsupervised conditions is discussed. The important problem of extracting the features of a pattern is also presented. Nonparametric approaches based on the Parzen or k-nearest neighbor estimates for the underlying density functions are also discussed.

In the second part of this chapter we discuss the identification of systems based on input and output measurements. We first consider the identification of the parameters of an autoregressive (AR) or autoregressive-moving average (ARMA) model of a stochastic process. Identification in signal models using state augmentation techniques is discussed. The maximum likelihood technique for identification in linear time-invariant models is also presented.

Three appendices are included. Appendix A provides a brief review of the bilateral Laplace and Z-transforms which are used widely in the text, Appendix B summarizes pertinent features of certain optimization techniques used in Chapter 4, and Appendix C covers useful vector and matrix operations, in particular, a useful matrix lemma.

While Chapters 3 through 6 are essential for an understanding of the remainder of the book, the text is structured such that the applications chapters can be covered selectively and in any order, depending on the interest and background of the intended audience. The exercises at the end of each chapter serve a dual purpose—to increase understanding of the material covered and to present new applications or results not covered in the text.

REFERENCES

1. S. Haykin, *Adaptive Filter Theory*, 2nd ed., Prentice Hall, Englewood Cliffs, NJ, 1991.
2. M. I. Skolnik, *Introduction to Radar Systems*, McGraw-Hill, New York, 1980.
3. F. E. Nathanson, *Radar Design Principles*, McGraw-Hill, New York, 1991.
4. J. C. Proakis, *Digital Communications*, McGraw-Hill, New York, 1989.
5. K. Fukunaga, *Introduction to Statistical Pattern Recognition*, 2nd ed., Academic Press, New York, 1990.
6. C. W. Therrien, *Decision, Estimation and Classification: An Introduction to Pattern Recognition and Related Topics*, Wiley, New York, 1989.
7. L. Ljung, *System Identification: Theory for the User*, Prentice Hall, Englewood Cliffs, NJ, 1987.
8. T. Soderstrom and P. Stoica, *System Identification*, Prentice Hall, Englewood Cliffs, NJ, 1989.
9. S. M. Kay, *Modern Spectral Estimation: Theory and Application*, Prentice Hall, Englewood Cliffs, NJ, 1988.
10. D. H. Johnson and D. E. Dudgeon, *Array Signal Processing: Concepts and Techniques*, Prentice Hall, Englewood Cliffs, NJ, 1993.

2

Signals and Systems

2.1 INTRODUCTION

In Chapter 1 we presented several examples in which statistical signal processing plays an important role. The role of the signal processor is to extract information from the observed signals and present it in a useful form. The proper design of the processor requires an appropriate description or characterization of the signals as well as of the systems generating these signals. For example, in radar tracking we would like to know the location and velocity of the object being tracked. The signal of interest in this case is the radar return, which is usually corrupted by various noises, such as thermal noise in the processor. A description of the signal can be given in terms of the equations of motion of the object being tracked while the noise is usually modeled as a sample function of a stochastic process.

In this chapter we review briefly models for the generation of random signals and discuss some representations of these signals. We will find these results to be of use in later chapters. Also included is a discussion of the concepts of likelihood and sufficiency, which are basic to statistical inference.

2.2 SYSTEM THEORY

We recall that a system represents a model of a physical object or collection of objects and is characterized by a relationship between an input $u(t)$ and an output $y(t)$ of the form

$$y(t) = \mathrm{T}\{u(t)\} \tag{2.2.1}$$

where T is an operator or function. The system is linear or nonlinear depending on whether T is linear or nonlinear [1,2].

Linear systems, which satisfy the superposition principle, can be characterized by the system impulse response $h(t, \tau)$, which is the response of the system to a unit impulse applied at time τ. The output $y(t)$ to an arbitrary input $u(t)$ is given by the convolution integral

$$y(t) = \int_{-\infty}^{\infty} h(t,\tau)u(\tau)\,d\tau \tag{2.2.2}$$

For a causal or nonanticipative system, $h(t,\tau) = 0$ for $t < \tau$.

If the system is time invariant, the convolution integral becomes

$$y(t) = \int_{-\infty}^{\infty} h(t-\tau)u(\tau)\,d\tau = \int_{-\infty}^{\infty} h(\tau)u(t-\tau)\,d\tau \tag{2.2.3}$$

If the system is causal, $h(t) = 0$ for $t < 0$.

For linear multivariable system with p inputs $u_1(t), u_2(t), \ldots, u_p(t)$ and q outputs $y_1(t), y_2(t), \ldots, y_q(t)$, let $h_{ij}(t)$ denote the response at time t at the ith output terminal due to a unit impulse applied at time τ at the jth input terminal. Then

$$y_i(t) = \sum_{j=1}^{p} \int_{-\infty}^{\infty} h_{ij}(t,\tau)u_j(\tau)\,d\tau \qquad \text{for } i = 1,2,\ldots,q \tag{2.2.4}$$

For linear time-invariant systems, we can define a transfer function (a transfer matrix for multivariable systems) as the Laplace transform, $H(s)$, or the Fourier transform, $H(\omega)$, of the system impulse response. We can then describe the system by the input–output relation

$$Y(s) = H(s)U(s) \tag{2.2.5a}$$

or

$$Y(\omega) = H(\omega)U(\omega) \tag{2.2.5b}$$

The impulse response provides a useful characterization of linear systems. For general (linear or nonlinear) systems, a useful characterization is in terms of state variables. Any set of variables that summarizes the past history of the system is said to be a *state* of the system. More precisely, we define the state of a system at time t_0 as the vector $x(t_0) = [x_1(t_0)\ \ x_2(t_0)\ \ \cdots\ \ x_n(t_0)]^T$, which together with the input $u(t)$ uniquely determines the system output $y(t)$ for all $t \geq t_0$. Here n denotes the order of the system. If the number of states is the minimum possible, the state vector has minimum order. If the number of states is finite, the system is a finite-dimensional system. In the sequel we assume that we are dealing with finite-dimensional systems.

We can write the *output equation* of the system in functional form as

$$y(t) = h(x_1(t)\cdots x_n(t), u(t), t) \tag{2.2.6}$$

For differential systems, the evolution of the states of the system with time can be described by the set of equations

$$\dot{x}_i(t) = f_i(x_1(t), x_2(t), \ldots, x_n(t), u(t), t),\ i = 1,2,\ldots n \tag{2.2.7}$$

These *state equations* form a set of simultaneous first-order differential equations.

The extension to multiple inputs and outputs is quite straightforward. Given a system with p inputs $u_1(t) \cdots u_p(t)$ and q outputs $y_1(t) \cdots y_q(t)$, we define as the state variables of the system any set of variables $x_1(t) \cdots x_n(t)$ which, together with a knowledge of the values of the inputs, is sufficient to determine uniquely the values of the outputs. We write the output and state equations as

$$y_i(t) = h_i(x_1(t) \cdots x_n(t), u_1(t) \cdots u_p(t), t) \qquad i = 1, \ldots, q \tag{2.2.8a}$$

$$\dot{x}_j(t) = f_j(x_1(t) \cdots x_n(t), u_1(t) \cdots u_p(t), t) \qquad j = 1, \ldots, n \tag{2.2.8b}$$

By defining the vectors

$$\mathbf{x}(t) = [x_1(t) \cdots x_n(t)]^T$$

$$\mathbf{y}(t) = [y_1(t) \cdots y_q(t)]^T$$

$$\mathbf{u}(t) = [u_1(t) \cdots u_p(t)]^T$$

we can write these equations compactly as

$$\mathbf{y}(t) = \mathbf{h}(\mathbf{x}(t), \mathbf{u}(t), t) \tag{2.2.9a}$$

$$\dot{\mathbf{x}}(t) = \mathbf{f}(\mathbf{x}(t), \mathbf{u}(t), t) \tag{2.2.9b}$$

where the vectors $\mathbf{h}(\cdot,\cdot,\cdot)$ and $\mathbf{f}(\cdot,\cdot,\cdot)$ have the obvious definitions. We will assume that the initial condition for solving Eq. (2.2.9) is specified at initial time t_0, $\mathbf{x}(t_0) = \mathbf{x}_0$. Once Eq. (2.2.9b) has been solved for $\mathbf{x}(t)$, the determination of $\mathbf{y}(t)$ from Eq. (2.2.9a) is fairly straightforward. Unfortunately, the analytical solution of Eq. (2.2.9b) is in general a formidable problem. In fact, such solutions can be obtained only for particular cases. If we can, however, establish that this equation has a unique solution, we can attempt to find this solution approximately by numerical integration of the equation.

State equations for linear continuous time systems. For linear systems, the functional dependence of the system outputs on the system inputs is linear, so that

$$y_i(t) = \sum_{j=1}^{n} c_{ij}(t)x_j(t) + \sum_{j=1}^{p} d_{ij}(t)u_j(t) \qquad i = 1, \ldots, q \tag{2.2.10}$$

Similarly, the state equations can be written as

$$\dot{x}_i(t) = \sum_{j=1}^{n} a_{ij}(t)x_j(t) + \sum_{j=1}^{p} b_{ij}(t)u_j(t) \qquad i = 1, \ldots, n \tag{2.2.11}$$

We can write these equations in vector-matrix notation as

$$\mathbf{y}(t) = \mathbf{C}(t)\mathbf{x}(t) + \mathbf{D}(t)\mathbf{u}(t) \tag{2.2.12a}$$

$$\dot{\mathbf{x}}(t) = \mathbf{A}(t)\mathbf{x}(t) + \mathbf{B}(t)\mathbf{u}(t) \tag{2.2.12b}$$

where the matrices $\mathbf{A}(t)$, $\mathbf{B}(t)$, $\mathbf{C}(t)$, and $\mathbf{D}(t)$ have the obvious definitions.

The solution to the state equations can be obtained in terms of the state transition matrix $\psi(t, \tau)$ as

$$\mathbf{x}(t) = \psi(t, t_0) + \int_{t_0}^{t} \psi(t, \tau)\mathbf{B}(\tau)\mathbf{u}(\tau)\, d\tau \qquad (2.2.13)$$

where $\psi(t, \tau)$ satisfies the following relations:

$$\frac{\partial\psi(t, \tau)}{\partial\tau} = \mathbf{A}(t)\psi(t, \tau) \qquad (2.2.14a)$$

$$\psi^{-1}(t, \tau) = \psi(\tau, t) \qquad (2.2.14b)$$

$$\psi(t, \xi)\psi(\xi, \tau) = \psi(t, \tau) \qquad \text{for any } t, \tau, \xi \qquad (2.2.14c)$$

From Eq. (2.2.12a) it follows that the output $\mathbf{y}(t)$ is given by

$$\begin{aligned}\mathbf{y}(t) &= \mathbf{C}(t)\mathbf{X}(t) + \mathbf{D}(t)\mathbf{u}(t)\\ &= \mathbf{C}(t)\left[\psi(t, t_0)\mathbf{x}(t_0) + \int_{t_0}^{t} \psi(t, \tau)\mathbf{B}(\tau)\mathbf{u}(\tau)\, d\tau\right] + \mathbf{D}(t)\mathbf{u}(t)\end{aligned} \qquad (2.2.15)$$

For time-invariant systems, the matrices **A**, **B**, **C**, and **D** are matrices of constants and the state equations can be written as

$$\mathbf{y}(t) = \mathbf{Cx}(t) + \mathbf{Du}(t) \qquad (2.2.16a)$$

$$\dot{\mathbf{x}}(t) = \mathbf{Ax}(t) + \mathbf{Bu}(t) \qquad (2.2.16b)$$

With initial time t_0 set equal to zero, the solution in this case is given by

$$\mathbf{x}(t) = \psi(t)\mathbf{x}(0) + \int_{0}^{t} \psi(t - \tau)\mathbf{Bu}(\tau)\, d\tau \qquad (2.2.17)$$

where $\psi(t)$ can be determined as

$$\psi(t) = e^{\mathbf{A}t} = \sum_{i=0}^{\infty} \frac{\mathbf{A}^i t^i}{i!} = \mathcal{L}^{-1}[s\mathbf{I} - \mathbf{A}]^{-1} \qquad (2.2.18)$$

The output is given by

$$\mathbf{y}(t) = \mathbf{C}\left[e^{\mathbf{A}t}\mathbf{x}(0) + \int_{0}^{t} \psi(t - \tau)\mathbf{Bu}(\tau)\, d\tau\right] + \mathbf{Du}(t) \qquad (2.2.19)$$

Taking Laplace transforms of both sides of Eq. (2.2.16b) with $\mathbf{x}(0) = 0$ and solving for $\mathbf{X}(s)$ gives

$$\mathbf{X}(s) = [s\mathbf{I} - \mathbf{A}]^{-1}\mathbf{BU}(s) \qquad (2.2.20)$$

Similarly transformation of Eq. (2.2.16a) gives

$$\mathbf{Y}(s) = \mathbf{CX}(s) + \mathbf{DU}(s) \qquad (2.2.21)$$

so that the transfer matrix relating $\mathbf{Y}(s)$ to $\mathbf{X}(s)$ is equal to

$$\mathbf{H}(s) = \mathbf{C}[s\mathbf{I} - \mathbf{A}]^{-1}\mathbf{B} + \mathbf{D} \qquad (2.2.22)$$

Discrete-time systems. The preceding discussions for continuous-time systems can easily be carried over for discrete-time systems. Given a linear discrete-time system whose input–output relationship is given by

$$\mathrm{T}\{u(k)\} = y(k) \tag{2.2.23}$$

let us denote by $h(k,j)$ the response of the system observed at time k to an impulse input applied at j. In terms of the impulse response $h(k,j)$, the output $y(k)$ to an arbitrary input $u(j)$ defined for j in $(-\infty, \infty)$ is given by the *convolution summation*

$$y(k) = \sum_{j=-\infty}^{\infty} h(k,j)u(j) \tag{2.2.24}$$

For time-invariant systems, if $h(k)$ denotes the response at time k due to the impulse input applied at time zero, the output due to an arbitrary input $u(\cdot)$ can be expressed as

$$y(k) = \sum_{j=-\infty}^{\infty} h(k-j)u(j) \tag{2.2.25}$$

We can define a transfer function for time-invariant, discrete-time systems in terms of the Z-transform. We thus write

$$H(z) = Z[h(k)] \tag{2.2.26}$$

so that Z-transformation of both sides of Eq. (2.2.25) yields

$$Y(z) = H(z)U(z) \tag{2.2.27}$$

The state-variable formulation for discrete-time systems can be obtained by writing the output $y(k)$ as

$$y(k) = f(x_1(k), x_2(k), \ldots, x_n(k), u(k), k) \tag{2.2.28}$$

where

$$\mathbf{x}(k) = [x_1(k) \cdots x_n(k)]^T$$

represents the state vector. The state equations can now be written as

$$x_i(k+1) = f_i(x_1(k) \cdots x_n(k), u(k), k) \qquad i = 1, \ldots, n \tag{2.2.29}$$

and form a set of n simultaneous first-order difference equations. For multiple-input multiple-output systems these equations can be written in vector notation as

$$\mathbf{x}(k+1) = \mathbf{f}(\mathbf{x}(k), \mathbf{u}(k), k) \tag{2.2.30a}$$

$$\mathbf{y}(k) = \mathbf{h}(\mathbf{x}(k), \mathbf{u}(k), k) \tag{2.2.30b}$$

where $\mathbf{u}(k)$ represents a p-vector of inputs and $\mathbf{y}(k)$ represents the q-vector of outputs.

For linear systems, the state equations become

$$\mathbf{x}(k+1) = \mathbf{A}(k)\mathbf{x}(k) + \mathbf{B}(k)\mathbf{u}(k) \tag{2.2.31a}$$

$$\mathbf{y}(k) = \mathbf{C}(k)\mathbf{x}(k) + \mathbf{D}(k)\mathbf{u}(k) \tag{2.2.31b}$$

and for time-invariant systems

$$\mathbf{x}(k+1) = \mathbf{A}\mathbf{x}(k) + \mathbf{B}\mathbf{u}(k) \tag{2.2.32a}$$

$$\mathbf{y}(k) = \mathbf{C}\mathbf{x}(k) + \mathbf{D}\mathbf{u}(k) \tag{2.2.32b}$$

in which the matrices are constant.

Solution of the state equations for linear discrete-time systems. The solution of the state equations for the linear discrete-time system of Eq. (2.2.31) is easily obtained by using the equations repeatedly. Thus given $\mathbf{x}(k_0)$ for any k_0, the solution $\mathbf{x}(k)$ for $k > k_0$ is obtained as follows:

$$\mathbf{x}(k) = \boldsymbol{\psi}(k, k_0)\mathbf{x}(k_0) + \sum_{j=k_0}^{k-1} \boldsymbol{\psi}(k, j+1)\mathbf{B}(j)\mathbf{u}(j) \tag{2.2.33}$$

where

$$\boldsymbol{\psi}(k, k) = \mathbf{I} \tag{2.2.34}$$

and for $k > j$

$$\boldsymbol{\psi}(k, j) = \prod_{i=j}^{k-1} \mathbf{A}(i) \tag{2.2.35}$$

The output is obtained as

$$\mathbf{y}(k) = \mathbf{C}(k)\left[\boldsymbol{\psi}(k, k_0)\mathbf{x}(k_0) + \sum_{j=k_0}^{k-1} \boldsymbol{\psi}(k, j+1)\mathbf{B}(j)\mathbf{u}(j)\right] + \mathbf{D}(k)\mathbf{u}(k) \tag{2.2.36}$$

By setting $\mathbf{x}(k_0) = 0$ and $\mathbf{u}(k) = \delta_{kl}$, we easily obtain the impulse response of the system as

$$\mathbf{H}(k, l) = \mathbf{C}(k)\boldsymbol{\psi}(k, l+1)\mathbf{B}(l) + \mathbf{D}(k)\delta_{kl} \tag{2.2.37}$$

For time-invariant systems, the state transition matrix becomes

$$\boldsymbol{\psi}(k, j) = \prod_{i=j}^{k-1} \mathbf{A}(i) = \mathbf{A}^{k-j} \tag{2.2.38}$$

so that the expression for the state vector is

$$\mathbf{x}(k) = \mathbf{A}^{k-k_0}\mathbf{x}(k_0) + \sum_{j=k_0}^{k-1} \mathbf{A}^{k-j-1}\mathbf{B}\mathbf{u}(j) \tag{2.2.39}$$

The output is given by

$$\mathbf{y}(k) = \mathbf{C}\left[\mathbf{A}^{k-k_0}\mathbf{x}(k_0) + \sum_{j=k_0}^{k-1} \mathbf{A}^{k-j-1}\mathbf{B}\mathbf{u}(j)\right] + \mathbf{D}\mathbf{u}(k) \tag{2.2.40}$$

We can obtain a frequency-domain description of the linear time-invariant system of Eq. (2.2.32). Taking the z-transform of both sides yields with $k_0 = 0$

$$z[\mathbf{X}(z) - \mathbf{x}(0)] = \mathbf{A}\mathbf{X}(z) + \mathbf{B}\mathbf{U}(z) \tag{2.2.41}$$

so that

$$\mathbf{X}(z) = z[z\mathbf{I} - \mathbf{A}]^{-1}\mathbf{x}(0) + [z\mathbf{I} - \mathbf{A}]^{-1}\mathbf{B}\mathbf{U}(z) \tag{2.2.42}$$

The output is

$$\mathbf{Y}(z) = \mathbf{C}[z(z\mathbf{I} - \mathbf{A})^{-1}\mathbf{x}(0) + (z\mathbf{I} - \mathbf{A})^{-1}\mathbf{B}\mathbf{U}(z)] + \mathbf{D}\mathbf{U}(z) \tag{2.2.43}$$

and the transfer function matrix is

$$\mathbf{H}(z) = \mathbf{C}(z\mathbf{I} - \mathbf{A})^{-1}\mathbf{B} + \mathbf{D} \tag{2.2.44}$$

2.3 STOCHASTIC PROCESSES

In this section we review some properties of stochastic processes that will be of use to us in subsequent discussions. We discuss independent increment processes and introduce the concept of a white noise process, which plays a useful role in many engineering applications. In section 2.5 we provide with a discussion of the representation of stochastic process in a set of complete orthonormal functions. Our treatment of these topics is necessarily brief and the reader is referred to [3–8] for detailed discussions of these and other results.

In our subsequent discussions in Section 2.6 and Chapters 3, 4, 5, and 7 we will be using uppercase letters to denote random variables and lowercase letters to denote specific values that these variables may take. However, in the case of both continuous- and discrete-time stochastic processes in Sections 2.3, 2.4, and 2.5 and Chapters 6, 8, and 9, we use lowercase letters for both random variables and specific realizations, to conform with the notation used in most engineering literature that discuss stochastic processes.

We recall that a stochastic process $x(t, \zeta)$ [or $x(t)$], defined over an interval T, can be characterized in terms of its joint distribution $P\{x(t_1) \le x_1, x(t_2) \le x_2, \ldots, x(t_n) \le x_n\}$ at a finite number of instants $t_1, t_2, \ldots, t_n$, and for some real numbers $x_1, x_2, \ldots, x_n$. The process is separable if, given the distribution for a countable set of t_i, the process can be defined for all t.

The process $x(t)$ is Markov of order k if

$$p(x_m \mid x_{m-1}, x_{m-2}, \ldots, x_1) = p(x_m \mid x_{m-1}, x_{m-2}, \ldots, x_{m-k}) \qquad k < m \tag{2.3.1}$$

We can use Bayes' rule to write the joint density function for a Markov-1 process in terms of the transition probability density $p(x_i \mid x_{i-1})$ as

$$p(x_1, x_2, \ldots, x_m) = p(x_m \mid x_{m-1})p(x_{m-1} \mid x_{m-2}) \cdots p(x_1) \tag{2.3.2}$$

$x(t)$ is a Gaussian process if the random variables $x(t_1), x(t_2), \ldots, x(t_n)$ are jointly normal for any n and any set of instants $t_1, t_2, \ldots, t_n$. That is, if we define the random vector $\mathbf{x}$ as

$$\mathbf{x} = [x(t_1)x(t_2) \cdots x(t_n)]^T$$

we have

$$p(\mathbf{x}) = \frac{1}{(2\pi)^{n/2}|\mathbf{V}|^{1/2}} \exp\left[-\frac{1}{2}(\mathbf{x} - \mathbf{m})^T \mathbf{V}^{-1}(\mathbf{x} - \mathbf{m})\right] \tag{2.3.3}$$

where $\mathbf{m}$ is the mean of $\mathbf{x}$ and $\mathbf{V}$ is the covariance matrix, defined by

$$\mathbf{V} = \mathrm{E}\{(\mathbf{x} - \mathbf{m})(\mathbf{x} - \mathbf{m})^T\} \tag{2.3.4}$$

A process that plays an important role in engineering applications is the white noise process, which can be considered to result from the formal differentiation of an independent increment process.

We say that a random process $x(t)$ is an independent increment process if for any instants $t_1 < t_2 < t_3 < t_4$, the increments $x(t_2) - x(t_1)$ and $x(t_4) - x(t_3)$ are independent of each other. We note that the increments are taken over nonoverlapping intervals.

Of special interest is the *Wiener process*, which is defined over the interval $t \in [0, \infty]$ and satisfies

1. $x(0, \xi) = 0$ for almost all ξ.
2. $x(t)$ is an independent increment process.
3. For $t_2 \geq t_1$,

$$P[x(t_2) - x(t_1) < \lambda] = \frac{1}{\sqrt{2\pi(t_2 - t_1)}} \int_{-\infty}^{\lambda} \exp\left[\frac{-u^2}{2(t_2 - t_1)}\right] du \tag{2.3.5}$$

Hence the increments of a Wiener process have a Gaussian distribution. It can be shown that the Wiener process is continuous almost everywhere. The process has mean zero and correlation function

$$\phi_x(t_1, t_2) = \min(t_1, t_2) \tag{2.3.6}$$

Even though the Wiener process is continuous, it is not differentiable. However, if we formally differentiate $x(t)$ to get the process $w(t) = \dot{x}(t)$, it follows from Eq. (2.3.6) that

$$\mathrm{E}\{w(t_1)\, w(t_2)\} = \delta(t_1 - t_2) \tag{2.3.7}$$

so that the spectrum of $w(t)$ is constant or "white" over all frequencies.

We can correspondingly define a nonstationary white process as one whose correlation function is

$$\mathrm{E}\{w(t_1)\, w(t_2)\} = V(t_1)\delta(t_1 - t_2) \tag{2.3.8}$$

where $V(t_1)$ is a nonnegative function.

Even though $w(t)$ does not correspond to a physical process, it plays an important role in many engineering applications. Given a wide-sense stationary process $x(t)$ with a prescribed spectrum $\Phi_x(s)$, we can model $x(t)$ as the output of a linear time-invariant system driven by a stationary white process $w(t)$ with unity spectrum. Then

$$\Phi_x(s) = H(s)H(-s)\Phi_w(s) = H(s)H(-s) \tag{2.3.9}$$

If $\Phi_x(s)$ is a rational function of s, we can determine $H(s)$ by making a spectral factorization of $\Phi_x(s)$ by collecting together all the poles and zeros into left half-plane and right half-plane factors $\Phi_x^+(s)$ and $\Phi_x^-(s)$, respectively, such that

$$\Phi_x(s) = \Phi_x^+(s)\Phi_x^-(s) \tag{2.3.10}$$

Since $\Phi_x(s)$ is a power spectrum, it can be shown that

$$\Phi_x^-(s) = \Phi_x^+(-s) \tag{2.3.11}$$

It follows from Eq. (2.3.9) that

$$H(s)H(-s) = \Phi_x^+(s)\Phi_x^+(-s) \tag{2.3.12}$$

We can obtain a causal model for $x(t)$ by choosing $H(s)$ as

$$H(s) = \Phi_x^+(s) \tag{2.3.13}$$

Conversely, if we pass the signal $x(t)$ through a filter $G(s)$ with transfer function $1/\Phi_x^+(s)$, the output will be white noise with unity spectrum. $G(s)$ is called a *whitening filter*. We note that $G(s)$ is also causal.

Analogous to our definition of a continuous white process, we can define a discrete white process as one for which the random variables $x(k)$ and $x(j)$ at two discrete time instants k and j are uncorrelated for any k and j. That is, if $x(\cdot)$ is a zero-mean process,

$$\phi_x(k,j) = E\{x(k)x(j)\} = \sigma^2(k)\delta_{kj} \tag{2.3.14}$$

If σ^2 is independent of k, Eq. (2.3.14) becomes

$$\phi_x(k-j) = \sigma^2\delta_{kj} \tag{2.3.15}$$

The power spectrum of the process obtained by taking the Z-transform of Eq. (2.3.15) is

$$\Phi_x(z) = \sigma^2 \tag{2.3.16}$$

and is constant for all z.

2.4 GAUSS–MARKOV MODELS

A useful characterization of a stochastic process modeled as the output of a linear system driven by white noise can be obtained if we use a state-space representation of the system. By imposing appropriate conditions on the model, it can be established that the output is a Markov process. If the input process is Gaussian, so will be the output process. Such models are therefore referred to as Gauss–Markov models. For discrete-time processes, we have

$$\mathbf{x}(k+1) = \mathbf{A}(k)\mathbf{x}(k) + \mathbf{B}(k)\mathbf{w}(k) \tag{2.4.1}$$

where $\mathbf{x}(k)$ is an n-vector representing the state of the model, $\mathbf{A}(k)$ and $\mathbf{B}(k)$ are matrices of appropriate dimension and $\mathbf{w}(k)$ is the system input, which is assumed to be white. Specifically, let $\mathbf{w}(k), k = 0, 1, \ldots$ be a p-dimensional Gaussian white sequence with mean

$$\mathrm{E}\{\mathbf{w}(k)\} = \mathbf{m}_{\mathbf{w}}(k) \tag{2.4.2}$$

and covariance

$$\mathrm{E}\{[\mathbf{w}(k) - \mathbf{m}_{\mathbf{w}}(k)][\mathbf{w}(j) - \mathbf{m}_{\mathbf{w}}(j)]^T\} = \mathbf{V}_{\mathbf{w}}(j)\delta_{kj} \tag{2.4.3}$$

The system output is assumed to be

$$\mathbf{y}(k) = \mathbf{C}(k)\mathbf{x}(k) \tag{2.4.4}$$

We assume further that the initial vector associated with the model of Eq. (2.4.1), $\mathbf{x}(0)$, is a Gaussian random vector with known mean

$$\mathrm{E}\{\mathbf{x}(0)\} = \mathbf{m}_{\mathbf{x}_0} \tag{2.4.5}$$

and covariance matrix

$$\mathrm{E}\{[\mathbf{x}(0) - \mathbf{m}_{\mathbf{x}_0}][\mathbf{x}(0) - \mathbf{m}_{\mathbf{x}_0}]^T\} = \mathbf{V}_{\mathbf{x}_0} \tag{2.4.6}$$

and independent of $\mathbf{w}(k)$ for all $k \geq 0$, such that

$$\mathrm{E}\{[\mathbf{x}(0) - \mathbf{m}_{\mathbf{x}_0}][\mathbf{w}(k) - \mathbf{m}_{\mathbf{w}}(k)]^T\} = \mathbf{0} \tag{2.4.7}$$

We now show that the process $\mathbf{x}(k)$ is Gauss–Markov. From Eq. (2.2.33), the solution to Eq. (2.4.1) is given by

$$\mathbf{x}(k) = \boldsymbol{\psi}(k,j)\mathbf{x}(j) + \sum_{i=j+1}^{k} \boldsymbol{\psi}(k,i)\mathbf{B}(i-1)\mathbf{w}(i-1) \tag{2.4.8a}$$

where

$$\boldsymbol{\psi}(k,j) = \mathbf{A}(k-1)\mathbf{A}(k-2)\cdots\mathbf{A}(j) \tag{2.4.8b}$$

From Eq. (2.4.8) it is clear that the conditional density function of $\mathbf{x}(k)$ given $\mathbf{x}(j), \mathbf{x}(j-1), \ldots, \mathbf{x}(1), \mathbf{x}(0)$ depends only on $\mathbf{x}(j)$ for any k and j. Therefore, $\mathbf{x}(k)$ is Markov. To show that $\mathbf{x}(k)$ is Gaussian, by setting $j = 0$ in Eq. (2.4.8), we see that $\mathbf{x}(k)$ is a linear combination of the Gaussian variables $\mathbf{x}(0), \mathbf{w}(0), \mathbf{w}(1), \ldots, \mathbf{w}(k-1)$ and hence is also Gaussian.

We now present two ways of describing the process. In the first description, we note that by using the chain rule, we can write the joint density function of the variables $\{\mathbf{x}(k+1), \mathbf{x}(k), \ldots, \mathbf{x}(0)\}$ as [see Eq. (2.3.2)]

$$p[\mathbf{x}(k+1), \mathbf{x}(k), \ldots, \mathbf{x}(0)] = p[\mathbf{x}(k+1) \mid \mathbf{x}(k)]p[\mathbf{x}(k) \mid \mathbf{x}(k-1)]\cdots p[\mathbf{x}(0)] \tag{2.4.9}$$

Since $p(\mathbf{x}(0))$ is known, it is enough to specify the transitional densities $p(\mathbf{x}(k+1)\,|\,\mathbf{x}(k))$. Since these are Gaussian densities, they are completely specified by the conditional mean

$$E[\mathbf{x}(k+1)\,|\,\mathbf{x}(k)] = \mathbf{A}(k)\mathbf{x}(k) + \mathbf{B}(k)\mathbf{m}_{\mathbf{w}}(k) \tag{2.4.10a}$$

and the conditional covariance

$$\begin{aligned} \mathbf{V}_{\mathbf{x}}(k+1\,|\,k) &= E\{[\mathbf{x}(k+1) - E\{\mathbf{x}(k+1)\,|\,\mathbf{x}(k)\}] \\ &\quad \cdot [\mathbf{x}(k+1) - E\{\mathbf{x}(k+1)\,|\,\mathbf{x}(k)\}]^T\} \\ &= \mathbf{B}(k)\mathbf{V}_{\mathbf{w}}(k)\mathbf{B}^T(k) \end{aligned} \tag{2.4.10b}$$

In the second description, we obtain equations describing the time propagation of the mean and variance of the process $\mathbf{x}(k)$. These will then provide a complete description of the process as it is Gaussian. Let

$$E\{\mathbf{x}(k)\} = \mathbf{m}_{\mathbf{x}}(k) \tag{2.4.11}$$

Taking expectations on both sides of Eq. (2.4.1) yields

$$\mathbf{m}_{\mathbf{x}}(k+1) = \mathbf{A}(k)\mathbf{m}_{\mathbf{x}}(k) + \mathbf{B}(k)\mathbf{m}_{\mathbf{w}}(k) \tag{2.4.12}$$

with initial condition $\mathbf{m}_{\mathbf{x}}(0) = \mathbf{m}_{\mathbf{x}_0}$.

To determine an equation for the covariance $\mathbf{V}_{\mathbf{x}}(k)$, define the normalized sequences

$$\bar{\mathbf{x}}(k) = \mathbf{x}(k) - \mathbf{m}_{\mathbf{x}}(k) \tag{2.4.13}$$

and

$$\bar{\mathbf{w}}(k) = \mathbf{w}(k) - \mathbf{m}_{\mathbf{w}}(k) \tag{2.4.14}$$

It can easily be verified that $\bar{\mathbf{x}}(k)$ and $\bar{\mathbf{w}}(k)$ are zero mean with covariances

$$\mathbf{V}_{\bar{\mathbf{x}}}(k) = \mathbf{V}_{\mathbf{x}}(k) \quad \text{and} \quad \mathbf{V}_{\bar{\mathbf{w}}}(k) = \mathbf{V}_{\mathbf{w}}(k) \tag{2.4.15}$$

Subtracting Eq. (2.4.12) from Eq. (2.4.1), we get

$$\bar{\mathbf{x}}(k+1) = \mathbf{A}(k)\bar{\mathbf{x}}(k) + \mathbf{B}(k)\bar{\mathbf{w}}(k) \tag{2.4.16}$$

The solution to this equation for $k > j$ can be obtained from Eq. (2.2.33) as

$$\bar{\mathbf{x}}(k) = \boldsymbol{\psi}(k,j)\bar{\mathbf{x}}(j) + \sum_{i=j+1}^{k} \boldsymbol{\psi}(k,i)\mathbf{B}(i-1)\bar{\mathbf{w}}(i-1) \tag{2.4.17}$$

with initial condition

$$\bar{\mathbf{x}}(0) = \mathbf{x}(0) - \mathbf{m}_{\mathbf{x}}(0) \tag{2.4.18}$$

Here $\boldsymbol{\psi}(k,j)$ is the state transition matrix of the system.

From Eq. (2.4.16) we can write

$$\begin{aligned} \mathbf{V}_{\bar{\mathbf{x}}}(k+1) &= E\{\bar{\mathbf{x}}(k+1)\bar{\mathbf{x}}^T(k+1)\} \\ &= E\{[\mathbf{A}(k)\bar{\mathbf{x}}(k) + \mathbf{B}(k)\bar{\mathbf{w}}(k)][\mathbf{A}(k)\bar{\mathbf{x}}(k) + \mathbf{B}(k)\bar{\mathbf{w}}(k)]^T\} \end{aligned} \tag{2.4.19}$$

To evaluate the cross-covariance matrix $E\{\bar{\mathbf{w}}(k)\bar{\mathbf{x}}^T(k)\}$ and its transpose, we set $j = 0$ in Eq. (2.4.17), postmultiply both sides by $\bar{\mathbf{w}}^T(k)$, and take expectations to get

$$E\{\bar{\mathbf{x}}(k)\bar{\mathbf{w}}^T(k)\} = \boldsymbol{\psi}(k,0)E(\bar{\mathbf{x}}(0)\bar{\mathbf{w}}^T(k)) + \sum_{i=1}^{k} \boldsymbol{\psi}(k,i)\mathbf{B}(i-1)E\{\bar{\mathbf{w}}(i-1)\bar{\mathbf{w}}^T(k)\} \tag{2.4.20}$$

From Eqs. (2.4.3) and (2.4.7) it follows that

$$E\{\bar{\mathbf{x}}(k)\bar{\mathbf{w}}^T(k)\} = 0 \tag{2.4.21}$$

Substitution in Eq. (2.4.19) then yields

$$\mathbf{V}_{\bar{\mathbf{x}}}(k+1) = \mathbf{A}(k)\mathbf{V}_{\bar{\mathbf{x}}}(k)\mathbf{A}^T(k) + \mathbf{B}(k)\mathbf{V}_{\bar{\mathbf{w}}}(k)\mathbf{B}^T(k) \tag{2.4.22}$$

with initial condition $\mathbf{V}_{\bar{\mathbf{x}}}(0) = \mathbf{V}_{\mathbf{x}}(0) = \mathbf{V}_{\mathbf{x}_0}$.

We now derive relations for finding $\mathbf{V}_{\bar{\mathbf{x}}}(k,j) = E\{\bar{\mathbf{x}}(k)\bar{\mathbf{x}}^T(j)\}$. For $k > j$, postmultiply Eq. (2.4.17) by $\bar{\mathbf{x}}^T(j)$ and take expectations to get

$$\mathbf{V}_{\bar{\mathbf{x}}}(k,j) = \boldsymbol{\psi}(k,j)\mathbf{V}_{\mathbf{x}}(j) + \sum_{i=j+1}^{k} \boldsymbol{\psi}(k,i)\mathbf{B}(i-1)E\{\bar{\mathbf{w}}(i-1)\bar{\mathbf{x}}^T(j)\} \tag{2.4.23}$$

We note that $\bar{\mathbf{w}}(i)$ is white. Since the system is causal, it follows that for $i \geq j$, $\bar{\mathbf{w}}(i)$ is independent of $\bar{\mathbf{x}}(j)$. Since it is also zero mean, it follows that the second term on the right side of Eq. (2.4.23) is zero. We therefore have

$$\mathbf{V}_{\bar{\mathbf{x}}}(k,j) = \boldsymbol{\psi}(k,j)\mathbf{V}_{\bar{\mathbf{x}}}(j) \qquad \text{for } k > j \tag{2.4.24}$$

For $k < j$, we can derive a similar result by writing

$$\begin{aligned}\mathbf{V}_{\bar{\mathbf{x}}}(k,j) &= [\mathbf{V}_{\bar{\mathbf{x}}}(j,k)]^T \\ &= [\boldsymbol{\psi}(j,k)\mathbf{V}_{\bar{\mathbf{x}}}(k)]^T \qquad \text{from Eq. (2.4.24)} \\ &= \mathbf{V}_{\bar{\mathbf{x}}}(k)\boldsymbol{\psi}^T(j,k)\end{aligned} \tag{2.4.25}$$

Finally, since $\mathbf{V}_{\bar{\mathbf{x}}}(k) = \mathbf{V}_{\mathbf{x}}(k)$, we can write from Eqs. (2.4.22), (2.4.24), and (2.4.25),

$$\mathbf{V}_{\mathbf{x}}(k+1) = \mathbf{A}(k)\mathbf{V}_{\mathbf{x}}(k)\mathbf{A}^T(k) + \mathbf{B}(k)\mathbf{V}_{\mathbf{w}}(k)\mathbf{B}^T(k) \tag{2.4.26}$$

with

$$\mathbf{V}_{\mathbf{x}}(0) = \mathbf{V}_{\mathbf{x}_0} \tag{2.4.27a}$$

and

$$\mathbf{V}_{\mathbf{x}}(k,j) = \begin{cases}\boldsymbol{\psi}(k,j)\mathbf{V}_{\mathbf{x}}(j) & \text{for } k > j \\ \mathbf{V}_{\mathbf{x}}(k)\boldsymbol{\psi}^T(j,k) & \text{for } k < j\end{cases} \tag{2.4.27b}$$

The covariance of the output is given by

$$\mathbf{V}_{\mathbf{y}}(k) = \mathbf{C}(k)\mathbf{V}_{\mathbf{x}}(k)\mathbf{C}^T(k)) \tag{2.4.28}$$

We now briefly consider the extension of these results to continuous-time models. For the most part, as indicated earlier, we will be performing the operations involved in a purely formal manner. We consider the linear system

$$\dot{\mathbf{x}}(t) = \mathbf{A}(t)\mathbf{x}(t) + \mathbf{B}(t)\mathbf{w}(t) \tag{2.4.29}$$

where $\mathbf{w}(t)$ is a white Gaussian process with mean $\mathbf{m_w}(t)$ and covariance

$$E\{[\mathbf{w}(t) - \mathbf{m_w}(t)][\mathbf{w}(t') - \mathbf{m_w}(t')]^T\} = \mathbf{V_w}(t)\delta(t - t') \tag{2.4.30}$$

We also assume that $\mathbf{x}(t_0)$ is a Gaussian random variable with mean $\mathbf{m_{x_0}}$ and covariance $\mathbf{V_{x_0}}$, and is independent of $\mathbf{w}(t)$ for all $t \geqslant t_0$. The system output given by

$$\mathbf{y}(t) = \mathbf{C}(t)\mathbf{x}(t) \tag{2.4.31}$$

The solution to Eq. (2.4.29) is given by

$$\mathbf{x}(t_m) = \boldsymbol{\psi}(t_m, t_{m-1})\mathbf{x}(t_{m-1}) + \int_{t_{m-1}}^{t_m} \boldsymbol{\psi}(t_m, \tau)\mathbf{B}(\tau)\mathbf{w}(\tau)\, d\tau \tag{2.4.32}$$

for any set of points t_m, t_{m-1}. Here $\boldsymbol{\psi}(t, \tau)$ is the state transition matrix corresponding to $\mathbf{A}(t)$. Thus given the values of $\mathbf{x}(t)$ for t equal to $t_0, t_1, \ldots, t_{m-1}, t_m$ where

$$t_0 < t_1 < \cdots < t_{m-1} < t_m$$

it follows from Eq. (2.4.32) that the density function of $\mathbf{x}(t_m)$ is dependent only on $\mathbf{x}(t_{m-1})$. Therefore, $\mathbf{x}(t)$ is Markov. It is a Gaussian process since $\mathbf{x}(t_m)$ is obtained as a linear functional of a Gaussian process. To determine expressions for the mean and variance of $\mathbf{x}(t)$, we again form the normalized zero mean processes

$$\overline{\mathbf{x}}(t) = \mathbf{x}(t) - \mathbf{m_x}(t) \tag{2.4.33a}$$

and

$$\overline{\mathbf{w}}(t) = \mathbf{w}(t) - \mathbf{m_w}(t) \tag{2.4.33b}$$

It can easily be verified that the variances are given by

$$\mathbf{V_{\overline{w}}}(t) = \mathbf{V_w}(t) \quad \text{and} \quad \mathbf{V_{\overline{x}}}(t) = \mathbf{V_x}(t) \tag{2.4.34}$$

To determine the mean or variance of $\overline{\mathbf{x}}(t)$ directly, we would need to solve for the transition matrix $\psi(t, \tau)$. Instead, we set up a differential equation to describe the evolution with time of these moments. Taking expectations on both sides of Eq. (2.4.29) yields

$$\dot{\mathbf{m}}_\mathbf{x}(t) = \mathbf{A}(t)\mathbf{m_x}(t) + \mathbf{B}(t)\mathbf{m_w}(t) \tag{2.4.35}$$

with

$$\mathbf{m_x}(t_0) = \mathbf{m_{x_0}}$$

Subtracting Eq. (2.4.35) from Eq. (2.4.29), we obtain the differential equation for $\overline{\mathbf{x}}(t)$ as

$$\dot{\overline{\mathbf{x}}}(t) = \mathbf{A}(t)\overline{\mathbf{x}}(t) + \mathbf{B}(t)\overline{\mathbf{w}}(t) \tag{2.4.36}$$

with

$$\overline{\mathbf{x}}(t_0) = \mathbf{x}(t_0) - \mathbf{m_{x_0}}$$

The solution to Eq. (2.4.36) is given by

$$\overline{\mathbf{x}}(t) = \boldsymbol{\psi}(t, t_0)\overline{\mathbf{x}}(t_0) + \int_{t_0}^{t} \boldsymbol{\psi}(t, \tau)\mathbf{B}(\tau)\overline{\mathbf{w}}(\tau)\, d\tau \tag{2.4.37}$$

Since

$$\mathbf{V}_{\bar{\mathbf{x}}}(t) = \mathrm{E}\{\bar{\mathbf{x}}(t)\bar{\mathbf{x}}^T(t)\} \tag{2.4.38}$$

we can write for its derivative

$$\frac{d}{dt}\dot{\mathbf{V}}_{\bar{\mathbf{x}}}(t) = \frac{d}{dt}\mathrm{E}\{\bar{\mathbf{x}}(t)\bar{\mathbf{x}}^T(t)\}$$

$$= \mathrm{E}\{\dot{\bar{\mathbf{x}}}(t)\bar{\mathbf{x}}^T(t)\} + \mathrm{E}\{\bar{\mathbf{x}}(t)\dot{\bar{\mathbf{x}}}^T(t)\} \tag{2.4.39}$$

Substitution of Eq. (2.4.36) in Eq. (2.4.39) yields

$$\begin{aligned}\frac{d}{dt}\mathbf{V}_{\bar{\mathbf{x}}}(t) &= \mathrm{E}\{[\mathbf{A}(t)\bar{\mathbf{x}}(t) + \mathbf{B}(t)\bar{\mathbf{w}}(t)]\bar{\mathbf{x}}^T(t)\} \\ &\quad + \mathrm{E}\{\bar{\mathbf{x}}(t)[\mathbf{A}(t)\bar{\mathbf{x}}(t) + \mathbf{B}(t)\bar{\mathbf{w}}(t)]^T\} \\ &= \mathbf{A}(t)\mathbf{V}_{\bar{\mathbf{x}}}(t) + \mathbf{V}_{\bar{\mathbf{x}}}(t)\mathbf{A}^T(t) + \mathbf{B}(t)\mathrm{E}\{\bar{\mathbf{w}}(t)\bar{\mathbf{x}}^T(t)\} \\ &\quad + \mathrm{E}\{\bar{\mathbf{x}}(t)\bar{\mathbf{w}}^T(t)\}\mathbf{B}^T(t)\end{aligned} \tag{2.4.40}$$

To evaluate the third and fourth terms in Eq. (2.4.40) we postmultiply Eq. (2.4.37) by $\bar{\mathbf{w}}^T(t)$ and take expectations to get

$$\mathrm{E}\{\mathbf{x}(t)\bar{\mathbf{w}}^T(t)\} = \psi(t, t_0)\mathrm{E}\{\bar{\mathbf{x}}(t_0)\bar{\mathbf{w}}^T(t)\} + \int_{t_0}^{t} \psi(t,\tau)\mathbf{B}(\tau)\mathrm{E}\{\bar{\mathbf{w}}(\tau)\bar{\mathbf{w}}^T(t)\}\, d\tau \tag{2.4.41}$$

The first term on the right-hand side of Eq. (2.4.41) is zero by virtue of the assumption that $\mathbf{w}(t)$ is independent of $\mathbf{x}(t_0)$ for all $t \geqslant t_0$. To evaluate the second term, we make use of Eq. (2.4.30) to write the second term as

$$\int_{t_0}^{t} \psi(t,\tau)\mathbf{B}(\tau)\mathbf{V}_{\mathbf{w}}(\tau)\delta(t-\tau)\, d\tau = \tfrac{1}{2}\mathbf{B}(t)\mathbf{V}_{\mathbf{w}}(t) \tag{2.4.42}$$

where the last step follows by assuming that the δ-function is symmetric.† Substitution of Eq. (2.4.42) in Eq. (2.4.41) yields

$$\mathrm{E}\{\bar{\mathbf{x}}(t)\bar{\mathbf{w}}^T(t)\} = \tfrac{1}{2}\mathbf{B}(t)\mathbf{V}_{\mathbf{w}}(t) \tag{2.4.43a}$$

It similarly follows that

$$\mathrm{E}\{\bar{\mathbf{w}}(t)\bar{\mathbf{x}}^T(t)\} = \tfrac{1}{2}\mathbf{V}_{\mathbf{w}}(t)\mathbf{B}^T(t) \tag{2.4.43b}$$

Substitution of Eqs. (2.4.43) in Eq. (2.4.40) yields for the variance $\mathbf{V}_{\mathbf{x}}(t)$

$$\frac{d}{dt}\mathbf{V}_{\bar{\mathbf{x}}}(t) = \mathbf{A}(t)\mathbf{V}_{\bar{\mathbf{x}}}(t) + \mathbf{V}_{\bar{\mathbf{x}}}(t)\mathbf{A}^T(t) + \mathbf{B}(t)\mathbf{V}_{\mathbf{w}}(t)\mathbf{B}^T(t) \tag{2.4.44}$$

† The δ-function is defined by $\int_a^b g(\tau)\delta(t-\tau)\,d\tau = \begin{cases} 0 & t < a \text{ or } t > b \\ \alpha g(a) & t = a \\ (1-\alpha)g(b) & t = b \\ g(t) & a < t < b \end{cases}$

where $0 \leqslant \alpha \leqslant 1$. We obtain the symmetric δ-function if $\alpha = \frac{1}{2}$.

Since $\mathbf{V_x}(t) = \mathbf{V_{\tilde{x}}}(t)$, we also have

$$\dot{\mathbf{V}}_{\mathbf{x}}(t) = \mathbf{A}(t)\mathbf{V_x}(t) + \mathbf{V_x}(t)\mathbf{A}^T(t) + \mathbf{B}(t)\mathbf{V_w}(t)\mathbf{B}^T(t) \tag{2.4.45}$$

with $\mathbf{V_x}(t_0) = \mathbf{V}_{\mathbf{x}_0}$.

The mean and the variance of the output $\mathbf{y}(t)$ are easily determined as

$$\mathbf{m_y}(t) = \mathbf{C}(t)\mathbf{m_x}(t) \tag{2.4.46}$$

and

$$\mathbf{V_y}(t) = \mathbf{C}(t)\mathbf{V_x}(t)\mathbf{C}^T(t) \tag{2.4.47}$$

Suppose that the system is time invariant, $\mathbf{w}(t)$ is stationary, and $t_0 = -\infty$. Then $\mathbf{x}(t)$ will also be stationary. In this case the matrices $\mathbf{A}$, $\mathbf{B}$, and $\mathbf{C}$ in Eqs. (2.4.29) and (2.4.31) [Eqs. (2.4.1) and (2.4.4) for the discrete-time case] will be constant. The covariance $\mathbf{V_x}(t)$ [$\mathbf{V_x}(k)$] can be obtained as the steady-state solution to the differential equation (2.4.45) [the difference equation (2.4.22) for the discrete case]. This solution can be found by solving the algebraic equation

$$\mathbf{A}\mathbf{V_x} + \mathbf{V_x}\mathbf{A}^T + \mathbf{B}\mathbf{V_w}\mathbf{B}^T = 0 \tag{2.4.48}$$

for the continuous-time case and

$$\mathbf{A}\mathbf{V_x}\mathbf{A}^T + \mathbf{B}\mathbf{V_w}\mathbf{B}^T = \mathbf{V_x} \tag{2.4.49}$$

for the discrete-time case.

We note that the equations for the propagation of the mean and variance of $\mathbf{x}(\cdot)$ and $\mathbf{y}(\cdot)$ hold even if the input process is not Gaussian but is only white. Of course, these two moments are no longer adequate to provide a complete statistical description of the output process.

2.5 REPRESENTATION OF STOCHASTIC PROCESSES

2.5.1 Karhunen–Loéve Expansion

It is often useful to obtain a representation, over an interval (t_0, t_f), of a second-order mean-square continuous stochastic process with $E\{x^2(t)\} < \infty$, in a series of the form

$$x(t) = \lim_{N\to\infty} \sum_{n=1}^{N} \beta_n g_n(t) \tag{2.5.1}$$

where the limit is in the mean-square sense.

The basis functions $g_n(t)$ are chosen to be orthonormal over the interval (t_0, t_f), so that

$$\int_{t_0}^{t_f} g_n(t) g_m(t)\, dt = \delta_{nm} \tag{2.5.2}$$

The coefficients β_n are random variables given by

$$\beta_n = \int_{t_0}^{t_f} x(t) g_n(t)\, dt \tag{2.5.3}$$

We would like to choose the basis functions $\{g_i(t)\}$ such that $\{\beta_n\}$ are a set of uncorrelated variables. If we assume that $E\{x(t)\} = 0$, then $E\{\beta_n\} = 0$. In this case, if the coefficients $\{\beta_n\}$ are orthogonal, they are also uncorrelated.

Let us choose the functions $g_n(t)$ as the solutions to the integral equation

$$\int_{t_0}^{t_f} \phi_x(t_1, t_2) g_n(t_2)\, dt_2 = \lambda_n g_n(t_1) \tag{2.5.4}$$

The kernel of this equation is the covariance function $\phi_x(t_1, t_2)$, and the values of $\{\lambda_n\}$ for which the equation is satisfied are its eigenvalues, with the functions $\{g_n(t)\}$ being the corresponding eigenfunctions. From the theory of linear integral equations, it follows that if $\phi_x(t_1, t_2)$ is positive definite, the eigenfunctions form a complete orthonormal set. If $\phi_x(t_1, t_2)$ is only positive semidefinite, the eigenfunctions will not form a complete orthonormal set. In this case we can augment the eigenfunctions with enough additional functions to form a complete set. In either case, the functions $\{g_n(t)\}$ will satisfy Eq. (2.5.1). We can therefore write, from Eq. (2.5.3),

$$\begin{aligned} E\{\beta_n \beta_m\} &= E\left\{\int_{t_0}^{t_f}\int_{t_0}^{t_f} x(t_1)x(t_2)g_n(t_1)g_m(t_2)\, dt_1\, dt_2\right\} \\ &= \int_{t_0}^{t_f}\left[\int_{t_0}^{t_f} \phi_x(t_1, t_2) g_m(t_2)\, dt_2\right] g_n(t_1)\, dt_1 \\ &= \int_{t_0}^{t_f} \lambda_m g_m(t_1) g_n(t_1)\, dt_1 \\ &= \lambda_m \delta_{nm} \end{aligned} \tag{2.5.5}$$

Thus with the basis functions chosen as the eigenfunctions of the integral equation (2.5.4), the coefficients β_n are uncorrelated. The resulting representation of $x(t)$ in Eq. (2.5.1) is known as the *Karhunen–Loéve expansion*.

It can be shown that since $\phi_x(t_1, t_2)$ is nonnegative definite, it can be uniformly expanded over $[t_0, t_f]$ in the series

$$\phi_x(t_1, t_2) = \sum_{n=1}^{\infty} \lambda_n g_n(t_1) g_n(t_2) \tag{2.5.6}$$

so that

$$E\left\{\int_{t_0}^{t_f} x^2(t)\, dt\right\} = \int_{t_0}^{t_f} \phi_x(t, t)\, dt = \sum_{n=1}^{\infty} \lambda_n \tag{2.5.7}$$

Finding the basis functions $\{g_n(t)\}$ by solving the integral equation (2.5.4) is in general a tedious task. Suppose that $x(t)$ is a white process with

$$\phi_x(t_1, t_2) = V(t_1)\delta(t_1 - t_2) \tag{2.5.8}$$

for some function $V(t_1) > 0$. Then substitution of Eq. (2.5.8) in Eq. (2.5.4) yields

$$\int_{t_0}^{t_f} V(t_1)\delta(t_1 - t_2)g_i(t_2)\,dt_2 = V(t_1)g_i(t_1) = \lambda g_i(t_1) \tag{2.5.9}$$

Thus we see that for a white process, the functions $g_i(t)$ can be chosen to be any arbitrary orthonormal set.

We point out that since $E\{x^2(t)\}$ is not bounded for a white noise process, the Karhunen–Loéve expansion is not strictly valid [9]. The expansion may be justified by assuming that $x(t)$ is a bandlimited process with a constant spectrum over its bandwidth W. If we now let W go to infinity, Eqs. (2.5.8) and (2.5.9) will be approximately satisfied. Moreover, when we use the Karhunen–Loéve expansion of a white process in our applications in subsequent chapters, we do obtain the correct results.

Finally, it follows from Eqs. (2.5.6), (2.5.8) and (2.5.9) that we can write

$$\delta(t_1 - t_2) = \sum_{n=1}^{\infty} g_n(t_1)g_n(t_2) \tag{2.5.10}$$

2.5.2 Complex Envelopes of Bandlimited Signals

In many applications involving radar and communication signals, the signal of interest, $x(t)$, is narrowband, with most of the signal energy being concentrated near a "center frequency" ω_0, so that the spectrum is of the form

$$\Phi_x(\omega) = 0 \quad \text{except for} \quad \omega_0 - \omega_c < |\omega| < \omega_0 + \omega_c \tag{2.5.11}$$

For such signals, it is often convenient to express operations on $x(t)$ in terms of an associated complex signal $\tilde{x}(t)$. We will assume that $x(t)$ is a zero-mean, wide-sense-stationary process.

Let $\check{x}(t)$ denote the *Hilbert transform* of $x(t)$:

$$\check{x}(t) = \int_{-\infty}^{\infty} \frac{x(\lambda)}{\pi(t - \lambda)}\,d\lambda \tag{2.5.12}$$

Thus $\check{x}(t)$ is the convolution of $x(t)$ with the function $h(t) = 1/\pi t$ whose Fourier transform $H(\omega)$ is given by

$$H(\omega) = \begin{cases} -j & \omega \geq 0 \\ j & \omega < 0 \end{cases} \tag{2.5.13}$$

Let us now define the complex signal $\tilde{x}(t)$ as

$$\tilde{x}(t) = [x(t) + j\check{x}(t)]\exp(-j\omega_0 t) \tag{2.5.14}$$

It then follows that

$$x(t) = \text{Re}[\tilde{x}(t)e^{j\omega_0 t}] \tag{2.5.15}$$

The signal $\tilde{x}(t)$ is the complex envelope of the (real) signal $x(t)$. $x(t)$ can be considered to be the result of amplitude modulating the envelope (baseband signal) $\tilde{x}(t)$ with a "carrier" $\exp(+j\omega_0 t)$.

The correlation function of $\tilde{x}(t)$ is given by

$$\begin{aligned}\phi_{\tilde{x}}(\tau) &= \mathrm{E}\{\tilde{x}(t)\tilde{x}^*(t+\tau)\}\\ &= [\phi_x(\tau) + j\phi_{\check{x}x}(\tau) - j\phi_{x\check{x}}(\tau) + \phi_{\check{x}}(\tau)]e^{j\omega_0\tau}\end{aligned} \tag{2.5.16}$$

It can then be shown using Eq. (2.5.13) that

$$\Phi_{\tilde{x}}(\omega) = \begin{cases} 4\Phi_x(\omega - \omega_0) & \omega - \omega_0 > 0 \\ 0 & \omega - \omega_0 < 0 \end{cases} \tag{2.5.17}$$

While in many situations a natural choice for ω_0 may exist, in other cases any convenient frequency can be chosen as the center frequency.

We now consider some useful results concerning the complex envelope $\tilde{x}(t)$ of the signal $x(t)$. Let $a(t)$ and $b(t)$ represent the real and imaginary parts of $\tilde{x}(t)$. We can use Eq. (2.5.14) to write

$$\begin{aligned} a(t) &= [x(t)\cos\omega_0 t + \check{x}(t)\sin\omega_0 t] \\ b(t) &= -[x(t)\sin\omega_0 t - \check{x}(t)\cos\omega_0 t] \end{aligned} \tag{2.5.18}$$

$a(t)$ and $b(t)$ are known as the *in-phase* and *quadrature* components of $x(t)$. We can easily show that $a(t)$ and $b(t)$ are lowpass processes with power spectra

$$\Phi_a(\omega) = \Phi_b(\omega) = \begin{cases} [\Phi_x(\omega + \omega_0) + \Phi_x(-\omega + \omega_0)] & |\omega| < \omega_0 \\ 0 & \text{otherwise} \end{cases} \tag{2.5.19}$$

It can also be shown that even though the processes $a(t)$ and $b(t)$ are correlated, for fixed t, the random variables $a(t)$ and $b(t)$ are uncorrelated.

We can write the bandlimited process $x(t)$ in terms of the real and imaginary parts of the complex envelope as

$$x(t) = a(t)\cos\omega_0 t - b(t)\sin\omega_0 t \tag{2.5.20}$$

This representation, known as *Rice's representation* [10], is particularly meaningful when $x(t)$ is a Gaussian process. An alternative representation, the so-called envelope and phase form, is obtained by writing Eq. (2.5.20) as

$$x(t) = \epsilon(t)\cos[\omega_0 t + \theta(t)] \tag{2.5.21}$$

with the obvious identifications

$$\epsilon(t) = [a^2(t) + b^2(t)]^{1/2} \tag{2.5.22a}$$

and

$$\theta(t) = \arctan\frac{b(t)}{a(t)} \tag{2.5.22b}$$

It is left as an exercise to the reader to show that if $x(t)$ is Gaussian distributed with mean zero, $\epsilon(t)$ is Rayleigh distributed and $\theta(t)$ is uniformly distributed in $[0, 2\pi)$.

Let $v(t)$ be a real zero-mean stationary Gaussian white process with two-sided power spectral density $N_0/2$ so that its autocorrelation function is

$$\phi_v(\tau) = \mathrm{E}\{v(t)v^*(t + \tau)\} = \frac{N_0}{2}\delta(\tau) \tag{2.5.23}$$

The autocorrelation function of its complex envelope $\tilde{v}(t)$ is given by [11]

$$\phi_{\tilde{v}}(\tau) = N_0\left[\delta(\tau) + j\frac{e^{-j\omega_0\tau}}{\pi\tau}\right] \tag{2.5.24}$$

It can be shown [11] that when operating with complex envelopes, the autocorrelation behaves essentially as if $\phi_{\tilde{v}}(\tau) = 2N_0\,\delta(\tau)$.

Let $\tilde{y}(t)$ represent the complex envelope of a deterministic signal $y(t)$ and let

$$Y = \int_0^T \tilde{v}(t)\tilde{y}^*(t)\,dt \tag{2.5.25}$$

Then

$$\mathrm{E}\{Y\} = \int_0^T \mathrm{E}\{\tilde{v}(t)\}\tilde{y}^*(t)\,dt = 0 \tag{2.5.26}$$

$$\begin{aligned}\sigma_Y^2 = \mathrm{E}\{YY^*\} &= \int_0^T\int_0^T \phi_{\tilde{v}}(t - \tau)\tilde{y}(\tau)\tilde{y}^*(t)\,d\tau\,dt \\ &= 2N_0\int_0^T |\tilde{y}(t)|^2\,dt \\ &= 2N_0,\end{aligned} \tag{2.5.27}$$

if $\tilde{y}(t)$ is normalized such that $\int_0^T |\tilde{y}(t)|^2\,dt = 1$. Hence Y is a complex Gaussian random variable with zero mean and variance $2N_0$. Its density is given by [12]

$$p(y) = \frac{1}{\pi\sigma_Y^2}\exp\left(-\frac{|y|^2}{\sigma_Y^2}\right) \tag{2.5.28}$$

Note that if Y_R and Y_I are the real and imaginary parts of Y, $p(y)$ corresponds to the joint density $p(y_R, y_I)$. That this density can be represented in terms of the single parameter σ_Y^2 is a consequence of the condition $\mathrm{E}\{\tilde{v}(t)\tilde{v}(t + \tau)\} = 0$ imposed by the stationarity of $v(t)$ [13].

Finally, let a bandpass process $x(t)$ be applied as input to a bandpass filter with impulse response $h(t)$. The output of the filter is then given by

$$y(t) = \int_{-\infty}^{\infty} h(t - \tau)x(\tau)\,d\tau \tag{2.5.29}$$

If we let $\tilde{h}(t)$ and $\tilde{x}(t)$ denote the complex envelopes of $h(t)$ and $x(t)$, respectively, it follows from Eq. (2.5.15) that

$$\begin{aligned}y(t) = \frac{1}{4}\int_{-\infty}^{\infty} &[\tilde{h}(t - \tau)e^{j\omega_0(t-\tau)} + \tilde{h}^*(t - \tau)e^{-j\omega_0(t-\tau)}] \\ &\cdot [\tilde{x}(\tau)e^{j\omega_0\tau} + \tilde{x}^*(\tau)e^{-j\omega_0\tau}]\,d\tau\end{aligned} \tag{2.5.30}$$

Since $\tilde{x}(t)$ and $\tilde{h}(t)$ are lowpass signals, we can neglect terms involving $\exp(\pm j2\omega_0\tau)$ in Eq. (2.5.30), as these terms integrate out to approximately zero. We can then write

$$y(t) = \frac{1}{4}\int_{-\infty}^{\infty} [\tilde{h}(t-\tau)\tilde{x}(\tau)e^{j\omega_0 t} + \tilde{h}^*(t-\tau)\tilde{x}^*(\tau)e^{-j\omega_0 t}]\,d\tau \qquad (2.5.31)$$

Comparison with Eq. (2.5.15) then shows that the complex envelope of $y(t)$ is given by

$$\tilde{y}(t) = \frac{1}{2}\int_{-\infty}^{\infty} \tilde{h}(t-\tau)\tilde{x}(\tau)\,d\tau \qquad (2.5.32)$$

This shows that for all practical purposes, the bandpass filtering operation of Eq. (2.5.29) can be replaced by the lowpass operation of Eq. (2.5.32). Thus the processing of bandpass signals by bandpass filters can be analyzed in terms of the lowpass filtering of the complex envelopes of bandpass signals.

2.6 LIKELIHOOD AND SUFFICIENCY

In statistical inference, the concepts of likelihood function and sufficient statistics play important roles. In our subsequent discussions on detection and estimation, we shall come across these concepts quite often. Several procedures for hypothesis testing or parameter estimation involve the likelihood function. Similarly, the notion of sufficient statistics is very useful in reducing the complexity of several problems, even though in some cases, its utility may be restricted. Because of their importance, we devote this section to a brief discussion of these concepts.

2.6.1 Likelihood Function

Consider a set of independent samples $\{x_1, x_2, \ldots, x_n\}$, denoted by $\mathbf{x}$, which are drawn from a population characterized by the density function $p(x \mid \theta)$. Here θ is a scalar or vector parameter of the density function. The likelihood function $L(\theta, \mathbf{x})$ is defined as

$$\begin{aligned} L(\theta, \mathbf{x}) &= p(x_1, x_2, \ldots, x_n \mid \theta) \\ &= \prod_{i=1}^{n} p(x_i \mid \theta) \end{aligned} \qquad (2.6.1)$$

Since the likelihood function is evaluated at a specific set of x_i, it is viewed as a function of the parameter θ with $\mathbf{x}$ fixed at the observed values. The concept of the likelihood function can be extended straightforwardly to multivariate densities.

The utility of the likelihood function arises from the fact that in many applications, we may know that the observed quantities are drawn from a density function of specific form, but certain parameters associated with the density may not be known completely. For example, we may model the observed samples as being from a Gaussian distribution whose mean and/or variance is unknown. A subsequent analysis may then be performed to make some inferences about θ.

Suppose that we want to determine whether a scalar parameter θ is positive or negative. We can hypothesize (assume) that the samples came from one of two possible distributions, the first corresponding to θ being positive and the second corresponding to θ being negative. Given the observations x_i, $1 \le i \le n$, the problem becomes one of deciding which hypothesis is correct and is hence called a hypothesis-testing problem.

If we are interested in finding the value of the parameter θ, the problem becomes one of estimation. Suppose that the right side of Eq. (2.6.1) is larger for a value of $\theta = \theta_1$ than it is for $\theta = \theta_2$. It is then more likely that the observed samples were drawn from the density function $p(x \mid \theta_1)$ than from the density $p(x \mid \theta_2)$. Given a set of observed values x_i, the estimate of θ can therefore be chosen as the value for which the likelihood function is maximum. The estimate is referred to as the *maximum likelihood estimate*, $\hat{\theta}_{ml}$. The reason for referring to $L(\theta, \mathbf{x})$ as the likelihood function is now clear—it provides a measure of how likely it is that the samples came from the density $p(x \mid \theta)$.

We note that although $L(\theta, \mathbf{x})$ is a positive function, it is not a density, since $\int L(\theta, \mathbf{x})\, d\theta$ may not be equal to unity. In general, an optimal procedure for making an inference on θ cannot be derived based solely on the likelihood function. One basic dilemma is whether to treat the likelihood function as a point function or as a set function. Either way, in many cases the results are not satisfactory and not logically consistent [14]. Nevertheless, many optimal statistical procedures designed to satisfy certain objectives turn out to depend on the observations only through likelihood functions. This alone justifies the importance assigned to these functions. We now consider some examples.

Example 2.6.1

Let $x_1, x_2, \ldots, x_n$ be random samples from a Poisson distribution with mean λ. Then

$$L(\lambda, \mathbf{x}) = \frac{e^{-n\lambda} \lambda^{\sum_{i=1}^{n} x_i}}{\prod_{i=1}^{n} x_i!} \tag{2.6.2}$$

Thus $L(\lambda, \mathbf{x})$ depends on $\mathbf{x}$ only through $\sum_{i=1}^{n} x_i$ and $\prod_{i=1}^{n} x_i!$. $L(\lambda, \mathbf{x})$ attains its maximum at $\lambda = (1/n) \sum_{i=1}^{n} x_i$ and hence $\hat{\lambda}_{ml} = (1/n) \sum_{i=1}^{n} x_i$.

Figure 2.6.1 shows a sketch of $L(\lambda, \mathbf{x})$ for $n = 3$, $x_1 = 2$, $x_2 = 0$, and $x_3 = 4$. The ML estimate is $\hat{\lambda}_{ml} = 2$.

Example 2.6.2

As a second example, consider the random samples to be from a Gaussian distribution with mean μ and variance σ^2. The likelihood function is given by

$$L(\mu, \sigma, \mathbf{x}) = \frac{1}{(\sqrt{2\pi}\sigma)^n} \exp\left[-\frac{\sum_{i=1}^{n} (x_i - \mu)^2}{2\sigma^2} \right] \tag{2.6.3}$$

The likelihood function is plotted in Fig. 2.6.2 as a surface plot for $x_1 = -1$, $x_2 = 2$, and $x_3 = 1.5$.

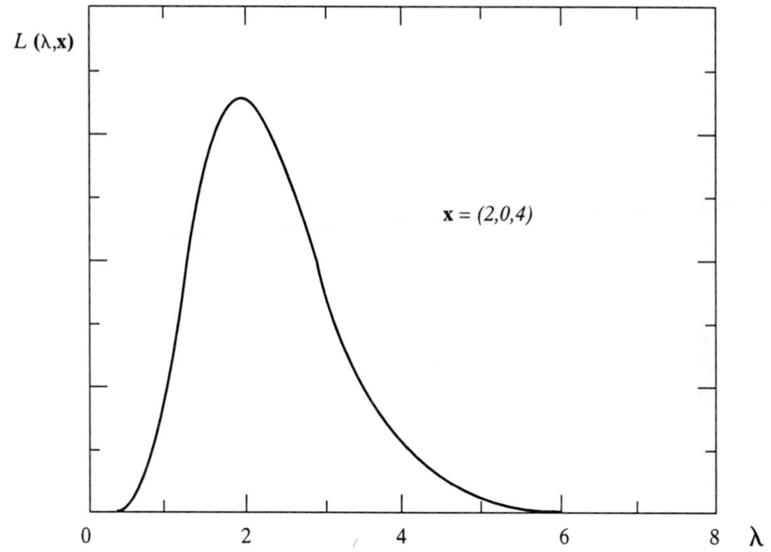

Figure 2.6.1 Likelihood function for Example 2.6.1.

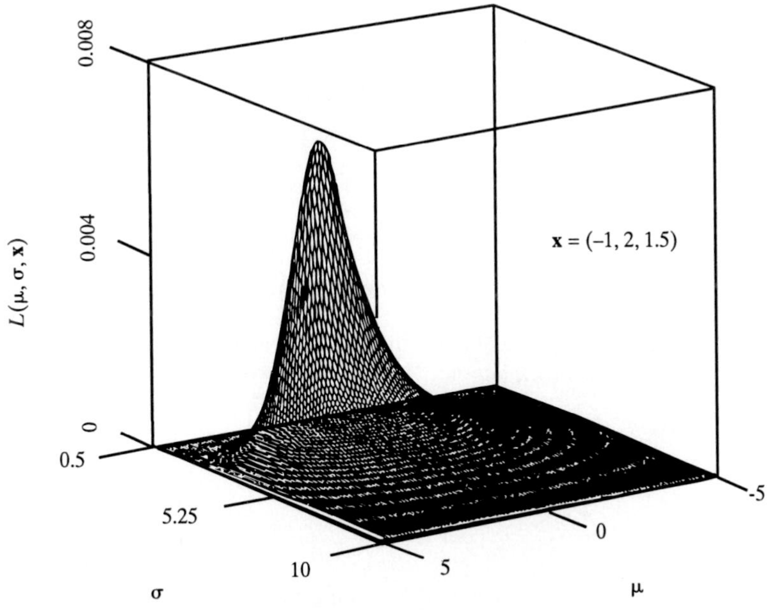

Figure 2.6.2 Likelihood function for Example 2.6.2.

Example 2.6.3

As a final example, consider a set of p-dimensional vectors $\mathbf{x}_i, 1 \le i \le n$, randomly drawn from a multivariate Gaussian density with mean vector μ and covariance matrix Σ. The likelihood function easily follows as

$$L(\mu, \Sigma, \mathbf{x}_1, \mathbf{x}_2, \ldots, \mathbf{x}_n) = \prod_{i=1}^{n} (2\pi)^{-p/2} |\Sigma|^{-1/2} \exp[-\tfrac{1}{2}(\mathbf{x}_i - \mu)^T \Sigma^{-1} (\mathbf{x}_i - \mu)] \tag{2.6.4}$$

In our subsequent discussions on detection, we will find that the *likelihood ratio* plays an important role. This is the ratio of the likelihood functions obtained by evaluating $L(\theta, \mathbf{x})$ at two values, $\theta = \theta_1$ and $\theta = \theta_2$.

Let $t(x_1, x_2, \ldots, x_n)$ be a function of only the observations. Then t is said to be a *statistic*. Note that given a specific set of observed values, the statistic is fully determined.

A family of densities $p(x \mid \theta)$ is said to have a *monotone likelihood ratio*

$$\Lambda = \frac{L(\theta_1, \mathbf{x})}{L(\theta_2, \mathbf{x})} \tag{2.6.5}$$

if there exists a statistic t such that Λ is either a nonincreasing or a nondecreasing function of t for every $\theta_1 < \theta_2$.

Example 2.6.4

Let $x_1, x_2, \ldots, x_n$ be a set of independent random variables uniformly distributed on $[0, \theta]$. Let $\theta_1 < \theta_2$. We have

$$\Lambda = \frac{L(\theta_1, \mathbf{x})}{L(\theta_2, \mathbf{x})} = \frac{(1/\theta_1^n) \prod_{i=1}^{n} I_{(0,\theta_1)}(x_i)}{(1/\theta_2^n) \prod_{i=1}^{n} I_{(0,\theta_2)}(x_i)} \tag{2.6.6}$$

where $I_{(\ldots)}(\cdot)$ is the indicator function defined by

$$I_{(0,\theta)}(x) = \begin{cases} 1 & 0 \le x \le \theta \\ 0 & \text{otherwise} \end{cases} \tag{2.6.7}$$

We choose as our statistic the quantity $t = \max\{x_1, x_2, \ldots, x_n\}$. t is referred to as the *maximum order statistic*. Now Λ can be written as

$$\begin{aligned} \Lambda &= \frac{(1/\theta_1^n) I_{(0,\theta_1)}(t)}{(1/\theta_2^n) I_{(0,\theta_2)}(t)} \\ &= \begin{cases} (\theta_2/\theta_1)^n & 0 \le t \le \theta_1 \\ 0 & \theta_1 < t \le \theta_2 \end{cases} \end{aligned} \tag{2.6.8}$$

Figure 2.6.3 is a sketch of Λ as a function of t for fixed θ_1 and θ_2. It is clear from the figure that the uniform density over $(0, \theta)$ has a monotone likelihood ratio.

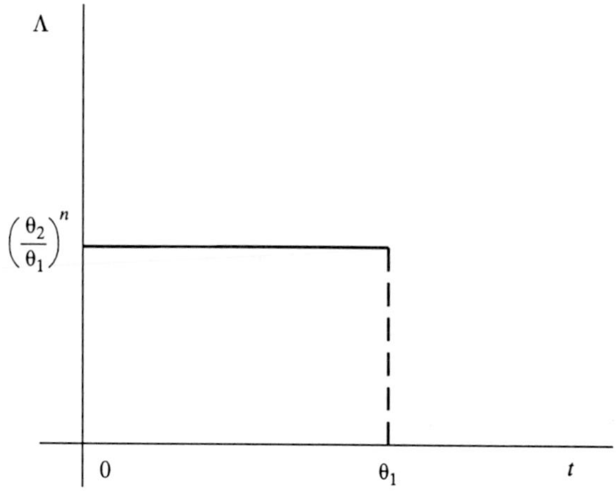

Figure 2.6.3 Likelihood ratio for Example 2.6.4.

2.6.2 Sufficiency

We have seen earlier that a statistic is a function that is totally determined by the observations. We have also seen that inferences regarding the statistical properties of $\mathbf{X}$, such as the value of a parameter θ associated with its density function, are made from a statistic $\mathbf{T}(\mathbf{X})$. In general, $\mathbf{T}(\mathbf{X})$ can be a k-dimensional vector $[T_1(\mathbf{X}), T_2(\mathbf{X}), \ldots, T_k(\mathbf{X})]$. If $k < n$, where n is the dimension of $\mathbf{X}$, making inferences about θ from $\mathbf{T}(\mathbf{X})$ may be easier than making them from $\mathbf{X}$. However, the mapping from $\mathbf{X}$ to $\mathbf{T}(\mathbf{X})$ is not one-to-one and not all the information about θ that is contained in $\mathbf{X}$ may be contained in $\mathbf{T}(\mathbf{X})$. It is therefore desirable to find a statistic $\mathbf{S}(\mathbf{X})$ of lower dimension that contains as much information as $\mathbf{X}$ about θ. Such a statistic is called a *sufficient statistic* for θ.

Formally, we say that a statistic $\mathbf{S}$ is a sufficient statistic (for θ) if the conditional density $p(x_1, x_2, \ldots, x_n \mid \mathbf{S}(x_1, x_2, \ldots, x_n) = \mathbf{s})$ is independent of θ. From the definition it can be seen that if $\mathbf{S}(\cdot)$ is a sufficient statistic, no further information about θ can be obtained from $\mathbf{X}$ than is provided by $\mathbf{S}(\cdot)$, and hence an inference about θ made from $\mathbf{S}(\cdot)$ will be equivalent to one made from the original observations $\mathbf{X}$.

Example 2.6.5

Suppose that we want to determine if a coin is biased by tossing it n times and observing the outcomes. Let the random variable X_i denote the outcome of the ith toss, with X_i being 1 or 0 depending on whether a head or a tail shows. Let θ denote the probability of a head showing in a random toss. Clearly, if the coin is unbiased, $\theta = \frac{1}{2}$.

The joint probability of the outcomes after n tosses is given by

$$P(x_1, x_2, \ldots, x_n) = \prod_{i=1}^{n} (\theta)^{x_i}(1 - \theta)^{(n-x_i)} \tag{2.6.9}$$

Let $s = \sum_{i=1}^{n} x_i$. Then

$$P(x_1, x_2, \ldots, x_n) = (\theta)^s(1 - \theta)^{(n-s)} \tag{2.6.10}$$

Suppose in a particular sequence of n tosses the outcomes are such that $\sum_{i=1}^{n} X_i = S = s$. Then

$$P(x_1, x_2, \ldots, x_n \mid S = s) = \frac{(\theta)^s(1-\theta)^{(n-s)}}{P(S = s)}$$
$$= \frac{(\theta)^s(1-\theta)^{(n-s)}}{\binom{n}{s}(\theta)^s(1-\theta)^{(n-s)}} = \frac{1}{\binom{n}{s}} \tag{2.6.11}$$

Since the conditional probability $P(\mathbf{x} \mid S)$ is independent of θ, $S = \sum_{i=1}^{n} X_i$ is a sufficient statistic. Note that S is just the number of times *heads* shows in a sequence of n tosses. Thus, to make an inference on θ, it is only necessary to count the number of times a head appears and it is not necessary to note down the outcome of each trial.

The quantity S/n can be used as an estimate of θ.

In Example 2.6.5 we made a guess as to what the sufficient statistic was and then established that it was indeed a sufficient statistic by verifying that it satisfied the formal definition given earlier. In general this is hard to do for arbitrary densities. Fortunately, there is a theorem called the *factorization theorem* which allows us to identify a sufficient statistic in most cases of practical interest. (The theorem is strictly valid only for the case of *dominated* densities.) The theorem states that if $p(x_1, x_2, \ldots, x_n \mid \theta)$ can be factored as

$$p(x_1, x_2, \ldots, x_n \mid \theta) = g(\mathbf{s}(x_1, x_2, \ldots, x_n), \theta)h(x_1, x_2, \ldots, x_n)m(\theta) \tag{2.6.12}$$

where $h(x_1, x_2, \ldots, x_n) > 0$, then $\mathbf{S}(X_1, X_2, \ldots, X_n)$ is a sufficient statistic (for θ). The proof of this result may be found in [15,16].

We note that a sufficient statistic is not unique and that any monotone one-to-one function of a sufficient statistic is also a sufficient statistic. For the coin-tossing problem of Example 2.6.5, the quantities $\sum_{i=1}^{n} x_i$, $(1/n)\sum_{i=1}^{n} x_i$, and $\sqrt{\sum_{i=1}^{n} x_i}$ are all sufficient.

Example 2.6.6

Let $x_i, 1 \le i \le n$, be randomly drawn from a density that is uniform over $(0, \theta)$ so that

$$p(\mathbf{x} \mid \theta) = \frac{1}{\theta^n} \prod_{i=1}^{n} I_{(0,\theta)}(x_i) \tag{2.6.13}$$

The right side of Eq. (2.6.13) can be factored as

$$p(\mathbf{x} \mid \theta) = \frac{1}{\theta^n}\left[\prod_{i=2}^{n-1} I_{(x(i-1), x(i+1))}(x_{(i)})\right] I_{(0, x(2))}(x_{(1)}) I_{(0,\theta)}(x_{(n)}) \tag{2.6.14}$$

where $x_{(1)}, x_{(2)}, \ldots, x_{(n)}$ are the ordered samples obtained by arranging $x_1, x_2, \ldots, x_n$ in increasing order of their values. Equation (2.6.14) satisfies the factorization theorem, with the sufficient statistic being the largest order statistic, $S(\mathbf{X}) = X_{(n)}$.

Example 2.6.7

Let x be a single observation from a zero-mean Gaussian density with variance σ^2. Let $\theta = \sigma^2$ so that

$$p(x \mid \theta) = \frac{1}{\sqrt{2\pi\theta}} \exp\left(-\frac{x^2}{2\theta}\right) \tag{2.6.15}$$

By letting $h(x) = 1, m(\theta) = 1/\sqrt{2\pi\theta}, g(s(x), \theta) = \exp(-x^2/2\theta)$, we see that Eq. (2.6.15) satisfies Eq. (2.6.12), so that the sufficient statistic is $S = X^2$. Clearly, $|X|$ is also a sufficient statistic, as it is a monotonic function of X^2. Further, since $|X|$ can be found from X, X is also a sufficient statistic for θ. However, since X cannot be recovered from $|X|$, we say that the latter is condensed.

Example 2.6.8

Let $x_i, 1 \leq i \leq n$, be drawn from a Gaussian density with mean μ and variance σ^2 so that

$$p(\mathbf{x}|\mu, \sigma^2) = \left(\frac{1}{\sqrt{2\pi}\,\sigma}\right)^n \exp\left[-\sum_{i=1}^{n} \frac{(x_i - \mu)^2}{2\sigma^2}\right]$$

$$= \exp\left(-\frac{1}{2\sigma^2}\sum_{i=1}^{n} x_i^2\right) \exp\left(\mu \frac{\sum_{i=1}^{n} x_i}{\sigma^2}\right) \left(\frac{1}{\sqrt{2\pi}\,\sigma}\right)^n \exp\left(-\frac{n\mu^2}{2\sigma^2}\right) \tag{2.6.16}$$

which is of the form of Eq. (2.6.12) with

$$g(\mathbf{s}(\mathbf{x})) = \exp\left(-\frac{1}{2\sigma^2}\sum_{i=1}^{n} x_i^2\right) \exp\left(\mu \frac{\sum_{i=1}^{n} x_i}{\sigma^2}\right) \tag{2.6.17}$$

$$h(\mathbf{x}) = 1$$

and

$$m(\mu, \sigma^2) = \left(\frac{1}{\sqrt{2\pi}\,\sigma}\right)^n \exp\left(-\frac{n\mu^2}{2\sigma^2}\right)$$

Thus the sufficient statistic is two-dimensional in this example and is given by

$$\mathbf{S} = [S_1 \quad S_2]^T = \left(\sum_{i=1}^{n} X_i^2 \quad \sum_{i=1}^{n} X_i\right)^T \tag{2.6.18}$$

In Example 2.6.8 both the mean and the variance were assumed to be unknown, and the sufficient statistic $\mathbf{S}(\mathbf{X})$ turned out to be a two-dimensional vector. The order in which the components of $\mathbf{S}(\cdot)$ were arranged was arbitrary and could as well have been reversed. Thus in the case where θ is a vector, while $\mathbf{S}$ is also a vector, this does not, however, imply that the first component of $\mathbf{S}$ is a sufficient statistic for the first component of θ, the second component of $\mathbf{S}$ is sufficient for the second component of θ, and so on.

If some of the parameters of the density function are known, the factorization condition must be applied with the known parameters being included in the function $h(\cdot)$. Thus, in Example 2.6.8, if the variance is known, it can be verified that $\sum_{i=1}^{n} X_i$ is a sufficient statistic for μ, whereas if the mean is known, $\sum_{i=1}^{n} X_i^2$ is a sufficient statistic for σ^2.

While the factorization theorem is helpful in identifying a sufficient statistic, the examples above show that the procedure is dependent on our ability to factorize the conditional density appropriately. If we are not able to perform the factorization, we may not be able to identify the sufficient statistic.

As we will see from our discussion in subsequent chapters, the concept of a sufficient statistic is useful only if there is a reduction in the complexity of a problem. Clearly, the set of samples $(X_1, X_2, \ldots, X_n)$ by definition constitutes a sufficient statistic. If the samples are from a continuous density, the order statistic $(X_{(1)}, X_{(2)}, \ldots, X_{(n)})$ can be shown to be sufficient [17]. Since the original samples cannot be recovered from the order statistic, the order statistic is more condensed than the original samples. As we have seen, a sufficient statistic condenses the data without any loss of information about the parameter. A sufficient statistic that provides the greatest condensation of the data may then be called a minimal sufficient statistic. Formally, a statistic is minimal sufficient if and only if it is a function of every other sufficient statistic. In Example 2.6.7, $|X|$ is minimal sufficient. If the dimension of the statistic $\mathbf{S}(\cdot)$ does not grow with sample size n, inferences regarding θ can be made with a fixed complexity without regard to n. Such a statistic is said to be a *fixed-dimensional sufficient statistic*.

Finally, as we shall see later, while the concept of a sufficient statistic is useful in many applications, in order to find optimal estimates of the parameter θ, $\theta \in \Theta$, we require the concept of completeness of density functions. Let $X_1, X_2, \ldots, X_n$ be a random sample from a density $p(\cdot|\theta)$, and let $T = t(X_1, X_2, \ldots, X_n)$ be a statistic. The family of densities of T is said to be complete if and only if $\mathrm{E}_\theta(Z(T)) = 0$ for all $\theta \in \Theta$ implies that $P_\theta(Z(T) = 0) = 1$ for some statistic $Z(T)$. In that case we say that the statistic T is also complete.

Example 2.6.9

Let $x_1, x_2, \ldots, x_n$ be a random sample drawn from a Bernoulli distribution. The statistic $T = X_1 - X_n$ is not complete since $\mathrm{E}_\theta(X_1 - X_2) = 0$, but $X_1 - X_n$ is not zero with probability 1.

Consider the statistic $T = \sum_{i=1}^n X_i$ and let $\mathrm{E}_\theta(Z(T)) = 0$ for some $Z(T)$ for all $\theta \in (0, 1)$. If T is complete, then $z(t) = 0$ for any $t \in (0, 1, 2, \ldots, n)$. Now

$$\begin{aligned} \mathrm{E}_\theta(Z(T)) &= \sum_{t=0}^{n} z(t)\tbinom{n}{t}\theta^t(1-\theta)^{n-t} \\ &= (1-\theta)^n \sum_{t=0}^{n} z(t)\tbinom{n}{t}\left(\frac{\theta}{1-\theta}\right)^t \end{aligned} \tag{2.6.19}$$

Hence $\mathrm{E}_\theta(Z(T)) = 0$ for all $0 \le \theta \le 1$ implies that

$$\sum_{t=0}^{n} z(t)\tbinom{n}{t}y^t = 0 \tag{2.6.20}$$

where $y = \theta/(1-\theta)$.

Since the left-hand side of Eq. (2.6.20) is a polynomial in y of degree n, it follows that the coefficients $z(t)\binom{n}{t}$ must be zero for each t. That is, $z(t) = 0$ for all $t = 0, 1, 2, \ldots, n$.

While we were able to demonstrate in Example 2.6.9 that $\sum_{i=1}^n X_i$ is complete, determining completeness is not easy in general.

In general, any complete statistic T does provide some form of data reduction. If T is also sufficient, it is said to be a complete and sufficient statistic.

2.6.3 Exponential Densities

As our previous discussion shows, finding a sufficient statistic for general density functions is not an easy task, and in fact a sufficient statistic of fixed dimension may not even exist. A family of densities for which a fixed-dimensional sufficient statistic exists is the exponential family. A one-parameter family of exponential densities characterized by the scalar θ can be expressed as

$$p(x \mid \theta) = a(\theta)b(x)\exp[c(\theta)\,d(x)] \tag{2.6.21}$$

where $a(\cdot)$, $b(\cdot)$, $c(\cdot)$, and $d(\cdot)$ are functions of their respective arguments.

A k-parameter exponential family of densities $p(x \mid \theta_1, \theta_2, \ldots, \theta_k)$ can be expressed as

$$p(x \mid \theta_1, \ldots, \theta_k) = a(\theta_1, \ldots, \theta_k)b(x)\exp\left[\sum_{i=1}^{k} c_i(\theta_1, \ldots, \theta_k)\,d_i(x)\right] \tag{2.6.22}$$

As an example, let us write the Gaussian density with mean μ and variance σ^2 as

$$p(x \mid \mu, \sigma) = \left[\frac{1}{\sqrt{2\pi}\,\sigma}\exp\left(-\frac{\mu^2}{2\sigma^2}\right)\right]\exp\left(-\frac{1}{2\sigma^2}x^2 + \frac{\mu}{\sigma^2}x\right) \tag{2.6.23}$$

Clearly, Eq. (2.6.23) is of the form of Eq. (2.6.22) with

$$\theta_1 = \mu,\ \theta_2 = \sigma^2 \qquad a(\theta_1, \theta_2) = \left[\frac{1}{\sqrt{2\pi}\,\sigma}\exp\left(-\frac{\mu^2}{2\sigma^2}\right)\right] \qquad b(x) = 1$$

$$c_1(\theta_1, \theta_2) = -\frac{1}{2\sigma^2} \qquad c_2(\theta_1, \theta_2) = \frac{\mu}{\sigma^2} \qquad d_1(x) = x^2 \qquad d_2(x) = x$$

Similarly, the beta density

$$p(x \mid \theta_1, \theta_2) = \frac{\Gamma(\theta_1 + \theta_2)}{\Gamma(\theta_1)\Gamma(\theta_2)}x^{\theta_1 - 1}(1 - x)^{\theta_2 - 1}I_{(0,1)}(x) \tag{2.6.24}$$

can be put in the form of Eq. (2.6.22) by setting

$$a(\theta_1, \theta_2) = \frac{\Gamma(\theta_1 + \theta_2)}{\Gamma(\theta_1)\Gamma(\theta_2)} \qquad b(x) = I_{(0,1)}(x)$$

$$c_1(\theta_1, \theta_2) = \theta_1 - 1 \qquad c_2(\theta_1, \theta_2) = \theta_2 - 1 \qquad d_1(x) = \ln x \qquad d_2(x) = \ln(1 - x)$$

While many standard densities, such as Poisson, simple exponential, gamma, Gaussian, and Bernoulli, belong to the exponential family, others, such as Laplace or uniform densities, do not. In fact, if the region of support for the density (the region over which the density is not zero) depends on θ, the density cannot be expressed in the form of Eq. (2.6.21) or (2.6.22) and hence does not belong to an exponential family of densities.

Given a set of random samples x_i from an exponential density, a sufficient statistic of fixed dimension exists. For the single-parameter family of Eq. (2.6.21), this statistic is given by $\sum_{i=1}^{n} d(X_i)$, while for the multiparameter family of Eq. (2.6.22) it is given by $\mathbf{S} = [\sum_{i=1}^{n} d_1(X_i) \quad \sum_{i=1}^{n} d_2(X_i) \quad \cdots \quad \sum_{i=1}^{n} d_k(X_i)]^T$.

This result can be verified by applying the factorization theorem. In fact, for the exponential family, the sufficient statistics identified by the factorization theorem are also minimal.

Finally, it has been shown that if the density from which a set of observations are drawn belongs to the one-parameter exponential family of Eq. (2.6.21), the sufficient statistic $T = \sum_{i=1}^{n} d(X_i)$, in addition to being minimal, is also complete. Also, if the parameter space corresponding to a k-parameter exponential family is k-dimensional, the family is complete [15,17].

2.7 SUMMARY

In this chapter we have reviewed briefly some results from systems theory and stochastic processes that will be of use to us in our discussions in subsequent chapters. Section 2.2 was concerned with the state-space representation of both continuous- and discrete-time systems and the solution of the state equations.

In Section 2.3 we discussed briefly a stochastic process of some importance, the Wiener process. Formal differentiation of this process yields the white noise process, which although not physically realizable, plays a useful role in the modeling of stochastic processes in many engineering applications. In Section 2.4 we discussed an important class of signal models, the Gauss–Markov models, and derived equations that describe the propagation of the mean and variance of the signal process.

In Section 2.5 we considered the Karhunen–Loéve expansion for wide-sense-stationary processes. The basis functions for the expansion are the eigenfunctions of an integral equation whose kernel is the covariance function of the process. The statistics of the random process can then be described in terms of the statistics of a set of uncorrelated random variables, the coefficients of the expansion. In this section we also considered the representation of narrowband stochastic processes in terms of their complex envelopes. Such processes arise naturally in modulation systems in which the message signal is used to modulate a carrier. Use of the complex envelope permits operations on these signals to be described in terms of the baseband signals.

In Section 2.6 we introduced the concepts of likelihood functions and sufficient statistics, which play important roles in statistical inference, and discussed their utility in applications. We concluded the section with a presentation of the family of exponential densities, which are important in that a fixed-dimensional sufficient statistic exists for this family.

REFERENCES

1. C. T. Chen, *Linear System Theory and Design*, Holt, Rinehart and Winston, New York, 1984.
2. T. Kailath, *Linear Systems*, Prentice Hall, Englewood Cliffs, NJ, 1980.
3. A. Papoulis, *Probability, Random Variables and Stochastic Processes*, 3rd ed., McGraw-Hill, New York, 1991.
4. J. L. Doob, *Stochastic Processes*, Wiley, New York, 1953.
5. E. Wong, *Stochastic Processes in Information and Dynamical Systems*, McGraw-Hill, New York, 1971.
6. K. Ito, *Lectures in Stochastic Processes*, Tata Institute of Fundamental Research, Bombay, India, 1961.
7. I. I. Gikhman and A. V. Skorokhod, *Introduction to the Theory of Random Processes*, W.B. Saunders, Philadelphia, 1969.
8. T. P. McGarty, *Stochastic Systems and State Estimation*, Wiley-Interscience, New York, 1974.
9. H. V. Poor, *An Introduction to Signal Detection and Estimation*, Springer-Verlag, New York, 1988.
10. S. O. Rice, Mathematical analysis of random noise, *Bell Syst. Tech. J.*, **23**, 282–332 (1944).
11. T. Kailath, The complex envelope of white noise, *IEEE Trans. Inf. Theory*, **IT-12**, No. 4, July, 397–398 (1966).
12. B. Picinbono, *Random Signals and Systems*, Prentice Hall, Englewood Cliffs, NJ, 1993.
13. C. W. Therrien, *Discrete Random Signals and Statistical Signal Processing*, Prentice Hall, Englewood Cliffs, NJ, 1992.
14. J. K. Ghosh, Ed., *Statistical Information and Likelihood: A Collection of Critical Essays by Dr. D. Basu*, Springer-Verlag, New York, 1988.
15. E. L. Lehmann, *Theory of Point Estimation*, Wiley, New York, 1983.
16. T. S. Ferguson, *Mathematical Statistics: A Decision Theoretic Approach*, Academic Press, New York, 1967.
17. S. Zacks, *Parametric Statistical Inference*, Pergamon Press, Oxford, 1981.

3

Detection Theory

3.1 INTRODUCTION

There are many situations in which we are required to choose among several possibilities. For example, in a radar detection problem, the radar return is observed and a decision has to be made as to whether a target was present or not. In a digital communication system, one of several possible waveforms is selected for transmission to the receiver. The waveforms may undergo some distortion in the transmission medium. Receiver noise contributes further to the distortion. The consequence is that at the receiver, it is no longer clear which of the signals was transmitted. The receiver has to make a decision as to which signal is actually present based on noisy observations. In pattern recognition problems a given pattern belongs to one of a fixed number of classes. For example, a given alphanumeric character belongs to the set of alphabets and numerals from 0 to 9. The pattern classifier must decide to which class (alphanumeric character) a given test pattern belongs. This decision has to be made by the classifier, which is an automatic machine, based on the noisy measurements and any statistical information regarding the various pattern classes.

In each case the solution involves making a decision among several choices. In this chapter we introduce the basic tools necessary for solving the decision problem, where we assume that the observables are a set of random variables. Such problems have been discussed by statisticians for many years and fall under the general category of statistical inference. In chapter 4 we extend these results to decision problems involving waveforms.

3.2 HYPOTHESIS TESTING

Signal detection or classification essentially belongs to a class of statistical problems dealing with hypothesis testing. Hypothesis testing is discussed in many books (see, for example, [1–7]). In a typical hypothesis-testing problem, given an observation or a set of observations, a decision has to be made regarding the source of the observations. We can think of a hypothesis as a statement of a possible source of observations. If the set of hypotheses is only two, the problem is a *binary* hypothesis-testing problem. If there are M hypotheses with $M > 2$, it is a *multiple-hypothesis* testing or M-ary detection problem. For example, in radar detection, the hypotheses are the statements "The observations correspond to target not being present" and "Observations correspond to target present," or simply, "Target not present" and "Target present." We label the two possible choices H_0 and H_1. We refer to H_0 as the *null hypothesis*.

Suppose that corresponding to each hypothesis we have an *observation*, a random variable that is generated according to some probabilistic law. The hypothesis-testing problem is one of deciding which hypothesis is the correct one, based on a single measurement, z, of this random variable. The range of values that z takes constitutes the *observation space*. The decision problem in the case of two hypotheses thus essentially consists in partitioning this one-dimensional space into two regions, Z_0 and Z_1, such that whenever z lies in Z_0, we decide that H_0 was the correct hypothesis, and whenever z lies in Z_1, we decide that H_1 was the correct hypothesis. The regions Z_0 and Z_1 are known as *decision regions*. Whenever a decision does not match the true hypothesis, we say that an *error* has occurred. The problem lies in choosing the decision regions such that we obtain the fewest errors with several realizations of Z.

More generally, we may have a set of observations, $\mathbf{z} = (z_1, z_2, \ldots, z_n)$. The observation space in this case is n-dimensional. As we shall see in Section 3.4, the hypothesis-testing problem with multiple measurements is essentially the same as with a single measurement, and many of the results we will derive for the case of a single observation carry over to multiple observations.

In most problems the observations $Z_1, Z_2, \ldots, Z_n$ are either continuous random variables with joint density functions $p(z_1, z_2, \ldots, z_n)$, or discrete random variables with joint probability mass functions $P(z_1, z_2, \ldots, z_n)$. In what follows we can simply extend any results that are developed for continuous random variables to discrete variables by replacing $p(\cdot)$ by $P(\cdot)$ and integrals by appropriate summations.

In this chapter we consider the parametric hypothesis-testing problem and defer until Chapter 7 a discussion of nonparametric tests. In the parametric problem, the density function under every hypothesis, $p(z_1, z_2, \ldots, z_n \mid H_i)$, is of a known form and characterized by a finite set of parameters, θ. In many problems of interest, θ is a scalar, although more generally, it could be vector-valued. The hypothesis is said to be a *simple hypothesis* if the parameter θ characterizing the hypothesis assumes a specific value. Thus, if under H_0, θ is equal to θ_0 while it is equal to θ_1 under H_1, the problem becomes one of testing two simple hypotheses against each other and is called *simple versus simple hypothesis testing*. On the other hand, if θ can belong to a set of values Θ_0 under H_0 and Θ_1 under H_1, where Θ_0 and Θ_1 are mutually exclusive, the problem is a *composite hypothesis-testing problem*. In many cases

H_0 will be simple while H_1 may be composite. Therefore, in general, a binary hypothesis-testing problem can be formulated as follows:

$$\begin{aligned} H_0&: Z_1, Z_2, \ldots, Z_n \sim p(z_1, z_2, \ldots, z_n \mid \Theta_0) \\ H_1&: Z_1, Z_2, \ldots, Z_n \sim p(z_1, z_2, \ldots, z_n \mid \Theta_1) \end{aligned} \tag{3.2.1}$$

or, equivalently, as

$$\begin{aligned} H_0&: \theta \in \Theta_0 \\ H_1&: \theta \in \Theta_1 \end{aligned} \tag{3.2.2}$$

Example 3.2.1

Let $Z_1, Z_2, \ldots, Z_n$ be independent, identically distributed (i.i.d.) Gaussian variables with mean θ and unity variance. The problem is to decide whether or not θ is zero. The two hypotheses become

$$\begin{aligned} H_0&: \theta = 0 \\ H_1&: \theta \neq 0 \end{aligned} \tag{3.2.3}$$

Thus Θ_0 consists of a single value, the origin, while Θ_1 consists of the entire real line minus the origin, so that H_0 is simple while H_1 is composite. The density functions of the observations under the two hypotheses are given by

$$\begin{aligned} H_0&: p(z_1, z_2, \ldots, z_n \mid \theta = 0) = \frac{1}{(\sqrt{2\pi})^n} \exp\left(-\frac{1}{2}\sum_{i=1}^{n} z_i^2\right) \\ H_1&: p(z_1, z_2, \ldots, z_n \mid \theta \neq 0) = \frac{1}{(\sqrt{2\pi})^n} \exp\left[-\frac{1}{2}\sum_{i=1}^{n} (z_i - \theta)^2\right] \end{aligned} \tag{3.2.4}$$

Example 3.2.2

Let $Z_1, Z_2, \ldots, Z_n$ be i.i.d. random variables with mean zero and unity variance. Let the density function of Z_i under H_0 be Gaussian, $f_G(z_i)$, while it is Laplacian, $f_L(z_i)$, under H_1. Clearly, the two hypotheses are both simple in this example. We can formulate the hypotheses in terms of a parameter θ by writing

$$p(z_1, z_2, \ldots, z_n \mid \theta) = \theta \prod_{i=1}^{n} f_L(z_i) + (1 - \theta) \prod_{i=1}^{n} f_G(z_i) \tag{3.2.5}$$

The hypotheses then become

$$\begin{aligned} H_0&: \theta = 0 \\ H_1&: \theta = 1 \end{aligned} \tag{3.2.6}$$

As stated earlier, the problem of deciding between two hypotheses is essentially one of partitioning the observation space into two appropriate regions. To quantify our judgment as to the best possible partitioning of this space, we first have to select a suitable criterion. While we will be discussing several such criteria in the sequel, let us consider a simple versus simple hypothesis-testing problem and adopt the following procedure. If $P(H_i \mid z), i = 0, 1$, denotes the probability that H_i was the true hypothesis given a particular value of the observation, we decide that the correct hypothesis is the one corresponding to the larger of the two probabilities. The decision rule will then be to choose H_0 if

$$P(H_0 \mid z) > P(H_1 \mid z) \tag{3.2.7}$$

and choose H_1 otherwise. We write the rule as

$$\frac{P(H_1 \mid z)}{P(H_0 \mid z)} \underset{H_0}{\overset{H_1}{\gtrless}} 1 \tag{3.2.8}$$

The criterion we have used is known as the *maximum a posteriori probability* (MAP) criterion, since we are choosing the hypothesis that corresponds to the maximum of the two posterior probabilities.

We can use Bayes rule to write the criterion in a more useful form. We thus write

$$P(H_i \mid z) = \frac{p(z \mid H_i)P(H_i)}{p(z)} \qquad i = 0, 1 \tag{3.2.9}$$

where $P(H_i)$ denotes the probability that source H_i was active, so that

$$\frac{P(H_1 \mid z)}{P(H_0 \mid z)} = \frac{p(z \mid H_1)}{p(z \mid H_0)} \frac{P(H_1)}{P(H_0)}$$

and the test becomes

$$\frac{p(z \mid H_1)}{p(z \mid H_0)} \underset{H_0}{\overset{H_1}{\gtrless}} \frac{P(H_0)}{P(H_1)} \tag{3.2.10}$$

The ratio $\Lambda(z) = p(z \mid H_1)/p(z \mid H_0)$ can be recognized as the likelihood ratio that we discussed in Section 2.6.1. The test therefore consists of comparing this ratio with a constant termed the threshold and is called a likelihood ratio test (LRT). We will see later that tests based on several other criteria also fall into the general class of likelihood ratio tests. But the thresholds will, in general, be different for different criteria.

Example 3.2.3

Consider a simple binary communication channel in which the source transmits over each T-second interval, either a signal pulse $y(t)$ of unit amplitude or no signal at all. Such a system is referred to as an *on–off keying* (OOK) *system* in digital communications. The communication channel adds noise $v(t)$ so that the received signal is either $y(t) + v(t)$ or $v(t)$ over each T-second interval. We observe the received signal at some instant during each signaling interval. The problem is to decide, on the basis of a single observation, whether or not the signal is present.

We will call the event that no signal was transmitted the *null hypothesis* and the *alternative hypothesis*, when a unit amplitude signal is transmitted, H_1. The received signal under the two hypotheses can then be written as

$$H_1\colon Z = 1 + V$$

$$H_0\colon Z = V$$

Let v be Gaussian, with zero mean and unit variance. The probability density of Z under each hypothesis follows easily as

$$p(z \mid H_0) = \frac{1}{\sqrt{2\pi}} \exp\left(-\frac{z^2}{2}\right)$$

and

$$p(z \mid H_1) = \frac{1}{\sqrt{2\pi}} \exp\left[-\frac{(z-1)^2}{2}\right]$$

The likelihood ratio is then given by

$$\Lambda(z) = \frac{p(z \mid H_1)}{p(z \mid H_0)} = \exp\left(z - \frac{1}{2}\right)$$

The decision rule is

$$\exp\left(z - \frac{1}{2}\right) \underset{H_0}{\overset{H_1}{\gtrless}} \frac{P(H_0)}{P(H_1)} \tag{3.2.11}$$

It is often more convenient to work with the natural logarithm of the likelihood ratio. Since the logarithm is a monotonically increasing function, the inequality in the decision rule will still hold and we can write the rule in terms of the log-likelihood ratio as

$$\ln \Lambda(z) = z - \frac{1}{2} \underset{H_0}{\overset{H_1}{\gtrless}} \ln \frac{P(H_0)}{P(H_1)}$$

or

$$z \underset{H_0}{\overset{H_1}{\gtrless}} \frac{1}{2} + \ln \frac{P(H_0)}{P(H_1)}$$

The decision rule is therefore to sample the received signal and compare it with a threshold. If the sample is greater than the threshold, it is decided that a "1" was sent; otherwise, it is decided that a "0" was sent. In general a test involves comparing a statistic with a threshold. We will denote the test statistic by $l(z)$.

Example 3.2.3 illustrates the basic components of the decision problem. The first component is the *source*. The second component is the *probabilistic mechanism*, which generates an observation z according to some probability law, depending on which hypothesis is true. The third component is the *observation space* Z, which could be multidimensional. In the preceding example, the probability densities of interest are $p(z \mid H_0)$ and $p(z \mid H_1)$, and the observation space is the entire real line from $-\infty$ to ∞. The decision rule then divides the observation space into the two regions Z_0 and Z_1 and assigns each point in the two regions to one of the two hypotheses. These concepts for the example are illustrated in Fig. 3.2.1.

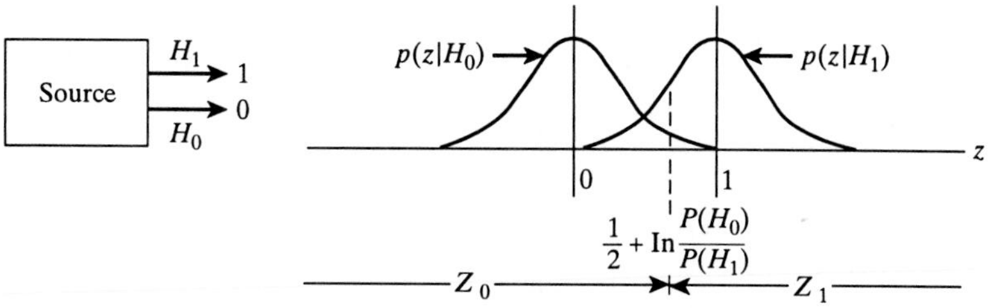

Figure 3.2.1 Source, prior probabilities, and decision regions for Example 3.2.3.

3.3 DECISION CRITERIA

In the simple example considered in Section 3.2 we divided the observation space into two regions in order to decide which hypothesis was true. It is clear that irrespective of where this point of division is chosen to be, we will occasionally make a wrong decision. Let us denote by D_i, our choice of H_i as the outcome of a test procedure.

In making a decision in any binary hypothesis-testing problem, we have four possibilities to consider: (1) H_0 is the true hypothesis, we decide D_0; (2) H_1 is the true hypothesis, we decide D_1; (3) H_0 is the true hypothesis, we decide D_1; and (4) H_1 is the true hypothesis, we decide D_0. The first two correspond to correct choices. The last two correspond to errors. In the statistical literature (3) is called a *type I error* and (4) a *type II error*. In radar terminology, these are referred to as *false alarm* (decide target is present when it is not) and *miss* (decide no target is present when it is). We will assume that the prior probabilities $P(H_0)$ and $P(H_1)$ of H_0 and H_1 occurring are known.

Given a test t, the probability that we reject H_0 is referred to in the statistical literature as the *power function* $\phi_t(\theta)$ of the test:

$$\phi_t(\theta) = P(\text{reject } H_0 | \theta) \tag{3.3.1}$$

The subscript t can be dropped if the reference to the test is clear from the context.

The probability of detection P_D is equal to 1 minus the probability of miss P_M and is equal to

$$P_D = \phi_t(\theta \in \Theta_1) = P(\text{reject } H_0 | \theta \in \Theta_1) \tag{3.3.2}$$

Note that P_D is a function of θ for $\theta \in \Theta_1$.

The probability of false alarm, P_F (or equivalently, the probability of a type I error, P_I), is defined as

$$P_F = \sup_{\theta \in \Theta_0} P(\text{reject } H_0 | \theta) \tag{3.3.3}$$

In the case of a simple null hypothesis, Θ_0 is a single point and the supremum in Eq. (3.3.3) is not required.

The purpose of a decision criterion is to attach some relative importance to the four possible choices. The consequences of one decision may not be the same as the consequences of another. For example, in the radar context the consequences of a miss are quite different from those of a false alarm. To reflect these differences we assign a cost to each of the decisions. We thus define C_{ij} to be the cost associated with making a decision D_i when the true hypothesis is H_j. We would now like to determine our decision rule so that the average cost or *risk* is minimized. This is known as the *Bayes criterion*. If we let $P(D_i, H_j)$ denote the joint probability that we decide H_i and that the true hypothesis is H_j, the average cost can be written as

$$\overline{C} = C_{00}P(D_0, H_0) + C_{10}P(D_1, H_0) + C_{01}P(D_0, H_1) + C_{11}P(D_1, H_1) \tag{3.3.4}$$

An application of Bayes rule yields

$$\begin{aligned}\overline{C} &= C_{00}P(D_0 \mid H_0)P(H_0) + C_{10}P(D_1 \mid H_0)P(H_0) \\ &\quad + C_{01}P(D_0 \mid H_1)P(H_1) + C_{11}P(D_1 \mid H_1)P(H_1)\end{aligned} \tag{3.3.5}$$

where $P(D_i \mid H_j)$ is the probability of deciding H_i when H_j is the true hypothesis.

As discussed earlier, for the binary decision problem, the decision rule divides the observation space into two regions, Z_0 and Z_1, by assigning each point in Z_0 to hypothesis H_0 and each point in Z_1 to H_1. The decision regions are shown in Fig. 3.3.1.

We can obtain the probabilities $P(D_i \mid H_j), i, j = 0, 1$, by integrating the conditional density function of z given hypothesis H_j, over Z_i, the decision space corresponding to hypothesis H_i:

$$P(D_i \mid H_j) = \int_{Z_i} p(z \mid H_j)\, dz \tag{3.3.6}$$

so that the average cost becomes

$$\begin{aligned}\overline{C} &= C_{00}P(H_0)\int_{Z_0} p(z \mid H_0)\, dz + C_{10}P(H_0)\int_{Z_1} p(z \mid H_0)\, dz \\ &\quad + C_{01}P(H_1)\int_{Z_0} p(z \mid H_1)\, dz + C_{11}P(H_1)\int_{Z_1} p(z \mid H_1)\, dz\end{aligned} \tag{3.3.7}$$

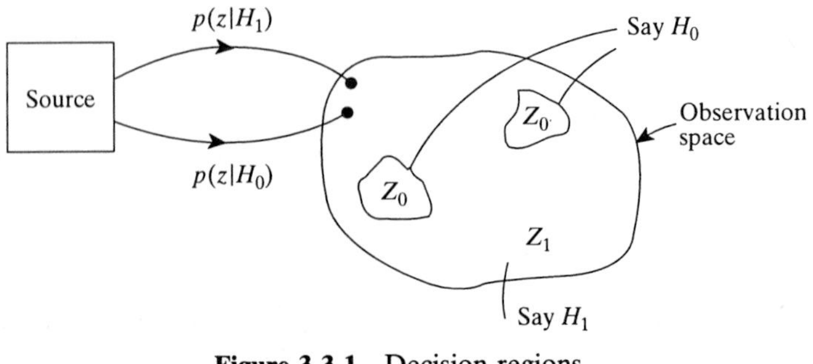

Figure 3.3.1 Decision regions.

It is reasonable at this point to assume that the cost of making an incorrect decision is higher than the cost of a correct decision. We will therefore assume that

$$C_{10} > C_{00} \quad \text{and} \quad C_{01} > C_{11} \tag{3.3.8}$$

Since $Z = Z_0 + Z_1 \underline{\Delta} Z_0 \cup Z_1$, and $Z_0 \cap Z_1 = \varnothing$, we can write

$$\begin{aligned} \int_{Z_1} p(z \mid H_i)\, dz &= \int_Z p(z \mid H_i)\, dz - \int_{Z_0} p(z \mid H_i)\, dz \\ &= 1 - \int_{Z_0} p(z \mid H_i)\, dz \qquad i = 0, 1 \end{aligned} \tag{3.3.9}$$

Consequently, Eq. (3.3.7) becomes

$$\begin{aligned} \overline{C} = C_{10} P(H_0) + C_{11} P(H_1) + \int_{Z_0} \{[P(H_1)(C_{01} - C_{11}) p(z \mid H_1)] \\ - [P(H_0)(C_{10} - C_{00}) p(z \mid H_0)]\}\, dz \end{aligned} \tag{3.3.10}$$

The only variable quantity in the preceding integral is the region Z_0. The assumption in Eq. (3.3.8) implies that the two terms in brackets within the integral are both positive. Therefore, all values of z such that the second term is larger should be assigned to Z_0. All values of z such that the second term is smaller should be excluded from Z_0, that is, assigned to Z_1. Points where the two terms are equal will be assigned arbitrarily to Z_1. The decision regions are, therefore, defined as follows:

If

$$P(H_1)(C_{01} - C_{11}) p(z \mid H_1) \geqslant P(H_0)(C_{10} - C_{00}) p(z \mid H_0) \tag{3.3.11}$$

assign z to Z_1 and consequently decide that H_1 is true. Otherwise, assign z to Z_0 and say that H_0 is true. In terms of the likelihood ratio

$$\Lambda(z) = \frac{p(z \mid H_1)}{p(z \mid H_0)} \tag{3.3.12}$$

and the threshold

$$\eta = \frac{P(H_0)(C_{10} - C_{00})}{P(H_1)(C_{01} - C_{11})} \tag{3.3.13}$$

the decision rule becomes

$$\Lambda(z) \underset{H_0}{\overset{H_1}{\gtrless}} \eta \tag{3.3.14}$$

Observe that the Bayes rule (3.3.14) is also a likelihood ratio test.

We note that the processing involved in computing $\Lambda(z)$ is independent of cost assignments and prior probabilities. In a practical implementation of the processor, we therefore leave the threshold as a variable quantity to accommodate changes in our cost assignments or prior probabilities.

In terms of the log-likelihood ratio, the decision rule becomes

$$\ln \Lambda(z) \underset{H_0}{\overset{H_1}{\gtrless}} \ln \eta \tag{3.3.15}$$

We note that if we choose $C_{10} - C_{00} = C_{01} - C_{11}$ in Eq. (3.3.13) the Bayes rule of Eq. (3.3.14) will be the same as was obtained using the MAP criterion.

We can give an alternative interpretation of the Bayes criterion as follows. We define the *conditional risk* or cost of making a decision in favor of hypothesis H_i as

$$C(H_i \mid z) = C_{i0} P(H_0 \mid z) + C_{i1} P(H_1 \mid z) \qquad i = 0, 1 \tag{3.3.16}$$

It can then easily be verified that the decision rule of Eq. (3.3.14) is equivalent to choosing that hypothesis for which the conditional risk is a minimum.

Example 3.3.1

We assume that under both hypotheses, the source output is a zero-mean Gaussian signal. Under H_1, the variance of the signal is σ_1^2, and under H_0, the variance is σ_0^2. We thus have

$$p(z \mid H_i) = \frac{1}{\sqrt{2\pi}\,\sigma_i} \exp\left(-\frac{z^2}{2\sigma_i^2}\right) \qquad i = 0, 1$$

$$\Lambda(z) = \frac{\sigma_0}{\sigma_1} \exp\left[\frac{z^2}{2}\left(\frac{1}{\sigma_0^2} - \frac{1}{\sigma_1^2}\right)\right]$$

so that the decision rule is

$$\frac{\sigma_0}{\sigma_1} \exp\left[\frac{z^2}{2}\left(\frac{1}{\sigma_0^2} - \frac{1}{\sigma_1^2}\right)\right] \underset{H_0}{\overset{H_1}{\gtrless}} \eta$$

In terms of the log-likelihood ratio, we can write this as

$$\frac{z^2}{2}\left(\frac{1}{\sigma_0^2} - \frac{1}{\sigma_1^2}\right) + \ln\frac{\sigma_0}{\sigma_1} \underset{H_0}{\overset{H_1}{\gtrless}} \ln \eta$$

Let $\sigma_1^2 > \sigma_0^2$. We can use $l(z) = z^2$ as the test statistic and write the test in terms of $l(z)$ as

$$l(z) \underset{H_0}{\overset{H_1}{\gtrless}} \frac{2\sigma_0^2\sigma_1^2}{\sigma_1^2 - \sigma_0^2} \ln\frac{\eta\sigma_1}{\sigma_0} \underline{\Delta} \gamma$$

The performance of the detector can be determined by evaluating the errors associated with making the decisions. The quantities of interest are the probabilities of the two types of error introduced earlier. We recall that a type I error (false alarm) corresponds to deciding H_1 when H_0 is true. We have, from Eq. (3.3.6),

$$P_{\mathrm{I}} = P_F = P(D_1 \mid H_0) = \int_{Z_1} p(z \mid H_0)\, dz \tag{3.3.17}$$

An error of the second kind (miss) corresponds to deciding H_0 when H_1 is true, so that

$$P_{II} = P_M = P(D_0 | H_1) = \int_{Z_0} p(z | H_1)\, dz \tag{3.3.18}$$

Similarly, the probability of detection P_D is given by

$$P_D = P(D_1 | H_1) = \int_{Z_1} p(z | H_1)\, dz = 1 - P_M \tag{3.3.19}$$

The average probability of error in decision making is then given by

$$P_e = P_F P(H_0) + P_M P(H_1) \tag{3.3.20}$$

We note that the regions Z_0 and Z_1 consist of those values of z for which the likelihood ratio $\Lambda(z)$ is, respectively, less than or greater than the threshold η. We can therefore write various error probabilities in terms of $\Lambda(z)$ as

$$P_F = \int_{\eta}^{\infty} p(\Lambda(z)|H_0)\, d\Lambda \tag{3.3.21}$$

$$P_M = \int_{0}^{\eta} p(\Lambda(z) | H_1)\, d\Lambda \tag{3.3.22}$$

and

$$P_D = \int_{\eta}^{\infty} p(\Lambda(z)|H_1)\, d\Lambda \tag{3.3.23}$$

Finally, we can write the average risk $\overline{C}$ in Eq. (3.3.10) in terms of P_F and P_M as

$$\begin{aligned} \overline{C} &= C_{00}(1 - P_F) + C_{10} P_F \\ &\quad + P(H_1)[(C_{11} - C_{00}) + (C_{01} - C_{11})P_M - (C_{10} - C_{00})P_F] \end{aligned} \tag{3.3.24}$$

3.3.1 Minimum Probability of Error Criterion

If we set $C_{00} = C_{11} = 0$ and $C_{10} = C_{01} = 1$, the expression for the risk in Eq. (3.3.7) becomes

$$\overline{C} = P(H_0) \int_{Z_1} p(z | H_0)\, dz + P(H_1) \int_{Z_0} p(z | H_1)\, dz \tag{3.3.25}$$

which is just the total probability of making an error. The criterion, therefore, yields the minimum probability of error. The decision rule becomes

$$\Lambda(z) \underset{H_0}{\overset{H_1}{\gtrless}} \frac{P(H_0)}{P(H_1)} \tag{3.3.26}$$

The receiver is then called an *ideal observer*. It can be seen that the decision rule is the same as for the MAP criterion.

Example 3.3.2

We assume that under the hypothesis H_1 we observe a constant signal of amplitude $m > 0$, corrupted by a zero-mean Gaussian noise V with variance σ^2; under the hypothesis H_0, we observe only the noise V. Given an observation Z, we need to decide whether the level was m or zero.

$$H_1\colon Z = m + V \qquad p(z \mid H_1) = \frac{1}{\sqrt{2\pi}\,\sigma} \exp\left[-\frac{(z-m)^2}{2\sigma^2}\right]$$

$$H_0\colon Z = V \qquad p(z \mid H_0) = \frac{1}{\sqrt{2\pi}\,\sigma} \exp\left(-\frac{z^2}{2\sigma^2}\right)$$

That is, we have to decide whether z is from a Gaussian distribution with mean m or with mean 0.

Assuming equally likely hypotheses and the minimum probability of error criterion, the likelihood ratio test is

$$\ln \frac{p(z \mid H_1)}{p(z \mid H_0)} \underset{H_0}{\overset{H_1}{\gtrless}} 0$$

This can be simplified as

$$z \underset{H_0}{\overset{H_1}{\gtrless}} \frac{m}{2}$$

We note that the observation itself is the sufficient (test) statistic. To determine the performance, it is convenient to normalize our observation to have a standard deviation of unity. Let $\bar{z} = z/\sigma$ denote the normalized observation, and let $d = m/\sigma$, a design parameter. The decision rule then becomes

$$\bar{z} \underset{H_0}{\overset{H_1}{\gtrless}} \frac{d}{2}$$

Note that

$$p(\bar{z} \mid H_1) \sim N(d, 1)$$

and

$$p(\bar{z} \mid H_0) \sim N(0, 1)$$

where $N(\mu, v)$ is a Gaussian random variable with mean μ and variance v.

The total probability of error P_e is given by

$$P_e = \tfrac{1}{2} P_{\mathrm{I}} + \tfrac{1}{2} P_{\mathrm{II}}$$

where

$$P_{\mathrm{I}} = \Pr\left\{\bar{z} > \frac{d}{2} \text{ given that } \bar{z} \text{ corresponded to } H_0\right\}$$

$$= \int_{d/2}^{\infty} p(\bar{z} \mid H_0)\, d\bar{z} = \int_{d/2}^{\infty} \frac{1}{\sqrt{2\pi}} \exp\left(-\frac{\bar{z}^2}{2}\right) d\bar{z}$$

Let $G(x)$ denote the cumulative distribution function for the standard Gaussian random variable with mean zero and unity variance,

$$G(x) = \frac{1}{\sqrt{2\pi}} \int_{-\infty}^{x} \exp\left(-\frac{x^2}{2}\right) dx$$

and let

$$Q(x) = 1 - G(x) = \frac{1}{\sqrt{2\pi}} \int_{x}^{\infty} \exp\left(-\frac{x^2}{2}\right) dx$$

Then

$$P_{\text{I}} = Q\left(\frac{d}{2}\right)$$

and

$$P_{\text{II}} = \Pr\left\{\bar{z} < \frac{d}{2} \text{ given that } \bar{z} \text{ corresponded to } H_1\right\}$$

$$= \int_{-\infty}^{d/2} p(\bar{z} \mid H_1)\, dz = \int_{-\infty}^{d/2} \frac{1}{\sqrt{2\pi}} \exp\left[\frac{(\bar{z} - d)^2}{2}\right] d\bar{z} = Q\left(\frac{d}{2}\right)$$

Therefore,

$$P_e = Q\left(\frac{d}{2}\right)$$

We now briefly discuss other criteria that also lead to LRTs.

3.3.2 Minimax Criterion

We have seen that the Bayes criterion requires that in addition to assigning costs to the various decisions, we must also assign prior probabilities to the two hypotheses. Frequently, not enough is known about the hypotheses which will enable us to determine the prior probabilities. In such a case, the Bayes criterion cannot be applied. One approach to obtaining the decision rule is to use the Bayes solution corresponding to the value of $P(H_1)$ for which the average cost is a maximum. The criterion is thus referred to as a *minimax criterion*.

For a fixed assignment of costs, the threshold of the Bayes test varies with $P(H_1)$. So does the average cost, which can therefore be written as $\overline{C}(P(H_1))$. If we now plot the minimum average cost $\overline{C}_{\min}$ (Bayes cost) as a function of $P(H_1)$, we obtain a curve somewhat like the one in Fig. 3.3.2. Typically, the curve will have a maximum at some value P_{I}^* of $P(H_1)$. If the likelihood ratio Λ is a continuous random variable with a cumulative distribution function that is strictly monotonic, then $\overline{C}_{\min}$ will be strictly concave downward, as in Fig. 3.3.2 [2]. In any case $\overline{C}_{\min}$ is always concave, although it may not have a derivative at every point [7].

Suppose we now assume that $P(H_1)$ is a fixed value P_1 and design the Bayes test accordingly. The decision regions are then fixed by the threshold of the test and

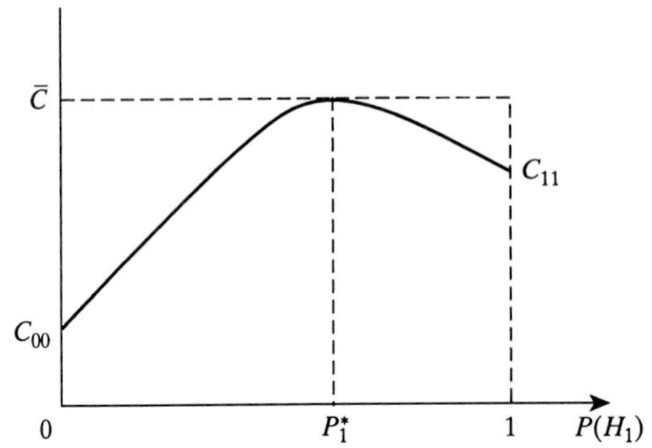

Figure 3.3.2 Bayes cost as a function of $P(H_1)$.

so are the error probabilities P_F and P_M. However, if the true value of $P(H_1)$ is some other value, the decision rule is no longer optimal and the average cost incurred will not be the Bayes cost but will be larger. In fact, as can be seen from Eq. (3.3.24), the variation in the average cost with $P(H_1)$ will be represented by a straight line. This straight line will of course be tangent to the $\bar{C}_{\min}$ curve at P_1, since the test is optimal for this value of $P(H_1)$. The actual cost incurred could thus be quite large depending on the true value of $P(H_1)$. However, if we were to design our test under the assumption that the prior probability was P_1^*, the tangent to the $\bar{C}_{\min}$ curve will be horizontal. We note that P_1^* represents the worst situation in that the Bayes cost is maximum at this point. We are then assured that irrespective of the true value of $P(H_1)$, the average cost will not exceed $\bar{C}_{\min}(P_1^*)$. This situation occurs when the partial derivative of $\bar{C}(P(H_1))$ with respect to $P(H_1)$ is zero and is given by

$$(C_{11} - C_{00}) + (C_{01} - C_{11})P_M - (C_{10} - C_{00})P_F = 0 \qquad (3.3.27)$$

Equation (3.3.27) is referred to as the *minimax equation*.

For the special case when $C_{00} = C_{11} = 0$, $C_{10} = C_{01} = 1$ the minimax equation reduces to

$$P_M = P_F \qquad (3.3.28)$$

and the minimax cost is the average probability of error. Even if $\bar{C}_{\min}$ is not differentiable everywhere on $(0, 1)$, the minimax test is still a Bayes test (see [7]).

Example 3.3.3

We consider the same problem as in Example 3.3.2 but assume that the prior probabilities $P(H_1)$ and $P(H_0)$ are not known. We would like to implement a minimax receiver in this case. For convenience, let us assume that $C_{ij} = 1 - \delta_{ij}$, $i, j = 0, 1$. Then we make use of Eq. (3.3.28) to obtain the threshold.

We can easily show that for this example

$$P_F = \int_\gamma^\infty p(z \mid H_0)\, dz = \int_\gamma^\infty \frac{1}{\sqrt{2\pi}\,\sigma} \exp\left(-\frac{z^2}{2\sigma^2}\right) dz = Q\left(\frac{\gamma}{\sigma}\right)$$

and

$$P_M = \int_{-\infty}^{\gamma} p(z \mid H_1)\, dz = \int_{-\infty}^{\gamma} \frac{1}{\sqrt{2\pi}\,\sigma} \exp\left[-\frac{(z-m)^2}{2\sigma^2}\right] dz$$
$$= G\left(\frac{\gamma - m}{\sigma}\right) = 1 - Q\left(\frac{\gamma - m}{\sigma}\right)$$

The optimum threshold, $\gamma = \gamma^*$, is obtained when $P_M = P_F$ according to Eq. (3.3.28). This results in the expression

$$Q\left(\frac{\gamma^*}{\sigma}\right) = 1 - Q\left(\frac{\gamma^* - m}{\sigma}\right)$$

from which

$$\gamma^* = \frac{m}{2}$$

Hence the minimax test is

$$z \underset{H_0}{\overset{H_1}{\gtrless}} \frac{m}{2}$$

The average probability of error is

$$P_e = P_F = Q\left(\frac{m}{2\sigma}\right) = Q\left(\frac{d}{2}\right)$$

In this particular example, the threshold for the minimax test is the same as for the minimum probability of error test.

3.3.3 Neyman–Pearson Tests

In many situations, not only the prior probabilities, but also cost assignments are difficult to make. For example, in radar detection, the cost of failing to detect a target cannot be easily determined. A simple procedure to bypass this difficulty is to work with the conditional probabilities P_F and P_D. In general, we would like to make P_F as small as possible and P_D as large as possible. However, these are usually conflicting objectives. It is usual, therefore, to determine a value of P_F that is acceptable and to seek a decision strategy that constrains P_F to this value while simultaneously maximizing P_D. Equivalently, we can seek to minimize P_M. The test is then said to be a *Neyman–Pearson test*. The value α chosen for P_F is called the *size* of the test. The Neyman–Pearson test is also called the *most powerful* (MP) *test* since it achieves the most power (largest P_D) among all the tests that have the same type I error probability, P_F.

We now obtain the decision rule by minimizing P_M subject to the constraint

$$P_F = \alpha$$

We thus construct the objective function (see Appendix B)

$$J = P_M + \lambda(P_F - \alpha) \tag{3.3.29}$$

where $\lambda \geqslant 0$ is a Lagrange multiplier. If we substitute for P_F and P_M from Eqs. (3.3.17) and (3.3.18), we get

$$\begin{aligned} J &= \int_{Z_0} p(z \mid H_1)\, dz + \lambda\left[\int_{Z_1} p(z \mid H_0)\, dz - \alpha\right] \\ &= \int_{Z_0} p(z \mid H_1)\, dz + \lambda\left[1 - \int_{Z_0} p(z \mid H_0)\, dz - \alpha\right] \\ &= \lambda(1 - \alpha) + \int_{Z_0} [p(z \mid H_1) - \lambda p(z \mid H_0)]\, dz \end{aligned} \tag{3.3.30}$$

It easily follows that a likelihood ratio test (LRT) will minimize J. The decision rule becomes

$$\Lambda(z) \underset{H_0}{\overset{H_1}{\gtrless}} \lambda \tag{3.3.31}$$

The threshold of the test, η, is equal to the Lagrange multiplier λ and is chosen to satisfy the constraint on P_F. We thus have

$$\alpha = P_F = \int_{Z_1} p(z \mid H_0)\, dz = \int_{\eta}^{\infty} p(\Lambda(z) \mid H_0)\, d\Lambda \tag{3.3.32}$$

An alternate derivation that does not employ Lagrange multipliers can be found in [1,4–7]. The existence and uniqueness of Neyman-Pearson tests are proved in standard tests on statistical theory [1,5,6].

Example 3.3.4

We consider the problem of Example 3.3.2 with $\sigma^2 = 1$ and construct a Neyman-Pearson receiver for the case where P_F is constrained to be 0.01. From Example 3.3.2, the likelihood ratio can be shown to be

$$\Lambda(z) = \exp\left[\left(z - \frac{m}{2}\right)m\right]$$

so that the decision rule is

$$z \underset{H_0}{\overset{H_1}{\gtrless}} \frac{\ln \eta}{m} + \frac{m}{2} \triangleq \gamma$$

To determine γ, we set $P_F = 0.01$ which gives

$$\int_{\gamma}^{\infty} \frac{1}{\sqrt{2\pi}} \exp\left[\frac{-z^2}{2}\right] dz = 0.01$$

from which $\gamma = 2.329$. The probability of detection is

$$P_D = P(D_1 | H_1) = \int_{\gamma}^{\infty} \frac{1}{\sqrt{2\pi}} \exp\left[-\frac{(z-m)^2}{2}\right] dz$$

For the case $m = 1$, this yields $P_D = 0.092$.

It may be noted that this is in general, an unacceptable level of P_D. For larger values of P_F, the threshold γ becomes lower, thus increasing P_D.

We have seen that all the three criteria discussed so far require the computation of a likelihood ratio. It is only the value of η with which this ratio is compared that varies with the criterion. The performance of any test is best analyzed in terms of a graph of the detection probability P_D against the probability of false alarm P_F. Such a curve is called a *receiver operating characteristic* (ROC). The ROC of a LRT depends only on the probability density functions of the observation under the two hypotheses. Figure 3.3.3 shows the ROC of an LRT for Example 3.3.4.

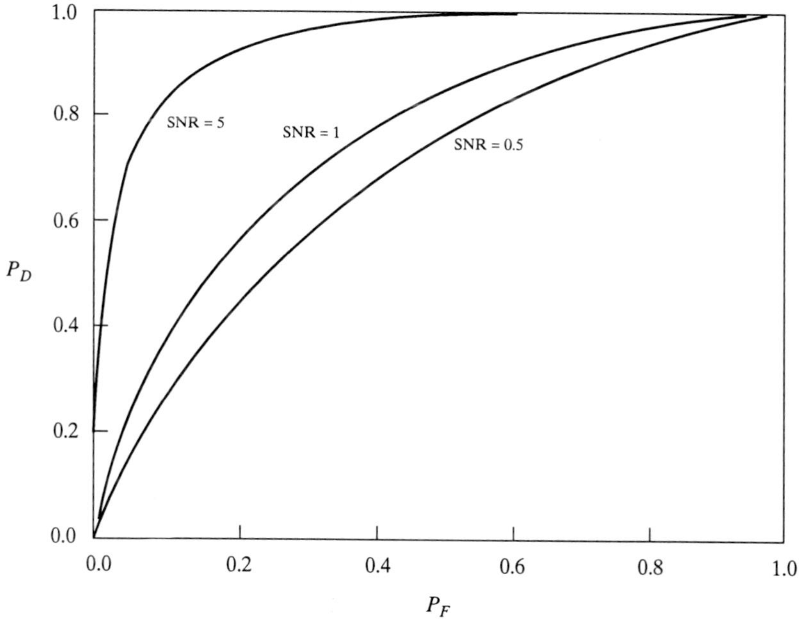

Figure 3.3.3 Receiver operating characteristics of a LRT for Example 3.3.4.

The three curves in the figure correspond to three different values of the ratio m^2/σ^2. Recall that in Example 3.3.4, σ^2 was fixed to be 1. The term m^2/σ^2 is called the *signal power/noise power ratio* or simply the *signal-to-noise ratio* (SNR). In detection problems in which the test involves the mean of a Gaussian density, a performance figure such as P_D is a function only of P_F and the SNR. As an example, if m and σ^2 are assumed to be arbitrary in Example 3.3.4, we have

$$P_D = Q\left(\frac{\gamma - m}{\sigma}\right) = Q\left(\frac{\gamma}{\sigma} - \sqrt{\frac{m^2}{\sigma^2}}\right)$$

so that

$$P_D = Q(Q^{-1}(P_F) - \sqrt{\text{SNR}})$$

In general, the higher the SNR, the larger will be the separation between the densities under the two hypotheses, and hence P_D for a fixed P_F will be larger. This can be clearly seen from the equation above, as well as from Fig. 3.3.3.

For the Neyman–Pearson test, the slope of the ROC corresponding to any specified value of P_F represents the critical value of the likelihood ratio. This easily follows from the definitions of P_D and P_F. From Eq. (3.3.23) we can write

$$\frac{dP_D}{d\eta} = \frac{d}{d\eta}\int_\eta^\infty p(\Lambda(z)\,|\,H_1)\,d\Lambda = -p(\eta\,|\,H_1) \tag{3.3.33a}$$

Similarly,

$$\frac{dP_F}{d\eta} = -p(\eta\,|\,H_0) \tag{3.3.33b}$$

However,

$$\begin{aligned} P_D(\eta) &= \Pr\{\Lambda(z) \geq \eta\,|\,H_1\} \\ &= \int_\eta^\infty p(\Lambda(z)\,|\,H_1)\,d\Lambda \\ &= \int_\eta^\infty \Lambda(z)p(\Lambda(z)\,|\,H_0)\,d\Lambda \end{aligned} \tag{3.3.34}$$

so that

$$\frac{dP_D(\eta)}{d\eta} = -\eta p(\eta\,|\,H_0) \tag{3.3.35}$$

Combining Eqs. (3.3.33) and (3.3.35) yields

$$\frac{p(\eta\,|\,H_1)}{p(\eta\,|\,H_0)} = \eta \tag{3.3.36}$$

from which it is clear that

$$\frac{dP_D}{dP_F} = \frac{dP_D/d\eta}{dP_F/d\eta} = \eta \tag{3.3.37}$$

For the Bayes criterion, the threshold η is determined by the costs and prior probabilities. The values of P_D and P_F corresponding to this threshold can be obtained by determining the point on the ROC at which the tangent has slope η.

It can easily be verified that the minimax equation represents a straight line in the P_D–P_F plane. Thus the probabilities P_D and P_F for this test can be determined from the point of intersection of the minimax line with the ROC. The slope of the tangent to the ROC at the point of intersection yields the threshold η. If the costs C_{ij} are such that either $C_{11} < C_{01} < C_{00} < C_{10}$ or $C_{00} < C_{10} < C_{11} < C_{01}$, the minimax line will not intersect the ROC in the region $0 < P_D, P_F < 1$. Thus the minimax solution does not exist in the usual sense. In the first case since C_{11} is less than C_{00}, the loss will always be greater when H_0 is true, so that the decision will be to choose H_0 always; whereas in the second case, the decision will be to choose H_1.

3.4 MULTIPLE MEASUREMENTS

The discussions of Section 3.3 can easily be extended to the more usual case in which the decision between the two hypotheses H_0 and H_1 is based on several observations. In this case we have N observations, $z_1, z_2, \ldots, z_N$, which may represent successive measurements of the same physical parameter or simultaneous measurement of several parameters or combinations thereof. The observations are described by the conditional density functions $p(z_1, z_2, \ldots, z_N \mid H_0)$ and $p(z_1, z_2, \ldots, z_N \mid H_1)$. We can derive the decision rules for this problem conveniently by considering the observations as a point in an N-dimensional space by defining the vector $\mathbf{z}$ as

$$\mathbf{z} = [z_1, z_2, \ldots, z_N]^T \tag{3.4.1}$$

The decision rule divides the observation space into two regions, Z_0 and Z_1, such that we decide H_0 when the observation vector $\mathbf{z}$ lies in Z_0 and decide H_1 when $\mathbf{z}$ lies in Z_1. The surface separating the two regions is known as the *decision surface*. The particular criterion used determines the location of the decision surface. For applying the Bayes criterion, we again must know the prior probabilities $P(H_0)$ and $P(H_1)$ and the costs C_{ij}. As in Eq. (3.3.4), the average risk (cost) is given by

$$\begin{aligned}\overline{C} = P(H_0)&\left[C_{00}\int_{Z_0} p(\mathbf{z}\mid H_0)\,d\mathbf{z} + C_{10}\int_{Z_1} p(\mathbf{z}\mid H_0)\,d\mathbf{z}\right] \\ &+ P(H_1)\left[C_{01}\int_{Z_0} p(\mathbf{z}\mid H_1)\,d\mathbf{z} + C_{11}\int_{Z_1} p(\mathbf{z}\mid H_1)\,d\mathbf{z}\right]\end{aligned} \tag{3.4.2}$$

where the integrals are N-fold integrals and $d\mathbf{z} = dz_1, dz_2, \ldots, dz_N$. We now choose our decision surface so as to minimize the average risk $\overline{C}$. Following a procedure similar to the one for the case of a single observation, we can easily show that the test requires the computation of the likelihood ratio

$$\Lambda(\mathbf{z}) = \frac{p(\mathbf{z}\mid H_1)}{p(\mathbf{z}\mid H_0)} = \frac{p(z_1, \ldots, z_N \mid H_1)}{p(z_1, \ldots, z_N \mid H_0)} \tag{3.4.3}$$

which is then compared with the threshold η in the usual manner, so that the decision rule is

$$\Lambda(\mathbf{z}) \underset{H_0}{\overset{H_1}{\gtrless}} \eta \tag{3.4.4}$$

If the costs are specified but the prior probabilities of the two hypotheses are unknown, we may use the minimax criterion. For each value of $P(H_1)$, we compute the minimum average risk $\overline{C}_{\min}(P(H_1))$. We then determine the probability P_1^* at which $\overline{C}_{\min}$ is maximum. The minimax strategy will then correspond to the Bayes rule assuming that $P(H_1) = P_1^*$ and the minimax cost is equal to $\overline{C}_{\min}(P_1^*)$. The error probabilities are given by

$$P_F = \int_{Z_1} p(\mathbf{z}|H_0)\,d\mathbf{z} = \int_{\eta}^{\infty} p(\Lambda(\mathbf{z})\,|\,H_0)\,d\Lambda \tag{3.4.5}$$

and

$$P_M = \int_{Z_0} p(\mathbf{z}|H_1)\,d\mathbf{z} = \int_{0}^{\eta} p(\Lambda(\mathbf{z})\,|\,H_1)\,d\Lambda \tag{3.4.6}$$

so that the minimax threshold can be obtained analogous to Eq. (3.3.27) as

$$C_{00}(1 - P_F) + C_{10}P_F = C_{01}P_M + C_{11}(1 - P_M) \tag{3.4.7}$$

In using the Neyman–Pearson criterion, we again specify an acceptable value for the false alarm probability P_F and seek to minimize the miss probability P_M. Use of a procedure similar to that for the single-observation case then yields the optimum decision rule as

$$\Lambda(\mathbf{z}) \underset{H_0}{\overset{H_1}{\gtrless}} \eta \tag{3.4.8}$$

where η is determined from the specified value α of P_F:

$$\int_{\eta}^{\infty} p(\Lambda(\mathbf{z})\,|\,H_0)\,d\Lambda = \alpha \tag{3.4.9}$$

Again as discussed in Section 3.3, the receiver operating characteristic can be used to evaluate the performance of various tests.

We illustrate the material in this section by means of two examples.

Example 3.4.1

We assume that we have N independent observations $z_i, i = 1, 2, \ldots, N$ of the received signal. The observations under each hypothesis are Gaussian independent, identically distributed random variables. Under H_1, we assume that the variables have positive mean m and variance σ^2, while under H_0, they are zero mean with variance σ^2. We will design a decision rule assuming the minimum probability of error criterion, and equally likely hypotheses.

For each i we can write

$$p(z_i | H_1) = \frac{1}{\sqrt{2\pi}\,\sigma} \exp\left[-\frac{(z_i - m)^2}{2\sigma^2} \right]$$

$$p(z_i | H_0) = \frac{1}{\sqrt{2\pi}\,\sigma} \exp\left(-\frac{z_i^2}{2\sigma^2} \right)$$

Since the z_i are assumed independent, their joint density function under each hypothesis is

$$p(\mathbf{z} | H_j) = \prod_{j=1}^{N} p(z_i | H_j) \qquad j = 0, 1$$

In terms of the vectors

$$\mathbf{z} = [z_1, z_2, \ldots, z_N]^T$$

and

$$\mathbf{m} = [m, m, \ldots, m]^T$$

we can write the joint density functions as

$$p(\mathbf{z} | H_1) = \frac{1}{(2\pi\sigma^2)^{N/2}} \exp\left[-\frac{1}{2\sigma^2} (\mathbf{z} - \mathbf{m})^T (\mathbf{z} - \mathbf{m}) \right]$$

and

$$p(\mathbf{z} | H_0) = \frac{1}{(2\pi\sigma^2)^{N/2}} \exp\left(-\frac{1}{2\sigma^2} \mathbf{z}^T \mathbf{z} \right)$$

so that the likelihood ratio is

$$\Lambda(\mathbf{z}) = \frac{p(\mathbf{z} | H_1)}{p(\mathbf{z} | H_0)} = \exp\left[\frac{1}{\sigma^2} \left(\mathbf{m}^T \mathbf{z} - \frac{1}{2} \mathbf{m}^T \mathbf{m} \right) \right]$$

The decision rule is

$$\exp\left[\frac{1}{\sigma^2} \left(\mathbf{m}^T \mathbf{z} - \frac{1}{2} \mathbf{m}^T \mathbf{m} \right) \right] \underset{H_0}{\overset{H_1}{\gtrless}} \eta$$

where the threshold η will be unity for the minimum probability of error criterion. Taking logarithms on both sides yields the more convenient form

$$\mathbf{m}^T \mathbf{z} \underset{H_0}{\overset{H_1}{\gtrless}} \tfrac{1}{2} \mathbf{m}^T \mathbf{m}$$

or explicitly,

$$\sum_{i=1}^{N} z_i \underset{H_0}{\overset{H_1}{\gtrless}} \frac{Nm}{2}$$

which can be rearranged as

$$\frac{1}{N}\sum_{i=1}^{N} z_i \underset{H_0}{\overset{H_1}{\gtrless}} \frac{m}{2} \triangleq \gamma$$

The test statistic in this case is $(1/N)\sum_{i=1}^{N} z_i$, which is the sample mean $\bar{z}$ of the observations. The receiver operation therefore consists of determining the sample mean of the observations and comparing it with the threshold γ. If the sample mean exceeds the threshold, we decide that H_1 is the true hypothesis; otherwise, we decide H_0.

We can easily determine P_F and P_D for this case from

$$P_F = \int_{\gamma}^{\infty} p(\bar{z} \mid H_0)\, d\bar{z}$$

and

$$P_D = \int_{\gamma}^{\infty} p(\bar{z} \mid H_1)\, d\bar{z}$$

Now $\bar{Z}$ is Gaussian under both hypotheses. Its mean and variance under H_1 are m and σ^2/N, respectively, whereas under H_0 it is zero mean but with the same variance, σ^2/N. It therefore follows that

$$P_D = Q\left(\frac{\gamma - m}{\sigma/\sqrt{N}}\right) \quad \text{and} \quad P_F = Q\left(\frac{\gamma}{\sigma/\sqrt{N}}\right)$$

Example 3.4.2

We consider the same problem as in Example 3.4.1 but now assume that the observations under the two hypotheses have mean zero but different variances σ_1^2 and σ_0^2, respectively, with $\sigma_1^2 > \sigma_0^2$. It can easily be verified that the conditional densities are given by

$$p(\mathbf{z} \mid H_1) = \frac{1}{(2\pi\sigma_1^2)^{N/2}} \exp\left(-\frac{1}{2\sigma_1^2}\mathbf{z}^T\mathbf{z}\right)$$

and

$$p(\mathbf{z} \mid H_0) = \frac{1}{(2\pi\sigma_0^2)^{N/2}} \exp\left(-\frac{1}{2\sigma_0^2}\mathbf{z}^T\mathbf{z}\right)$$

The likelihood ratio is

$$\Lambda(\mathbf{z}) = \left(\frac{\sigma_0^2}{\sigma_1^2}\right)^{N/2} \exp\left(\mathbf{z}^T\mathbf{z}\frac{\sigma_1^2 - \sigma_0^2}{2\sigma_0^2\sigma_1^2}\right)$$

and the decision rule becomes

$$\left(\frac{\sigma_0^2}{\sigma_1^2}\right)^{N/2} \exp\left(\mathbf{z}^T\mathbf{z}\frac{\sigma_1^2 - \sigma_0^2}{2\sigma_0^2\sigma_1^2}\right) \underset{H_0}{\overset{H_1}{\gtrless}} 1$$

We can arrive at a simple form for the rule by taking logarithms on both sides and rearranging terms:

$$\mathbf{z}^T\mathbf{z} \underset{H_0}{\overset{H_1}{\gtrless}} \frac{2\sigma_0^2\sigma_1^2}{\sigma_1^2 - \sigma_0^2} N \ln \frac{\sigma_1}{\sigma_0} \underline{\Delta} \gamma$$

We see that the test statistic in this case is $l(\mathbf{z}) = \mathbf{z}^T\mathbf{z} = \sum_{i=1}^N z_i^2$. Under both hypotheses, it is the sum of the squares of N independent, identically distributed Gaussian random variables. Within a scaling factor $l(\mathbf{z})$ is distributed as a chi-square with N degrees of freedom. Specifically, the densities of l under the two hypotheses are

$$p(l \mid H_i) = \begin{cases} \dfrac{l^{(N/2)-1}}{2^{N/2}\sigma_i^N \Gamma(N/2)} \exp\left(-\dfrac{l}{2\sigma_i^2}\right) & l \geq 0, \quad i = 0, 1 \\ 0 & \text{otherwise} \end{cases}$$

The expressions for P_D and P_F are given by

$$P_D = \int_\gamma^\infty \left[2^{N/2}\sigma_1^N \Gamma\left(\frac{N}{2}\right)\right]^{-1} l^{(N/2)-1} \exp\left(-\frac{l}{2\sigma_1^2}\right) dl$$

and

$$P_F = \int_\gamma^\infty \left[2^{N/2}\sigma_0^N \Gamma\left(\frac{N}{2}\right)\right]^{-1} l^{(N/2)-1} \exp\left(-\frac{l}{2\sigma_0^2}\right) dl$$

If N is even, by making the change of variables $M = (N/2) - 1$ and $\gamma_i = \gamma/2\sigma_i^2$, $i = 0$, 1, we can express P_D and P_F in terms of the incomplete gamma function

$$I(u, M) \underline{\Delta} \int_0^{u\sqrt{M+1}} \frac{l^M}{M!} \exp(-l)\, dl$$

which has been tabulated [8]. We then obtain

$$P_D = 1 - I\left(\frac{\gamma_1}{\sqrt{M+1}}, M\right)$$

and

$$P_F = 1 - I\left(\frac{\gamma_0}{\sqrt{M+1}}, M\right)$$

In both of the preceding examples, the observation space was N-dimensional. However, the solution to the decision problem could be obtained explicitly in terms of the test statistic $l(\mathbf{z})$, which is a scalar quantity. $l(\mathbf{z})$, which is equal to $(1/N)\sum_{i=1}^N z_i$ in the first example and to $\sum_{i=1}^N z_i^2$ in the second example is, in fact, sufficient in both these examples. Thus, according to the definition of sufficiency, $l(\mathbf{z})$ contains all the information about the parameter θ (which corresponds to the mean m in the first example and the variance σ^2 in the second) for making a decision regarding the correct hypothesis. Therefore, the fact that an optimum test concerning θ is based solely on $l(\mathbf{z})$ is not surprising.

Also, when a sufficient statistic of fixed, reduced dimension exists, as in the previous two examples, it is possible to recast the problem in terms of the distributions of the sufficient statistic under the two hypotheses. We illustrate this for Example 3.4.2 but reiterate that this approach will work only if a sufficient statistic of fixed dimension exists for the original problem which involves N independent, identically distributed samples.

Example 3.4.2 (revisited)

It is clear that the example involves testing whether the samples $z_i, i = 1, \ldots, N$ are from a zero-mean Gaussian density with variance $\sigma_0^2(H_0)$ or one with variance $\sigma_1^2(H_1)$. Since the test corresponds to an inference regarding a Gaussian density with known mean, it follows from Eq. (2.6.3) that $l = \sum_{i=1}^N z_i^2$ is sufficient for σ^2. l carries all the information about σ^2 that the original samples z_i have. Therefore, the hypothesis-testing problem can be recast using the density functions of l from Example 3.4.2.

It follows that for the minimum probability of error criterion, the LRT is of the form

$$\frac{[l^{(N/2)-1}\exp(-l/2\sigma_1^2)]/[2^{(N/2)}\sigma_1^N\Gamma(N/2)]}{[l^{(N/2)-1}\exp(-l/2\sigma_0^2)]/[2^{(N/2)}\sigma_0^N\Gamma(N/2)]} \underset{H_0}{\overset{H_1}{\gtrless}} 1$$

Upon simplification and noting that $\sigma_1^2 > \sigma_0^2$, the test can be written as

$$l \underset{H_0}{\overset{H_1}{\gtrless}} \frac{2\sigma_0^2\sigma_1^2}{\sigma_1^2 - \sigma_0^2} N \ln\frac{\sigma_1}{\sigma_0}$$

This is exactly the same result that was obtained earlier based on the joint density functions of the N i.i.d. samples z_i under the two hypotheses.

3.5 MULTIPLE-HYPOTHESIS TESTING

We have so far considered the problem of deciding between two hypotheses. In many situations the source has several outputs and we will have to decide which of the several hypotheses corresponding to these outputs is the correct one. The multiple-hypothesis testing problem is treated in selected texts, such as [2,3,7]. With M outputs, there are a total of M alternatives that may occur during each experiment. We apply the Bayes criterion for this problem exactly as we did for the simple binary hypothesis testing problem. We thus assume that the prior probabilities $P(H_0), P(H_1), \ldots, P(H_{M-1})$ are known. We associate a cost C_{ij} with choosing H_i when the correct hypothesis is H_j.

The average cost is then

$$\overline{C} = \sum_{i=0}^{M-1}\sum_{j=0}^{M-1} C_{ij}P(D_i \mid H_j)P(H_j) \tag{3.5.1}$$

where $P(D_i \mid H_j)$ is the probability of choosing hypothesis H_i when H_j is true. We again divide the observation space Z into several mutually exclusive subspaces

$Z_0, Z_1, \ldots, Z_{M-1}$ by assigning the points in each subspace to the corresponding hypothesis. The objective is to find the decision surfaces so that the average cost $\overline{C}$ is minimized. We thus write

$$Z = Z_0 + Z_1 + \cdots + Z_{M-1} \underline{\Delta}\, Z_0 \cup Z_1 \cup \cdots \cup Z_{M-1} \tag{3.5.2}$$

In terms of these decision regions, the probability $P(D_i \mid H_j)$ can be written as

$$P(D_i \mid H_j) = \int_{Z_i} p(z \mid H_j)\, dz \tag{3.5.3}$$

Substitution of Eq. (3.5.3) in Eq. (3.5.1) yields

$$\begin{aligned} \overline{C} &= \sum_{i=0}^{M-1} \sum_{j=0}^{M-1} C_{ij} P(H_j) \int_{Z_i} p(z \mid H_j)\, dz \\ &= \sum_{i=0}^{M-1} C_{ii} P(H_i) \int_{Z_i} p(z \mid H_i)\, dz + \sum_{i=0}^{M-1} \sum_{\substack{j=0 \\ j \neq i}}^{M-1} C_{ij} P(H_j) \int_{Z_i} p(z \mid H_j)\, dz \end{aligned} \tag{3.5.4}$$

We can put this in a more convenient form by noting that in the first summation Z_i can be written as $Z - \bigcup_{j=0, j\neq i}^{M-1} Z_j$. Then, since $\int_Z p(z \mid H_i)\, dz = 1$, we obtain for the average cost

$$\overline{C} = \sum_{i=0}^{M-1} C_{ii} P(H_i) + \sum_{i=0}^{M-1} \int_{Z_i} \sum_{\substack{j=0 \\ j \neq i}}^{M-1} P(H_j)(C_{ij} - C_{jj}) p(z \mid H_j)\, dz \tag{3.5.5}$$

As before, the first term represents the fixed cost and the integrals represent the variable cost that depends on our choice of the decision regions. Clearly, we assign each observation z to the region in which the value of the integral is the smallest. Thus if we let

$$I_i(z) = \sum_{\substack{j=0 \\ j \neq i}}^{M-1} P(H_j)(C_{ij} - C_{jj}) p(z \mid H_j) \tag{3.5.6}$$

our decision rule would be to choose that hypothesis which corresponds to the minimum value of $I_i(z)$ as the correct one.

We can rewrite the decision rule in terms of the likelihood ratios by defining

$$\Lambda_i(z) = \frac{p(z \mid H_i)}{p(z \mid H_0)} \qquad i = 0, \ldots, M-1 \tag{3.5.7}$$

and

$$J_i(z) = \frac{I_i(z)}{p(z \mid H_0)} = \sum_{\substack{j=0 \\ j \neq i}}^{M-1} P(H_j)(C_{ij} - C_{jj}) \Lambda_j(z) \qquad i = 0, 1, \ldots, M-1 \tag{3.5.8}$$

The decision rule is then to choose that hypothesis for which $J_i(z)$ is minimum. It can be seen that the decision rules correspond to hyperplanes in the $(M-1)$-dimensional space of the likelihood ratios $\Lambda_1(z) \cdots \Lambda_{M-1}(z)$.

Of particular interest is the case where $C_{ij} = 1$ for $i \neq j$ and $C_{ii} = 0$. The criterion in this case is the minimum probability of error. Then Eq. (3.5.6) reduces to

$$\begin{aligned} I_i(z) &= \sum_{\substack{j=0\\ j\neq i}}^{M-1} P(H_j)p(z\,|\,H_j) = \sum_{\substack{j=0\\ j\neq i}}^{M-1} P(H_j\,|\,z)p(z) \\ &= [1 - P(H_i\,|\,z)]p(z) \end{aligned} \tag{3.5.9}$$

For this particular assignment of costs, therefore, minimizing $I_i(z)$ is equivalent to maximizing $P(H_i\,|\,z)$. This is the posterior probability of hypothesis H_i given the observations z and the processor for the minimum probability of error is the maximum a posteriori probability computer.

An obvious extension is to the case when all the hypotheses are equally likely:

$$P(H_0) = P(H_1) = P(H_2) = \cdots = P$$

In this case Eq. (3.5.9) can be written as

$$I_i(z) = \sum_{\substack{j=0\\ j\neq i}}^{M-1} p(z\,|\,H_j)P = p(z) - p(z\,|\,H_i)P \tag{3.5.10}$$

The decision rule is then to choose the hypothesis H_i for which $p(z\,|\,H_i)$ is maximum. Such a rule is called the maximum likelihood rule since $p(z\,|\,H_i)$ is nothing but a likelihood function (see Section 2.6.1). An example where this occurs is in a digital communication system using M-ary alphabets where each signal is transmitted with equal probability.

Example 3.5.1

Based on sampled observations of a signal, we are required to decide the correct hypothesis out of three possibilities ($M = 3$). Under each hypothesis H_i, the received signal is normally distributed with variance σ^2 and mean m_i. The means m_i are as follows:

$$m_1 = 1 \qquad m_2 = 2 \qquad m_3 = -1$$

The probabilities of occurrence of the hypotheses are equal. Further, we assume minimum probability of error criterion (Bayes criterion with $C_{ij} = 1$ for $i \neq j$ and $C_{ij} = 0$ for $i = j$) to make our decision.

From our previous discussions we see that the decision rule is to choose the hypothesis corresponding to the maximum of $p(\mathbf{z}\,|\,H_i), i = 1,2,3$. If there are N independent sampled observations, we can write $p(\mathbf{z}\,|\,H_i)$ as

$$p(\mathbf{z}\,|\,H_i) = \frac{1}{(2\pi\sigma^2)^{N/2}} \exp\left[-\frac{1}{2}\sum_{j=1}^{N}\frac{(z_j - m_i)^2}{\sigma^2}\right] \qquad i = 1,2,3$$

The preceding density function can be rewritten as

$$\begin{aligned} p(\mathbf{z}\,|\,H_i) &= \frac{1}{(2\pi\sigma^2)^{N/2}} \exp\left(-\frac{1}{2\sigma^2}\sum_{j=1}^{N} z_j^2\right) \\ &\quad \cdot \exp\left[\frac{1}{2\sigma^2}\sum_{j=1}^{N}(2z_j m_i - m_i^2)\right] \qquad i = 1,2,3 \end{aligned}$$

It is obvious that we choose that hypothesis H_i for which

$$\frac{1}{N}\sum_{j=1}^{N}(2z_j m_i - m_i^2) = \left(\frac{2}{N}\sum_{j=1}^{N} z_j m_i\right) - m_i^2$$

is maximum. That is, we compute the following quantities:

$$\left(\frac{2}{N}\sum_{j=1}^{N} z_j\right) - 1 \qquad \left(\frac{4}{N}\sum_{j=1}^{N} z_j\right) - 4 \qquad \left(\frac{-2}{N}\sum_{j=1}^{N} z_j\right) - 1$$

based on the sampled observations, $\mathbf{z}$. If the second quantity exceeds the other two, hypothesis H_2 is chosen, and so on. It is left as a simple exercise for the reader to show that the test statistic is the sample mean $\hat{m} = (1/N)\sum_{j=1}^{N} z_j$, of the received signal, and the decision regions are as follows:

$$\begin{aligned} \hat{m} &\geq \tfrac{3}{2}, && \text{choose } H_2 \\ 0 \leq \hat{m} &< \tfrac{3}{2}, && \text{choose } H_1 \\ \hat{m} &< 0, && \text{choose } H_3 \end{aligned}$$

3.5.1 General Gaussian Problem

The multivariate Gaussian distribution arises in many signal-processing applications as it is often used as the underlying model for signals. As an example of its application in hypothesis testing, let us consider the following binary hypothesis problem, in which the observation $\mathbf{z} = (z_1, z_2, \ldots, z_n)^T$ under the two hypotheses is given by

$$\begin{aligned} H_0&: \mathbf{Z} \sim N(\mathbf{m}_0, \Sigma_0) \\ H_1&: \mathbf{Z} \sim N(\mathbf{m}_1, \Sigma_1) \end{aligned} \tag{3.5.11}$$

where $\mathbf{m}_i, \Sigma_i, i = 0, 1$, are the mean vectors and covariance matrices of the observation under the two hypotheses. Let us restrict ourselves to the case of equally likely hypotheses and minimum error criterion, since other cases can be handled in a similar fashion. The likelihood ratio test is given by

$$\Lambda = \frac{|\Sigma_0|^{1/2}\exp[-\frac{1}{2}(\mathbf{z}-\mathbf{m}_1)^T\mathbf{Q}_1(\mathbf{z}-\mathbf{m}_1)]}{|\Sigma_1|^{1/2}\exp[-\frac{1}{2}(\mathbf{z}-\mathbf{m}_0)^T\mathbf{Q}_0(\mathbf{z}-\mathbf{m}_0)]} \underset{H_0}{\overset{H_1}{\gtrless}} 1 \tag{3.5.12}$$

where $\mathbf{Q}_i = \Sigma_i^{-1}$ and $|\Sigma_i|$ is the determinant of Σ_i.

We can write the test equivalently as

$$\tfrac{1}{2}[(\mathbf{z}-\mathbf{m}_0)^T\mathbf{Q}_0(\mathbf{z}-\mathbf{m}_0) - (\mathbf{z}-\mathbf{m}_1)^T\mathbf{Q}_1(\mathbf{z}-\mathbf{m}_1)] \underset{H_0}{\overset{H_1}{\gtrless}} -\ln(|\Sigma_0|/|\Sigma_1|)^{1/2} \tag{3.5.13}$$

This can be further simplified for special cases. For example, if the covariances under the two hypotheses are equal, the test can be written as

$$\mathbf{z}^T \mathbf{Q}(\mathbf{m}_1 - \mathbf{m}_0) \underset{H_0}{\overset{H_1}{\gtrless}} \tfrac{1}{2}[\mathbf{m}_1^T \mathbf{Q} \mathbf{m}_1 - \mathbf{m}_0^T \mathbf{Q} \mathbf{m}_0] \tag{3.5.14}$$

The test statistic is linear in the observation $\mathbf{z}$.

If the observation under H_0 is zero mean, $\mathbf{m}_0 = 0$, the problem can be considered as testing signal present (hypothesis H_1) versus no signal present (hypothesis H_0) condition. If, in addition, the observations are independent and identically distributed, we have $\Sigma = \sigma^2 \mathbf{I}$, so that the test simplifies to

$$\sum_{k=1}^{n} z_k m_{1k} \underset{H_0}{\overset{H_1}{\gtrless}} \tfrac{1}{2} \sum_{k=1}^{n} m_{1k}^2 \tag{3.5.15}$$

Since the left side is a discrete-time correlation statistic, the receiver is called a discrete-time matched filter. (The matched filter is discussed in detail in Chapter 4.) Further results on the general Gaussian problem can be found in Chapter 9 and in [2].

3.6 COMPOSITE HYPOTHESIS TESTING

In previous sections we have considered the case where the parameter characterizing a hypothesis assumed only a known single value. In many situations, however, this parameter can take an unknown value on a range of values. For example, if the signal is a sine wave, its phase could be an unknown that lies between 0 and 2π. If the signal is a realization of a stochastic process, its mean or variance could lie on a range of values. As stated in Section 3.2, such hypotheses are called *composite hypotheses* as distinct from the simple hypothesis, in which the characterizing parameter is fixed and known. We confine our discussion in this section to the two-hypothesis case, for reasons of clarity. The extensions of the ideas presented to the multiple-hypothesis case is quite straightforward.

We can view the set of parameters characterizing a hypothesis as a vector θ in parameter space χ_θ. The probability density governing the mapping from the parameter space to the observation space, $p(\mathbf{z} \mid \theta)$ is assumed known for all values of θ. Depending on how we characterize θ, we can proceed to derive rules for deciding between various hypotheses.

We may characterize θ either as a random variable or as an unknown constant. If θ is a random variable with known probability densities under the two hypotheses, the extension of the LRT test to the composite-hypothesis case is quite straightforward. We can compute the likelihood ratio as

$$\Lambda(\mathbf{z}) = \frac{\displaystyle\int_{\Theta_1} p(\mathbf{z} \mid \theta) p(\theta \mid H_1)\, d\theta}{\displaystyle\int_{\Theta_0} p(\mathbf{z} \mid \theta) p(\theta \mid H_0)\, d\theta} \tag{3.6.1}$$

The known density on θ under the two hypotheses has enabled us to compute the probability density of the observations over the two hypotheses by integrating over θ. When Θ is a random variable with an unknown density, the test procedure to be used is not clearly specified. In this case we could seek a minimax test, that is, a test that minimizes the maximum average cost (risk) as the prior density varies over the assumed class.

If θ is a nonrandom variable, a Bayes test is not meaningful since θ has no probability density function over which we can average $p(\mathbf{z} \mid H_i, \theta)$. We therefore use a Neyman–Pearson test. If there exists a test that maximizes the power $P(D_1 \mid H_1)$ for every $\theta \in \Theta_1$ subject to the constraint that the size of the test $\sup_{\theta \in \Theta_0} P(D_1 \mid H_0)$ is a specified value, such a test is called a *uniformly most powerful* (UMP) test. The phrase uniformly refers to the fact that the test maximizes the power for every $\theta \in \Theta_1$.

If a UMP test does not exist, we may try estimating the values of θ under the two hypotheses H_1 and H_0 and use these estimates as the true values in an LRT. For nonrandom parameters, the usual estimate is the *maximum likelihood estimate*,† obtained by maximizing the conditional density function $p(z \mid \theta)$. We thus construct a *generalized likelihood ratio* and write the test as

$$\Lambda_g(z) = \frac{\max\limits_{\theta \in \Theta_1} p(\mathbf{z} \mid \theta)}{\max\limits_{\theta \in \Theta_0} p(\mathbf{z} \mid \theta)} \underset{H_0}{\overset{H_1}{\gtrless}} \eta \tag{3.6.2}$$

Example 3.6.1

The signals under the two hypotheses are Gaussian with variance σ^2. Under the null hypothesis, the mean is assumed to be zero, whereas under hypothesis H_1 the mean is a constant ($m_0 \leqslant m \leqslant m_1$). For a single observation the conditional densities are given by

$$p(z \mid H_1, m) = \frac{1}{\sqrt{2\pi}\,\sigma} \exp\left[-\frac{(z-m)^2}{2\sigma^2}\right] \qquad m_0 \leqslant m \leqslant m_1$$

and

$$p(z \mid H_0) = \frac{1}{\sqrt{2\pi}\,\sigma} \exp\left(-\frac{z^2}{2\sigma^2}\right)$$

For fixed m the likelihood ratio is

$$\Lambda(z) = \exp\left(\frac{mz - m^2/2}{\sigma^2}\right)$$

Using the log-likelihood ratio yields the decision rule as

$$L(z) = mz - \frac{m^2}{2} \underset{H_0}{\overset{H_1}{\gtrless}} \gamma$$

or

$$mz \underset{H_0}{\overset{H_1}{\gtrless}} \gamma' \underline{\Delta} \gamma + \frac{m^2}{2}$$

†This estimate is mentioned in Section 2.6 and is defined in Chapter 5.

If $m_0 > 0$, then m can take on only positive values and the test can be written as

$$z \underset{H_0}{\overset{H_1}{\gtrless}} \gamma^+ = \frac{\gamma}{m} + \frac{m}{2}$$

For a Neyman–Pearson test with P_F equal to some specified value α, we determine the threshold γ^+ from

$$P_F = \int_{\gamma^+}^{\infty} \frac{1}{\sqrt{2\pi}\,\sigma} \exp\left(\frac{-z^2}{2\sigma^2}\right) dz = \alpha$$

Note that the threshold γ^+ may be negative. If $m_1 \leqslant 0$, m takes on only nonpositive values and the test becomes

$$z \underset{H_0}{\overset{H_1}{\lesseqgtr}} \gamma^- = \frac{\gamma}{m} + \frac{m}{2}$$

In this case, the threshold γ^- is determined from

$$P_F = \int_{-\infty}^{\gamma^-} \frac{1}{\sqrt{2\pi}\,\sigma} \exp\left(\frac{-z^2}{2\sigma^2}\right) dz = \alpha$$

For either of the two possibilities preceding, we see that the test (includes the threshold) can be completely designed without knowledge of the true value of m. Thus a UMP test exists in both cases. Some general procedures to find UMP tests for one-sided hypotheses are discussed in Section 3.6.1. If, however, m can take on both positive and negative values, a UMP test does not exist. In a situation such as this, if we were to determine the decision rule assuming that the mean is positive (negative) whereas it actually is negative (positive), very poor performance results. To demonstrate this, assume that the type I error $P(D_1 \mid H_0)$ has been fixed at some value α and the test has been designed as though the mean were positive. The situations resulting when the true mean is either positive or negative are shown in Fig. 3.6.1.

The resulting $P(D_1 \mid H_1)$ as a function of the true value of the mean is shown in Fig. 3.6.2. Similar results hold when the test is designed assuming that the mean is negative. Note that when the test is designed assuming that $m > 0$ ($m < 0$) for negative (positive) values of m, P_D is less than P_F.

Since a UMP test does not exist in the case when m can take on either positive or negative values, a reasonable choice in this case would be to use a two-sided test as shown in Fig. 3.6.3. The decision rule is to choose H_1 if $|z| > \gamma$.

This test obviously will not perform as well as the one-sided tests discussed earlier with correct assumptions but is much better in the case of incorrect assumptions.

Example 3.6.2

We will determine a generalized LRT test for Example 3.6.1. Assume that N independent observations are available. The relevant probability densities are

$$p(\mathbf{z} \mid m, H_1) = \prod_{i=1}^{N} \frac{1}{\sqrt{2\pi}\,\sigma} \exp\left[-\frac{(z_i - m)^2}{2\sigma^2}\right]$$

$$p(\mathbf{z} \mid H_0) = \prod_{i=1}^{N} \frac{1}{\sqrt{2\pi}\,\sigma} \exp\left(-\frac{z_i^2}{2\sigma^2}\right)$$

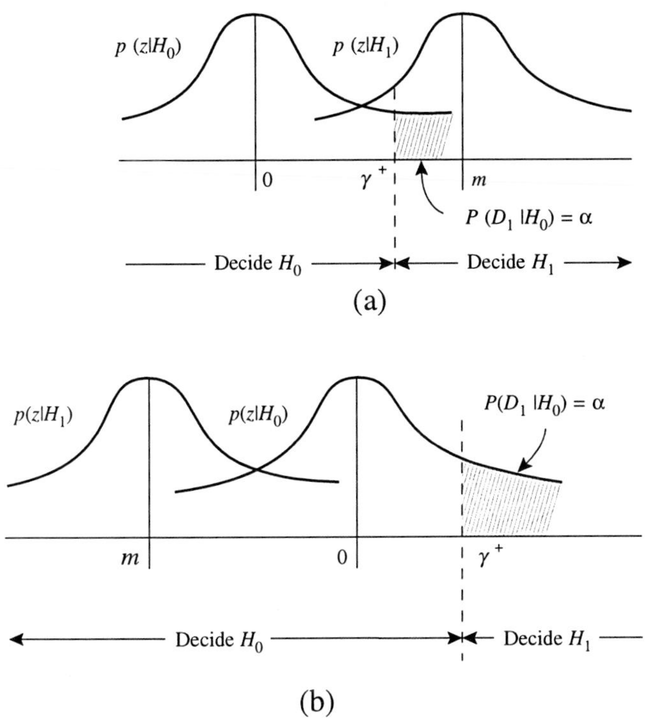

Figure 3.6.1 Decision regions for Example 3.6.1, assuming that mean is positive: (a) true mean positive; (b) true mean negative.

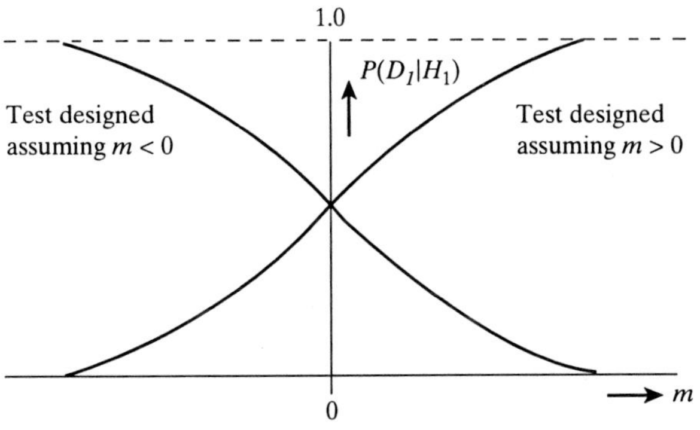

Figure 3.6.2 Performance of tests for Example 3.6.1, $P(D_1 | H_0)$ fixed.

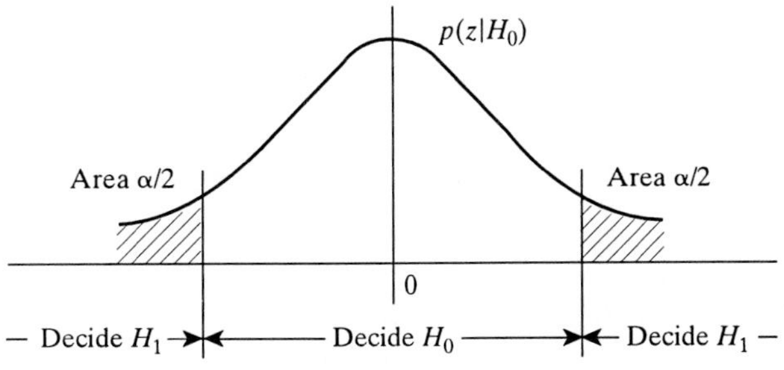

Figure 3.6.3 Two-sided test for Example 3.6.1.

The maximum likelihood estimate of the parameter m is given by the sample mean as (see Chapter 5)

$$\hat{m} = \frac{1}{N}\sum_{i=1}^{N} z_i$$

and the LRT is

$$\Lambda_g(z) = \frac{\prod_{i=1}^{N} \frac{1}{\sqrt{2\pi}\,\sigma} \exp\left[-\left(z_i - 1/N\sum_{i=1}^{N} z_i\right)^2 / 2\sigma^2\right]}{\prod_{i=1}^{N} \frac{1}{\sqrt{2\pi}\,\sigma} \exp\left[-\left(\frac{z_i^2}{2\sigma^2}\right)\right]} \underset{H_0}{\overset{H_1}{\gtrless}} \gamma$$

Canceling common terms and taking the logarithm yields

$$\frac{1}{2\sigma^2 N}\left(\sum_{i=1}^{N} z_i\right)^2 \underset{H_0}{\overset{H_1}{\gtrless}} \ln \gamma$$

An equivalent test is to choose

$$\left(\sum_{i=1}^{N} z_i\right)^2 \underset{H_0}{\overset{H_1}{\gtrless}} \gamma_1^2 \underline{\Delta} 2\sigma^2 N \ln \gamma$$

That is,

$$\left|\sum_{i=1}^{N} z_1\right| \underset{H_0}{\overset{H_1}{\gtrless}} \gamma_1$$

which for $N = 1$ is just the two-sided test discussed in Example 3.6.1.

3.6.1 One-Sided Hypotheses and UMP Tests

Consider the one-sided hypotheses testing problem in which, given random samples $z_1, z_2, \ldots, z_n$ from a density $p(z \mid \theta)$, we want to test the hypothesis H_0: $\theta \leq \theta_0$ against H_1: $\theta > \theta_0$ or H_0: $\theta = \theta_0$ against H_1: $\theta > \theta_0$. A UMP test may be identified by using one of the following two results. We simply state these results without proofs.

Suppose that $p(z \mid \theta)$ belongs to a one-parameter exponential family of the form of Eq. (2.6.21). Let $t(\mathbf{z}) = \sum_{i=1}^{n} d(z_i)$. Let c^* denote the critical region for the test, that is, the region where we decide H_1. If $c(\theta)$ is a monotone increasing function in θ, and if there exists a constant k^* such that $P(T(\mathbf{Z}) > k^* \mid \theta_0) = \alpha$, the test with $c^* = \{\mathbf{z}: t(\mathbf{z}) > k^*\}$ is a UMP test of size α. (If $c(\theta)$ is monotone decreasing, the critical region is given by $c^* = \{\mathbf{z}: t(\mathbf{z}) < k^*\}$) [1,4].

Example 3.6.3

Consider the problem of Example 3.6.1 with $m_0 > 0$. The observations $Z_i, i = 1, 2, \ldots, N$ are Gaussian and hence belong to a one-parameter exponential family. By writing

$$p(\mathbf{z} \mid H_1) = \exp\left(-\frac{Nm^2}{2\sigma^2}\right)\left[\frac{1}{\sqrt{2\pi}\,\sigma} \exp\left(\frac{\sum_{i=1}^{N} z_i^2}{2\sigma^2}\right)\right] \exp\left(\frac{m}{\sigma^2}\sum_{i=1}^{N} z_i\right)$$

it easily follows that $t(\mathbf{z}) = d(\mathbf{z}) = \sum_{i=1}^{N} z_i$ and we immediately obtain the UMP test

$$\sum_{i=1}^{N} z_i \underset{H_0}{\overset{H_1}{\gtrless}} \gamma^+$$

where γ^+ is determined from

$$P\left(\sum_{i=1}^{N} Z_i > \gamma^+ \,|\, H_0\right) = \alpha$$

Example 3.6.4

Let the observations $z_1, z_2, \ldots, z_n$ be from an exponential density with parameter λ, so that

$$p(\mathbf{z}\,|\,\lambda) = \lambda \exp\left(-\lambda \sum_{i=1}^{n} z_i\right)$$

Let the two hypotheses be H_0: $\lambda = \lambda_0$ and H_1: $\lambda > \lambda_0$. Since $d(z_i) = z_i$, it follows that $t(\mathbf{z}) = \sum_{i=1}^{n} z_i$ and the test

$$\text{Reject } H_0 \text{ if } \sum_{i=1}^{n} z_i < k^*$$

with

$$P\left(\sum_{i=1}^{n} Z_i < k^* \,|\, \lambda_0\right) = \alpha$$

is a UMP test of size α.

For example, if $\lambda_0 = 1, \alpha = 10^{-2}, n = 2, \sum_{i=1}^{2} Z_i$ has a gamma $(2, 1)$ density and k^* can be determined from

$$1 - \sum_{j=0}^{n-1} \frac{e^{-k^*}(k^*)^j}{j!} = 10^{-2}$$

An iterative solution yields $k^* = 0.148$.

A similar result about the existence of a UMP test holds for one-sided hypotheses, even when $p(\mathbf{z}\,|\,\theta)$ does not belong to the exponential family but nevertheless has a monotone likelihood ratio in the statistic $t(\mathbf{z})$. That is, if the monotone likelihood ratio as defined in (2.6.5) with $\mathbf{x}$ replaced by $\mathbf{z}$, is nondecreasing in $t(\mathbf{z})$, and if there exists a constant k^* such that $P(T(\mathbf{Z}) < k^* \,|\, \theta_0) = \alpha$, the test with $c^* = \{\mathbf{z}: t(\mathbf{z}) < k^*\}$ is a UMP of size α for a test of H_0: $\theta \leq \theta_0$ versus H_1: $\theta > \theta_0$. (If the monotone likelihood ratio is nonincreasing in $t(\mathbf{z})$, then $c^* = \{\mathbf{z}: t(\mathbf{z}) > k^*\}$) [1,4].

Example 3.6.5

Let $z_1, z_2, \ldots, z_n$ be random samples from a uniform distribution over $[0, \theta]$ and let the two hypotheses be H_0: $\theta = \theta_0$ and H_1: $\theta > \theta_0$. We have seen in Example 2.6.4 that the uniform distribution has a nonincreasing monotone likelihood ratio Λ in the maximum order statistic t. Thus the test

$$t \underset{H_0}{\overset{H_1}{\gtrless}} k^*$$

is a UMP test. k^* is determined from

$$P(T > k^* \mid \theta_0) = \alpha$$

or, equivalently,

$$P(Z_1 < k^*, Z_2 < k^*, \ldots, Z_n < k^* \mid \theta_0) = 1 - \alpha$$

which gives

$$k^* = \theta_0[1 - \alpha]^{1/n}$$

In rare cases, a UMP test for a two-sided hypotheses may exist (see [9]). Other applications of these results are given in the exercises.

3.7 ASYMPTOTIC ERROR RATE OF LRT FOR SIMPLE HYPOTHESIS TESTING

As the number of observations n approaches infinity, we may expect that any reasonable test will tend toward making perfect inferences of the corresponding hypotheses. That is, the probability of error of the test should approach zero. In the case of simple versus simple hypotheses, we can determine the rate at which the probability of error of an LRT approaches zero as n approaches infinity. This rate depends on the *Kullback–Liebler* (KL) *information number*, which is a measure for discriminating between two densities. The KL number for the two densities $p_0(\mathbf{z}) = p(\mathbf{z} \mid H_0)$ and $p_1(\mathbf{z}) = p(\mathbf{z} \mid H_1)$ is defined as [5]

$$\delta = \int p_0(\mathbf{z}) \ln \frac{p_0(\mathbf{z})}{p_1(\mathbf{z})} d\mathbf{z} \tag{3.7.1}$$

It can be shown that δ is greater than or equal to zero with equality holding only if $p_0(\mathbf{z}) = p_1(\mathbf{z})$ almost everywhere. We assume that for densities of interest, $\delta < \infty$ [5].

Consider a set of random samples $z_1, z_2, \ldots, z_n$ from either $p_0(z)(H_0)$ or $p_1(z)(H_1)$. Suppose that we design a sequence of Neyman–Pearson likelihood ratio tests of the form

$$\Lambda_n = \frac{\prod_{i=1}^{n} p_1(z_i)}{\prod_{i=1}^{n} p_0(z_i)} \underset{H_0}{\overset{H_1}{\gtrless}} \lambda_n \tag{3.7.2}$$

with α_n being the specified probability of false alarm and β_n the corresponding probability of miss. Then it can be shown [10,11] that if the sequence of tests is such that α_n tends to a constant $\alpha, 0 < \alpha \leq 1$ as n tends to infinity, then

$$\lim_{n \to \infty} \frac{1}{n} \ln \beta_n = -\delta$$

That is, the probability of miss, β_n, approaches zero exponentially as $\exp(-n\delta)$. The rate at which it approaches zero is determined by the KL number δ.

If, instead of the Neyman–Pearson test, we used a sequence of Bayesian tests with the threshold of the tests chosen to be $\lambda_n = 1$ (this corresponds to the minimum probability of error criterion for equally likely hypotheses), it can be shown that both α_n and β_n approach zero as n approaches infinity [10].

3.8 CFAR DETECTION

In radar target detection, the term *constant false alarm rate* (CFAR) is widely used. It refers to a test of signal (H_1) versus no signal (H_0) hypotheses in which the rate (probability) of false alarm is required to be constant, independent of any unknown (nuisance) parameters that may be present in the density functions of the observations under either hypothesis. A CFAR detector need not be optimal in any sense, but it does provide the required false alarm rate under the assumed conditions. To illustrate this approach, we consider the following example of a *cell-averaging* radar detector, in which we are required to determine whether or not a target is present in a specific range cell.

Let us denote the energy of the detector output of the test cell under examination by z_0. It can be shown that under the assumption that the observation noise is narrowband Gaussian, with no signal present, Z_0 has an exponential density function with parameter, say, λ_0. If the signal is present, and if the target is a Rayleigh fluctuating target (see Chapter 8), it can be shown that Z_0 still has an exponential density function but with parameter $\lambda_1, \lambda_1 < \lambda_0$. λ_0 is usually unknown, and the test with a fixed constant threshold t,

$$z_0 \underset{H_0}{\overset{H_1}{\gtrless}} t \tag{3.8.1}$$

cannot maintain a constant probability of false alarm irrespective of the value of λ_0. Therefore, to obtain a CFAR test, several range cells adjacent to the test cell are sampled to produce energy outputs $Z_1, Z_2, \ldots, Z_n$. The Z_i values are statistically independent of Z_0. Assuming that the noise level in these adjacent cells is identical to the noise level in the test cell, and assuming that these cells do not contain reflections from other interfering targets, the Z_i are independent, exponentially distributed with parameter λ_0. The test

$$z_0 \underset{H_0}{\overset{H_1}{\gtrless}} t\sum_{i=1}^{n} z_i \tag{3.8.2}$$

is CFAR since it is equivalent to

$$\lambda_0 z_0 \underset{H_0}{\overset{H_1}{\gtrless}} t\sum_{i=1}^{n} (\lambda_0 z_i) \tag{3.8.3}$$

and $\lambda_0 z_i, i = 1, 2, \ldots, n$ are independent, identically exponentially distributed random variables with parameter equal to unity, and under H_0, $\lambda_0 z_0$ is also exponential with parameter 1. Therefore, the probability of false alarm given by

$$P_F = P\left(Z_0 > t\sum_{i=1}^{n} Z_i \,|\, H_0\right) \tag{3.8.4}$$

is independent of λ_0. The test in Eq. (3.8.2) is called a *cell-averaging CFAR*.

The CFAR property is destroyed if there are interfering targets appearing within the adjacent range cells or if the noise/clutter levels in adjacent cells are different from that in the test cell.

Other CFAR tests can be designed for the above-stated problem. Some of these are based on the order statistics of the samples [12].

3.9 SEQUENTIAL DETECTION: WALD'S TEST

In previous sections we have considered decision problems in which the total number of observations is fixed. In many practical situations, however, the observations are sequential in nature, and more information becomes available as time progresses. In such cases we may desire to make a decision by processing the samples sequentially as they occur. Even though the Bayes procedure may be used in this case, the resulting test is fairly complicated. Instead, we focus our attention on a modified Neyman–Pearson type of test, referred to as *Wald's sequential test* or the *sequential probability ratio test* (SPRT) [4,13,14]. These tests have found application in a number of areas, such as pattern recognition, communications, and radar [14,15]. For purposes of clarity, we again restrict our discussion to the simple binary hypothesis-testing problem of Section 3.3.

We recall that in our discussion of the Neyman–Pearson rule in Section 3.3, we computed the likelihood ratio and compared it with a threshold η which was determined by the specified value of P_F. In the SPRT, at each stage of the decision procedure, the likelihood ratio is compared with two thresholds, η_0 and η_1. If the likelihood ratio is greater than or equal to η_1, we decide in favor of H_1. If it is less than η_0, we decide that H_0 was the correct hypothesis. If its value lies between η_0 and η_1, we conclude that we do not have enough information to decide between the two hypotheses and hence take another observation.

The key problem in the design of the SPRT is to determine the thresholds η_1 and η_0 that will yield, approximately, the specified values of P_F and P_M. We develop the necessary relations in this section for determining these thresholds. For a given level of performance (that is, for given values of P_F and P_M), this test will terminate, on the average, in fewer samples than the fixed sample size tests [16]. This can be claimed as the principal virtue of the method because of the costs involved in making additional measurements.

Let $z_1, z_2, \ldots, z_j, \ldots$ represent the observation samples and let $\mathbf{z}_N$ be the vector defined as

$$\mathbf{z}_N = [z_1, z_2, \ldots, z_N]^T$$

The likelihood ratio based on N samples is then given by

$$\Lambda(\mathbf{z}_N) = \frac{p(\mathbf{z}_N \mid H_1)}{p(\mathbf{z}_N \mid H_0)} \tag{3.9.1}$$

Thus, in general, in computing the likelihood ratio $\Lambda(\mathbf{z}_j)$ at each stage (that is, for each value of j), we need to know the joint density functions of the samples $z_1, z_2, \ldots, z_j$ up to that stage. If, however, these samples are independent, we can easily compute $\Lambda(\mathbf{z}_j)$ recursively as follows:

$$\Lambda(\mathbf{z}_j) = \frac{p(\mathbf{z}_j \mid H_1)}{p(\mathbf{z}_j \mid H_0)} = \prod_{i=1}^{j} \frac{p(z_i \mid H_1)}{p(z_i \mid H_0)} = \frac{p(z_j \mid H_1)}{p(z_j \mid H_0)} \prod_{i=1}^{j-1} \frac{p(z_i \mid H_1)}{p(z_i \mid H_0)}$$

That is,

$$\Lambda(\mathbf{z}_j) = \Lambda(z_j)\Lambda(\mathbf{z}_{j-1}) \tag{3.9.2}$$

with the initial condition

$$\Lambda(\mathbf{z}_1) = \Lambda(z_1)$$

Let us assume that the error probabilities are specified as

$$P_F = \alpha \quad \text{and} \quad P_M = \beta \tag{3.9.3}$$

We now derive a pair of inequalities for the thresholds η_0 and η_1 in terms of α and β. We recall from Eqs. (3.3.17) and (3.3.19) that the false alarm and detection probabilities are given by

$$P_F = P(D_1 \mid H_0) = \int_{Z_1} p(\mathbf{z}_j \mid H_0)\, d\mathbf{z}_j \tag{3.9.4}$$

and

$$\begin{aligned} P_D = P(D_1 \mid H_1) &= \int_{Z_1} p(\mathbf{z}_j \mid H_1)\, d\mathbf{z}_j \\ &= \int_{Z_1} \Lambda(\mathbf{z}_j) p(\mathbf{z}_j \mid H_0)\, d\mathbf{z}_j \end{aligned} \tag{3.9.5}$$

where Z_1 represents the decision region for hypothesis H_1. When H_1 is the true hypothesis and the decision is also H_1, we must have $\Lambda(\mathbf{z}_j) \geqslant \eta_1$, so that Eq. (3.9.5) can be rewritten as

$$P_D = P(D_1 \mid H_1) \geqslant \eta_1 \int_{Z_1} p(\mathbf{z}_j \mid H_0)\, d\mathbf{z}_j \tag{3.9.6}$$

From Eq. (3.9.4), the integral on the right-hand side represents P_F, which is assumed to have a value α. Since $P_D = 1 - P_M = 1 - \beta$, Eq. (3.9.6) yields the inequality

$$1 - \beta \geqslant \eta_1 \alpha$$

so that the threshold η_1 satisfies

$$\eta_1 \leqslant \frac{1 - \beta}{\alpha} \tag{3.9.7}$$

Using a similar argument, it follows that

$$\eta_0 \geq \frac{\beta}{1-\alpha} \tag{3.9.8}$$

Often, we use the log-likelihood ratio because of its computational advantages. Taking the logarithm of Eq. (3.9.2), we obtain the recursive relation for updating $L(\mathbf{z}_j) \underline{\Delta} \ln \Lambda(\mathbf{z}_j)$ as

$$L(\mathbf{z}_j) = L(\mathbf{z}_{j-1}) + L(z_j) \tag{3.9.9}$$

The corresponding thresholds are given by ln η_1 and ln η_0. When the increments $L(z_j)$ in Eq. (3.9.9) are small, $L(\mathbf{z}_j)$ will at best exceed (or fall below) the thresholds by only a small quantity. In such a case, it is not unreasonable to treat the inequalities in Eqs. (3.9.7) and (3.9.8) as equalities to obtain the simple design rule:

$$\ln \eta_1 \simeq \ln \frac{1-\beta}{\alpha} \tag{3.9.10}$$

and

$$\ln \eta_0 \simeq \ln \frac{\beta}{1-\alpha} \tag{3.9.11}$$

We now calculate the average number of observations required to make a decision under each hypothesis. To do this, let us assume that the test terminates in N samples, which implies that $L(\mathbf{z}_N)$ approximately assumed either of the values ln η_0 or ln η_1. If the true hypothesis is H_0 and $L(\mathbf{z}_N) \geq \ln \eta_1$, we have a false alarm with probability α. In this case, the probability that $L(\mathbf{z}_N)$ is less than or equal to ln η_0 will be $1 - \alpha$. Similarly, if the true hypothesis is H_1 and $L(\mathbf{z}_N) \leq \ln \eta_0$, we have miss with probability β, and that the probability that $L(\mathbf{z}_N)$ is greater than or equal to ln η_1 is $1 - \beta$. The expected values of $L(\mathbf{Z}_N)$ under the two hypotheses are therefore approximately equal to

$$E[L(\mathbf{Z}_N)\,|\,H_1] = (1-\beta)\ln \eta_1 + \beta \ln \eta_0 \tag{3.9.12}$$

and

$$E[L(\mathbf{Z}_N)\,|\,H_0] = \alpha \ln \eta_1 + (1-\alpha)\ln \eta_0 \tag{3.9.13}$$

Let us define the binary variable K_m such that

$$K_m = \begin{cases} 1 & \text{if no decision was made up to } (m-1)\text{st stage} \\ 0 & \text{if a decision was made at an earlier stage} \end{cases} \tag{3.9.14}$$

Since the test is assumed to terminate after N samples, we can use Eqs. (3.9.9) and (3.9.14) to write

$$L(\mathbf{z}_N) = \sum_{m=1}^{N} L(z_m) = \sum_{m=1}^{\infty} K_m L(z_m) \tag{3.9.15}$$

The conditional expectation of $L(\mathbf{Z}_N)$ given hypothesis H_i is

$$\begin{aligned} E(L(\mathbf{Z}_N)\,|\,H_i\} &= \sum_{n=1}^{\infty} E(L(\mathbf{Z}_N)\,|\,H_i, N = n)\, P(N = n\,|\,H_i) \\ &= \sum_{n=1}^{\infty} E\left(\sum_{m=1}^{\infty} K_m L(Z_m)\,|\,H_i, N = n\right) P(N = n\,|\,H_i) \end{aligned} \tag{3.9.16}$$

We note that K_m depends only on $Z_1, Z_2, \ldots, Z_{m-1}$ and not on Z_m so that K_m and $L(Z_m)$ are independent. Hence we can write Eq. (3.9.16) as

$$E(L(\mathbf{Z}_N\,|\,H_i)) = \sum_{n=1}^{\infty} P(N = n\,|\,H_i) \sum_{m=1}^{\infty} E(L(Z_m\,|\,H_i))E(K_m\,|\,H_i, N = n) \tag{3.9.17}$$

If we now assume that the observations under each hypothesis are identically distributed so that

$$E(L(Z_m\,|\,H_i)) = E(L(Z\,|\,H_i)) \tag{3.9.18}$$

we can write

$$\begin{aligned} E(L(Z_N\,|\,H_i)) &= E(L(Z)\,|\,H_i) \sum_{n=1}^{\infty} P(N = n\,|\,H_i) E\left(\sum_{m=1}^{\infty} K_m\,|\,H_i, N = n\right) \\ &= E(L(Z)\,|\,H_i) \sum_{n=1}^{\infty} P(N = n\,|\,H_i) n \end{aligned} \tag{3.9.19}$$

Or

$$E(L(\mathbf{Z}_N)\,|\,H_i) = E(L(Z)\,|\,H_i)E(N\,|\,H_i) \tag{3.9.20}$$

Substitution of Eq. (3.9.20) in Eq. (3.9.12) yields the average number of samples when H_1 is the correct hypothesis as

$$E[N\,|\,H_1] = \frac{(1 - \beta) \ln \eta_1 + \beta \ln \eta_0}{E[L(z)\,|\,H_1]} \tag{3.9.21}$$

Similarly, Eq. (3.9.13) yields

$$E[N\,|\,H_0] = \frac{\alpha \ln \eta_1 + (1 - \alpha) \ln \eta_0}{E[L(z)\,|\,H_0]} \tag{3.9.22}$$

We can easily show that the test terminates with probability 1. That is,

$$\lim_{n\to\infty} P(N \geqslant n) = 0 \tag{3.9.23}$$

We note that we take another sample if $\ln \eta_0 < L(z) < \ln \eta_1$. If we let $\delta = \ln \eta_1 - \ln \eta_0$, in general, $P(|L(z_i)| \leqslant \delta) = p < 1$. Thus the probability that the first n observations are all such that $|L(z_i)| < \delta$ is equal to p^n, since the observations are independent. This probability tends to zero as n tends to infinity, thus establishing Eq. (3.9.23).

It has also been shown by Wald and Wolfowitz [16] that for specified values of P_F and P_M, the SPRT minimizes the average number of observations $E(N\,|\,H_1)$ and $E(N\,|\,H_0)$. We omit the proof since it is somewhat involved.

Our development of the sequential method for decision making assumed the following: (1) The samples z_j are independent and identically distributed; (2) the probabilities P_F and P_D were constant for the entire interval, resulting in constant threshold given by Eqs. (3.9.10) and (3.9.11). But neither of the assumptions is required in general. In the absence of independent samples, we cannot obtain the recursive relations of Eqs. (3.9.2) and (3.9.9). Consequently, we must compute $\Lambda(\mathbf{z}_j)$ from the joint density functions $p(\mathbf{z}_j \mid H_i), i = 0, 1$, and for each j. This, in general, is computationally difficult. Removal of assumption (2) yields varying thresholds, $\eta_1(j)$ and $\eta_0(j)$, and it is no longer simple to calculate the average number of samples for decision making.

Although the SPRT terminates with probability 1, in many cases it is preferable to fix an upper limit N^* on the total number of samples we can afford. This is called a *truncated SPRT*. At the end of N^* samples, a fixed sample size test is performed to obtain a decision. We conclude this section with an example.

Example 3.7.1

We consider the same problem as in Example 3.2.3, but assume that the samples are available in a sequence. From Example 3.2.3, we have the likelihood ratio at the jth stage as

$$\Lambda(\mathbf{z}_j) = \exp\left(\sum_{i=1}^{j} z_i - \frac{j}{2}\right)$$

and the log-likelihood ratio is

$$L(\mathbf{z}_j) = \sum_{i=1}^{j} z_i - \frac{j}{2}$$

Let us assume that our acceptable level of performance is

$$P_F = \alpha = 0.1 \quad \text{and} \quad P_M = \beta = 0.1$$

Using Eqs. (3.9.10) and (3.9.11), we obtain the two thresholds as

$$\ln \eta_1 = 2.197 \quad \text{and} \quad \ln \eta_0 = -2.197$$

The decision rule, therefore, is

$$L(\mathbf{z}_j) \geqslant 2.197, \qquad \text{decide } H_1$$

$$L(\mathbf{z}_j) \leqslant -2.197 \qquad \text{decide } H_0$$

and

$$-2.197 \leqslant L(\mathbf{z}_j) \leqslant 2.197 \qquad \text{take another sample } z_{j+1}$$

The average number of samples for the test to terminate under each hypothesis is obtained using Eqs. (3.9.21) and (3.9.22). Since

$$\mathrm{E}[L(Z) \mid H_1] = \mathrm{E}\left[\left(Z - \frac{1}{2}\right) \middle| H_1\right] = \frac{1}{2}$$

and

$$E[L(Z)\,|\,H_0] = E\left[\left(Z - \frac{1}{2}\right)\middle|\,H_0\right] = -\frac{1}{2}$$

we obtain

$$E[N\,|\,H_1] = 3.5 \text{ samples}$$

and

$$E[N\,|\,H_0] = 3.5 \text{ samples}$$

The preceding results indicate that a sample size of 4 may yield the desired performance.

There have been some criticisms directed against the classical theories of Neyman–Pearson (and consequently, Wald's sequential test) and the Bayesian approach to hypotheses testing. In the case of the Neyman–Pearson test, the criticism is directed toward the requirement for a proper choice of the size of the type I error (P_F). Recall that in a Neyman–Pearson test, P_F determines the threshold of the likelihood ratio test. Given a set of observations, if tests are designed for two different choices of P_F, the decisions regarding the hypotheses could be contradictory! A choice of 0.05 as the value for the type I error is common in medical and social sciences, whereas a value in the range 10^{-6} to 10^{-3} is common in radar target detection. Certainly, a judicious choice of P_F, based perhaps on judgment or past experience, is required. Similarly, the Bayesian approach has been criticized from the perspective of the requirement of costs and prior probabilities [17]. The costs and prior probabilities will again have to be determined based on the user's judgment and past experience. In either case, the determination of these required quantities will be only approximate, and reasonable guesses in general.

3.10 SUMMARY

In this chapter we have developed the basic ideas of statistical hypothesis testing. The components of the problem were introduced in Section 3.2 and the basic decision rule for testing between two hypotheses was developed in Section 3.3. The test involves computation of the likelihood ratio $\Lambda(z)$ and comparing it with a threshold in order to make a decision. The decision space is thus a one-dimensional space. The threshold of the test depends on the particular criterion chosen. The Bayes test requires the specification of costs and prior probabilities. In the absence of prior probabilities, we can use the minimax test, which has its origins in the theory of games [18]. The Neyman–Pearson test can be used when assignment of neither costs nor prior probabilities can be made. We also determined the performance of the tests by computing the probabilities of the two types of error. The performance is most conveniently analyzed by consideration of the receiver operating characteristic, which is a plot of detection probability P_D versus false alarm probability P_F.

These concepts were extended to the case of multiple observations in Section 3.4. While the observation space is multidimensional for this problem, the decision space is of reduced dimension. We have also shown that an optimal test is based on a sufficient statistic.

Section 3.5 considered the M-ary hypothesis-testing problem. The decision rule involves the computation of $M - 1$ likelihood ratios from the density functions $p(\mathbf{z} \mid H_i)$. The decision boundaries are hyperplanes in the $(M - 1)$-dimensional space of the likelihood ratios. For the special case of minimum probability of error criterion, the test reduces to choosing the hypothesis corresponding to the largest of the posterior probabilities $p(H_i \mid \mathbf{z})$. While the optimum test is fairly straightforward to determine, the computation of the error probabilities is frequently difficult in real world problems.

The composite hypothesis-testing problem, in which the parameters θ characterizing the hypotheses are not known, is considered in Section 3.6. The case in which the unknown parameter is random with known probability density is fairly straightforward. If the density function for θ is not known, we may try integrating over several assumed densities and pick the least favorable one. Frequently, the test structure is insensitive to the choice of density function. If θ is a nonrandom variable, we may consider applying the Neyman–Pearson test. This is useful if a UMP test exists. Otherwise, we may try estimating the values of θ under the two hypotheses and use these estimates as the true values in a likelihood ratio test. Section 3.7 contains a brief discussion of the asymptotic error rate of a likelihood ratio test, while the notion of a CFAR test was introduced in Section 3.8. We concluded the chapter with a discussion of the sequential probability ratio test in Section 3.9.

The development of the discussions in this chapter have assumed that the density functions of the observations under each hypothesis are either known or can be determined. In Chapter 7 we consider problems in which the underlying density functions are not known completely. An approach to solving this class of problems is the *nonparametric* or *distribution-free* approach.

EXERCISES

3.1. Derive a likelihood ratio test for the following case. Under hypothesis H_1, the observation Z is Gaussian with mean m_1 and variance σ_1^2. Under H_0, Z is Gaussian with mean m_0 and variance σ_0^2. Find the decision regions as a function of the LRT threshold.

3.2. Consider the binary symmetric channel shown in Fig. E.3.1, where the transmitted symbols are 0 and 1 and the received symbols are a and b. ϵ represents the crossover probability [i.e., the probability that the channel output is $a(b)$ when the input is 1(0) and is small]. Obtain a decision rule to minimize the total probability of error if the prior probability of a 1 being transmitted is p. What is the probability of error? Assume that $\epsilon < \frac{1}{2}$.

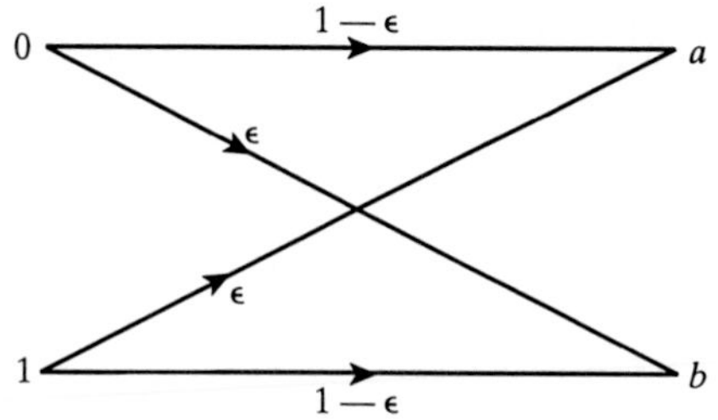

Figure E.3.1

3.3. Consider the hypothesis-testing problem in which

$$p(\mathbf{z}|H_1) = \frac{1}{\sqrt{2\pi}} \exp\left(-\frac{z^2}{2}\right)$$

$$p(\mathbf{z}|H_0) = \frac{1}{2} \exp(-|z|)$$

(a) Set up LRT and determine the decision regions. Plot the ROC.
(b) For $C_{00} = C_{11} = 0$ and $C_{01} = C_{10} = 1$, find P_F and P_D for the Bayes test if $P(H_1) = \frac{3}{4}$.
(c) For the same assignment of costs as in part (b), find the performance of a minimax test.
(d) Set up the Neyman–Pearson test for $P_F = 0.2$.

3.4. Consider the following two-channel problem, in which the observations under the two hypotheses are

$$H_1\colon Z_1 = V_1 + 1$$
$$Z_2 = 0.5V_2 + 0.5$$
$$H_0\colon Z_1 = V_1 - 1$$
$$Z_2 = 0.5V_2 - 0.5$$

where V_1 and V_2 are independent, zero-mean Gaussian variables with variance σ^2. Find the minimum probability of error receiver if both hypotheses are equally likely. Simplify the receiver structure.

3.5. The observed random variable Z is Gaussian distributed on the three hypotheses,

$$p(\mathbf{z}|H_i) = \frac{1}{\sqrt{2\pi}\,\sigma_i} \exp\left[-\frac{(z - m_i)^2}{2\sigma_i^2}\right] \qquad i = 0, 1, 2$$

where $m_0 = m_2 = 0$, $m_1 = 1$, and $\sigma_0 = \sigma_1 = 1$. Find the decision rules for the minimum probability of error criterion assuming equal prior probabilities. What is the probability of error if $\sigma_2 = 2$?

3.6. In a binary hypothesis-testing problem, the received signals under the two hypotheses are

$$H_1\colon Z = X_1^2 + X_2^2$$
$$H_2\colon Z = X_1^2$$

where X_1 and X_2 are independent, identically Gaussian distributed with zero mean and unity variance. Obtain the optimum decision rule for the Bayes criterion. Assume that $P_0 = P_1 = \frac{1}{2}$, $C_{00} = C_{11} = 0$, $C_{10} = C_{01} = 1$.

3.7. Obtain a minimum probability of error decision rule to discriminate between two hypotheses H_0 and H_1 based on the observation Z. The distributions of Z under the two hypotheses are as shown in Fig. E.3.2.

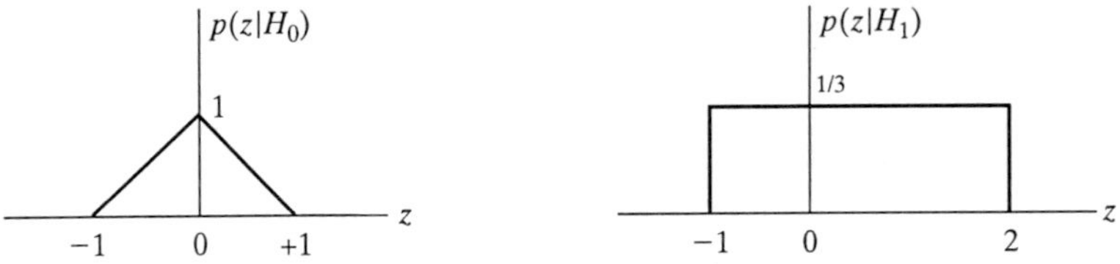

Figure E.3.2

3.8. In many instances the observations under the two hypotheses are discrete random variables. We can derive a likelihood ratio test for this case by using probability mass functions. Let us assume that the observation under the two hypotheses have the Poisson distributions

$$P(Z = z \,|\, H_1) = \frac{m_1^z}{z!}\exp(-m_1) \qquad z = 0, 1, 2, \ldots$$

$$P(Z = z \,|\, H_0) = \frac{m_0^z}{z!}\exp(-m_0) \qquad z = 0, 1, 2, \ldots$$

with $m_1 > m_0$.

(a) Show that the LRT test is

$$z \underset{H_0}{\overset{H_1}{\gtrless}} \frac{\ln \eta + m_1 - m_0}{\ln m_1 - \ln m_0}$$

(b) Because z takes on only integer values, it is more convenient to write the decision rule as

$$z \underset{H_0}{\overset{H_1}{\gtrless}} \gamma' \qquad \gamma' = 0, 1, 2 \ldots$$

Show that for $\gamma' \geqslant 1$, the error probabilities are given by

$$P_F = 1 - \exp(-m_0) \sum_{n=0}^{\gamma'-1} \frac{(m_0)^n}{n!}$$

and

$$P_M = \exp(-m_1) \sum_{n=0}^{\gamma'-1} \frac{(m_1)^n}{n!}$$

Plot the ROC for this problem assuming that $m_0 = 1$ and $m_1 = 2$.

3.9. In Exercise 3.8, P_F can assume only certain discrete values. Suppose that we want to design a Neyman–Pearson rule for some intermediate value of P_F. It is clear that an ordinary LRT will not give this performance. We will have to use a *randomized decision* rule in which instead of assigning a point $z \in Z_1$ to H_1, we assign it to H_1 with a certain probability $P_a(z)$ and to H_0 with probability $1 - P_a(z)$. We do this by combining the results of two LRTs in a probabilistic manner. Let us assume that for Exercise 3.8 we want

$$P_F = 1 - \tfrac{1}{2}\exp(-m_0).$$

(a) Show that a randomized rule combines two LRTs, one with a threshold $\gamma' = 0$ and the other with threshold $\gamma' = 1$. Find the test explicitly.

(b) Find P_D.

3.10. In a binary hypothesis-testing problem, the observation Z is Rayleigh distributed under both hypotheses with different parameters, that is,

$$p(z \mid H_i) = \frac{z}{\sigma_i^2}\exp\left(-\frac{z^2}{2\sigma_i^2}\right) \qquad z \geqslant 0, \quad i = 0, 1$$

Obtain the likelihood ratio test for the Bayes criterion. Extend your results to N independent observations for the minimum probability of error criterion. Assume that $P(H_1) = P(H_0) = \frac{1}{2}$. Derive the expressions for the resulting probability of error.

3.11. In a binary hypothesis-testing problem, the observed samples, Z_i, are independent and identically distributed under each hypothesis. The distributions are

$$p(z_i \mid H_j) = \frac{1}{\sqrt{2\pi}\,\sigma_j}\exp\left[-\frac{(z_i - m_j)^2}{2\sigma_j^2}\right] \qquad \begin{array}{l} j = 0, 1 \\ i = 1, \ldots, N \end{array}$$

where $\sigma_0 = 1$, $\sigma_1 = 2$, $m_0 = -1$, and $m_1 = 1$. Find the LRT for this problem. Express your results in terms of the quantities $l_1 = \sum_{i=1}^N z_i$ and $l_2 = \sum_{i=1}^N z_i^2$ and determine the decision regions in the l_1–l_2 plane.

3.12. We are given K independent observations:

$$H_1\colon Z_k = V_k \qquad k = 1, 2, \ldots, K$$

$$H_0\colon Z_k = 1 + V_k \qquad k = 1, 2, \ldots, K$$

where V_k is a zero-mean Gaussian random variable with variance σ^2.

(a) Compute the likelihood ratio and the threshold for the optimum receiver. Assume that $C_{00} = C_{11} = 0$, $C_{01} = 2$, $C_{10} = 1$ and that $P(H_0) = 0.7$, $P(H_1) = 0.3$.

(b) Find P_F and P_M.

(c) Plot the ROC for $K = 1, \sigma^2 = 2$.

(d) Find the value of K for which the risk is no more than one-half of the risk for $K = 1$.

3.13. Plot the ROC for the problem of Example 3.4.2. Assume that $N = 2, \sigma_1^2 = 4\sigma_0^2$.

3.14. Consider M-ary hypothesis testing with the Bayes criterion. Prove that the optimum decision criterion is to choose the hypothesis for which the conditional risk $C(H_i \mid z)$ is smallest, where

$$C(H_i \mid z) = \sum_{j=0}^{M-1} C_{ij} P(H_j \mid z) \qquad i = 0, 1, \ldots, M-1$$

3.15. The observation Z is the output under one of the three hypotheses, $H_i, i = 0, 1, 2$. The probability density functions under the hypotheses are shown in Fig. E.3.3. What is the decision rule that minimizes the probability of error? Find the minimum probability of error. Assume equal prior probabilities.

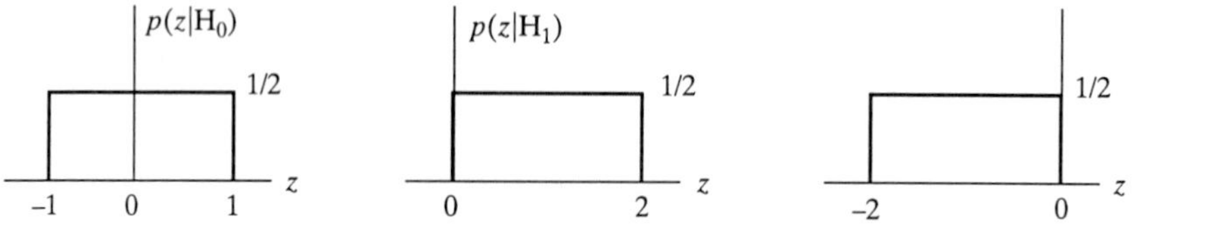

Figure E.3.3

3.16. Consider the following binary hypothesis-testing problem:

$$H_0\colon \mathbf{Z} = \mathbf{m}_0 + V$$

$$H_1\colon \mathbf{Z} = \mathbf{m}_1 + V$$

where $\mathbf{Z} = [Z_1 \quad Z_2]^T$, $\mathbf{m}_0 = [m_{10} \quad m_{20}]^T$, $\mathbf{m}_1 = [m_{11} \quad m_{21}]^T$, and $\mathbf{V} = [V_1 \quad V_2]^T$, with V_1 and V_2 being independent Gaussian random variables with mean zero and variance σ^2.

(a) Find the minimum probability of error receiver for equally likely hypotheses if the vectors $\mathbf{m}_1$ and $\mathbf{m}_0$ are known constants.

(b) How does the test simplify if $\mathbf{m}_1 = [1 \quad 0]^T$ and $\mathbf{m}_0 = [0 \quad 1]^T$?

3.17. Consider the problem of Exercise 3.16 with the modification that the vectors $\mathbf{m}_i$ are not completely known. It is, however, known that the vectors lie on a circle with radius r_0 under H_0 and radius r_1 under H_1. Since the vectors can lie anywhere on the appropriate circles, it is reasonable to write $m_{10} = r_0 \cos\theta$, $m_{20} = r_0 \sin\theta$ and $m_{11} = r_1 \cos\phi$, $m_{21} = r_1 \sin\phi$, where θ and ϕ are assumed to be independent random variables uniformly distributed in $(0, 2\pi]$. Thus the problem can be formulated as one of testing with unknown parameters.

(a) Formulate the likelihood ratio test for this problem and simplify it to the extent possible.

(b) It seems reasonable to use $z_1^2 + z_2^2$ as a test statistic. For what range of values of r_0^2/σ^2 and r_1^2/σ^2 can the likelihood ratio test be approximated by a test based on this test statistic?

(This problem is somewhat similar to a problem in incoherent detection of ASK signals discussed in [19].)

3.18. Consider the following binary hypothesis testing problem:

$$H_1\colon \mathbf{Z} = \mathbf{y}_1 + \mathbf{V}$$

$$H_2\colon \mathbf{Z} = \mathbf{y}_2 + \mathbf{V}$$

where $\mathbf{Z}$, $\mathbf{y}_1$, $\mathbf{y}_2$, and $\mathbf{V}$ are vectors of dimension N and $\mathbf{V}$ is Gaussian with mean zero and covariance matrix Σ. For equally likely hypotheses and a minimum probability of error criterion, find the optimum decision rule. Find the receiver structure and the probability of error for the case $\Sigma = \sigma^2\mathbf{I}$.

3.19. Design a likelihood ratio test to choose between the hypotheses H_1 and H_0, where

$$p(z \mid H_1) = \frac{1}{\sqrt{2\pi}\,\sigma}\exp\left(-\frac{z^2}{2\sigma^2}\right) \quad \text{and} \quad p(z \mid H_0) = \begin{cases} \frac{1}{2} & -1 < z < 1 \\ 0 & \text{otherwise} \end{cases}$$

but σ is unknown. Consider that σ is **(a)** nonrandom and lies in $[0, 1]$, and **(b)** random and is uniformly distributed between 0 and 1.

3.20. Let $X_1, X_2, \ldots, X_m$ represent a random sample from an exponential distribution with parameter θ_1 and let $Y_1, Y_2, \ldots, Y_n$ represent a random sample from an exponential distribution with parameter θ_2. Assume that X_i are independent of Y_j.

(a) Find the generalized likelihood ratio test (GLRT) for testing H_0: $\theta_1 = \theta_2$ versus H_1: $\theta_1 \neq \theta_2$.

(b) Show that the GLRT can be expressed in terms of the statistic $T = \sum_i X_i/(\sum_i X_i + \sum_j Y_j)$.

3.21. Let $X_1, X_2, \ldots, X_n$ be a random sample from the Poisson distribution

$$P(X = x \mid \theta) = \frac{e^{-\theta}\theta^x}{x!} \qquad x = 0, 1, 2, \ldots$$

Let $\theta_0 > 0$ be a specific value of θ.

(a) Find the uniformly most powerful test for deciding between H_0: $\theta = \theta_0$ and H_1: $\theta > \theta_0$.

(b) Find the generalized LRT for testing H_0: $\theta = \theta_0$ against H_1: $\theta \neq \theta_0$.

3.22. Let $X_1, X_2, \ldots, X_m$ represent a random sample from a truncated exponential density

$$p(x \mid \theta, b) = \begin{cases} \theta e^{-\theta(x-b)} & x > b \quad b \geqslant 0 \\ 0 & \text{otherwise} \end{cases}$$

Show that the minimum order statistic $Y_1 = \min(X_1, X_2, \ldots, X_n)$ is sufficient for b. Find the uniformly most powerful test for H_0: $b = 0$ versus H_1: $b > 0$ at $P_F = \alpha$. Assume that θ is known.

3.23. A Cauchy density with location parameter θ is given by

$$p(x \mid \theta) = \frac{1}{\pi[1 + (x - \theta)^2]}$$

Let X_1, be a random sample from the foregoing density.

(a) For equally likely hypotheses, find the minimum probability of error test for H_0: $\theta = \theta_0$ versus H_1: $\theta = \theta_1$.

(b) Find the Neyman–Pearson test for $P_F = \alpha$ for θ_0 versus θ_1.

(c) Does a UMP test exist for H_0: $\theta = \theta_0$ versus H_1: $\theta > \theta_0$?

3.24. Consider a binary hypothesis-testing problem involving H_0: $\theta = \theta_0$ versus H_1: $\theta = \theta_1$. Suppose that there are two sensors providing two sets of observations $\{X_1, X_2, \ldots, X_n\}$ and $\{X_{n+1}, X_{n+2}, \ldots, X_{2n}\}$ for making the decision. We assume that the observations under each hypotheses are independent, identically distributed with density $p(x \mid H_j), j = 0, 1$, with

$$p(x \mid H_j) = \theta_j \exp(-\theta_j x)u(x)$$

and $\theta_0 = 1, j = 0, 1$. Here $u(x)$ is the unit step function. We can consider two schemes for processing the observations in order to arrive at a decision regarding which hypothesis is correct:

1. A centralized scheme where the $2n$ observations from the two sensors are used together to make the decision.
2. A decentralized scheme in which at each sensor site, a decision is made using only the observations available at that site.

Let $u_k, k = 1, 2$ be the decision variable at site k, with u_k being 1 if the decision is H_1 and 0 if the decision is H_0. The final decision u_0 can be made by combining the decisions u_1 and u_2. Two possible rules are:

(i) *AND rule:* $u_0 = 1$ if and only if both u_1 and u_2 are 1.

(ii) *OR rule:* $u_0 = 1$ if either u_1 or u_2 or both are 1.

(a) Find the optimal centralized rule that maximizes P_D with $P_F = \alpha$.

(b) Find the optimal decision rules for sites 1 and 2 that maximize the probability of detection at each site with prescribed probability of false alarms. That is, find the rule at site k for maximizing $P_{D_k} = P(u_k = 1 \mid H_1)$ subject to $P_{F_k} = P(u_k = 1 \mid H_0) = \alpha_k, k = 1, 2$.

(c) For AND rule, $u_0 = u_1 \; u_2$. (i) Show that the probability of false alarm under this rule is $\alpha_0 = P(u_0 = 1 \mid H_0) = \alpha_1 \; \alpha_2$. (ii) Find the probability of detection $P_{D_0} = P(u_0 = 1 \mid H_1)$ as a function of α_0, α_1, and θ_1. (iii) Choose α_1 to maximize P_{D_0} at a given α_0, for $n = 1$.

(d) Repeat (i), (ii) and (iii) in part (c) for the OR rule. For the OR rule, $u_0 = 1 - (1 - u_1)(1 - u_2)$.

(e) Compare the probabilities of detection obtained with the decentralized schemes with that obtained for the centralized scheme for the case $n = 1$.

This and the following problem are examples of decentralized decision-making systems. For a discussion of related problems, see [20].

3.25. In decentralized detection problems, it is not uncommon to have an optimal solution that is asymmetrical even though the corresponding centralized optimal rule exhibits symmetry [21]. To illustrate this, consider the following problem in which, given an observation from each of two sensors, we are required to make a decision between the two equally likely hypotheses H_1 and H_0. Let the observations Y_1 and Y_2 at the two sensors be identically distributed discrete random variables with the following probability mass functions under the two hypotheses:

	Y = 1	2	3
H_1	$\frac{4}{5}$	$\frac{1}{5}$	0
H_0	$\frac{1}{3}$	$\frac{1}{3}$	$\frac{1}{3}$

Assume that Y_1 and Y_2 given H_i are statistically independent. A centralized test would employ $y_1 + y_2$ as the test statistic, thus exhibiting symmetry with respect to the two sensors. Show that the optimal decentralized rule requires the two sensors to employ nonidentical tests by explicitly finding the local decision rules u_1 and u_2 based on the observations y_1 and y_2 and the fusion rule, which combines u_1 and u_2 to form the final decision u_0 if the criterion is minimum probability of error. That is, design the test so as to minimize

$$P_e = \tfrac{1}{2}[P(U_0 = 1 \mid H_0) + P(U_0 = 0 \mid H_1)]$$

3.26. In composite-hypothesis testing, we can obtain an upper bound on the test performance by assuming that the receiver first measures the unknown parameters perfectly and then design the optimum likelihood ratio test. Such a bound is called a *perfect measurement bound*. The ROC of any test will be bounded by the ROC of this fictitious perfect measurement test. Given that under H_0, the observation Z is $N(0, \sigma_0^2)$ and under H_1,

Z is $N(0, \sigma_1^2)$ with $\sigma_1 > \sigma_0$, find the upper bound on the ROC for a P_F of 10^{-1} and determine if a UMP test exists.

3.27. Consider the following composite hypothesis-testing problem. The observations are

$$H_1: Z_k = Y_k + V_k \qquad k = 1, 2, \ldots, K$$

$$H_0: Z_k = V_k \qquad k = 1, 2, \ldots, K$$

with Y_k and V_k independent, identically distributed random variables with densities $N(0, \sigma_y^2)$ and $N(0, 1)$, where σ_y is unknown.

(a) Does a UMP test exist?

(b) If a UMP test does not exist, find a generalized LRT.

3.28. *CFAR* test: Consider the following hypotheses concerning a test cell sample X_0:

$$H_0: X_0 \sim p(x \mid \theta_0)$$

$$H_1: X_0 \sim p(x \mid \theta) \qquad \theta \neq \theta_0$$

with θ being the scale parameter of the density function. That is, the density of X_0/θ, which is $\theta p(\theta x \mid \theta)$, is independent of θ. Let $X_1, X_2, \ldots, X_n$ be independent, identically distributed samples from the adjacent cells, with density function $p(x \mid \theta_0)$, and with each X_i being independent of X_0. Let $Z = f(X_1, X_2, \ldots, X_n)$. Show that the test

$$X_0 \underset{H_0}{\overset{H_1}{\gtrless}} tZ$$

is a CFAR test if θ_0 is the scale parameter of the density function of Z. [Examples satisfying this condition are: (i) Z is an order statistic of $X_1, \ldots, X_n$, and (ii) $Z = \sum_{i=n_1}^{n_2} X_{(i)}, 1 \leq n_1 \leq n \leq n_2$.]

3.29. Derive a sequential ratio test when the observations are

$$H_1: Z_k = Y_k^1 \qquad k = 1, 2, \ldots$$

$$H_0: Z_k = Y_k^0 \qquad k = 1, 2, \ldots$$

where Y_k^1 and Y_k^0 are independent, identically distributed Gaussian random variables with zero mean and variances σ_0^2 and σ_1^2, respectively. Assume that $\sigma_1 > \sigma_0$ and that $P_F = 0.2$ and $P_M = 0.1$. Find the average number of samples for the test to terminate if $\sigma_0 = 1, \sigma_1 = 2$, and $P(H_0) = P(H_1) = \frac{1}{2}$.

3.30. Consider the binary hypothesis-testing problem with

$$H_0: Z_k = 1 + V_k \qquad k = 1, 2, \ldots$$

$$H_1: Z_k = -1 + V_k \qquad k = 1, 2, \ldots$$

Obtain a SPRT for this case if V_k is zero mean, white, and Gaussian with variance σ^2.

REFERENCES

1. E. L. Lehmann, *Testing Statistical Hypotheses*, Wiley, New York, 1986.
2. H. L. Van Trees, *Detection, Estimation and Modulation Theory*, Part I, Wiley, New York, 1968.

3. A. D. Whalen, *Detection of Signals in Noise*, Academic Press, New York, 1971.
4. A. M. Mood, F. A. Graybill, and D. C. Boes, *Introduction to the Theory of Statistics*, McGraw-Hill, New York, 1974.
5. S. Zacks, *Parametric Statistical Inference*, Pergamon Press, Oxford, 1981.
6. T. S. Ferguson, *Mathematical Statistics: A Decision Theoretic Approach*, Academic Press, New York, 1967.
7. H. V. Poor, *An Introduction to Signal Detection and Estimation*, Springer-Verlag, New York, 1988.
8. K. Pearson, *Tables of the Incomplete Γ-Function*, Cambridge University Press, Cambridge, 1934, reissued, 1951.
9. D. Kazakos and P. Papantoni-Kazakos, *Detection and Estimation*, Computer Science Press, New York, 1990.
10. C. R. Rao, *Linear Statistical Inference and Its Applications*, 2nd ed., Wiley, New York, 1973.
11. R. R. Bahadur, *Some Limit Theorems in Statistics*, SIAM, Philadelphia, 1971.
12. P. P. Gandhi and S. A. Kassam, Analysis of CFAR processors in nonhomogeneous background, *IEEE Trans. Aerosp. Electron. Syst.*, July, 427–445 (1988).
13. A. Wald, *Sequential Analysis*, Wiley, New York, 1959.
14. K. S. Fu, *Sequential Methods in Pattern Recognition and Machine Learning*, Academic Press, New York, 1968.
15. R. E. Ziemer and R. L. Peterson, *Digital Communications and Spread Spectrum Systems*, Macmillan, New York, 1985.
16. A. Wald and J. Wolfowitz, Optimum character of the sequential probability ratio test, *Ann. Math. Stat.*, **19**, 326–339 (1948).
17. J. O. Berger, *Statistical Decision Theory and Bayesian Analysis*, 2nd ed., Springer-Verlag, New York, 1985.
18. D. Blackwell and M. A. Grischik, *Theory of Games and Statistical Decision*, Wiley, New York, 1954.
19. E. Arthurs and H. Dym, On the optimum detection of digital signals in the presence of white Gaussian noise—a geometric interpretation and a study of three basic data transmission systems, *IRE Trans. Commun. Syst.*, December, 336–372 (1962).
20. J. N. Tsitsiklis, Decentralized detection, in *Advances in Statistical Signal Processing*, Vol. 2, *Signal Detection*, H. V. Poor and J. B. Thomas, Eds., JAI Press, Greenwich, CT, 1990.
21. J. N. Tsitsiklis, Decentralized detection by a large number of sensors, *Math. Controls Signals Syst.*, **1**, 167–182 (1988).

4

Detection of Signals in Noise

4.1 INTRODUCTION

In Chapter 3 we introduced results required for the detection of signals in noise. In this chapter we discuss the design of optimum receivers which will detect the presence of a signal or distinguish between different signals in the presence of noise. The term *receiver* is used to indicate the mathematical operation that is to be performed on the noisy observations. In order to design an optimum receiver we first specify a criterion of performance as in Chapter 3 and then design a receiver that is the best with respect to the chosen criterion. It should be noted that a receiver designed to optimize a specific criterion may perform poorly with respect to other criteria, or even with respect to the chosen criterion if the underlying assumptions change.

In designing communication receivers, we use primarily the minimum probability of error criterion. As indicated in Chapter 3 for radar and sonar applications, the Neyman–Pearson criterion is usually preferred. In both cases, the procedure involves the computation of appropriate likelihood ratios.

We start with a discussion of the detection of known signals in additive noise. That is, the parameters characterizing the signal, such as amplitude, frequency, phase, and time of arrival, are assumed known. While such an assumption may seem idealistic, in some practical systems these quantities are indeed known. For such situations at least, the ideal system represents a fairly good approximation. In any case, this discussion will serve to clarify the concepts used and will be extended in later sections to the problem of detection of signals with unknown parameters.

4.2 DETECTION OF KNOWN SIGNALS IN WHITE NOISE: THE CORRELATION RECEIVER

Consider a simple binary communication system in which the transmitter sends out one of two signals whose waveforms are completely known. The signal at the receiver is observed over an interval $[t_0, t_f]$ and is assumed to be corrupted by additive noise $v(t)$. We are required to design a receiver that operates on the received signal $z(t)$ and choose one of the two possible hypotheses:

$$H_0\colon z(t) = y_0(t) + v(t) \qquad t_0 \leq t \leq t_f$$

$$H_1\colon z(t) = y_1(t) + v(t) \qquad t_0 \leq t \leq t_f$$

We assume that the noise is Gaussian, white, with mean zero and spectral density $N_0/2$.

An obvious approach to solving this problem is to try to use our results from Chapter 3. We recall, however, that the results of Chapter 3 were obtained for a discrete set of observations. To apply those results to the present problem, where the signal waveform is a continuous function of time, let us assume that we sample the signal so that N samples are available in the interval $[t_0, t_f]$ and later let N tend to infinity without limit [1]. In doing this, we are assuming that all the signals are separable.

Let z_k, y_{ik} $(i = 0, 1)$, and v_k denote the sample values of $z(t)$, $y_i(t)$, and $v(t)$ at time $t = t_k$. The received samples under the two hypotheses in the interval $[t_0, t_f]$ are then given by

$$\begin{aligned} H_0\colon Z_k &= y_{0k} + V_k \qquad 1 \leq k \leq N \\ H_1\colon Z_k &= y_{1k} + V_k \qquad 1 \leq k \leq N \end{aligned} \tag{4.2.1}$$

Let us first assume that the noise samples are independent and Gaussian with mean zero and variance σ^2. The observations under the two hypotheses are also independent, Gaussian, with

$$\mathrm{E}\{Z_k \mid H_i\} = y_{ik}$$

and

$$\operatorname{var}\{Z_k \mid H_i\} = \operatorname{var}\{v_k\} = \sigma^2 \qquad i = 0, 1 \tag{4.2.2}$$

Let us define the vectors $\mathbf{y}_0$, $\mathbf{y}_1$, and $\mathbf{z}$ as

$$\mathbf{y}_0 = [y_{01}, y_{02}, \ldots, y_{0N}]^T$$

$$\mathbf{y}_1 = [y_{11}, y_{12}, \ldots, y_{1N}]^T$$

and

$$\mathbf{z} = [z_1, z_2, \ldots, z_N]^T$$

It follows from Eq. (4.2.2) that the likelihood ratio is

$$\begin{aligned} \Lambda(\mathbf{z}) &= \frac{p(\mathbf{z} \mid H_1)}{p(\mathbf{z} \mid H_0)} = \frac{\displaystyle\prod_{k=1}^{N} \frac{1}{\sqrt{2\pi}\,\sigma} \exp\left[-\frac{(z_k - y_{1k})^2}{2\sigma^2}\right]}{\displaystyle\prod_{k=1}^{N} \frac{1}{\sqrt{2\pi}\,\sigma} \exp\left[-\frac{(z_k - y_{0k})^2}{2\sigma^2}\right]} \\ &= \exp\left[-\frac{1}{2}\sum_{k=1}^{N}\left(\frac{2z_k y_{0k}}{\sigma^2} - \frac{2z_k y_{1k}}{\sigma^2} - \frac{y_{0k}^2 - y_{1k}^2}{\sigma^2}\right)\right] \end{aligned} \tag{4.2.3}$$

The decision rule is

$$\Lambda(\mathbf{z}) \underset{H_0}{\overset{H_1}{\gtrless}} \eta \tag{4.2.4}$$

Substituting Eq. (4.2.3) into Eq. (4.2.4), taking logarithms, and rearranging terms yields the decision rule as

$$\sum_{k=1}^{N} z_k y_{1k} - \sum_{k=1}^{N} z_k y_{0k} \underset{H_0}{\overset{H_1}{\gtrless}} \sigma^2 \ln \eta + \tfrac{1}{2} \sum_{k=1}^{N} (y_{1k}^2 - y_{0k}^2) \tag{4.2.5}$$

In terms of the vectors $\mathbf{y}_0$, $\mathbf{y}_1$, and $\mathbf{z}$, the decision rule becomes

$$\mathbf{z}^T(\mathbf{y}_1 - \mathbf{y}_0) \underset{H_0}{\overset{H_1}{\gtrless}} \sigma^2 \ln \eta + \tfrac{1}{2}(\mathbf{y}_1^T \mathbf{y}_1 - \mathbf{y}_0^T \mathbf{y}_0) \tag{4.2.6}$$

The key step in implementing the rule is to obtain the quantity $\mathbf{z}^T(\mathbf{y}_1 - \mathbf{y}_0)$. The other terms in the rule do not depend on the observations and can be precomputed. We note that the sufficient statistic for this problem is $\mathbf{z}^T(\mathbf{y}_1 - \mathbf{y}_0) = \sum_{k=1}^{N} z_k(y_{1k} - y_{0k})$, which is a Gaussian random variable under both hypotheses.

To relate these results to our original problem of deciding between the two waveforms $y_0(t)$ and $y_1(t)$, let us assume that the noise process $v(t)$ is bandlimited, so that

$$\Phi_v(\omega) = \begin{cases} \dfrac{N_0}{2} & |\omega| < \Omega \\ 0 & \text{otherwise} \end{cases} \tag{4.2.7a}$$

with correlation function

$$\phi_v(\tau) = \frac{N_0 \Omega}{2\pi} \frac{\sin \Omega\tau}{\Omega\tau} \tag{4.2.7b}$$

Thus, for the noise samples to be independent, we must sample $z(t)$ at intervals $\tau = k\pi/\Omega, k = 1, 2, 3$, and so on. The number of such samples in the interval $[t_0, t_f]$ will be $N = \Omega(t_f - t_0)/\pi$. The variance of the noise samples is obtained from Eq. (4.2.7b) as $\sigma^2 = N_0 \Omega/2\pi$.

We may now seek to obtain the decision rule directly in terms of the original waveforms by letting the samples become dense. In the limit as Ω approaches infinity, $v(\cdot)$ becomes a true white process with the noise variance becoming an impulse of intensity $N_0/2$. We then have from Eq. (4.2.3)

$$\begin{aligned} \ln \Lambda[z(t)] &= \lim_{N \to \infty} \ln \Lambda(\mathbf{z}) \\ &= \frac{2}{N_0} \int_{t_0}^{t_f} z(t) y_1(t)\, dt - \frac{2}{N_0} \int_{t_0}^{t_f} z(t) y_0(t)\, dt \\ &\quad + \frac{1}{N_0} \int_{t_0}^{t_f} [y_0^2(t) - y_1^2(t)]\, dt \end{aligned} \tag{4.2.8}$$

so that the decision rule can be written as

$$\int_{t_0}^{t_f} z(t)y_1(t)\,dt - \int_{t_0}^{t_f} z(t)y_0(t)\,dt \underset{H_0}{\overset{H_1}{\gtrless}} \gamma \tag{4.2.9}$$

where the threshold is

$$\gamma = \tfrac{1}{2}N_0 \ln \eta + \tfrac{1}{2}\int_{t_0}^{t_f} [y_1^2(t) - y_0^2(t)]\,dt \tag{4.2.10}$$

The preceding derivation for obtaining the decision rule becomes involved when the observation noise is not white but is correlated (colored). In this case, the arguments used in going to the limit as N approaches infinity tend to become somewhat involved, since the correlation between samples must be taken into account. We present instead an alternative approach to the detection of signal waveforms observed in noise, which uses a representation of the noise in terms of a set of complete orthonormal functions. Basically, the approach is again to convert the problem of deciding between two waveforms into one of deciding between two countable sets of random variables. However, these variables now are the coefficients of the expansion of the observed waveform in terms of the basis functions. If the basis functions are chosen appropriately, the coefficients will be uncorrelated random variables and the derivation of the decision rule will be straightforward. We will initially use a finite number K of these variables and later let K increase without limit.

Thus let $\{g_j(t)\}$, $j = 1, 2, \ldots$ denote a complete set of orthonormal functions over the interval $[t_0, t_f]$ such that (see 2.5.1)

$$\int_{t_0}^{t_f} g_j(t)g_k(t)\,dt = \delta_{jk} \tag{4.2.11}$$

We defer until later consideration of the choice of these functions. Let

$$v_K(t) = \sum_{j=1}^{K} V_j g_j(t) \tag{4.2.12}$$

represent the K-term approximation of $v(t)$ in terms of the functions $\{g_j(t)\}$, where

$$V_j = \int_{t_0}^{t_f} v(t)g_j(t)\,dt \qquad j = 1, \ldots, K \tag{4.2.13}$$

Let us similarly define $z_K(t)$, $y_{1K}(t)$ and $y_{0K}(t)$ as the K-term approximations of $z(t)$, $y_1(t)$, and $y_0(t)$, respectively. Then

$$z(t) = \lim_{K\to\infty} z_K(t) = \lim_{K\to\infty} \sum_{j=1}^{K} Z_j g_j(t) \tag{4.2.14}$$

where

$$Z_j = \int_{t_0}^{t_f} z(t)g_j(t)\,dt \qquad j = 1, \ldots, K \tag{4.2.15}$$

It is clear that Z_j, $j = 1 \cdots K$ constitute a set of Gaussian random variables under each hypothesis.

The first two moments of these variables can be determined as follows. Under H_1 we have

$$\begin{aligned} Z_j &= \int_{t_0}^{t_f} [y_1(t) + v(t)] g_j(t)\, dt \\ &= y_{1j} + V_j \end{aligned} \tag{4.2.16}$$

so that

$$\mathrm{E}\{Z_j \mid H_1\} = y_{1j} \tag{4.2.17a}$$

$$\operatorname{var}\{Z_j \mid H_1\} = \mathrm{E}\{V_j^2\} \tag{4.2.17b}$$

$$\begin{aligned} \mathrm{E}\{V_j^2\} &= \mathrm{E}\left[\int_{t_0}^{t_f} v(t) g_j(t)\, dt \int_{t_0}^{t_f} v(\sigma) g_j(\sigma)\, d\sigma\right] \\ &= \int_{t_0}^{t_f}\int_{t_0}^{t_f} \phi_v(t-\sigma) g_j(t) g_j(\sigma)\, dt\, d\sigma = \frac{N_0}{2} \end{aligned} \tag{4.2.17c}$$

The last step follows from the fact that $v(t)$ is assumed to be white with spectral density $N_0/2$ and the orthogonality of the basis functions. Thus

$$\operatorname{var}\{Z_j \mid H_1\} = \frac{N_0}{2} \tag{4.2.18}$$

Similarly, the first two moments under H_0 are given by

$$\mathrm{E}\{Z_j \mid H_0\} = y_{0j}$$

and

$$\operatorname{var}\{Z_j \mid H_0\} = \frac{N_0}{2} \tag{4.2.19}$$

From Chapter 2 we know that the variables Z_j, $j = 1, \ldots, K$ are uncorrelated, and being Gaussian, are independent under each hypothesis. We can therefore write the likelihood ratio for this problem as

$$\begin{aligned} \Lambda[z(t)] &= \lim_{K\to\infty} \Lambda[z_K(t)] = \lim_{K\to\infty} \frac{p[z_k(t) \mid H_1]}{p[z_K(t) \mid H_0]} \\ &= \lim_{K\to\infty} \frac{p(z_1, z_2, \ldots, z_K \mid H_1)}{p(z_1, z_2, \ldots, z_K \mid H_0)} \end{aligned} \tag{4.2.20}$$

Use of Eqs. (4.2.17) and (4.2.19) enables us to write the likelihood ratio $\Lambda[z_K(t)]$ as

$$\Lambda[z_k(t)] = \frac{\prod_{j=1}^{K} \frac{1}{\sqrt{\pi N_0}} \exp\left[-\frac{1}{2}\frac{(z_j - y_{1j})^2}{N_0/2}\right]}{\prod_{j=1}^{K} \frac{1}{\sqrt{\pi N_0}} \exp\left[-\frac{1}{2}\frac{(z_j - y_{0j})^2}{N_0/2}\right]} \tag{4.2.21}$$

Taking the logarithm and canceling common terms yields

$$\ln \Lambda[z_K(t)] = \frac{2}{N_0}\sum_{j=1}^{K} z_j(y_{1j} - y_{0j}) + \frac{1}{N_0}\sum_{j=1}^{K}(y_{0j}^2 - y_{1j}^2) \tag{4.2.22}$$

Now

$$\begin{aligned}\int_{t_0}^{t_f} z_K(t)y_{iK}(t)\,dt &= \int_{t_0}^{t_f}\left[\sum_{j=1}^{K} z_j g_j(t)\right]\left[\sum_{k=1}^{K} y_{ik} g_k(t)\right]dt \\ &= \sum_{j=1}^{K}\sum_{k=1}^{K} z_j y_{ik}\int_{t_0}^{t_f} g_j(t)g_k(t)\,dt \\ &= \sum_{j=1}^{K} z_j y_{ij} \qquad i = 0, 1\end{aligned} \tag{4.2.23}$$

Similarly

$$\sum_{j=1}^{K} y_{ij}^2 = \int_{t_0}^{t_f} y_{iK}^2(t)\,dt \qquad i = 0, 1 \tag{4.2.24}$$

Substitution of Eqs. (4.2.23) and (4.2.24) in Eq. (4.2.22) yields

$$\begin{aligned}\ln \Lambda[z_K(t)] &= \frac{2}{N_0}\int_{t_0}^{t_f} z_K(t)[y_{1K}(t) - y_{0K}(t)]\,dt \\ &\quad - \frac{1}{N_0}\int_{t_0}^{t_f}[y_{1K}^2(t) - y_{0K}^2(t)]\,dt\end{aligned} \tag{4.2.25}$$

The log-likelihood ratio $\ln \Lambda(t)$ obtained by taking the limit of Eq. (4.2.25) as K tends to infinity is

$$\ln \Lambda(t) = \frac{2}{N_0}\int_{t_0}^{t_f} z(t)[y_1(t) - y_0(t)]\,dt - \frac{1}{N_0}\int_{t_0}^{t_f}[y_1^2(t) - y_0^2(t)]\,dt \tag{4.2.26}$$

and the decision rule is

$$\ln \Lambda(t) \underset{H_0}{\overset{H_1}{\gtrless}} \ln \eta \tag{4.2.27}$$

Substitution of Eq. (4.2.26) in Eq. (4.2.27) yields the decision rule as

$$\int_{t_0}^{t_f} z(t)[y_1(t) - y_0(t)]\,dt \underset{H_0}{\overset{H_1}{\gtrless}} \frac{N_0}{2}\ln \eta + \frac{1}{2}\int_{t_0}^{t_f}[y_1^2(t) - y_0^2(t)]\,dt \tag{4.2.28}$$

which is the same as was obtained earlier [see Eqs. (4.2.9) and (4.2.10)]. The structure of the receiver is shown in Fig. 4.2.1. The receiver is called a *correlation receiver*, since the received signal $z(t)$ is cross correlated with the signals $y_0(t)$ and $y_1(t)$. If the signals are of equal energy, the threshold is signal independent.

We now comment briefly on the choice of the basis functions $\{g_i(t)\}$. We recall from our discussion of series representations of stochastic signals in Chapter 2 that

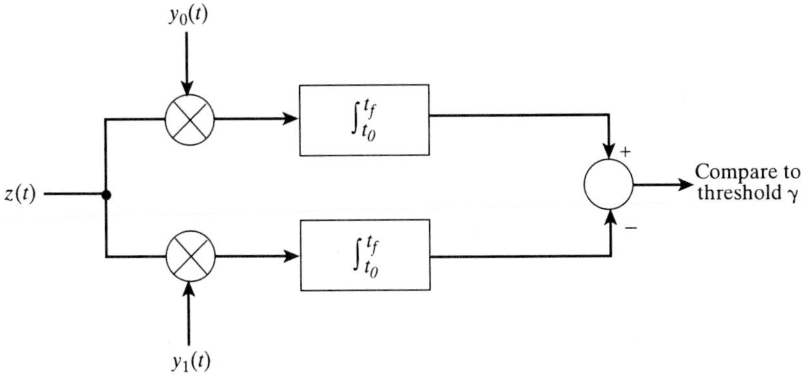

Figure 4.2.1 Correlation receiver for binary decision problem.

for white processes, *any* set of orthonormal functions can be used for decomposing the process. As we have seen, the coefficients of the expansion will be independent irrespective of the choice of basis functions. For colored (nonwhite) processes, our choice of basis functions is of course more restricted. In the next section we discuss the use of the Karhunen–Loéve expansion for this case.

The nonuniqueness of the basis functions for the white observation noise problem enables us to give a simple derivation of the likelihood ratio test in terms of a sufficient statistic. If we transform the observation vector into two components, one of which is dependent on which hypothesis is true and the other is independent of the hypothesis, the decision can be based entirely on the first component. The transformation involves choosing an appropriate set of coordinate or basis functions for the decision space [3]. We will use a complete set of orthonormal functions chosen as follows. Let ϵ_1 and ϵ_0 denote the energies in the signals $y_1(t)$ and $y_0(t)$:

$$\epsilon_1 = \int_{t_0}^{t_f} y_1^2(t)\,dt \quad \text{and} \quad \epsilon_0 = \int_{t_0}^{t_f} y_0^2(t)\,dt \tag{4.2.29}$$

We choose as our first basis function

$$g_1(t) = \frac{y_1(t)}{\sqrt{\epsilon_1}} \qquad t_0 \leq t \leq t_f \tag{4.2.30}$$

The second function $g_2(t)$ is obtained from the signal $y_0(t)$ by subtracting the component correlated with $g_1(t)$ and normalizing. Thus $g_2(t)$ is chosen as

$$g_2(t) = \frac{1}{\sqrt{\epsilon_0\epsilon_1 - \rho^2}}\left[\sqrt{\epsilon_1}y_0(t) - \frac{\rho}{\sqrt{\epsilon_1}}y_1(t)\right] \qquad t_0 \leq t \leq t_f \tag{4.2.31}$$

where the correlation ρ is given by

$$\rho = \int_{t_0}^{t_f} y_0(t)y_1(t)\,dt \tag{4.2.32}$$

The remaining orthonormal basis functions $g_i(t)$ can be arbitrary. In terms of the basis functions, we write the received waveform as

$$z(t) = \sum_{i=1}^{\infty} Z_i g_i(t) \qquad t_0 \le t \le t_f \tag{4.2.33}$$

where

$$Z_i = \int_{t_0}^{t_f} z(t) g_i(t)\,dt \qquad i = 1, 2, \ldots \tag{4.2.34}$$

It is clear that only Z_1 and Z_2 depend on which hypothesis is true. Thus the decision region is two-dimensional.

The mean and variance of Z_i under each hypothesis are given by

$$\mathrm{E}\{Z_i \mid H_0\} = \int_{t_0}^{t_f} y_0(t) g_i(t)\,dt = y_{0i} \qquad i = 1, 2 \tag{4.2.35a}$$

$$\mathrm{E}\{Z_i \mid H_1\} = \int_{t_0}^{t_f} y_1(t) g_i(t)\,dt = y_{1i} \qquad i = 1, 2 \tag{4.2.35b}$$

and

$$\mathrm{var}\{Z_i \mid H_j\} = \int_{t_0}^{t_f}\int_{t_0}^{t_f} \phi_v(t-s) g_i(t) g_i(s)\,dt\,ds = \frac{N_0}{2} \qquad j = 1, 2 \text{ and } i = 1, 2 \tag{4.2.36}$$

Following an argument similar to that used in setting up Eq. (4.2.22) we can easily show that the decision rule becomes

$$\sum_{i=1}^{2} (2z_i - y_{1i}) \frac{y_{1i}}{N_0} - \sum_{i=1}^{2} (2z_i - y_{0i}) \frac{y_{0i}}{N_0} \underset{H_0}{\overset{H_1}{\gtrless}} \ln \eta \tag{4.2.37}$$

or

$$\sum_{i=1}^{2} z_i (y_{1i} - y_{0i}) \underset{H_0}{\overset{H_1}{\gtrless}} \frac{N_0}{2} \ln \eta + \frac{1}{2}\left(\sum_{i=1}^{2} y_{1i}^2 - \sum_{i=1}^{2} y_{0i}^2\right) \tag{4.2.38}$$

The implementation of the receiver is shown in Fig. 4.2.2.

The receiver is again a correlation receiver in which the observation $z(t)$ is correlated with the two basis functions $g_1(t)$ and $g_2(t)$. With the use of Eqs. (4.2.30), (4.2.31), and (4.2.35a,b) and a little algebraic manipulation, it can be shown that the structure of Fig. 4.2.2 is equivalent to that of Fig. 4.2.1.

For equally likely hypotheses and minimum probability of error criterion, η is 1. We can obtain an alternative interpretation of the receiver structure in this case by proceeding as follows. By adding $\sum_{i=1}^{2} z_i^2$ to both sides of the inequality in Eq. (4.2.38) and rearranging terms, the decision rule becomes

$$\sum_{i=1}^{2} (z_i - y_{1i})^2 \underset{H_0}{\overset{H_1}{\gtrless}} \sum_{i=1}^{2} (z_i - y_{0i})^2 \tag{4.2.39}$$

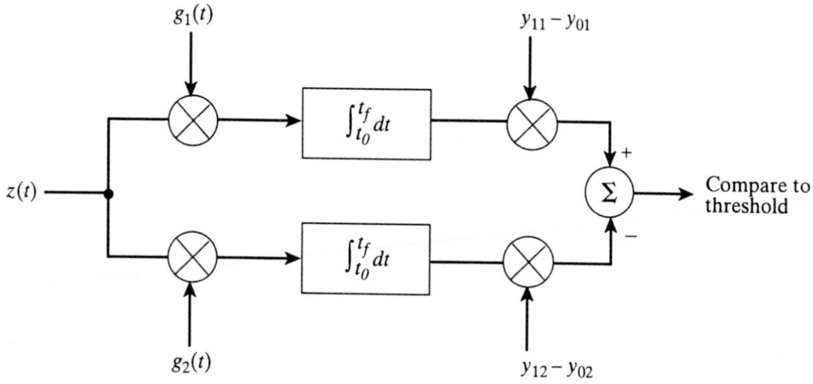

Figure 4.2.2 Alternative implementation of correlation receiver for binary problem.

The left side of the inequality above represents the Euclidean distance between the observation vector $\mathbf{z} = [z_1 \quad z_2]^T$ and the signal vector $\mathbf{y}_1 = [y_{11} \quad y_{12}]^T$ while the right side is the distance between the vector $\mathbf{z}$ and the signal vector $\mathbf{y}_0 = [y_{01} \quad y_{02}]^T$. Thus the decision rule is to choose that hypothesis for which the distance between $\mathbf{z}$ and the corresponding signal vector is minimum. This is illustrated in Fig. 4.2.3, which shows the observation vector, the signal vectors, and the corresponding decision regions.

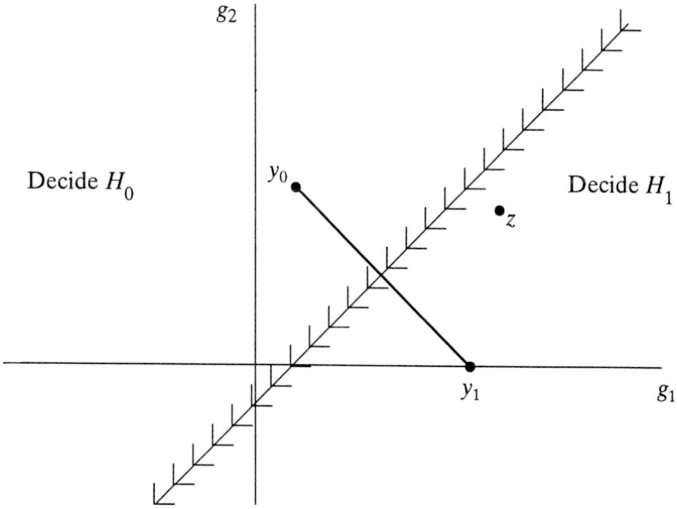

Figure 4.2.3 Interpretation of binary decision rule as a minimum distance classifier.

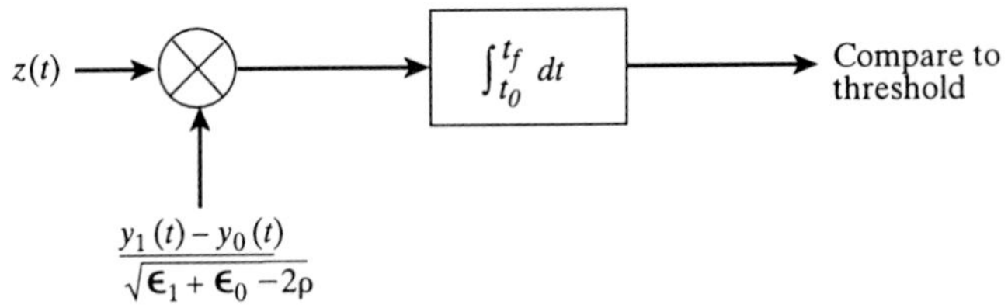

Figure 4.2.4 Implementation of a binary receiver with a single correlator.

Finally, we can reduce the structure of Fig. 4.2.2 to one that requires only a single correlation operation as shown in Fig. 4.2.4. This is equivalent to choosing as our first basis function

$$g_1(t) = \frac{y_1(t) - y_0(t)}{\sqrt{\epsilon_1 + \epsilon_0 - 2\rho}} \tag{4.2.40}$$

with the other basis functions being arbitrary.

If we again write $z(t)$ as in Eq. (4.2.35), only z_1 will depend on which hypothesis is true and the decision region becomes one-dimensional. Thus z_1 is a sufficient statistic for this problem.

4.2.1 Receiver Performance

To determine the performance of the receiver, let us rewrite the decision rule in the following equivalent form:

$$\int_{t_0}^{t_f} z(t)y_1(t)\,dt - \int_{t_0}^{t_f} z(t)y_0(t)\,dt + \tfrac{1}{2}\int_{t_0}^{t_f} [y_0^2(t) - y_1^2(t)]\,dt \underset{H_0}{\overset{H_1}{\gtrless}} \frac{N_0}{2} \ln \eta \triangleq \gamma \tag{4.2.41}$$

We recall that the threshold η depends on the particular criterion chosen. Denote the left side of the inequality in Eq. (4.2.41) by I. The probabilities of error are then given by

$$P_F = P(D_1 | H_0) = \int_{\gamma}^{\infty} p(i | H_0)\,di$$

and

$$P_M = P(D_0 | H_1) = \int_{-\infty}^{\gamma} p(i | H_1)\,di \tag{4.2.42}$$

We note that I is a Gaussian variable under both hypotheses. Under H_0, the received signal is $z(t) = y_0(t) + v(t)$, so that

$$\begin{aligned} I = &\int_{t_0}^{t_f} [y_0(t) + v(t)]y_1(t)\,dt - \int_{t_0}^{t_f} [y_0(t) + v(t)]y_0(t)\,dt \\ &+ \tfrac{1}{2}\int_{t_0}^{t_f} [y_0^2(t) - y_1^2(t)]\,dt \end{aligned} \tag{4.2.43}$$

Then we have

$$E\{I \mid H_0\} = -\tfrac{1}{2}\int_{t_0}^{t_f} [y_0(t) - y_1(t)]^2\, dt \tag{4.2.44}$$

where we have used the fact that $v(t)$ is zero mean.

The variance of I can also be calculated as

$$\begin{aligned}
\text{var}[I \mid H_0] &= E\{[I - E(I \mid H_0)]^2 \mid H_0\} \\
&= E\left\{\int_{t_0}^{t_f} v(t)[y_1(t) - y_0(t)]\, dt \right. \\
&\quad \left. \times \int_{t_0}^{t_f} v(\tau)[y_1(\tau) - y_0(\tau)]\, d\tau\right\} \qquad (4.2.45) \\
&= \int_{t_0}^{t_f}\int_{t_0}^{t_f} E\{v(t)v(\tau)\}[y_1(t) - y_0(t)][y_1(\tau) - y_0(\tau)]\, dt\, d\tau \\
&= \tfrac{1}{2}N_0 \int_{t_0}^{t_f} [y_1(t) - y_0(t)]^2\, dt
\end{aligned}$$

since

$$E\{v(t)v(\tau)\} = \frac{N_0}{2}\delta(t - \tau)$$

It may similarly be shown that under H_1,

$$E\{I \mid H_1\} = \tfrac{1}{2}\int_{t_0}^{t_f} [y_1(t) - y_0(t)]^2\, dt \tag{4.2.46}$$

and

$$\text{var}\{I \mid H_1\} = \frac{N_0}{2}\int_{t_0}^{t_f} [y_0(t) - y_1(t)]^2\, dt \tag{4.2.47}$$

We denote the average energy of the two signals by

$$\epsilon = \tfrac{1}{2}\int_{t_0}^{t_f} [y_0^2(t) + y_1^2(t)]\, dt = \tfrac{1}{2}(\epsilon_0 + \epsilon_1) \tag{4.2.48}$$

and define the (normalized) correlation coefficient

$$\bar{\rho} = \frac{1}{\epsilon}\int_{t_0}^{t_f} y_0(t)y_1(t)\, dt \tag{4.2.49}$$

We note that $|\bar{\rho}| \leq 1$. Then it can easily be verified that

$$E\{I \mid H_0\} = -\epsilon(1 - \bar{\rho})$$
$$E\{I \mid H_1\} = \epsilon(1 - \bar{\rho})$$

and

$$\text{var}\{I \mid H_0\} = \text{var}\{I \mid H_1\} = N_0\,\epsilon(1 - \bar{\rho}) \tag{4.2.50}$$

Let γ^+ and γ^- be given by

$$\gamma^+ \underline{\Delta} \frac{\gamma + \epsilon(1 - \bar{\rho})}{\sqrt{N_0 \epsilon(1 - \bar{\rho})}} \qquad \gamma^- \underline{\Delta} \frac{\gamma - \epsilon(1 - \bar{\rho})}{\sqrt{N_0 \epsilon(1 - \bar{\rho})}} \tag{4.2.51}$$

It easily follows that the error probabilities are given by

$$P_F = \int_{\gamma^+}^{\infty} \frac{1}{\sqrt{2\pi}} e^{-u^2/2}\, du = Q(\gamma^+) \tag{4.2.52}$$

and

$$P_M = \int_{-\infty}^{\gamma^-} \frac{1}{\sqrt{2\pi}} e^{-u^2/2}\, du = 1 - Q(\gamma^-) \tag{4.2.53}$$

The performance of the receiver thus depends only on the three parameters ϵ, $\bar{\rho}$, and N_0 and is independent of the particular waveform used. For equally likely hypotheses and the minimum probability of error criterion, γ equals zero. Thus $\gamma^+ = -\gamma^-$ and the two error probabilities become equal. The total probability of error is just P_F or P_M and decreases as $\epsilon(1 - \bar{\rho})/N_0$ increases. For fixed N_0, therefore, the optimum system is one for which $\bar{\rho} = -1$ or $y_0(t) = -y_1(t)$.

Example 4.2.1

We will determine the performance of three commonly used communication systems in which the binary signals are sine waves. We will assume that the signal phase at the receiving end is known so that we can normalize the phase to be zero. Such systems are referred to as *coherent systems*. We will use the minimum probability of error criterion in all cases.

In coherent phase shift keying (CPSK) the signals are

$$y_0(t) = A \sin \omega_c t \quad \text{and} \quad y_1(t) = -A \sin \omega_c t \qquad t_0 \leq t \leq t_f$$

Since $\bar{\rho} = -1$, this is an ideal binary system and the probability of error is obtained from 4.2.51 and 4.2.52 as

$$P_e = Q\left(\sqrt{\frac{2\epsilon}{N_0}}\right)$$

In coherent frequency shift keying (CFSK) the binary signals are

$$y_0(t) = A \sin \omega_0 t \quad \text{and} \quad y_1(t) = A \sin \omega_1 t \qquad t_0 \leq t \leq t_f$$

ω_1 and ω_0 are chosen so that the two signals are orthogonal, that is, $\bar{\rho} = 0$. In this case the error probability is given by

$$P_e = Q\left(\sqrt{\frac{\epsilon}{N_0}}\right)$$

In on–off keying (OOK) the binary signals are

$$y_0(t) = 0 \quad \text{and} \quad y_1(t) = A \sin \omega_1 t \qquad t_0 \leq t \leq t_f$$

Clearly, $\bar{\rho} = 0$. Since $\epsilon_0 = 0$, we have $\epsilon = \epsilon_1/2$. The error probability is

$$P_e = Q\left(\sqrt{\frac{\epsilon_1}{2N_0}}\right) = Q\left(\sqrt{\frac{\epsilon}{N_0}}\right)$$

Example 4.2.2

The probabilities of error for the binary signal problem can be determined from Eqs. (4.2.52) and (4.2.53). However, for equally likely hypotheses and minimum probability of error criterion, in some cases we can exploit the symmetry properties of the decision regions in signal space to determine these probabilities. This is particularly useful in M-ary signal detection (see Exercise 4.6). We illustrate this approach for the binary signal detection problem when the two signals are orthogonal, with equal energy, E, and the observation noise is additive, Gaussian, white, with mean zero and spectral density $N_0/2$.

For this problem, the basis functions can be chosen to be the transmitted signals so that the optimum decision regions will be as shown in Fig. 4.2.5a. Let $P_D(y_i), i = 1,2$ denote the probability of making a correct decision when the transmitted signal is $y_i(t)$. Then the probability of correct detection is given by

$$P_D = \tfrac{1}{2}P_D(y_1) + \tfrac{1}{2}P_D(y_2)$$

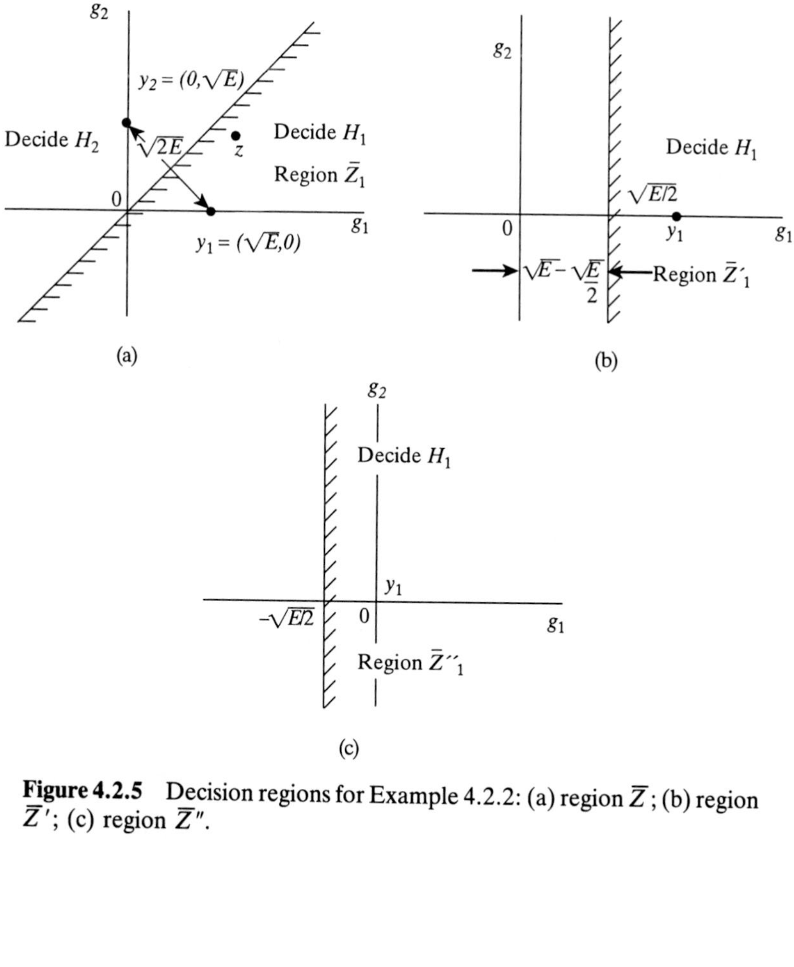

Figure 4.2.5 Decision regions for Example 4.2.2: (a) region $\bar{Z}$; (b) region $\bar{Z}'$; (c) region $\bar{Z}''$.

It is clear from Fig. 4.2.5a that

$$P_D(y_1) = P(Z_2 < Z_1)$$

where Z_1 and Z_2 are independent, Gaussian random variables with mean $\sqrt{E}$ and variance $N_0/2$.

Therefore, to compute $P_D(y_1)$, we need to integrate over the two-dimensional region $\overline{Z}_1$ in the signal space, which is shown hatched in the figure. The computation is greatly simplified by exploiting some properties of the signal space. First we note that the probability that the point $\mathbf{z}$ is in the region $\overline{Z}_1$ is invariant to rotation of the region about the point $y_1 = (\sqrt{E}, 0)$. This follows from the fact that under H_1, $\mathbf{z} = \mathbf{y}_1 + \mathbf{v}$, so that $P(\mathbf{z} \in \overline{Z}_1 | \mathbf{y}_1) = P(\mathbf{v} \in \overline{Z}_1 - \mathbf{y}_1 | \mathbf{y}_1)$. Since the density of $\mathbf{v}$ is circularly symmetric, it follows that $P(\mathbf{z} \in \overline{Z}_1 | \mathbf{y}_1) = P(\mathbf{z} \in \overline{Z}_1' | \mathbf{y}_1)$, where the region $\overline{Z}_1'$ obtained after rotation of $\overline{Z}_1$ is shown in Figure 4.2.5b.

Let $\overline{Z}_1''$ denote the region obtained when $\overline{Z}_1'$ is translated to the left by $\mathbf{y}_1 = (\sqrt{E}, 0)$, $\overline{Z}_1'' = \overline{Z}_1' - \mathbf{y}_1$. Since $P(\mathbf{z} \in \overline{Z}_1' | \mathbf{y}_1) = P(\mathbf{z} - \mathbf{y}_1 \in \overline{Z}_1'' | \mathbf{y}_1)$, it follows that the probability that the point $\mathbf{z}$ is in the region $\overline{Z}_1'$ is unaltered if the region is translated. The region $\overline{Z}_1''$ is shown in Fig. 4.2.5c. We therefore have

$$P_D(y_1) = P(\mathbf{z}_1 \in \overline{Z}_1'' | y_1 = 0)$$

and is the probability that the zero-mean Gaussian random variable Z_1 is larger than $-E/2$. We thus have

$$P_D(y_1) = Q\left(\frac{-\sqrt{E/2}}{\sqrt{N_0/2}}\right) = Q\left(-\sqrt{\frac{E}{N_0}}\right)$$

Because of the symmetry of the decision regions in Fig. 4.2.5a, this probability also equals $P_D(y_2)$ and is hence also equal to P_D. The probability of error is given by

$$P_e = 1 - Q\left(-\sqrt{\frac{E}{N_0}}\right) = Q\left(\sqrt{\frac{E}{N_0}}\right)$$

This example illustrates the fact that exploiting the symmetry and invariance properties of the decision regions greatly simplifies the calculation of error probabilities in some cases.

Matched filter implementation of receiver. The implementation of the decision rule of Eq. (4.2.28) involves computation of the integrals

$$\int_{t_0}^{t_f} z(t)y_i(t)\,dt \qquad i = 0, 1 \tag{4.2.54}$$

and requires one multiplier and integrator for determining each integral (see Fig. 4.2.1).

We can obtain an alternative implementation as follows. Suppose that we apply $z(t)$ to a linear-time-invariant filter with impulse response $h_i(t)$. Then the output of the filter at time τ is given by

$$r(\tau) = \int_{t_0}^{t_f} h_i(\tau - t)z(t)\,dt \tag{4.2.55}$$

Comparison with Eq. (4.2.54) shows that if we let $h_i(\tau - t) = y_i(t)$ for $t_0 \leqslant t \leqslant t_f$

(or equivalently $h_i(t) = y_i(\tau - t)$ for $\tau - t_f \leqslant t \leqslant \tau - t_0$) the output $r(\tau)$ of the filter will be equal to the correlation between the signals $z(t)$ and $y_i(t)$. Such an implementation is referred to as a *matched filter*. The matched filter implementation is usually preferred in practice and is shown in Fig. 4.2.6. We note if $\tau < t_f$, the filter in Eq. (4.2.55) is noncausal (not physically realizable) since the output at time τ depends on values of the input over the interval $[\tau, t_f]$. If, however, τ is chosen to be equal to t_f the filter will be causal with impulse response

$$h_i(t) = y_i(t_f - t) \qquad \text{for} \quad 0 \leqslant t \leqslant t_f - t_0$$

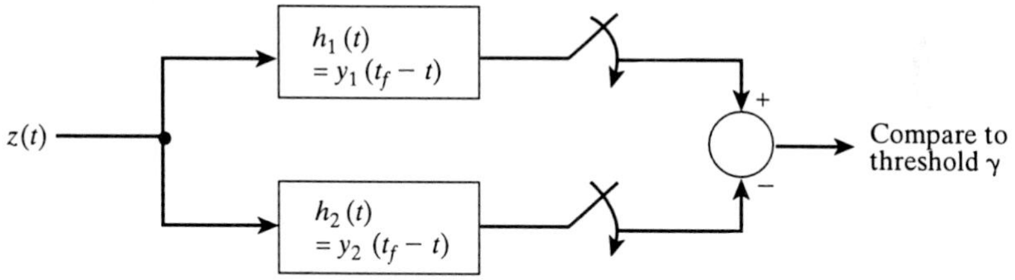

Figure 4.2.6 Matched filter implementation of correlation receiver.

4.2.2 Optimum Detection of *M*-ary Signals in White Gaussian Noise

We now consider the extension of the previous results to the case where we have to decide which one of M possible signal waveforms $\{y_1(t), y_2(t), \ldots, y_M(t)\}$ has been transmitted. We define the received signal under hypothesis $H_i, i = 1, 2, \ldots, M$ as

$$H_i\colon z(t) = y_i(t) + v(t) \qquad t_0 \leq t \leq t_f \tag{4.2.56}$$

where $v(t)$ is a zero-mean, Gaussian white noise.

We can obtain a solution to this problem in a manner somewhat similar to our solution in Section 4.2, page 88. It may be recalled that there we expanded the signals $z(t)$, $y_i(t)$, and $v(t)$ in a (possibly infinite) set of orthonormal functions $g_i(t)$. However, it can be shown that we can expand any set of M finite-energy waveforms in a *finite* set of orthonormal signals $g_1(t), g_2(t), \ldots, g_N(t)$ with $N \leqslant M$. The orthonormal signals $g_i(t)$ can be generated from the signal waveforms as in the binary signal case by using the Gram–Schmidt orthogonalization procedure [5].

Let us write

$$y_i(t) = \sum_{k=1}^{N} y_{ik} g_k(t) \qquad 1 \leq i \leq M \tag{4.2.57}$$

where it easily follows that

$$y_{ik} = \int_{t_0}^{t_f} y_i(t) g_k(t) \qquad 1 \leq k \leq N \tag{4.2.58}$$

Thus the signal $y_i(t)$ can be described by the signal vector $y_i = [y_{i1}, y_{i2}, \ldots, y_{iN}]^T$.

We have seen that to obtain a series representation of $v(t)$, in general, we will have to use an infinite number of orthonormal functions in the expansion. However, $v(t)$ is white, so any orthonormal set of functions can be used. We will therefore choose the first N functions of this set to be exactly the same as $g_1(t), g_2(t), \ldots, g_N(t)$, with the functions $g_{N+1}(t), g_{N+2}(t), \ldots$ being arbitrary orthonormal waveforms. We can therefore write

$$v(t) = \sum_{k=1}^{\infty} V_k g_k(t) \tag{4.2.59}$$

where the V_k are determined as in Eq. (4.2.13). If we now write the observations under hypothesis H_i as

$$z(t) = \sum_{k=1}^{\infty} Z_k g_k(t) \tag{4.2.60}$$

where

$$Z_k = \int_{t_0}^{t_f} z(t) g_k(t)\, dt \tag{4.2.61}$$

it follows from Eqs. (4.2.58) and (4.2.59) that

$$Z_k = \begin{cases} y_{ik} + V_k & 1 \le k \le N \\ V_i & k \ge N + 1 \end{cases} \tag{4.2.62}$$

Since the coefficients $Z_k, k \ge N + 1$ do not depend on which hypothesis is correct, we can make a decision based solely on the first N components. Hence $\mathbf{Z} = [Z_1 \quad Z_2 \quad \cdots \quad Z_N]^T$ is a sufficient statistic for the M-ary detection problem.

As in Section 4.2.1, it can easily be established that the variables Z_k are Gaussian distributed and are independent under each hypothesis, with

$$\mathrm{E}\{Z_k \mid H_i\} = y_{ik} \tag{4.2.63}$$

and

$$\mathrm{var}\{Z_k \mid H_i\} = \frac{N_0}{2}$$

so that

$$p(\mathbf{z} \mid H_i) = \frac{1}{\sqrt{\pi N_0}} \exp\left[-\frac{1}{2} \frac{\sum_{k=1}^{N} (z_k - y_{ik})^2}{N_0/2} \right] \tag{4.2.64}$$

For equally likely signals, that is, $P(H_i) = 1/M, i = 1, 2, \ldots, M$, and minimum probability of error criterion, it follows from our discussions in Chapter 3 that the decision rule is to choose that hypothesis for which $p(\mathbf{z} \mid H_i)$ is maximum. Equivalently, from Eq. (4.2.64), it is clear that the decision rule can be stated as: Decide H_i if

$$\sum_{k=1}^{N} (z_k - y_{jk})^2, \; j = 1, 2, \ldots, M \tag{4.2.65}$$

is minimum for $j = i$.

Again the sum in Eq. (4.2.65) is the Euclidean distance between the observation $\mathbf{z}$ and the signal vector $\mathbf{y}_j$, so that the rule chooses the signal vector closest to the received vector $\mathbf{z}$.

If the signals $y_i(t)$ all have equal energy, then $\sum_{k=1}^{N} y_{ik}^2 = \epsilon$ for each i. In this case, minimizing the expression in Eq. (4.2.65) is equivalent to maximizing over j, the quantity

$$I_j = \sum_{k=1}^{N} z_k y_{jk} \tag{4.2.66}$$

Since it can easily be verified that $I_j = \int_{t_0}^{t_f} z(t) y_j(t)\, dt$, it is clear that the receiver may again be implemented either as a correlation receiver or as a matched filter.

Example 4.2.3

We can find an expression for the probability of error for the problem of detection of equal energy, orthogonal signals, observed in additive, white Gaussian noise. In this case, the signals $y_i(t)$ are such that

$$\int_{t_0}^{t_f} y_i(t) y_j(t)\, dt = \begin{cases} \epsilon & i = j \\ 0 & k \neq j \end{cases} \tag{4.2.67}$$

It easily follows from Eq. (4.2.66) that

$$E\{I_j \mid H_i\} = \begin{cases} \epsilon & i = j \\ 0 & i \neq j \end{cases} \tag{4.2.68}$$

$$E\{I_i I_j \mid H_k\} = 0$$

$$E\{I_j^2 \mid H_k\} = \begin{cases} \epsilon^2 + \dfrac{N_0 \epsilon}{2} & j = k \\ \dfrac{N_0 \epsilon}{2} & j \neq k \end{cases}$$

Therefore, the variance of I_j assuming hypothesis H_i is true is given by

$$\operatorname{var}\{I_j \mid H_i\} = \frac{N_0 \epsilon}{2} = \sigma^2 \qquad \text{for all } i \text{ and } j \tag{4.2.69}$$

Let us assume that H_i is the true hypothesis. The probability that no error is made is then equal to the probability that I_i is the maximum of the random variables $I_1, I_2, \ldots, I_M$. Suppose that I_i has a particular value x. Then the probability of no error is equal to the probability that I_j is less than x for $j \neq i$:

$$P(\text{no error} \mid I_i = x, H_i) = P\left(\bigcap_{\substack{j=1 \\ j \neq i}}^{M} \{I_j < x\} \,\middle|\, H_i \right)$$

Since the variables I_j are Gaussian and uncorrelated, they are also independent, so that we can write

$$\begin{aligned} P(\text{no error} \mid I_i = x, H_i) &= [P(I_j < x \mid H_i)]^{M-1} \\ &= \left[\int_{-\infty}^{x} \frac{1}{\sqrt{2\pi}\,\sigma} \exp\left(-\frac{u^2}{2\sigma^2} \right) du \right]^{M-1} \end{aligned} \tag{4.2.70}$$

The probability of error is then given by

$$
\begin{aligned}
P_e &= 1 - \int_{-\infty}^{\infty} P(\text{no error} \mid I_i = x, H_i) P(I_i = x \mid H_i)\, dx \\
&= 1 - \int_{-\infty}^{\infty} \frac{1}{\sqrt{2\pi}\,\sigma} \exp\left[-\frac{(x-\epsilon)^2}{2\sigma^2}\right] \\
&\quad \times \left[\int_{-\infty}^{x} \frac{1}{\sqrt{2\pi}\,\sigma} \exp\left(-\frac{u^2}{2\sigma^2}\right) du\right]^{M-1} dx
\end{aligned}
\tag{4.2.71}
$$

This expression cannot in general be integrated analytically. However, values of P_e for certain values of $\sqrt{\epsilon/N_0}$ and M can be numerically evaluated.

Example 4.2.4

We consider a widely used scheme in digital communications known as *quadrature amplitude modulation* (QAM) in which one of four waveforms corresponding to one of four possible symbols is transmitted over each signaling interval $[0, T]$. The waveforms are

$$y_1(t) = \sqrt{\frac{E}{T}} \cos \omega_c(t) + \sqrt{\frac{E}{T}} \sin \omega_c(t)$$

$$y_2(t) = -\sqrt{\frac{E}{T}} \cos \omega_c(t) + \sqrt{\frac{E}{T}} \sin \omega_c(t)$$

$$y_3(t) = -\sqrt{\frac{E}{T}} \cos \omega_c(t) - \sqrt{\frac{E}{T}} \sin \omega_c(t)$$

$$y_4(t) = +\sqrt{\frac{E}{T}} \cos \omega_c(t) - \sqrt{\frac{E}{T}} \sin \omega_c(t)$$

We assume that the signals are observed in zero-mean additive Gaussian white noise with spectral density $N_0/2$.

If we choose as our basis functions $g_1(t) = \sqrt{2/T} \cos \omega_c(t)$ and $g_2(t) = \sqrt{2/T} \sin \omega_c t$, the four signal vectors will be

$$\mathbf{y}_1 = \left[\sqrt{\frac{E}{2}} \quad \sqrt{\frac{E}{2}}\right]^T \qquad \mathbf{y}_2 = \left[-\sqrt{\frac{E}{2}} \quad +\sqrt{\frac{E}{2}}\right]^T$$

$$\mathbf{y}_3 = \left[-\sqrt{\frac{E}{2}} \quad -\sqrt{\frac{E}{2}}\right]^T \qquad \mathbf{y}_4 = \left[+\sqrt{\frac{E}{2}} \quad -\sqrt{\frac{E}{2}}\right]^T$$

as shown in Fig. 4.2.7. The decision regions for this problem are also shown in the figure.

Let us denote the probability of making a correct decision when the transmitted signal is $y_i(t)$ as $P_D(\mathbf{y}_i)$. Because of the symmetry of the decision regions, this probability is the same value, say P_D, for each of the transmitted waveforms. We can therefore find this probability by finding for example $P_D(\mathbf{y}_1)$:

$$P_D = P_D(\mathbf{y}_1) = P(Z_1 > 0, Z_2 > 0 \mid H_1)$$

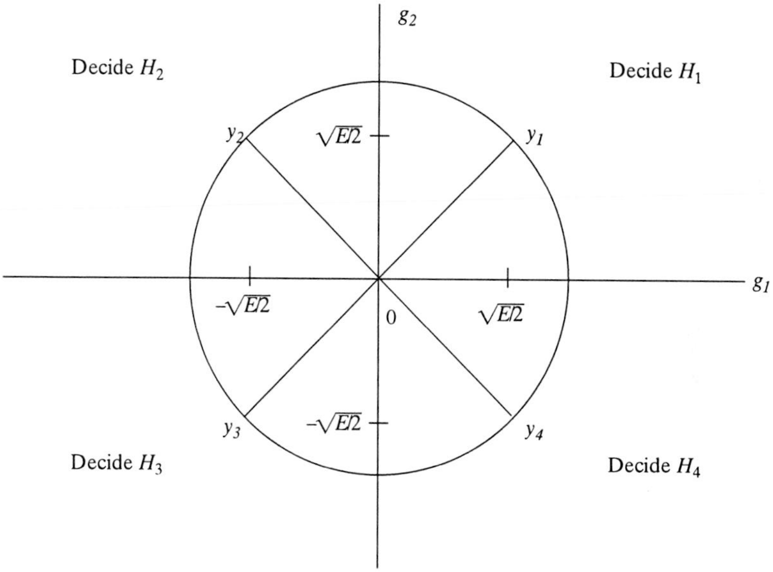

Figure 4.2.7 Signal space and decision regions for Example 4.2.4.

Since under hypothesis H_1, Z_1 and Z_2 are independent, Gaussian random variables with mean $\sqrt{E/2}$ and variance $N_0/2$, it follows that the probability of error

$$P_e = 1 - P(Z_1 > 0)P(Z_2 > 0)$$
$$= Q^2\left(\sqrt{\frac{E}{N_0}}\right)$$

4.3 DETECTION OF KNOWN SIGNALS IN COLORED NOISE

In Section 4.2 we assumed that the observation noise $v(t)$ was Gaussian, white, with mean zero. We now extend our results to the case when the noise is colored with covariance function $\phi_v(t, u)$. We can obtain the solution in exactly the same manner as in Section 4.2 by expanding the observations in a set of orthonormal basis functions. The functions are again chosen so that the coefficients of the expansion are uncorrelated. Specifically, we will use the Karhunen–Loéve expansion, in which the orthogonal functions are the eigenfunctions of the integral equation

$$\int_{t_0}^{t_f} \phi_v(t, u)g_j(u)\,du = \lambda_j g_j(t) \qquad t_0 \leq t \leq t_f \tag{4.3.1}$$

We can then write the expansion of $v(t)$ as

$$v(t) = \lim_{K\to\infty} \sum_{j=1}^{K} V_j g_j(t) \tag{4.3.2}$$

It follows from our discussion of the Karhunen–Loéve expansion in Chapter 2 that

$$E\{V_j\} = 0 \quad \text{and} \quad E\{V_j V_k\} = \lambda_j \delta_{jk} \tag{4.3.3}$$

We can similarly expand $z(t)$ in terms of $g_j(t)$ as

$$z(t) = \lim_{K\to\infty} \sum_{j=1}^{K} Z_j g_j(t) \tag{4.3.4}$$

where

$$Z_j = \int_{t_0}^{t_f} z(t) g_j(t)\, dt \tag{4.3.5}$$

and is hence Gaussian.

Under hypothesis H_1, we have

$$\begin{aligned} Z_j &= \int_{t_0}^{t_f} [y_1(t) + v(t)] g_j(t)\, dt \\ &= y_{1j} + V_j \end{aligned} \tag{4.3.6}$$

We can therefore write

$$E\{Z_j \mid H_1\} = y_{1j} \quad \text{and} \quad \text{var}\{Z_j \mid H_1\} = \lambda_j \tag{4.3.7}$$

where the last step follows from Eq. (4.3.3). Similarly, under hypothesis H_0, we have

$$E\{Z_j \mid H_0\} = y_{0j} \quad \text{and} \quad \text{var}\{Z_j \mid H_0\} = \lambda_j \tag{4.3.8}$$

It can easily be established by virtue of Eq. (4.3.3) that the variables Z_j are independent under each hypothesis. Similar to Eq. (4.2.21), we can write the likelihood ratio for a K-term expansion as

$$\Lambda(z_K(t)) = \frac{\prod_{j=1}^{K} \frac{1}{\sqrt{2\pi\lambda_j}} \exp\left[-\frac{1}{2\lambda_j}(z_j - y_{1j})^2\right]}{\prod_{j=1}^{K} \frac{1}{\sqrt{2\pi\lambda_j}} \exp\left[-\frac{1}{2\lambda_j}(z_j - y_{0j})^2\right]} \tag{4.3.9}$$

Canceling common terms, taking the logarithm, and letting K tend to infinity yields the decision rule as

$$\ln[\Lambda(z(t))] = \sum_{j=1}^{\infty} \frac{z_j}{\lambda_j}(y_{1j} - y_{0j}) + \frac{1}{2}\sum_{j=1}^{\infty} \frac{1}{\lambda_j}(y_{0j}^2 - y_{1j}^2) \underset{H_0}{\overset{H_1}{\gtrless}} \ln \eta \tag{4.3.10}$$

We can rewrite the likelihood ratio in terms of $z(t)$ and the signals $y_1(t)$ and $y_0(t)$ by using Eqs. (4.3.5) and (4.3.6). We then obtain the decision rule as

$$\begin{aligned} &\int_{t_0}^{t_f}\int_{t_0}^{t_f} [y_1(v) - y_0(v)] z(u) \left[\sum_{j=1}^{\infty} \frac{g_j(v) g_j(u)}{\lambda_j}\right] dv\, du \\ &\quad + \frac{1}{2}\int_{t_0}^{t_f}\int_{t_0}^{t_f} [y_1(v) y_1(u) - y_0(v) y_0(u)] \\ &\quad \times \left[\sum_{j=1}^{\infty} \frac{g_j(v) g_j(u)}{\lambda_j}\right] dv\, du \underset{H_0}{\overset{H_1}{\gtrless}} \ln \eta \end{aligned} \tag{4.3.11}$$

The decision rule of Eq. (4.3.11) can be interpreted as a correlation detector as follows. Let us define the function $h(\cdot)$ by

$$\begin{aligned} h(u) &= \int_{t_0}^{t_f} [y_1(v) - y_0(v)]\left[\sum_{j=1}^{\infty} \frac{1}{\lambda_j} g_j(v) g_j(u)\right] dv \\ &= \sum_{j=1}^{\infty} \frac{y_{1j} - y_{0j}}{\lambda_j} g_j(u) \end{aligned} \tag{4.3.12}$$

The second term on the left side of Eq. (4.3.11) is an integral of known functions. Hence it can be combined with the ln η term on the right side to form a new threshold γ. Use of Eq. (4.3.12) then yields the decision rule as

$$\int_{t_0}^{t_f} z(u) h(u)\, du \underset{H_0}{\overset{H_1}{\gtrless}} \gamma \tag{4.3.13}$$

As in the case of white observation noise, we see that the receiver for colored observation noise can be interpreted as a correlation detector. We shall now show that the signal $h(\cdot)$ with which the observation $z(\cdot)$ is correlated satisfies an integral equation. To do so, let us multiply Eq. (4.3.12) by $\phi_v(t, u)$ and integrate over the interval (t_0, t_f). We then get

$$\begin{aligned} \int_{t_0}^{t_f} \phi_v(t,u) h(u)\, du &= \int_{t_0}^{t_f} \phi_v(t,u)\left[\sum_{j=1}^{\infty} \frac{y_{1j} - y_{0j}}{\lambda_j} g_j(u)\right] du \\ &= \sum_{j=1}^{\infty} (y_{1j} - y_{0j}) g_j(t) \\ &= y_1(t) - y_0(t) \end{aligned} \tag{4.3.14}$$

Thus $h(u)$ is obtained as the solution to the preceding integral equation.

4.3.1 Whitening Approach

An alternative approach to obtaining the decision rule for the colored noise problem is to seek to convert it first into a problem in which the results of Section 4.2 can be applied directly. We do this by "whitening" the observation noise. We thus do a preliminary processing of the observations by passing them through a linear time-varying filter with known impulse response. As can be expected, the filter response will depend on the noise characteristics. At the output of the whitening filter, we will have a known signal in additive white noise and the problem thus reduces to the one considered in Section 4.2 [2,4].

The output of the whitening filter can be written as

$$\begin{aligned} z_w(t) &= \int_{t_0}^{t_f} h_w(t,u) z(u)\, du \\ &= \int_{t_0}^{t_f} h_w(t,u) y_i(u)\, du + \int_{t_0}^{t_f} h_w(t,u) v(u)\, du \\ &= y_{wi}(t) + v_w(t) \qquad t \geq t_0, \quad i = 0, 1 \end{aligned} \tag{4.3.15}$$

where $h_w(t, u)$ represents the impulse response of the whitening filter, $y_{wi}(t)$ is the output of this filter due to input $y_i(t)$, and $v_w(t)$ is the output due to input $v(t)$.

We want to choose $h_w(\cdot,\cdot)$ so that

$$\phi_{v_w}(t, u) = \mathrm{E}\{v_w(t)v_w(u)\} = \delta(t - u) \qquad t_0 \leqslant t,\, u \leqslant t_f$$

where we have arbitrarily specified a unity spectral height for the noise output. Assuming for the moment that such a filter has been found, we can use Eq. (4.2.28) directly to determine the decision rule as

$$\int_{t_0}^{t_f} z_w(t)y_{w_1}(t)\,dt - \int_{t_0}^{t_f} z_w(t)y_{w_0}(t)\,dt \underset{H_0}{\overset{H_1}{\gtrless}} \gamma \tag{4.3.16}$$

where the threshold is

$$\gamma = \ln \eta - \tfrac{1}{2}\int_{t_0}^{t_f} [y_{w0}^2(t) - y_{w1}^2(t)]\,dt \tag{4.3.17}$$

In terms of the original waveforms, the decision rule can be written as

$$\int_{t_0}^{t_f} dt \int_{t_0}^{t_f} h_w(t, u)z(u)\,du \int_{t_0}^{t_f} h_w(t, v)y_1(v)\,dv$$

$$- \int_{t_0}^{t_f} dt \int_{t_0}^{t_f} h_w(t, u)z(u)\,du \int_{t_0}^{t_f} h_w(t, v)y_0(v)\,dv \underset{H_0}{\overset{H_1}{\gtrless}} \gamma \tag{4.3.18}$$

The structure of the receiver is shown in Fig. 4.3.1.

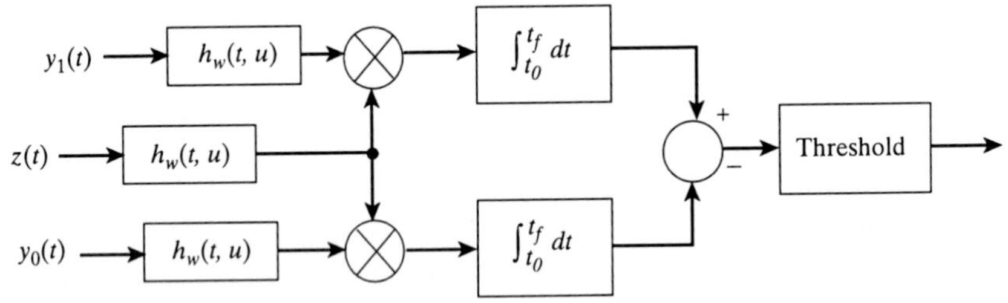

Figure 4.3.1 Receiver structure for colored noise.

All that remains is to determine the whitening filter $h_w(t, u)$. We can obtain a formal solution for $h_w(t, u)$ in terms of the eigenfunctions of $\phi_v(t, \sigma)$. To do so, let us expand $h_w(t, u)$ in terms of the orthogonal functions $g(t)$ as follows. We write

$$h_w(t, u) = \sum_{j=1}^{\infty} h_j g_j(t) g_j(u) \tag{4.3.19}$$

The noise output from the whitening filter is obtained from Eq. (4.3.15) as

$$v_w(t) = \int_{t_0}^{t_f} h_w(t, u)v(u)\,du \tag{4.3.20}$$

so that

$$\int_{t_0}^{t_f}\int_{t_0}^{t_f} h_w(t,u)h_w(s,v)\phi_v(u,v)\,du\,dv = \mathrm{E}\{v_w(t)v_w(s)\} = \delta(t-s) \tag{4.3.21}$$

We now substitute for $h_w(\cdot,\cdot)$ from Eq. (4.3.19) in the left side of Eq. (4.3.21) and make use of Eq. (4.3.1) and the orthogonality of the eigenfunctions to obtain

$$\sum_{j=1}^{\infty} h_j^2 \lambda_j g_j(t) g_j(s) = \delta(t-s) \tag{4.3.22}$$

We recall from Chapter 2 that we can formally express the impulse function as a series

$$\delta(t-s) = \sum_{j=1}^{\infty} g_j(t) g_j(s) \tag{4.3.23}$$

Comparison of Eqs. (4.3.22) and (4.3.23) then yields

$$h_j = \frac{1}{\sqrt{\lambda_j}} \tag{4.3.24}$$

so that the whitening filter is given by

$$h_w(t,u) = \sum_{j=1}^{\infty} \frac{1}{\sqrt{\lambda_j}} g_j(t) g_j(u) \tag{4.3.25}$$

Finally, we can establish the equivalence of the decision rules of Eqs. (4.3.11) and (4.3.18) by substituting Eq. (4.3.25) in Eq. (4.3.18) and making use of the orthogonality of the eigenfunctions.

In going through the preceding formal derivation, we have ignored certain mathematical niceties. For example, if the correlation function $\phi_i(t,u)$ is not positive definite, the eigenfunctions may not form a complete orthonormal set. We can overcome this problem by assuming that the noise has a white component in addition to the correlated component. The thermal noise in the receiver can usually be used to account for this white component.

The determination of the whitening filter is, in general, fairly tedious. The expression for $h_w(\cdot,\cdot)$ in Eq. (4.3.25) is in the form of an infinite series and in general cannot be put in a closed form. Furthermore, the determination of the eigenfunctions involves solving the integral equation of (4.3.1). This in itself is usually a pretty formidable task. We discuss solution techniques for some special cases in Section 4.5.

One case of practical importance for which the whitening filter may be obtained in a straightforward fashion is that of stationary noise with rational spectrum and semi-infinite observation interval, that is, $t_0 = -\infty$. In this case, the whitening filter may be chosen to be time invariant. If $\Phi_v(s)$ is the power spectrum of the noise, we perform a spectral factorization and write

$$\Phi_v(s) = \Phi_v^+(s)\Phi_v^-(s) \tag{4.3.26}$$

where $\Phi_v^+(\cdot)$ consists of all poles and zeros in the left half-plane and $\Phi_v^-(\cdot)$ consists of all poles and zeros in the right half-plane. Then the transfer function of the whitening filter is given by

$$H_w(s) = \frac{1}{\Phi_v^+(s)} \tag{4.3.27}$$

4.3.2 Receiver Performance

We can determine the performance of the receiver of Eq. (4.3.16) from the results in Section 4.2 for the white observation noise problem by replacing $y_0(t)$ and $y_1(t)$ by $y_{w0}(t)$ and $y_{w1}(t)$, respectively, and setting $N_0 = 2$. For the minimum probability of error criterion and equal prior probabilities, we can use Eqs. (4.2.52) with $\gamma = 0$ to write the probability of error as

$$P_e = \int_d^\infty \frac{1}{\sqrt{2\pi}} e^{-z^2/2}\, dz \tag{4.3.28}$$

where

$$d^2 = \frac{1}{4}\int_{t_0}^{t_f} [y_{w0}(t) - y_{w1}(t)]^2\, dt \tag{4.3.29}$$

In terms of the original signals this becomes

$$d^2 = \frac{1}{4}\int_{t_0}^{t_f}\left[\int_{t_0}^{t_f} h_w(t,u)[y_0(u) - y_1(u)]\, du\right]^2 dt \tag{4.3.30}$$

The calculation of d^2 requires solving integral equations and is at best tedious. We also note that the performance of the detector is no longer independent of the signal shape.

It is clear from Eq. (4.3.28) that the probability of error decreases monotonically as d^2 increases. To optimize the performance of the communication system, therefore, the signals $y_0(t)$ and $y_1(t)$ must be chosen so as to maximize d^2 subject to the constraint

$$\frac{1}{2}\int_{t_0}^{t_f} [y_1^2(t) + y_0^2(t)]\, dt = \text{constant} = \epsilon \tag{4.3.31}$$

We thus form the objective function

$$\begin{aligned} J &= d^2 - 2\mu\epsilon \\ &= \int_{t_0}^{t_f}\left[\int_{t_0}^{t_f} h_w(t,u)[y_0(u) - y_1(u)]\, du\right]^2 dt \\ &\quad - \mu\int_{t_0}^{t_f} [y_1^2(t) + y_0^2(t)]\, dt \end{aligned} \tag{4.3.32}$$

where μ is a Lagrange multiplier. It then follows by using the calculus of variations that the optimum signals $y_1(t)$ and $y_0(t)$ are given by the relations (see Appendix B)

$$\int_{t_0}^{t_f}\int_{t_0}^{t_f} [y_0(u) - y_1(u)]h_w(t,u)h_w(t,v)\, dt\, du = \mu y_0(v) = -\mu y_1(v) \tag{4.3.33}$$

We thus note that in the optimum system

$$y_1(v) = -y_0(v) \qquad t_0 \leq v \leq t_f \tag{4.3.34}$$

This is the same result as that found for the optimum binary system with white noise and corresponds to the case $\bar{\rho} = -1$ in Eq. (4.2.49). Substitution of Eqs. (4.3.33) and (4.3.34) in Eq. (4.3.30) yields the optimal value for d^2 as

$$d^2 = \mu \int_{t_0}^{t_f} y_0^2(t)\, dt = \mu \int_{t_0}^{t_f} y_1^2(t)\, dt \tag{4.3.35}$$

Since the signal energies are fixed, d^2 will be maximum (P_e will be minimum) when μ is maximum.

4.3.3 Signal Design

We now consider the problem of finding the optimum signals for transmission. Substitution of Eq. (4.3.34) in Eq. (4.3.33) yields

$$\int_{t_0}^{t_f}\int_{t_0}^{t_f} y_1(u)h_w(t,u)h_w(t,v)\, dt\, du = \frac{\mu}{2} y_1(v) \tag{4.3.36}$$

Let us now multiply both sides of Eq. (4.3.21) by $h_w(t,\lambda)$ and integrate over $t_0 < t < t_f$ to get

$$\int_{t_0}^{t_f} h_w(t,\lambda) \int_{t_0}^{t_f}\int_{t_0}^{t_f} h_w(t,u)h_w(s,v)\phi_v(u,v)\, du\, dv\, dt = \int_{t_0}^{t_f} \delta(t-s)h_w(t,\lambda)\, dt = h_w(s,\lambda) \tag{4.3.37}$$

or equivalently

$$\int_{t_0}^{t_f} h_w(s,v)\left[\int_{t_0}^{t_f}\int_{t_0}^{t_f} h_w(t,u)h_w(t,\lambda)\phi_v(u,v)\, dt\, du\right] dv = h_w(s,\lambda) \tag{4.3.38}$$

It therefore follows that

$$\int_{t_0}^{t_f}\int_{t_0}^{t_f} h_w(t,u)h_w(t,\lambda)\phi_v(u,v)\, dt\, du = \delta(v-\lambda) \tag{4.3.39}$$

Multiplying both sides of Eq. (4.3.36) by $\phi_v(v,\sigma)$ and integrating over $t_0 < v < t_f$ gives

$$\int_{t_0}^{t_f} y_1(u) \int_{t_0}^{t_f}\int_{t_0}^{t_f} h_w(t,u)h_w(t,v)\phi_v(v,\sigma)\, dt\, dv\, du = \frac{\mu}{2}\int_{t_0}^{t_f} y_1(v)\phi_v(v,\sigma)\, dv \tag{4.3.40}$$

Use of Eq. (4.3.39) then enables us to write the previous equation as

$$\int_{t_0}^{t_f} y_1(v)\phi_v(v,\sigma)\, dv = \frac{2}{\mu} y_1(\sigma) \tag{4.3.41}$$

We see that the optimum signal waveform is an eigenfunction of the covariance kernel $\phi_v(v,\sigma)$ corresponding to the eigenvalue $2/\mu$. Since d^2 is maximum when $2/\mu$

is minimum, we choose as the optimum signal $y_1(t)$ the eigenfunction corresponding to the *smallest* eigenvalue.

Finally, we note that if the noise is white, the eigenvalues are all equal (see Section 2.3) and it makes no difference which waveforms we choose as long as $y_1(t) = -y_0(t)$. This is in agreement with Section 4.2.

We also point out that perfect detection (probability of error is zero) occurs when $d^2 = \infty$. From Eq. (4.3.35) we see that this happens when the smallest eigenvalue of the kernel $\phi_v(u, v)$ is zero. In this case, for arbitrarily small time intervals and arbitrarily small energy levels, we achieve perfect detection. Such a situation is referred to as *singular detection*. Since this kind of performance cannot correspond to an actual physical situation, it is important that we make our mathematical model realistic enough to eliminate the possibility of singular detection. Again, a simple way of doing this is to assume that the observation noise has a white component.

4.4 DETECTION OF KNOWN SIGNALS IN NOISE: MAXIMUM SNR CRITERION

In communication problems, a measure of the efficiency of the system is obtained by the signal-to-noise ratio (SNR), which is the ratio of the signal power to the noise power. In fact, we have seen that the probability of error in the binary communication problem of Section 4.3 was dependent on this ratio ϵ/N_0 and that the larger this ratio, the smaller the error probability. In a number of situations, the SNR and the probability of error are monotonically related. Even though this need not be the case in general, we may still use the signal-to-noise ratio as a criterion of the efficacy of the system and seek to design a receiver that maximizes this ratio [6].

The analysis that follows is applicable in many cases other than with the binary communication problem considered in Section 4.3: for example, in radar systems where it is required to detect the presence of a signal echo in noise. The criterion is useful in applications where we are not interested in maintaining the fidelity of the transmitted pulse but are interested only in recognizing the presence of a signal in noise. In applications such as pulse-position modulation, in which the time of occurrence is of prime importance, it is necessary to retain the fidelity or sharpness of pulse rise time. In such systems, therefore, the signal-to-noise ratio is not a very meaningful measure of the performance of the system.

We consider the following problem. The received signal $z(t)$ consists of known signal $y(t)$ [defined over the interval (t_0, t_f)] and additive noise $v(t)$. The noise is assumed to be at least wide-sense stationary with zero mean and covariance $\phi_v(\tau)$. We seek to design a linear filter so as to maximize the output signal-to-noise ratio at some time T.

Let the impulse response of the filter be $h(t)$. Then the signal and noise components at the output at time T are

$$y_0(T) = \int_{t_0}^{t_f} h(T - \tau)y(\tau)\, d\tau \tag{4.4.1}$$

$$V_0(T) = \int_{t_0}^{t_f} h(T - \tau)v(\tau)\, d\tau \tag{4.4.2}$$

The noise power is given by

$$\begin{aligned} \mathrm{E}\{V_0^2(T)\} &= \mathrm{E}\left\{\int_{t_0}^{t_f} h(T - \tau)v(\tau)\, d\tau \int_{t_0}^{t_f} h(T - \sigma)v(\sigma)\, d\sigma\right\} \\ &= \int_{t_0}^{t_f}\int_{t_0}^{t_f} h(T - \tau)h(T - \sigma)\phi_v(\tau - \sigma)\, d\sigma\, d\tau \end{aligned} \tag{4.4.3}$$

The ratio of signal to noise power is then given by $y_0^2(T)/\mathrm{E}\{V_0^2(T)\}$. To make the maximization of this quantity meaningful, we assume that $y_0(T)$ is constrained to be a constant. The problem of maximizing the signal-to-noise ratio is then equivalent to that of minimizing the quantity $\mathrm{E}\{V_0^2(T)\}$ subject to the constraint that $y_0(T)$ is a constant. We thus form the objective function (see Appendix B)

$$J = \mathrm{E}\{V_0^2(T)\} - 2\mu y_0(T) \tag{4.4.4}$$

where μ is a Lagrange multiplier. Substituting from Eqs. (4.4.1) and (4.4.3), we can write Eq. (4.4.4) as

$$\begin{aligned} J &= \int_{t_0}^{t_f}\int_{t_0}^{t_f} h(T - \tau)h(T - \sigma)\phi_v(\tau - \sigma)\, d\sigma\, d\tau \\ &\quad - 2\mu \int_{t_0}^{t_f} h(T - \tau)y(\tau)\, d\tau \\ &= \int_{t_0}^{t_f} h(T - \tau)\left[\int_{t_0}^{t_f} h(T - \sigma)\phi_v(\tau - \sigma)\, d\sigma - 2\mu y(\tau)\right] d\tau \end{aligned} \tag{4.4.5}$$

We now use the calculus of variations (Appendix B) to minimize J with respect to the filter function $h(t)$. The first variation in the cost function can easily be shown to be

$$\begin{aligned} \delta J = \int_{t_0}^{t_f}\int_{t_0}^{t_f} &[\delta h(T - \tau)h(T - \sigma) \\ &+ \delta h(T - \sigma)h(T - \tau)]\phi_v(\tau - \sigma)\, d\tau\, d\sigma \\ &- 2\mu \int_{t_0}^{t_f} \delta h(T - \tau)y(\tau)\, d\tau \end{aligned} \tag{4.4.6}$$

which can be written as

$$\delta J = \int_{t_0}^{t_f} 2\delta h(T - \tau)\left\{\left[\int_{t_0}^{t_f} h(T - \sigma)\phi_v(\tau - \sigma)\, d\sigma\right] - \mu y(\tau)\right\} d\tau \tag{4.4.7}$$

For a minimum, δJ must be zero. Since $\delta h(\cdot)$ is arbitrary, it follows that the inner integral must be identically zero over the interval $t_0 \leqslant \tau \leqslant t_f$. Thus the optimum filter must satisfy the equation

$$\int_{t_0}^{t_f} h(T - \sigma)\phi_v(\tau - \sigma)\, d\sigma = \mu y(\tau) \qquad t_0 \leqslant \tau \leqslant t_f \tag{4.4.8}$$

Since the constant multiplier μ changes only the filter gain and does not affect the signal-to-noise ratio, we can arbitrarily set it equal to unity. Therefore, the filter function should satisfy the integral equation

$$\int_{t_0}^{t_f} h(T-\sigma)\phi_v(\tau-\sigma)\,d\sigma = y(\tau) \qquad t_0 \leqslant \tau \leqslant t_f \tag{4.4.9}$$

Determination of the optimum filter therefore requires that we use information about the signal $y(t)$ over the entire observation interval. If $T < t_f$, the optimum system corresponds to a physically nonrealizable (noncausal) system: to make a decision at time T, we need information about the signal in the future, that is, on the interval $[T, t_f]$. If the received signal needs to be processed in real time, we will have to make a decision at time T based only on the data available up to that time. Thus the receiver structure will not be truly optimum. However if the processing can be carried out in nonreal time on stored or recorded data, we can arrive at a decision based on the signal-to-noise ratio at time T using the values of the received signal over the entire observation interval. The processing delay in this case is equal to $t_f - T$.

The maximum signal-to-noise ratio (SNR) obtained by substituting the optimum filter function in Eqs. (4.4.1) and (4.4.3) is given by

$$\begin{aligned}\text{SNR}_{\max} &= \frac{\left[\displaystyle\int_{t_0}^{t_f} h(T-\tau)y(\tau)\,d\tau\right]^2}{\displaystyle\int_{t_0}^{t_f} h(T-\tau)\left[\int_{t_0}^{t_f} h(T-\sigma)\phi_v(\tau-\sigma)\,d\sigma\right]d\tau} \\ &= \int_{t_0}^{t_f} h(T-\tau)y(\tau)\,d\tau\end{aligned} \tag{4.4.10}$$

where we have substituted Eq. (4.4.9) in the denominator.

To determine the optimum filter explicitly, we need to solve Eq. (4.4.9). This is a Fredholm integral equation of the first kind [7]. In general, the solution of such equations is not straightforward. We consider some solution techniques for integral equations in Section 4.5. A special case, in which the solution can be obtained easily, is when the noise signal is white with autocorrelation $(N_0/2)\delta(\tau)$. For this case we easily see that the solution to Eq. (4.4.9) is given by

$$h(T-\tau) = \frac{2}{N_0}\,y(\tau) \qquad t_0 \leqslant \tau \leqslant t_f \tag{4.4.11}$$

or, equivalently,

$$h(\tau) = \frac{2}{N_0}\,y(T-\tau) \qquad T - t_f \leqslant \tau \leqslant T - t_0 \tag{4.4.12}$$

which is precisely the matched filter discussed in Section 4.2. The maximum signal-to-noise ratio is obtained from Eq. (4.4.10) as

$$\text{SNR}_{\max} = \int_{t_0}^{t_f} \frac{2}{N_0} y^2(\tau)\, d\tau = \frac{2\epsilon}{N_0} \tag{4.4.13}$$

where ϵ is the signal energy. We now consider the frequency-domain representation of the filter in Eq. (4.4.12).

By taking the (bilateral) Laplace transform of Eq. (4.4.12), we obtain

$$H(s) = \frac{2}{N_0} Y(-s)e^{-sT} \tag{4.4.14}$$

If we assume that $y(t) = 0$ for $t > T$, then $y(T - \tau)$ in Eq. (4.4.12) is zero for $\tau < 0$. It therefore follows that $h(\tau)$ is a causal (physically realizable) function and that its transform, $H(s)$, has poles only in the left half-plane.

If $T < t_f$ but $y(t)$ is not zero for $t > T$, $y(T - \tau)$ is not zero for $\tau < 0$ and the optimum filter will not be physically realizable. As discussed earlier, we can overcome this problem by operating on stored data. If we want to restrict ourselves to causal filters, we can proceed as follows. Let $\{Y(-s)e^{-sT}\}_+$ and $\{Y(-s)e^{-sT}\}_-$ represent the transforms of the causal and anticausal parts of $y(T - \tau)$. The causal filter is then obtained as

$$H_c(s) = \frac{2}{N_0} \{Y(-s)e^{-sT}\}_+ \tag{4.4.15}$$

When the observation noise is not white, we may proceed in a manner similar to that in Section 4.3. That is, we first pass the observations through a whitening filter which whitens the observation noise and then design a filter matched to the signal output from the whitening filter. Specifically, if the impulse response of the whitening filter is $h_1(t, u)$, the output will be

$$\begin{aligned} z_1(t) &= \int_{t_0}^{t_f} h_1(t, u) z(u)\, du \\ &= \int_{t_0}^{t_f} h_1(t, u) y(u)\, du + \int_{t_0}^{t_f} h_1(t, u) v(u)\, du \\ &= y_1(t) + v_1(t) \end{aligned} \tag{4.4.16}$$

Since $v_1(t)$ is white, we can use the results of Eq. (4.4.12) to obtain the optimum matched filter $h_2(t)$. The filter $h_2(t)$ has to be matched to the signal $y_1(t)$ so that

$$h_2(t) = y_1(T - t) \qquad T - t_f \leqslant t \leqslant T - t_0 \tag{4.4.17}$$

Again the basic problem is in determining the whitening filter. As discussed earlier, if $t_0 = -\infty$, this filter is easily determined in terms of the spectrum of the noise signal $v(t)$ as

$$H_1(s) = \frac{1}{\Phi_v^+(s)} \tag{4.4.18}$$

which represents a causal system. Transformation of Eq. (4.4.17) yields the transfer function of the filter $h_2(t)$ as

$$H_2(s) = [Y_1(-s)e^{-sT}] \tag{4.4.19}$$

Since

$$\begin{aligned} Y_1(s) &= H_1(s)Y(s) \\ &= \frac{1}{\Phi_v^+(s)} Y(s) \end{aligned} \tag{4.4.20}$$

it follows that

$$H_2(s) = \frac{Y(-s)e^{-sT}}{\Phi_v^-(s)} \tag{4.4.21}$$

The realizable part of $H_2(s)$ is thus given by

$$H_{2c}(s) = \left[\frac{Y(-s)e^{-sT}}{\Phi_v^-(s)}\right]_+ \tag{4.4.22}$$

The optimum causal filter obtained by cascading $H_1(s)$ and $H_{2c}(s)$ is thus given by

$$\begin{aligned} H(s) &= H_1(s)H_{2c}(s) \\ &= \frac{1}{\Phi_v^+(s)}\left[\frac{Y(-s)e^{-sT}}{\Phi_v^-(s)}\right]_+ \end{aligned} \tag{4.4.23}$$

We conclude this section with an example.

Example 4.4.1

Consider the problem of designing a matched filter to detect the signal

$$y(t) = \begin{cases} e^{-t/2} - e^{-t} & t \geqslant 0 \\ 0 & t < 0 \end{cases}$$

The noise spectral density is

$$\Phi_v(s) = \frac{1}{1 - s^2}$$

The signal-to-noise ratio is to be maximized at some specified time $T > 0$.

The noise spectrum may be factored as

$$\Phi_v(s) = \Phi_v^+(s)\Phi_v^-(s) = \frac{1}{1+s}\frac{1}{1-s}$$

so that the whitening filter has transfer function

$$H_1(s) = \frac{1}{\Phi_\nu^+(s)} = 1 + s$$

The signal output from the whitening filter is

$$Y_1(s) = Y(s)H_1(s) = \left(\frac{2}{1+2s} - \frac{1}{1+s}\right)(1+s)$$

or

$$Y_1(s) = \frac{1}{1+2s}$$

The corresponding time function is

$$y_1(t) = \tfrac{1}{2}e^{-t/2} \qquad \text{for} \quad t \geq 0$$

and exists in white noise of spectral height unity. Since $y_1(t)$ is nonzero for all $t \geq 0$, the optimum matched filter in this case is unrealizable and is given by

$$H_2(s) = Y_1(-s)e^{-sT} = \frac{1}{1-2s}e^{-sT}$$

with impulse response

$$h_2(t) = \begin{cases} \tfrac{1}{2}e^{(t-T)/2} & -\infty < t \leq T \\ 0 & t > T \end{cases}$$

We take as the realizable filter, the causal part of the preceding

$$h_{2c}(t) = \begin{cases} \tfrac{1}{2}e^{(t-T)/2} & 0 \leq t \leq T \\ 0 & \text{elsewhere} \end{cases}$$

with transfer function

$$\begin{aligned} H_{2c}(s) &= \int_0^T \tfrac{1}{2}e^{(t-T)/2}\,e^{-st}\,dt \\ &= \frac{1}{1-2s}(e^{-sT} - e^{-T/2}) \end{aligned}$$

The overall matched filter is given by

$$\begin{aligned} H(s) = H_1(s)H_{2c}(s) &= \frac{1+s}{1-2s}(e^{-sT} - e^{-T/2}) \\ &= \left(-\frac{1}{2} + \frac{3/2}{1-2s}\right)(e^{-sT} - e^{-T/2}) \end{aligned}$$

The impulse response of the optimal filter is given by

$$h(t) = -\tfrac{1}{2}\delta(t-T) + \tfrac{1}{2}e^{-T/2}\delta(t) + \tfrac{3}{4}e^{(t-T)/2}[e^{-T/2}u(T-t) - u(-t)]$$

where $u(\cdot)$ denotes the unit step function. The various signals are illustrated in Fig. 4.4.1.

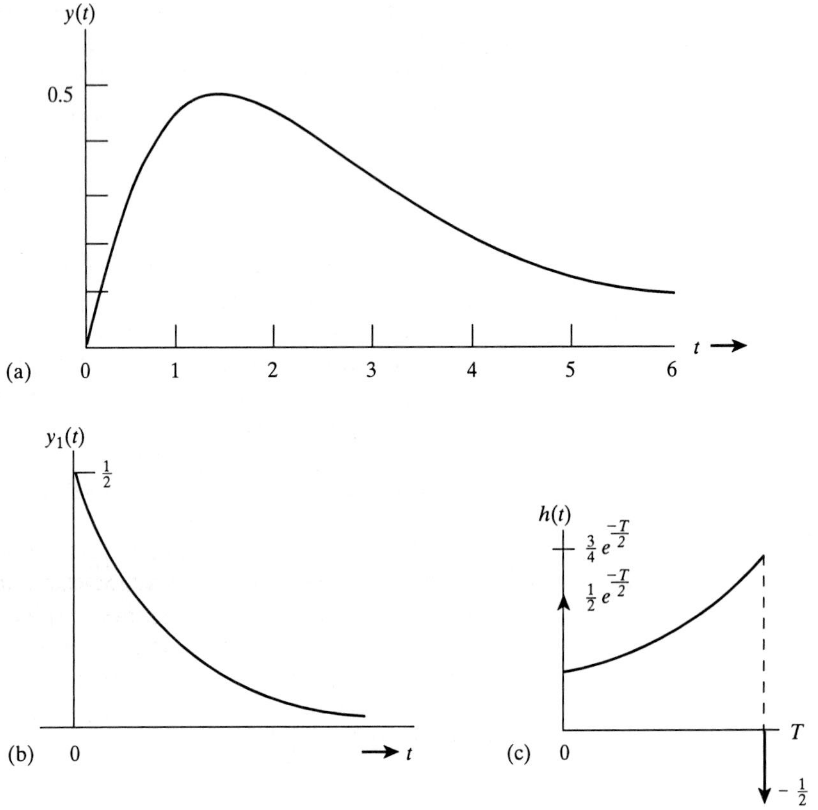

Figure 4.4.1 Matched filter for Example 4.4.1: (a) signal input; (b) signal output from whitening filter; (c) impulse response of optimum filter.

4.5 SOLUTION OF INTEGRAL EQUATIONS

We now briefly discuss solution techniques for some of the integral equations we have encountered in previous sections and work some typical examples [1,2,7–11]. The first equation of interest that we encountered in the detection of known signals was of the form

$$\int_{t_0}^{t_f} \phi_v(t,\sigma)h(\sigma)\,d\sigma = y(t) \qquad t_0 \leq t \leq t_f \tag{4.5.1}$$

in which $y(t)$ and $\phi_v(t,\sigma)$ are known and we are required to solve for $h(t)$.

If there is no white noise present, the kernel $\phi_v(t,\sigma)$ will have no singularities. Then Eq. (4.5.1) represents a *Fredholm equation of the first kind*. The solution for $h(t)$ will in general contain singularity functions (impulses and their derivatives) at the endpoints of the observation interval.

If there is a white noise component in the additive noise, we may write

$$\phi_v(t,\sigma) = \frac{N_0}{2}\delta(t-\sigma) + \phi_c(t,\sigma) \tag{4.5.2}$$

where $\phi_c(t,\sigma)$ is a continuous square-integrable function. In this case Eq. (4.5.1) becomes

$$\frac{N_0}{2}h(t) + \int_{t_0}^{t_f} \phi_c(t,\sigma)h(\sigma)\,d\sigma = y(t) \tag{4.5.3}$$

This is called a *Fredholm equation of the second kind*.

The solutions to these equations are at best tedious and in many cases they do not appear to be solvable. When the kernel is of certain special types, however, straightforward procedures for solving Eqs. (4.5.1) and (4.5.3) are available. One type occurs when the noise $v(t)$ is the steady-state response of a linear time-invariant system excited with white noise. In this case the noise output is stationary and the noise power spectrum is the ratio of two polynomials in s^2. Let us write the spectrum as

$$\Phi_v(s) = \int_{-\infty}^{\infty} \phi_v(\tau)e^{-s\tau}\,d\tau \,\underline{\Delta}\, \frac{N(s^2)}{D(s^2)} \tag{4.5.4}$$

For the noise to have finite power, the denominator $D(s^2)$ must be at least two degrees higher than the numerator. Kernels whose transforms satisfy Eq. (4.5.4) are called *rational* kernels.

The basic approach in solving an integral equation with rational kernel is first to find an equivalent differential equation. Because of the form of the kernel, the differential equation will be linear with constant coefficients and hence its solution can be easily determined. To obtain the differential equation, assume that

$$\Phi_v(s) = \frac{N(s^2)}{D(s^2)} \tag{4.5.5}$$

Multiplying both sides by $D(s^2)$ yields

$$D(s^2)\Phi_v(s) = N(s^2) \tag{4.5.6}$$

Recalling that multiplication by s in the frequency domain corresponds to differentiation with respect to t, we can write Eq. (4.5.6) as

$$D(p^2)\phi_v(t-u) = N(p^2)\delta(t-u) \tag{4.5.7}$$

where $p \,\underline{\Delta}\, d/dt$. Now the equation of interest is

$$y(t) = \int_{t_0}^{t_f} \phi_v(t-\sigma)h(\sigma)\,d\sigma \qquad t_0 \leq t \leq t_f \tag{4.5.8}$$

Operating on both sides by $D(p^2)$, we obtain

$$D(p^2)y(t) = \int_{t_0}^{t_f} D(p^2)\phi_v(t-\sigma)h(\sigma)\,d\sigma \qquad t_0 \leq t \leq t_f \tag{4.5.9}$$

Substitution of Eq. (4.5.7) in the right side of Eq. (4.5.9) then yields

$$D(p^2)y(t) = N(p^2)h(t) \qquad t_0 \leqslant t \leqslant t_f \tag{4.5.10}$$

The solution consists of three parts. The homogeneous solution is found by solving

$$N(p^2)h(t) = 0$$

and will consist of exponential terms. The particular solution is found by solving Eq. (4.5.10) directly without regard to the limits $[t_0, t_f]$ or equivalently by using transform techniques.

For example, transforming both sides of Eq. (4.5.10) yields

$$H(s) = \frac{D(s^2)Y(s)}{N(s^2)} = \frac{Y(s)}{\Phi_v(s)}$$

so that the particular solution is

$$h_p(t) = \frac{1}{2\pi j}\int_{c-j\infty}^{c+j\infty} \frac{Y(s)}{\Phi_v(s)} e^{st}\, ds \qquad t_0 \leqslant t \leqslant t_f$$

The total solution is found by adding the homogeneous and particular solutions to terms of the form

$$\sum_{i=0}^{p-q-1} a_i\, \delta^{(i)}(t - t_0) + b_i\, \delta^{(i)}(t - t_f)$$

where $\delta^{(i)}(t)$ is the ith derivative of the delta function and where p and q refer to the order of $D(s^2)$ and $N(s^2)$, respectively. The coefficients a_i and b_i are evaluated by substituting the assumed solution into the integral equation.

Example 4.5.1

We consider the problem of designing a correlation receiver for a known signal observed over the interval $[0, T]$ in the presence of noise with spectral density and autocorrelation function

$$\Phi_v(s) = \frac{2\alpha\beta}{-s^2 + \beta^2} \quad \text{and} \quad \phi_v(\tau) = \alpha e^{-\beta|\tau|}$$

The relevant differential equation corresponding to Eq. (4.5.10) is easily seen to be

$$2\alpha\beta h(t) = \beta^2 y(t) - \ddot{y}(t) \qquad 0 \leqslant t \leqslant T$$

Thus the equation reduces to a simple algebraic one for this example. We have

$$h(t) = \frac{1}{2\alpha\beta}\,[\beta^2 y(t) - \ddot{y}(t)]$$

Since $p - q - 1 = 0$, we modify the solution by including delta functions at the endpoints but not derivatives of the delta functions. The total solution is therefore of the form

$$h(t) = \frac{1}{2\alpha\beta}\,[\beta^2 y(t) - \ddot{y}(t)] + a\delta(t) + b\delta(t - T)$$

If we substitute this solution into the integral Eq. (4.5.1), we obtain

$$y(t) = \int_0^T \alpha e^{-\beta|t-\sigma|}\left[\frac{1}{2\alpha\beta}[\beta^2 y(\sigma) - \ddot{y}(\sigma)] + a\delta(\sigma) + b\delta(\sigma - T)\right]d\sigma \qquad 0 \leq t \leq T$$

Carrying out the integration on the right-hand side yields

$$\begin{aligned} y(t) = y(t) &+ e^{-\beta t}\left[a\alpha + \frac{\dot{y}(0)}{2\beta} - \frac{y(0)}{2}\right] \\ &+ e^{\beta t}e^{-\beta T}\left[\alpha b - \frac{\dot{y}(T)}{2\beta} - \frac{y(T)}{2}\right] \qquad 0 \leq t \leq T \end{aligned}$$

Since this equation must be satisfied for all t in the interval $[0, T]$, this implies that

$$a = \frac{1}{2\alpha\beta}[\beta y(0) - \dot{y}(0)] \quad \text{and} \quad b = \frac{1}{2\alpha\beta}[\beta y(T) + \dot{y}(T)]$$

The solution for $h(t)$ is thus

$$\begin{aligned} h(t) = \frac{1}{2\alpha\beta}\{&[\beta y(0) - \dot{y}(0)]\delta(t) + [\beta y(T) + \dot{y}(T)]\delta(t - T) \\ &+ \beta^2 y(t) - \ddot{y}(t)\} \qquad 0 \leq t \leq T \end{aligned}$$

The output of the processor is

$$\int_0^T z(t)h(t)\,dt = az(0) + bz(T) + \frac{1}{2\alpha\beta}\int_0^T z(t)[\beta^2 y(t) - \ddot{y}(t)]\,dt$$

The solution to the Fredholm equation of the second kind proceeds in exactly the same way except that we do not need to add singularity functions at the endpoints.

For example, let us assume that there is also a white component in the noise spectrum so that

$$\Phi_v(s) = \frac{2\alpha\beta}{-s^2 + \beta^2} + \frac{N_0}{2} = \frac{N_0/2(-s^2 + \gamma^2)}{-s^2 + \beta^2} = \frac{N(s^2)}{D(s^2)}$$

where $\gamma^2 = \beta^2 + 4\alpha\beta/N_0$.

The optimum filter $h(t)$ is obtained by solving the integral equation

$$\int_0^T \phi_v(t - \sigma)h(\sigma)\,d\sigma = y(t)$$

which in this case reduces to [see Eq. (4.5.3)]

$$\frac{N_0}{2}h(t) + \int_0^T \alpha e^{-\beta|t-\sigma|}h(\sigma)\,d\sigma = y(t)$$

The corresponding differential equation is obtained from Eq. (4.5.10) as

$$\frac{N_0}{2}[-\ddot{h}(t) + \gamma^2 h(t)] = -\ddot{y}(t) + \beta^2 y(t)$$

The complementary solution is easily seen to be

$$h_c(t) = \alpha_1 e^{\gamma t} + \alpha_2 e^{-\gamma t}$$

To obtain the particular solution, we note that

$$H(s) = \frac{2}{N_0}\frac{-s^2 + \beta^2}{-s^2 + \gamma^2}Y(s)$$

The particular solution can thus be obtained as

$$h_p(t) = \frac{1}{2\pi j}\int_{c-j\infty}^{c+j\infty} \frac{2}{N_0}\frac{-s^2 + \beta^2}{-s^2 + \gamma^2}Y(s)e^{st}\,ds$$

The total solution is obtained as

$$h(t) = h_p(t) + \alpha_1 e^{\gamma t} + \alpha_2 e^{-\gamma t} \qquad 0 \leq t \leq T$$

We conclude this section with a consideration of integral equations with separable kernels. A kernel is said to be *separable* if it can be written as

$$\phi_v(t, \sigma) = \sum_{i=1}^{N} \lambda_i g_i(t) g_i(\sigma) \qquad t_0 \leq t,\, u \leq t_f \tag{4.5.11}$$

where λ_i and $g_i(t)$ are the eigenvalues and eigenfunctions of the equation

$$\int_{t_0}^{t_f} \phi_v(t, \sigma) g_i(\sigma)\, d\sigma = \lambda_i g_i(t) \tag{4.5.12}$$

Note that N is assumed to be finite in Eq. (4.5.11), since if N were allowed to be infinite, all kernels could be considered separable. We can easily show in this case that

$$h(t) = \begin{cases} \dfrac{2}{N_0}\left[y(t) - \displaystyle\sum_{i=1}^{N} \frac{y_i \lambda_i}{N_0/2 + \lambda_i} g_i(t)\right] & t_0 \leq t \leq t_f \\ 0 & \text{elsewhere} \end{cases}$$

where

$$y(t) = \sum_{i=1}^{N} y_i g_i(t)$$

Some results for the solution of integral equations have used a state-variable model for the generation of the covariance kernel. The technique again is to obtain a set of differential equations from the integral equation which are then solved to obtain the required solution to the integral equation. The details are quite involved and the reader is referred to the literature [12–14].

4.6 DETECTION OF SIGNALS WITH UNKNOWN PARAMETERS

We have so far assumed in this chapter that the signals of concern were completely known. The only uncertainty was caused by additive noise. In many practical situations, however, this assumption is not valid. For example, in a radar tracking problem, the reflected pulse acquires a random phase. In such cases the signals would be functions of an unknown parameter vector. Thus for the binary detection problem considered in this chapter, the received signals under the two hypotheses would be given by

$$\begin{aligned} H_1&: z(t) = y_1(t, \boldsymbol{\theta}_1) + v(t) \qquad t_0 \leqslant t \leqslant t_f \\ H_0&: z(t) = y_0(t, \boldsymbol{\theta}_0) + v(t) \qquad t_0 \leqslant t \leqslant t_f \end{aligned} \tag{4.6.1}$$

where $\boldsymbol{\theta}_0$ and $\boldsymbol{\theta}_1$ denote the unknown parameters. We assume that the signals $y_1(t, \boldsymbol{\theta}_1)$ and $y_0(t, \boldsymbol{\theta}_0)$ are conditionally deterministic. The problem is the waveform counterpart to the composite hypothesis-testing problem in Chapter 3. Here we very briefly discuss the approach to solving such problems and defer until Chapter 8 some of the details.

The parameters $\boldsymbol{\theta}_i$ may be either random variables or nonrandom variables. In the former case, the associated probability density functions may or may not be known. The first case of known density functions is the one most frequently encountered in practice. The solution in this case is straightforward. We again expand the observations in a set of orthogonal basis functions and denote the coefficients of the K-term approximation by the K-dimensional vector $\mathbf{z}_K$. We then construct the likelihood ratio and let $K \rightarrow \infty$. To construct the likelihood ratio, we need the density functions $p(\mathbf{z}_K | H_1)$ and $p(\mathbf{z}_K | H_0)$. If the prior densities for $\boldsymbol{\theta}_i$ are known, we can calculate these two densities as

$$p(\mathbf{z}_K | H_1) = \int_{\chi_{\boldsymbol{\theta}_1}} p(\mathbf{z}_K | \boldsymbol{\theta}_1, H_1) p(\boldsymbol{\theta}_1 | H_1) \, d\boldsymbol{\theta}_1 \tag{4.6.2}$$

and

$$p(\mathbf{z}_K | H_0) = \int_{\chi_{\boldsymbol{\theta}_0}} p(\mathbf{z}_K | \boldsymbol{\theta}_0, H_0) p(\boldsymbol{\theta}_0 | H_0) \, d\boldsymbol{\theta}_0 \tag{4.6.3}$$

where $\chi_{\boldsymbol{\theta}_i}, i = 0, 1$ denotes the space of the parameter $\boldsymbol{\theta}_i$.

The likelihood ratio is then given as

$$\Lambda[z(t)] = \lim_{K \to \infty} \frac{p(\mathbf{z}_K | H_1)}{p(\mathbf{z}_K | H_0)} \tag{4.6.4}$$

While the procedure to be used is conceptually straightforward, the actual computation of the integrals depends on the form of the functions to be integrated.

If $\boldsymbol{\theta}_0$ and $\boldsymbol{\theta}_1$ are nonrandom variables, we may again try estimating the values of $\boldsymbol{\theta}_i$ under the two hypothesis H_1 and H_0, and use these estimates in the LRT. Using maximum likelihood estimates† of the parameters yields the generalized likelihood ratio

$$\Lambda_g[z(t)] = \lim_{K \to \infty} \frac{\max\limits_{\boldsymbol{\theta}_1} p(\mathbf{z}_K | \boldsymbol{\theta}_1)}{\max\limits_{\boldsymbol{\theta}_0} p(\mathbf{z}_K | \boldsymbol{\theta}_0)} \underset{H_0}{\overset{H_1}{\gtrless}} \gamma \tag{4.6.5}$$

which can be compared with a threshold γ to make a decision.

† See Section 5.4.

Example 4.6.1

Consider the on–off keying system of Example 3.2.1 in which a digital 1 is conveyed by transmitting a sinusoidal waveform of known amplitude and frequency and a digital zero by no transmission. The receiver observes the transmitted signal in the presence of additive zero-mean Gaussian white noise $v(t)$ of spectral density $N_0/2$. In the case of transmission over fading channels, the attenuation in the received signal can be modeled as a random variable. Further, the transmission undergoes a random change in phase. The observations under the two hypotheses can thus be modeled as

$$H_1\colon z(t) = \sqrt{\frac{2}{T}} A \cos(\omega_c t + \theta) + v(t) \qquad 0 \leq t \leq T$$

$$H_0\colon z(t) = v(t) \qquad 0 \leq t \leq T$$

where A and θ are assumed to be random variables with known prior distributions. We assume that there are an integral number of cycles corresponding to ω_c in the interval $[0, T]$.

Since the observations under H_0 do not depend on A and θ, instead of finding $p(\mathbf{z}_k \mid H_i)$ as in Eqs. (4.6.2) and (4.6.3), we can first obtain the conditional likelihood ratio $\Lambda(z \mid A, \theta)$ and average it over the joint density function of A and θ to find the likelihood ratio $\Lambda(z(t))$:

$$\Lambda(z(t)) = \int_{\chi_E}\int_{\chi_\theta} \Lambda(z(t) \mid a, \theta) p(a, \theta)\, da\, d\theta$$

where χ_A and χ_θ are the ranges of A and θ, respectively. To derive the conditional likelihood ratio, we expand $z(t)$ in an orthonormal basis. We choose as our first basis function

$$g_1(t) = \sqrt{\frac{2}{T}} \cos \omega_c t$$

The other orthonormal functions can be chosen arbitrarily. It is clear that only the coefficient z_1 in the expansion will depend on which hypothesis is true. Hence

$$Z_1 = \int_0^T z(t) g_1(t)\, dt$$

is a sufficient statistic. The conditional likelihood ratio can thus be written as

$$\Lambda(z(t) \mid a, \theta) = \frac{p(z_1 \mid a, \theta, H_1)}{p(z_1 \mid a, \theta, H_0)}$$

From our assumptions, it is clear that Z_1 is Gaussian under both hypotheses with moments given by

$$\mathrm{E}\{Z_1 \mid a, \theta, H_1\} = a \cos \theta \quad \text{and} \quad E\{Z_1 \mid a, \theta, H_0\} = 0$$

$$\operatorname{var}\{Z_1 \mid a, \theta, H_1\} = \operatorname{var}\{Z_1 \mid a, \theta, H_0\} = \frac{N_0}{2}$$

so that

$$\Lambda(z(t) \mid a, \theta) = \exp\left(\frac{2}{N_0} z_1 a \cos \theta\right) \exp\left(-\frac{1}{N_0} a^2 \cos^2 \theta\right)$$

We now average this conditional likelihood function over the joint density function $p(a, \theta)$. A frequently used channel model is the Rayleigh fading channel in which A and

θ are assumed to be independent with A having a Rayleigh density function and θ being uniformly distributed in $(0, 2\pi)$. Thus

$$p(a,\theta) = \begin{cases} \dfrac{1}{2\pi}\dfrac{2a}{\epsilon}\exp\left(\dfrac{-a^2}{\epsilon}\right) & a \geqslant 0, \quad 0 \leqslant \theta \leqslant 2\pi \\ 0 & \text{otherwise} \end{cases}$$

where ϵ is the average energy, $\epsilon = \mathrm{E}(A^2)$. To evaluate the likelihood function, we make the change of variables $L \underline{\Delta} A \cos\theta$. It then follows that

$$p(l) = \frac{1}{\sqrt{\pi\epsilon}}\exp\left(-\frac{l^2}{\epsilon}\right) \qquad -\infty < l < \infty$$

The likelihood ratio conditioned on the variable l is

$$\Lambda(z(t)\,|\,l) = \exp\left(\frac{2}{N_0}z_1 l - \frac{1}{N_0}l^2\right)$$

so that

$$\begin{aligned}\Lambda(z(t)) &= \int_{-\infty}^{\infty}\Lambda(z(t)\,|\,l)p(l)\,dl \\ &= \int_{-\infty}^{\infty}\frac{1}{\sqrt{\pi\epsilon}}\exp\left(-l^2\frac{\epsilon + N_0}{\epsilon N_0} + \frac{2}{N_0}z_1 l\right)dl \\ &= \frac{N_0}{\epsilon + N_0}\exp\left(\frac{\epsilon/N_0}{\epsilon + N_0}z_1^2\right)\end{aligned}$$

The decision rule is

$$\frac{N_0}{\epsilon + N_0}\exp\left(\frac{\epsilon/N_0}{\epsilon + N_0}z_1^2\right) \underset{H_0}{\overset{H_1}{\gtrless}} \gamma$$

where γ is a threshold depending on the criterion chosen. Taking logarithms and rearranging yields the decision rule in terms of a new threshold η as

$$z_1^2 \underset{H_0}{\overset{H_1}{\gtrless}} \eta$$

Thus the receiver can be implemented as a correlator followed by a square-law detector as shown in Fig. 4.6.1. The performance of the receiver can be derived following the procedure of Section 4.2 and is left as an exercise for the reader.

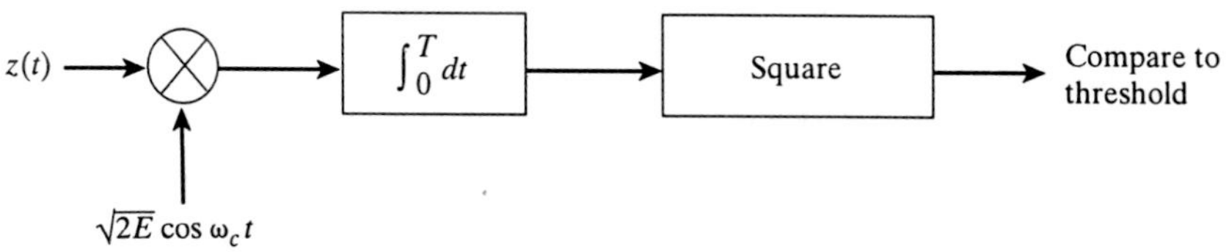

Figure 4.6.1 Receiver structure for Example 4.6.1.

Decision rules for the other communication schemes of Example 4.2.1 and their performance are discussed in the wider context of digital communications in Chapter 8. A similar situation also arises in detecting the presence of a fluctuating radar target and is discussed in Chapter 8.

4.7 SUMMARY

In this chapter we have extended our results on hypothesis testing to the problem of detecting waveforms in the presence of noise. In Section 4.2 we considered the problem of deciding between two known signal waveforms when the observation noise is additive white Gaussian. Since the results of Chapter 3 were for discrete random variables, to apply those results to the present case, the observations were expanded in a set of orthonormal basis functions. Since the observations can now be expressed in terms of the coefficients of the expansion, the likelihood ratio can be computed easily. The decision rule consists in correlating the observations with each of the signal waveforms and comparing the difference with a threshold. The correlation operation can be replaced by a matched filter whose impulse response is matched to the signal waveform. The performance of the receiver is independent of signal shape and depends only on ϵ, the average energy of the two signals, ρ, the correlation between the signals and N_0, the spectral density of the observation noise. For the minimum probability of error criterion, the performance is optimum if $\rho = -1$, that is, if the signals are maximally negatively correlated. The extension of these results to M-ary detection in white noise is straightforward. However, the performance of the receiver cannot be determined analytically and recourse must be had to numerical tables.

In Section 4.3 we considered the case of colored observation noise with known covariance function $\phi_\nu(t, u)$. One approach to obtaining the decision rule in this case is to pass the observations through a whitening filter whose impulse response can be predetermined. The function of the whitening filter is to convert the colored observation noise to white noise so that the results of Section 4.2 become applicable. The determination of the whitening filter, however, is in most cases, difficult. A formal solution can be obtained in terms of the eigenvalues and eigenfunctions of an integral equation with kernel $\phi_\nu(t, u)$. The solution of the integral equation is in many cases a formidable task.

A case in which the solution is determined in a relatively straightforward fashion is when the noise is stationary, with rational spectrum and the observation interval is semi-infinite. The whitening filter is then obtained from a spectral factorization of the noise spectrum.

The evaluation of the receiver performance is again not easy. For minimum probability of error, the correlation ρ is again -1. The receiver performance is, however, not independent of signal shape in this case. In fact, the optimum signal is the eigenfunction corresponding to the smallest eigenvalue.

A different criterion, the maximum output signal-to-noise ratio, was used in Section 4.4 to obtain the optimum linear time-invariant filter for the detection of known signals in noise. The receiver is again a filter matched to the signal.

While Section 4.5 was concerned with solution techniques for some integral equations of interest, in Section 4.6 we discussed the detection of signals with unknown parameters. Such problems arise, for example, in communication systems in which there is fading over the channel or in analyzing radar returns for target ranging and tracking. Many problems in pattern classification also fall in this category. The problem is the waveform counterpart of the composite hypothesis-testing problem of Chapter 3 and the techniques can be extended to this problem.

One topic that we have not discussed is the detection of stochastic signals in noise. In many applications, the observations, instead of consisting of a known signal observed in additive noise, must be modeled as a random signal in noise. Such instances arise, for example, in multipath transmission over a number of randomly varying (fading) paths. At the receiver the effect of these multiple transmissions can be modeled as a narrowband Gaussian signal. In radio astronomy or spectroscopy, the received signals will be very narrowband random signals. While we can obtain the receiver structure for this case by following an approach similar to the known signal case, the structure of the receiver is difficult to determine explicitly [1,2,14–16].

We recall that in the case of the detection of a known signal in white noise, the receiver correlates the observations with the known signal to generate the sufficient statistic. For the detection of a random signal, we might intuitively expect that the sufficient statistic may be generated by correlating an *estimate* of the signal with the observation. While we will discuss the estimation of random waveforms in Chapter 6, we note that for the detection of Gaussian signals in Gaussian white noise, the receiver does indeed have an estimator–correlator structure and that the estimate to be used is the minimum-mean-square estimate [16,17].

EXERCISES

4.1. Design a receiver to choose between the hypotheses

$$H_1\colon z(t) = y_1(t) + v(t)$$
$$H_0\colon z(t) = y_0(t) + v(t)$$

if the signals $y_1(t)$ and $y_0(t)$ are as shown in Fig. E.4.1. Use the minimum probability of error criterion. Assume that $v(t)$ is zero mean white Gaussian with spectral density $N_0/2$. Plot the probability of error as a function of $2\epsilon/N_0$. Assume equal prior probabilities for the two hypotheses.

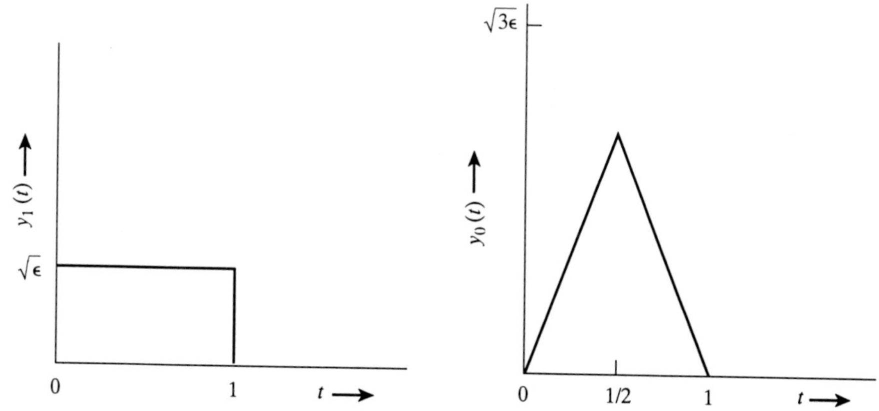

Figure E.4.1

4.2. In a ternary communication system, the transmitted signals over an interval $0 \leq t \leq T$ are as follows:

$$H_0: y_0(t) = 0$$

$$H_1: y_1(t) = A_1 \sin \omega_0 t$$

$$H_2: y_2(t) = A_2 \sin \omega_0 t,$$

where $\omega_0 = \dfrac{2\pi}{T}$. The signals are observed in white additive noise of spectral density $N_0/2$.

(a) Design a receiver that minimizes the probability of error.
(b) Find the probabilities of detection for the three hypotheses. Hence find the probability of a correct decision. Assume equal prior probabilities for the hypotheses.

4.3. One of two equally likely, equal-energy ($= E$) orthogonal signals is transmitted over an interval of $[0, T]$, over an AWGN channel of spectral density $N_0/2$. A minimum probability of error receiver is to be designed for detecting the signals.
(a) Find the probability of error in terms of E and N_0.

Due to a fault in the transmitter, when the signal to be transmitted is $y_i(t)$, $i = 1, 2$, the signal that is actually transmitted is found to be

$$y_{a_i}(t) = \epsilon y_j(t) + (1 - \epsilon) y_i(t) \qquad j \neq i, \quad 0 < t \leq T$$

Let us assume that it is known that the transmitter is faulty and that the receiver is designed for the actual transmitted signals $y_{a_i}(t)$.
(b) Find the energies in the actual transmitted signals $y_{a_i}(t)$, $i = 1, 2$, and the correlation coefficient between these two signals.
(c) Draw the block diagram of the minimum probability of error receiver. Find the probability of error in terms of E, ϵ, and N_0. What happens when $\epsilon = 0.5$?

4.4. One of two equally likely, equal-energy, orthogonal signals is transmitted over an interval of $[0, T]$, over an AWGN channel of spectral density $N_0/2$. A minimum probability of error receiver is to be designed for detecting the signals. Assume that $T = 1$ and that the two orthogonal signals are as shown in Fig. E.4.2.

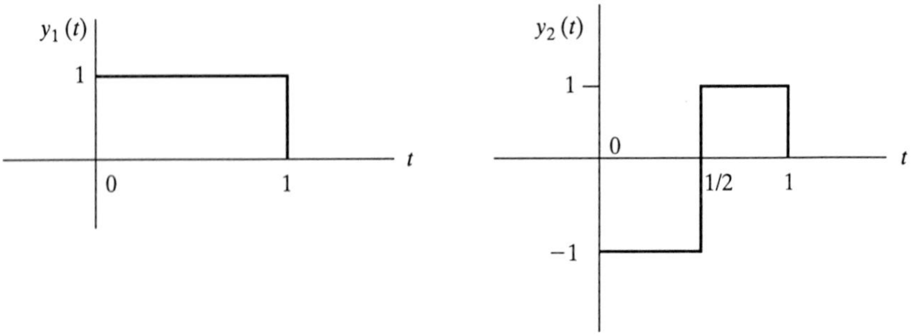

Figure E.4.2

(a) Find the receiver structure and write down the expression for the probability of error.
(b) Due to equipment failure, the actual transmission of each symbol does not last for the entire signaling interval of 1 second, but ends abruptly τ seconds earlier than

scheduled, where τ $(0 < \tau < 1)$ is assumed to be known (see Fig. E.4.3). If the receiver of part (a) is used, calculate the probability of error as a function of τ.

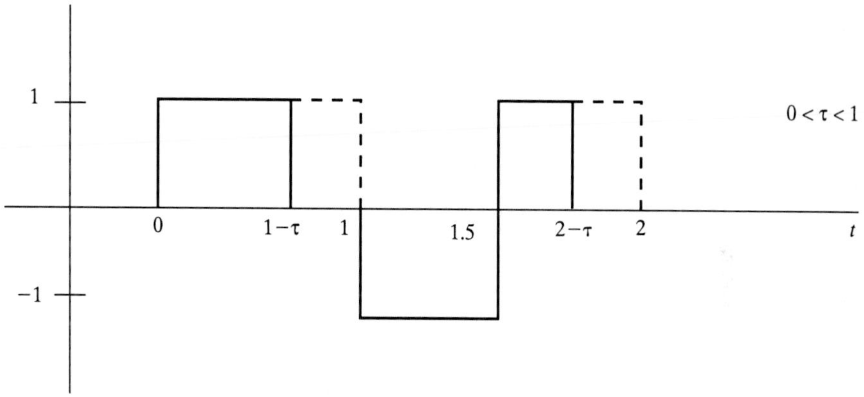

Figure E.4.3

(c) Assume that the receiver is designed based on the knowledge that each symbol signal is nonzero for $1 - \tau$ seconds and is zero after that. Calculate the probability of error for this receiver as a function of τ.

(d) Compare your answers in parts (b) and (c). Are the results what you might reasonably expect?

4.5. The signal space for a seven-signal set transmitted over an additive white Gaussian noise (AWGN) channel with equal prior probabilities is shown in Fig. E.4.4.

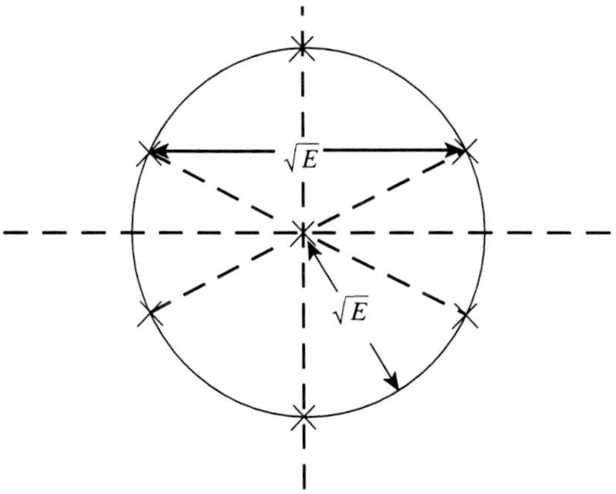

Figure E.4.4

(a) Determine the optimum decision regions if the criterion used is minimum probability of error.

(b) Find the probability of error.

4.6. One of eight equally likely signals is to be communicated over an AWGN channel with spectral density $N_0/2$. The vector representation of the signal set is as shown in Fig. E.4.5.

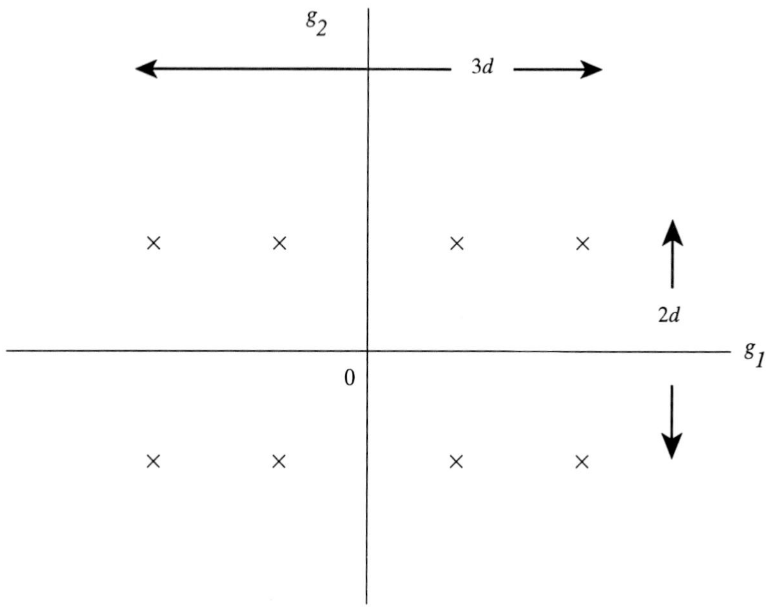

Figure E.4.5

(a) Draw the optimum decision regions for the minimum probability of error receiver.

(b) Find the corresponding probability of error. (*Hint:* To compute the probability of error, utilize the property that the conditional probability of a correct decision is not affected by a translation of the decision regions along with the transmitted signal.)

4.7. Design a filter that maximizes the output signal-to-noise ratio when the transmitted signal $y(t)$ is observed in additive white noise of spectral density $N_0/2$. The signal $y(t)$ is given by

$$y(t) = \begin{cases} e^{-t/2} - e^{3t/2} & 0 \leqslant t \leqslant T \\ 0 & \text{otherwise} \end{cases}$$

What is the maximum output signal-to-noise ratio? Assume that

$$\int_0^T y^2(t)\,dt = 1$$

4.8. Instead of using the optimum filter, we want to use a simplified filter for the detection problem of Exercise 4.7. The filter has impulse response

$$h(t) = \begin{cases} e^{-\alpha t} & t \geqslant 0 \\ 0 & t < 0 \end{cases}$$

(a) Find T so that the energy of $y(t) = 1$. Choose the parameter α to maximize the output signal-to-noise ratio.

(b) Compare the performance of this receiver with the results of Exercise 4.7.

(c) By how much should the transmitter energy be increased to match the performance of the optimum filter?

4.9. In the problem of Exercise 4.7, the signal $y(t)$ is observed in the presence of colored noise with the spectrum

$$\Phi_v(s) = N_0 \frac{1 - s^2}{2 - s^2}$$

Find the optimum matched filter and determine the maximum output signal-to-noise ratio.

4.10. Repeat Exercise 4.9 when the transmitted signal is

$$y(t) = \begin{cases} e^{-3t} & t \geq 0 \\ 0 & \text{otherwise} \end{cases}$$

and it is desired to maximize the output signal-to-noise ratio at some time $T > 0$. Determine the performance of the receiver as a function of T.

4.11. Consider the detection of a known signal in a mixture of white and colored noise. The observations are thus modeled as

$$H_1\text{: } z(t) = y(t) + v_1(t) + v_2(t) \qquad t_0 \leq t \leq t_f$$

$$H_0\text{: } z(t) = v_1(t) + v_2(t) \qquad t_0 \leq t \leq t_f$$

where $v_1(t)$ is zero-mean, Gaussian, white with spectral density $N_0/2$ and $v_2(t)$ is zero-mean, Gaussian with correlation function $\phi_{v_2}(\tau)$. Show that the test statistic can be chosen as

$$l = \int_{t_0}^{t_f} [z(t) - \tfrac{1}{2}y(t)]h(t)\,dt$$

where $h(t)$ satisfies the integral equation

$$h(t) + \frac{2}{N_0}\int_{t_0}^{t_f} h(\tau)\phi_{v_2}(t - \tau)\,d\tau = \frac{2}{N_0}y(t) \qquad t_0 \leq t \leq t_f$$

4.12. Consider the integral equation

$$\int_0^T h(\tau)\phi_v(\tau - \sigma)\,d\tau = g(\sigma) \qquad 0 \leq t \leq T$$

where

$$\phi_v(\tau) = \alpha e^{-\alpha|\tau|}$$

Let $g(t)$ be such that

$$\dot{g}(0) = \alpha g(0) \quad \text{and} \quad \dot{g}(T) = -\alpha g(T)$$

Find the solution to the integral equation.

4.13. Find the eigenvalues and eigenfunctions corresponding to the covariance kernel

$$\phi_v(t, u) = \sigma^2 \min(t, u)$$

by solving the integral equation

$$\lambda g(t) = \int_0^T \phi_x(t, u)g(u)\,du$$

Note that $x(t)$ is a Wiener process.

4.14. Derive Eq. (4.3.33) for optimum signal design.

4.15. In a binary communication system, the received signals are observed in additive noise with spectral density

$$\Phi_v(s) = \frac{a}{a - s^2} \qquad a > 0$$

Design the transmitted signals to yield minimum probability of error. What is P_e?

4.16. Assume that in Exercise 4.15, the noise has in addition a white component of spectral density $N_0/2$. Design the optimum signals and obtain P_e for this case.

4.17. Our discussions in Section 4.4 showed that for the detection of known signals in colored noise, the probability of error P_e was not independent of signal shape and that by appropriately choosing the transmitted waveforms, we could optimize P_e. In radar problems where a Neyman–Pearson criterion is used, we seek to maximize P_D for a given P_F level. Show how to choose the transmitted signals to maximize P_D. The problem of detection in colored noise in radar systems occurs when there is unwanted interference (clutter) present in the observations.

4.18. In certain applications involving band-limited signals, it is convenient to use the complex representation of signals discussed in Chapter 2. We consider the problem of detecting a known signal in the presence of bandlimited, white Gaussian noise so that the observations under the two hypotheses are

$$H_1\colon z(t) = y(t) + v(t)$$

$$H_0\colon z(t) = v(t)$$

In terms of the respective complex envelopes, we can write

$$H_1\colon \tilde{z}(t) = \tilde{y}(t) + \tilde{v}(t)$$

$$H_0\colon \tilde{z}(t) = \tilde{v}(t)$$

where

$$z(t) = \mathrm{Re}[\tilde{z}(t)\exp(j\omega_0 t)]$$

and similar relations hold for $\tilde{y}(t)$ and $\tilde{v}(t)$. Here ω_0 is the center frequency of the bandlimited signals. Show that a sufficient statistic for the decision problem is

$$\tilde{y}_1 = \int_0^T \tilde{z}(t)\tilde{y}(t)\,dt$$

Obtain the decision rule and determine P_e.

4.19. Extend the results of Exercise 4.18, to the colored observation noise case.

4.20. In many digital communication systems, orthogonal signals are preferred because of various reasons. Consider a binary system in which

$$\int_0^T y_1(t)y_0(t)\,dt = 0$$

and

$$\int_0^T y_i^2(t)\,dt = \epsilon \qquad i = 0, 1$$

The additive noise is colored with spectral density $\Phi_v(s)$. How will you design the signals for minimizing P_e?

4.21. For the digital communication problem discussed in Example 4.6.1, obtain the probability of error expression assuming equal prior probabilities.

4.22. In Example 4.6.1, assume that the phase θ is exactly known. Derive the optimum receiver and its performance.

4.23. In Example 4.6.1, we assumed a Rayleigh fading channel. This is also a valid model in certain radar situations where the aspect of the target is changing. Derive the P_F and P_D expressions and sketch the ROC.

4.24. In Example 4.6.1 we assumed additive Gaussian white noise. Suppose that the noise is colored with spectral density

$$\Phi_v(s) = \frac{-s^2}{1 - s^2}$$

Can you derive the optimum receiver? What is its performance?

4.25. In radar applications the transmitted waveform is often a segment of a sinusoidal signal of known frequency. The frequency of the return signal from the target, however, is shifted by an amount called the *Doppler frequency*, which is approximately equal to $\omega_d = 2v\omega_0/c$, where v is the radial velocity of the target and c is the velocity of propagation of the electromagnetic wave. For targets of unknown velocity, ω_d can be modeled as a random variable with known density function. The received observations corresponding to whether a target is present or not will thus be

$$H_1: z(t) = \sqrt{2E}\,\sin(\omega_d t + \theta) + v(t) \qquad 0 \leqslant t \leqslant T$$

$$H_0: z(t) = v(t) \qquad 0 \leqslant t \leqslant T$$

where E and θ are assumed known. Determine the likelihood ratio test for this problem under the assumption that $v(t)$ is zero mean, Gaussian, white.

4.26. The range (distance from transmitter) of a radar target is determined from the delay between the transmitted and return signals. For targets of unknown range, the delay will be a random variable. The decision problem in this case can be modeled as

$$H_1: z(t) = y(t - \tau) + v(t) \qquad 0 \leqslant t \leqslant T$$

$$H_0: z(t) = v(t) \qquad 0 \leqslant t \leqslant T$$

where $y(t)$ is a known waveform and τ is the unknown delay. Derive the likelihood ratio test for this problem under the assumption that $v(t)$ is zero mean, white, Gaussian, and τ is uniformly distributed in $[\tau_0, \tau_1]$.

REFERENCES

1. C. W. Helstrom, *Statistical Theory of Signal Detection*, Pergamon Press, Oxford, 1968.
2. H. L. Van Trees, *Detection, Estimation and Modulation Theory*, Part I, Wiley, New York, 1968.
3. A. D. Whalen, *Detection of Signals in Noise*, Academic Press, New York, 1971.

4. H. W. Bode and C. E. Shannon, A simplified derivation of linear least-square smoothing and prediction theory, *Proc. IRE*, **38**, April, 417–425 (1950).
5. J. M. Wozencraft and I. M. Jacobs, *Principles of Communication Engineering*, Wiley, New York, 1965.
6. D. O. North, Analysis of the factors which determine signal/noise discrimination in radar, *Report PTR-6C*, RCA Laboratories, Princeton, NJ (June 1943); also in *Proc. IEEE*, **51**, July (1963).
7. R. Courant and D. Hilbert, *Methods of Mathematical Physics*, Vol. I, English Translation, Interscience, New York, 1953.
8. F. Smithies, *Integral Equations*, Cambridge University Press, London, 1965.
9. L. A. Zadeh and J. R. Ragazzini, Optimum filters for the detection of signals in noise, *Proc. IRE*, **40**, 1233 (1952).
10. K. S. Miller and L. A. Zadeh, Solution of an integral equation occurring in the expansion of second-order stationary random functions, *IRE Trans. Inf. Theory*, **IT-2**, 72 (1956).
11. D. Youla, The solution of a homogeneous Wiener–Hopf integral equation occurring in the expansion of second-order stationary random functions, *IRE Trans. Inf. Theory*, **IT-3**, 187 (1957).
12. A. Baggeroer, A state variable approach to the solution of Fredholm integral equations, *IEEE Trans. Inf. Theory*, **IT-15**, No. 5, September 556–569 (1969).
13. H. L. Van Trees, *Detection, Estimation and Modulation Theory*, Part II, Wiley, New York, 1971.
14. D. Middleton, On the detection of stochastic signals in additive normal noise, *IRE Trans. Inf. Theory*, **IT-3**, No. 2, June, 86 (1957).
15. T. Kailath, Solution of an integral equation occurring in multipath communication problems, *IRE Trans. Inf. Theory*, **IT-6**, No. 3, June, 412 (1960).
16. T. T. Kadota and L. A. Shepp, On the best finite set of linear observables for discriminating two Gaussian signals, *IEEE Trans. Inf. Theory*, **IT-13**, No. 2, April, 278–284 (1967).
17. T. Kailath, Likelihood ratios for Gaussian processes. *IEEE Trans. Inf. Theory*, **IT-16**, No. 3, May, 276–288 (1970).

5

Estimation Theory

5.1 INTRODUCTION

In Chapters 3 and 4, we were concerned primarily with the detection of signals. We wished to know whether or not a signal was present and were not explicitly concerned with determining the signal waveform or parameters associated with it. In many instances, however, we need to know these parameters. For example, in radar tracking, the time at which the return pulse arrives at the transmitter is of interest in determining the range of the target. In a gun control problem, it is necessary not only to estimate the present position and heading of the target, but also to predict its future position.

As in Chapters 3 and 4, we assume that we have available a received waveform which consists of the transmitted signal (which may be distorted in the transmission medium) and local receiver noise. Based on either a finite number of samples or a continuous observation of the received signal, we wish to estimate either the transmitted signal or parameters associated with it. We start with a discussion of the estimation of parameters in this chapter and defer the estimation of waveforms until Chapter 6.

5.2 ESTIMATION OF PARAMETERS

We start our discussion by considering a simple example.

Example 5.2.1

A generator is available in which the voltage θ is some value between 0 and 10 volts. We wish to determine this value based on a measurement that is corrupted by additive noise. The observed variable is thus

$$Z = \theta + V$$

where V is modeled as an independent zero-mean Gaussian random variable. The problem is to observe z and estimate θ.

The probability density governing the observation process is

$$p(z \mid \theta) = \frac{1}{\sqrt{2\pi}\,\sigma_v} \exp\left[-\frac{(z-\theta)^2}{2\sigma_v^2}\right]$$

There are several ways in which an estimate may be found. If we treat θ as a random variable, then, as we did in Chapter 3, we may choose as our basis for determining the estimate, the maximum a posteriori (MAP) criterion. The approach in this case would be to form the posterior density function $p(\theta \mid z)$ and determine the value of θ that maximizes this density.

On the other hand, if we treat θ as a constant but unknown value, we may seek to determine θ by maximizing the likelihood function $p(z \mid \theta)$. As mentioned in Chapter 2, such an estimate is called a *maximum likelihood estimate* (the estimate can be interpreted as that value of θ which most likely gave rise to the observed quantity z).

The example brings out the structure of the estimation problem in the following manner. The first component is the source, whose output is a parameter that can be regarded as a point in parameter space. The parameter may be either a random variable or a nonrandom, unknown quantity. The second component is the probabilistic transition mechanism that governs the effect of the parameter on the observation. The third component is the observation space. The estimation rule then provides a mapping of the observation space into an estimate. This is shown in Fig. 5.2.1.

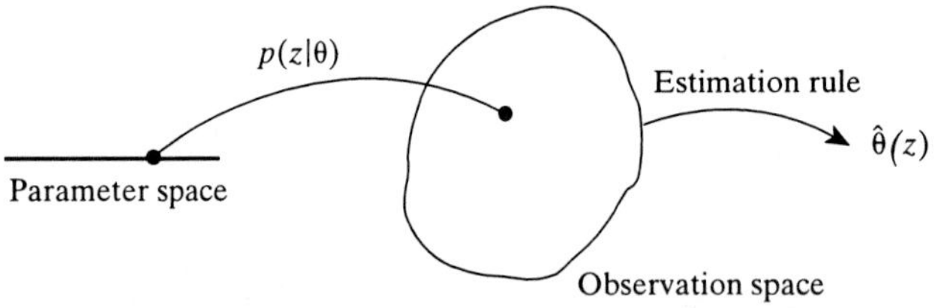

Figure 5.2.1 Estimation model.

We note that the first three components are similar to the decision problem. We develop suitable estimation rules in this chapter for both random and nonrandom variables and discuss some of the properties of such estimators. We start with the random parameter case.

5.3 RANDOM PARAMETERS: BAYES ESTIMATES

In the detection problem we saw that we had to specify two quantities, the set of costs C_{ij} and the prior probabilities $P(H_i)$. For the estimation problem, if we designate the estimate of a parameter θ by $\hat{\theta}(z)$, we have to assign a nonnegative cost to all pairs $[\theta, \theta(z)]$ over the range of interest. For most problems of interest, it is usual to choose the cost $C(\theta, \hat{\theta}(z))$ as a function of the parameter estimation error $\tilde{\theta}(z)$ defined as

$$\tilde{\theta}(z) = \theta - \hat{\theta}(z) \tag{5.3.1}$$

so that we can write the cost as $C(\tilde{\theta}(z))$.

In the preceding discussion we have denoted the estimate $\hat{\theta}$ and the error $\tilde{\theta}$ as functions of z since they are explicitly dependent on the observations. Normally, in the sequel we suppress this dependence and write $\hat{\theta}$ and $\tilde{\theta}$ except where clarity requires it.

Some typical cost functions are shown in Fig. 5.3.1. They correspond to the cases

$$
\begin{aligned}
C(\theta, \hat{\theta}) &= (\theta - \hat{\theta})^2 && \text{squared error} \\
C(\theta, \hat{\theta}) &= |\theta - \hat{\theta}| && \text{absolute value of error} \\
C(\theta, \hat{\theta}) &= \begin{cases} 1 & |\theta - \hat{\theta}| \geqslant \dfrac{\Delta}{2} \\ 0 & |\theta - \hat{\theta}| < \dfrac{\Delta}{2} \end{cases} && \text{uniform cost}
\end{aligned}
\tag{5.3.2}
$$

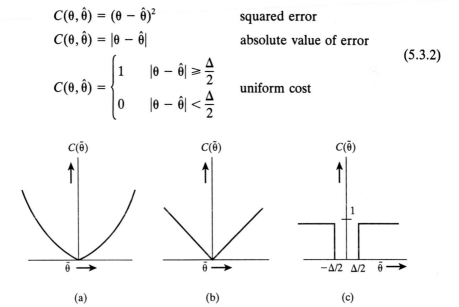

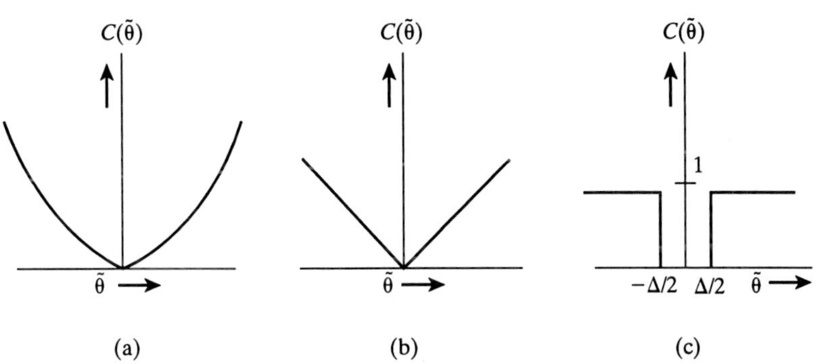

Figure 5.3.1 Cost functions: (a) squared error; (b) absolute value; (c) uniform cost.

In a given situation, we choose a cost function on the basis of the use to which the results will be put and on the tractability of the resulting problem. The first criterion is somewhat subjective and the choice of the cost function, as with all optimization problems, is usually a compromise between the goals set by the designer and the resulting complexity in obtaining the solution. Fortunately, in many problems of interest the optimum estimate is usually insensitive to the choice of the criterion. The same estimate will be optimum for a large class of cost functions.

Corresponding to the prior probabilities in the detection problem, we have a prior density $p(\theta)$ for the random parameter estimation problem. The density function $p(\theta)$ is assumed to be known.

Once we have specified the cost function and the prior density, we may write an expression for the average cost or risk as

$$\overline{C} = \int_{-\infty}^{\infty}\int_{-\infty}^{\infty} C(\theta - \hat{\theta}(z))p(z, \theta)\, dz\, d\theta \tag{5.3.3}$$

We want to choose $\hat{\theta}$ so as to minimize the average cost. We do this by making use of Bayes rule to write $\overline{C}$ in terms of two integrals only one of which depends on $\hat{\theta}$ as

$$\overline{C} = \int_{-\infty}^{\infty} \left[\int_{-\infty}^{\infty} C(\theta - \hat{\theta}(z)) p(\theta \mid z) \, d\theta \right] p(z) \, dz \tag{5.3.4}$$

It is clear in the preceding equation that the inner integral and $p(z)$ are nonnegative. Therefore, we can minimize $\overline{C}$ by minimizing the inner integral with respect to $\hat{\theta}$. Thus the optimal estimate is obtained by minimizing the conditional risk

$$\overline{C}(\hat{\theta} \mid z) = \int_{-\infty}^{\infty} C(\theta - \hat{\theta}(z)) p(\theta \mid z) \, d\theta \tag{5.3.5}$$

Let us now determine the estimates for the typical cost functions of Fig. 5.3.1. For the squared error cost, if we denote the conditional risk as $\overline{C}_{\text{ms}}$, Eq. (5.3.5) becomes

$$\overline{C}_{\text{ms}}(\hat{\theta} \mid z) = \int_{-\infty}^{\infty} (\theta - \hat{\theta})^2 p(\theta \mid z) \, d\theta \tag{5.3.6}$$

We find the optimum estimate $\hat{\theta}_{\text{ms}}$ by differentiating Eq. (5.3.6) with respect to $\hat{\theta}$ and setting the result to zero:

$$\begin{aligned} &\frac{d}{d\hat{\theta}} \int_{-\infty}^{\infty} (\theta - \hat{\theta})^2 p(\theta \mid z) \, d\theta \\ &= -2 \int_{-\infty}^{\infty} \theta p(\theta \mid z) \, d\theta + 2\hat{\theta}(z) \int_{-\infty}^{\infty} p(\theta \mid z) \, d\theta \\ &= 0 \end{aligned} \tag{5.3.7}$$

We therefore have

$$\hat{\theta}_{\text{ms}}(z) = \int_{-\infty}^{\infty} \theta p(\theta \mid z) \, d\theta \tag{5.3.8}$$

where we have used the fact that $\int_{-\infty}^{\infty} p(\theta \mid z) \, d\theta$ equals 1.

This represents a unique minimum since the second derivative is positive. The term on the right side of Eq. (5.3.8) represents the mean $\text{E}(\theta \mid z)$ of the posterior density. The mean-square estimate is therefore sometimes referred to as the conditional mean estimate.

To find the Bayes estimate for the absolute value cost function, we write Eq. (5.3.5) as

$$\begin{aligned} \overline{C}_{\text{abs}}(\hat{\theta} \mid z) &= \int_{-\infty}^{\infty} |\tilde{\theta}(z)| p(\theta \mid z) \, d\theta \\ &= \int_{-\infty}^{\hat{\theta}(z)} (-\theta + \hat{\theta}(z)) p(\theta \mid z) \, d\theta + \int_{\hat{\theta}(z)}^{\infty} (-\hat{\theta}(z) + \theta) p(\theta \mid z) \, d\theta \end{aligned} \tag{5.3.9}$$

Differentiating both sides with respect to $\hat{\theta}(z)$ and setting the result to zero gives

$$\int_{-\infty}^{\hat{\theta}_{\text{abs}}(z)} p(\theta \mid z) \, d\theta = \int_{\hat{\theta}_{\text{abs}}(z)}^{\infty} p(\theta \mid z) \, d\theta \tag{5.3.10}$$

Thus the estimate $\hat{\theta}_{abs}(z)$ is the median of the conditional density $p(\theta \mid z)$. For the uniform cost function, the conditional risk becomes

$$\overline{C}_{unf}(\hat{\theta} \mid z) = 1 - \int_{\hat{\theta}_{unf}-\Delta/2}^{\hat{\theta}_{unf}+\Delta/2} p(\theta \mid z)\, d\theta \tag{5.3.11}$$

To minimize the conditional risk, we need to maximize the integral on the right-hand side. For small Δ we see that the best choice for $\hat{\theta}$ is the value of θ at which the posterior density has its maximum. Thus the estimate corresponds to the maximum posterior† (MAP) estimate discussed in Section 5.2 and is given by the mode of the posterior density.

For most densities of interest, we can find $\hat{\theta}_{map}$ as a solution to the equation

$$\left. \frac{\partial \ln p(\theta \mid z)}{\partial \theta} \right|_{\theta = \hat{\theta}_{map}(z)} = 0 \tag{5.3.12}$$

Equation (5.3.12) is referred to as the *MAP equation*. We may rewrite Eq. (5.3.12) to separate the role of the observation z and the prior knowledge. We thus write

$$p(\theta \mid z) = \frac{p(z \mid \theta) p(\theta)}{p(z)} \tag{5.3.13}$$

Taking logarithms on both sides yields

$$\ln p(\theta \mid z) = \ln p(z \mid \theta) + \ln p(\theta) - \ln p(z)$$

Differentiating with respect to θ yields the MAP equation

$$\left. \frac{\partial \ln p(z \mid \theta)}{\partial \theta} \right|_{\theta = \hat{\theta}(z)} + \left. \frac{\partial \ln p(\theta)}{\partial \theta} \right|_{\theta = \hat{\theta}(z)} = 0 \tag{5.3.14}$$

The first term gives the dependence on z and the second corresponds to the prior knowledge.

We have seen that given an observation z, the Bayes estimators corresponding to the squared error, the absolute error, and the uniform cost function are, respectively, the mean, the median, and the mode of the posterior density $p(\theta \mid Z = z)$. The posterior density combines the prior information about the parameter θ contained in $p(\theta)$ with the information contained in the sample (observation) z in the form of the likelihood function $p(z \mid \theta)$. If a sufficient statistic $t(z)$ exists, then by an application of the factorization theorem (see Chapter 2), we can see that the posterior density $p(\theta \mid Z = z)$ depends on the observation only through the sufficient statistic $t(z)$. Therefore, any of the Bayes estimators depends on the observation only through $t(z)$. That is, a Bayes estimator is a function of the sufficient statistic.

Bayes estimation is treated in several statistics and electrical engineering texts (see, for example, [1–4]). We shall be concerned primarily with mean square and MAP in the sequel.

† Maximum posterior and maximum a posteriori both refer to the same estimate.

Example 5.3.1

We consider the problem of estimating a parameter Θ observed in additive noise. The observations are given by

$$Z_i = \Theta + V_i \qquad i = 1, \ldots, N$$

We assume that the V_i are independent, identically distributed Gaussian random variables with zero mean and variance σ_v^2. We also assume that Θ is Gaussian, zero mean, and with variance σ_θ^2.

It is clear from our previous discussions that the first step in determining the estimate is to find the relevant probability density function $p(\theta \mid \mathbf{z})$, where $\mathbf{z}$ denotes the set of observations $\{z_1, z_2, \ldots, z_N\}$. Now

$$p(\theta \mid \mathbf{z}) = \frac{p(\mathbf{z} \mid \theta) p(\theta)}{p(\mathbf{z})} = \frac{p(\mathbf{z} \mid \theta) p(\theta)}{\int_{\chi_\theta} p(\mathbf{z} \mid \theta) p(\theta)\, d\theta}$$

Because of the form of the observation equations, we have

$$p(\mathbf{z} \mid \theta) = \prod_{i=1}^{N} p(z_i \mid \theta) = \prod_{i=1}^{N} \frac{1}{\sqrt{2\pi}\,\sigma_v} \exp\left(-\frac{(z_i - \theta)^2}{2\sigma_v^2}\right)$$

Also,

$$p(\theta) = \frac{1}{\sqrt{2\pi}\,\sigma_\theta} \exp\left(-\frac{\theta^2}{2\sigma_\theta^2}\right)$$

The computation of $p(\mathbf{z})$ can tend to be algebraically tedious. We note, instead, that $p(\mathbf{z})$ is just a normalizing constant, which causes

$$\int_{-\infty}^{\infty} p(\theta \mid \mathbf{z})\, d\theta = 1$$

Substituting for $p(\mathbf{z} \mid \theta)$ and $p(\theta)$ thus yields

$$p(\theta \mid \mathbf{z}) = \frac{\left(\prod_{i=1}^{N} \frac{1}{\sqrt{2\pi}\,\sigma_v}\right) \frac{1}{\sqrt{2\pi}\,\sigma_\theta}}{p(\mathbf{z})} \exp\left\{-\frac{1}{2}\left[\sum_{i=1}^{N} \left(\frac{z_i - \theta}{\sigma_v}\right)^2 + \frac{\theta^2}{\sigma_\theta^2}\right]\right\}$$

Let

$$\sigma_m^2 = \frac{\sigma_\theta^2 \sigma_v^2}{N\sigma_\theta^2 + \sigma_v^2}$$

We can complete the square in the exponent by multiplying with $\exp\left[\pm \frac{1}{2}\left(\frac{\sigma_m^2}{\sigma_v^2} \sum_{i=1}^{N} z_i\right)^2\right]$. We can then write $p(\theta \mid \mathbf{z})$ as

$$p(\theta \mid \mathbf{z}) = k(\mathbf{z}) \exp\left\{-\frac{1}{2\sigma_m^2}\left[\theta - \frac{\sigma_m^2}{\sigma_v^2}\left(\sum_{i=1}^{N} z_i\right)\right]^2\right\}$$

where $k(\mathbf{z})$ denotes all the terms not explicitly involving θ. We see that $p(\theta \mid \mathbf{z})$ is just a Gaussian density. Thus the mean, mode, and the median of the density are the same, implying that

$$\hat{\theta}_{ms}(\mathbf{z}) = \hat{\theta}_{MAP}(\mathbf{z}) = \hat{\theta}_{abs}(\mathbf{z})$$

The estimate is given by

$$\hat{\theta}_{ms} = \frac{\sigma_m^2}{\sigma_v^2}\left(\sum_{i=1}^{N} z_i\right)$$

$$= \frac{\sigma_\theta^2}{\sigma_\theta^2 + \sigma_v^2/N}\left(\frac{1}{N}\sum_{i=1}^{N} z_i\right)$$

We note that the observations enter into the posterior density only through their sum. From our results on exponential densities in Chapter 2, it is clear that

$$l(\mathbf{z}) = \sum_{i=1}^{N} z_i$$

is a sufficient statistic.

Example 5.3.2

Suppose now our observations are of the form

$$Z_i = y(\Theta) + V_i$$

where $y(\cdot)$ is a nonlinear function of θ. As before, we will have

$$p(\theta \mid \mathbf{z}) = \frac{\left(\prod_{i=1}^{N} \frac{1}{\sqrt{2\pi}\,\sigma_v}\right)\frac{1}{\sqrt{2\pi}\,\sigma_\theta}}{p(\mathbf{z})} \exp\left\{-\frac{1}{2}\left[\sum_{i=1}^{N}\frac{[z_i - y(\theta)]^2}{\sigma_v^2} + \frac{\theta^2}{\sigma_\theta^2}\right]\right\}$$

The determination of the conditional mean in this case will be quite complicated since the posterior density is no longer Gaussian. In this case $\hat{\theta}_{MAP}$ will not in general be the same as $\hat{\theta}_{ms}$.

Example 5.3.3

As a final example in this section, we consider a binary communication problem. We assume that the signals corresponding to the two hypotheses H_1 and H_0 are $+1$ and -1, respectively. We are required to estimate the value of the signal based on a single observation. We thus have

$$Z = Y + V$$

where V is assumed $N(0, \sigma_v^2)$. If we assume the prior probabilities for the two hypotheses to be the same, we have

$$p(y) = 0.5\delta(y - 1) + 0.5\delta(y + 1)$$

Since

$$p(z \mid y) = \frac{1}{\sqrt{2\pi}\,\sigma_v}\exp\left[-\frac{(z - y)^2}{2\sigma_v^2}\right]$$

we can compute $p(z)$ as

$$p(z) = \int_{-\infty}^{\infty} p(z \mid y) p(y)\, dy$$

$$= \int_{-\infty}^{\infty} \frac{1}{\sqrt{2\pi}\,\sigma_v} \exp\left[-\frac{(z-y)^2}{2\sigma_v^2}\right] [0.5\delta(y-1) + 0.5\delta(y+1)]\, dy$$

$$= \frac{1}{2\sqrt{2\pi}\,\sigma_v}\left[\exp\left(-\frac{(z-1)^2}{2\sigma_v^2}\right) + \exp\left(-\frac{(z+1)^2}{2\sigma_v^2}\right)\right]$$

The posterior density is given by

$$p(y \mid z) = \frac{p(z \mid y) p(y)}{p(z)}$$

$$= \frac{\exp\left[-\dfrac{(z-y)^2}{2\sigma_v^2}\right]\{\delta(y-1) + \delta(y+1)\}}{\exp\left[-\dfrac{(z-1)^2}{2\sigma_v^2}\right] + \exp\left[-\dfrac{(z+1)^2}{2\sigma_v^2}\right]}$$

Obviously, to maximize $p(y \mid z)$, we have to choose $\hat{y} = \pm 1$, as otherwise the density is zero. Since minimizing $|z - y|$ maximizes the density, we choose $\hat{y}$ to be the value closest to z. We therefore have

$$\hat{y}_{\text{MAP}} = \operatorname{sgn} z$$

where

$$\operatorname{sgn} z = \begin{cases} +1 & \text{if } z \geqslant 0 \\ -1 & \text{if } z < 0 \end{cases}$$

To find the mean-square estimate, we find the mean of the posterior density. Thus

$$\hat{y}_{\text{ms}} = \int_{-\infty}^{\infty} y p(y \mid z)\, dy$$

$$= \int_{-\infty}^{\infty} y \frac{\exp\left[-\dfrac{(z-y)^2}{2\sigma_v^2}\right]\{\delta(y-1) + \delta(y+1)\}}{\exp\left[-\dfrac{(z-1)^2}{2\sigma_v^2}\right] + \exp\left[-\dfrac{(z+1)^2}{2\sigma_v^2}\right]}\, dy$$

$$= \frac{\exp\left(\dfrac{z}{\sigma_v^2}\right) - \exp\left(-\dfrac{z}{\sigma_v^2}\right)}{\exp\left(\dfrac{z}{\sigma_v^2}\right) + \exp\left(\dfrac{-z}{\sigma_v^2}\right)} = \tanh \frac{z}{\sigma_v^2}$$

Thus the MAP and minimum-mean-square estimates are not equivalent.

Exercise 5.3 considers the extension of these results to multiple observations.

5.3.1 Invariance of Estimators

We have seen from the preceding examples that different cost criteria yield different estimates. If an estimate could be found that was optimum for several cost functions, this would be desirable because of the subjective judgments used in choosing $C(\tilde{\theta})$.

As we shall establish in the sequel, the minimum-mean-square estimate possesses this property under certain conditions [3, 5].

We will assume that the cost function we are considering is a symmetric convex-upward function and that the posterior density is symmetric about its conditional mean. Hence we have

$$C(\tilde{\theta}) = C(-\tilde{\theta}) \qquad \text{symmetry} \tag{5.3.15a}$$

$$C(b\theta_1 + (1-b)\theta_2) \leqslant bC(\theta_1) + (1-b)C(\theta_2) \qquad \text{convexity} \tag{5.3.15b}$$

for any $b \in (0, 1)$ and for all θ_1 and θ_2

$$p(\theta - \hat{\theta}_{\text{ms}} \,|\, z) = p(\hat{\theta}_{\text{ms}} - \theta \,|\, z) \qquad \text{symmetry about conditional mean} \tag{5.3.15c}$$

We obtain the estimate for this cost function by minimizing the conditional risk

$$\overline{C}(\hat{\theta} \,|\, z) = \mathrm{E}\{C(\theta - \hat{\theta}) \,|\, z\} \tag{5.3.16}$$

Let $\delta = \theta - \hat{\theta}_{\text{ms}}$. Then we can write

$$\begin{aligned} \overline{C}(\hat{\theta} \,|\, z) &= \int_{-\infty}^{\infty} C(\delta + \hat{\theta}_{\text{ms}} - \hat{\theta}) p(\delta \,|\, z)\, d\delta \\ &= \int_{-\infty}^{\infty} C(-\delta - \hat{\theta}_{\text{ms}} + \hat{\theta}) p(\delta \,|\, z)\, d\delta \qquad \text{by (5.3.15a)} \\ &= \int_{-\infty}^{\infty} C(\delta - \hat{\theta}_{\text{ms}} + \hat{\theta}) p(\delta \,|\, z)\, d\delta \qquad \text{by (5.3.15c)} \end{aligned} \tag{5.3.17}$$

We can therefore, by using the first and third expressions on the right side of Eq. (5.3.17), write for the conditional risk

$$\begin{aligned} \overline{C}(\hat{\theta} \,|\, z) &= \int_{-\infty}^{\infty} [\tfrac{1}{2}C(\delta + \hat{\theta}_{\text{ms}} - \hat{\theta}) + \tfrac{1}{2}C(\delta - \hat{\theta}_{\text{ms}} + \hat{\theta})] p(\delta \,|\, z)\, d\delta \\ &\geqslant \int_{-\infty}^{\infty} C(\tfrac{1}{2}(\delta + \hat{\theta}_{\text{ms}} - \hat{\theta}) + \tfrac{1}{2}(\delta - \hat{\theta}_{\text{ms}} + \hat{\theta})) p(\delta \,|\, z)\, d\delta \\ &\qquad\qquad \text{by Eq. (5.3.15b)} \\ &= \overline{C}[\hat{\theta}_{\text{ms}} \,|\, z] \end{aligned} \tag{5.3.18}$$

Equality will be achieved if $\hat{\theta} = \hat{\theta}_{\text{ms}}$.

We have thus established that the minimum-mean-square estimate $\hat{\theta}_{\text{ms}}$ is also the optimal estimate for all cost functions satisfying Eq. (5.3.15).

For nonconvex cost functions we require the further condition that

$$\lim_{x \to \infty} C(x) p(x \,|\, z) = 0 \tag{5.3.19}$$

The proof is similar to the one for the case of convex cost functions.

5.4 ESTIMATION OF NONRANDOM PARAMETERS

If the parameter we are seeking to estimate is a nonrandom but unknown constant, we would not be able to define a prior probability density for the parameter and hence could not use a Bayes estimator. In this section we consider two types of estimates for nonrandom parameters. The first is a *point estimate*, a function of the observation vector, which assumes a specific value for any given set of observations. The specific estimate we consider is the maximum likelihood estimate. The second, an *interval estimate*, defines an interval in which the parameter is expected to lie.

5.4.1 Maximum Likelihood Estimation

We recall that the maximum likelihood estimate is obtained by maximizing the likelihood function $p(z \mid \theta)$ and can be interpreted as that value of θ that most likely gave rise to the observation z. This estimate can also be used when the parameter is random but has uniform prior density (called the *noninformative prior*).

In cases where the likelihood function is smooth, the MLE can be obtained as a solution to the likelihood equation

$$\left.\frac{\partial \ln p(\mathbf{z} \mid \theta)}{\partial \theta}\right|_{\theta=\hat{\theta}_{\mathrm{ml}}} = 0 \tag{5.4.1}$$

In our discussion of the maximum posterior estimate, we found that we had to solve the equation

$$\left.\frac{\partial p(\theta \mid z)}{\partial \theta}\right|_{\theta=\hat{\theta}_{\mathrm{MAP}}} = 0 \tag{5.4.2}$$

Since

$$p(\theta \mid z) = \frac{p(z \mid \theta)\, p(\theta)}{p(z)} \tag{5.4.3}$$

we required the prior density $p(\theta)$ to be available. In terms of the logarithms of the density functions, the MAP equation was given by

$$\left[\frac{\partial \ln p(z \mid \theta)}{\partial \theta} + \frac{\partial \ln p(\theta)}{\partial \theta}\right]_{\theta=\hat{\theta}_{\mathrm{MAP}}} = 0 \tag{5.4.4}$$

If the density function $p(\theta)$ is uniform, the second term on the left side of Eq. (5.4.4) will be zero. Since the first term on the left side of Eq. (5.4.4) is the same as the term on the left side of Eq. (5.4.1), it is clear that in this case the MAP and the maximum likelihood estimates will be the same.

Example 5.4.1

We determine a maximum likelihood estimator for the problem of Example 5.3.1, where we have

$$p(\mathbf{z} \mid \theta) = \prod_{i=1}^{N} \frac{1}{\sqrt{2\pi}\,\sigma_v} \exp\left[-\frac{(z_i - \theta)^2}{2\sigma_v^2}\right]$$

Taking logarithms and differentiating we have

$$\frac{\partial \ln p(\mathbf{z}|\theta)}{\partial \theta} = \frac{1}{\sigma_v^2}\left(\sum_{i=1}^{N} z_i - N\theta\right)$$

Thus

$$\hat{\theta}_{\text{ml}} = \frac{1}{N}\sum_{i=1}^{N} z_i$$

The estimate is the sample mean of the observations.

Example 5.4.2

We consider the problem of Example 5.3.2. The likelihood equation for this problem is

$$\left.\frac{\partial \ln p(\mathbf{z}|\theta)}{\partial \theta}\right|_{\theta=\hat{\theta}_{\text{ml}}} = \frac{1}{\sigma_v^2}\sum_{i=1}^{N}[z_i - y(\theta)]\left.\frac{\partial y(\theta)}{\partial \theta}\right|_{\theta=\hat{\theta}_{\text{ml}}} = 0$$

or

$$\left[\frac{\partial y(\theta)}{\partial \theta}\frac{1}{\sigma_v^2}\right]\left[\frac{1}{N}\sum_{i=1}^{N} z_i - y(\theta)\right]_{\theta=\hat{\theta}_{\text{ml}}} = 0$$

If $y^{-1}(\cdot)$ exists and if $(1/N)\sum_{i=1}^{N} z_i$ is in the range of $y(\cdot)$, we can solve the preceding to obtain

$$\hat{\theta}_{\text{ml}} = y^{-1}\left(\frac{1}{N}\sum_{i=1}^{N} z_i\right)$$

We note that the MLE commutes over nonlinear operation [6]. That is, if the estimate of θ is $\hat{\theta}_{\text{ml}}$, the estimate of $y(\theta)$ is $y(\hat{\theta}_{\text{ml}})$. This is not true for minimum mean square estimate or MAP estimate.

Example 5.4.3

Consider the problem of estimating the amplitudes and frequencies of M sinusoids observed in additive zero-mean Gaussian white noise with unknown variance σ_v^2 [7,8], from measurements of the form

$$z(n) = \sum_{m=1}^{M} \alpha_m \cos(n\omega_m + \phi_m) + v(n) \qquad n = 0, 1, \ldots, N-1$$

The interest in this problem arises from the mathematical similarity of the observation model of the preceding equation to the observation model in certain applications involving the tracking of targets. We illustrate this similarity by considering the following example. Suppose that a target is emitting a narrowband electromagnetic signal with center frequency ω_0, which is picked up by an array of sensors which are spatially located along a straight line (a linear array) equidistance from one another. If the target is in the *far field* of the array, the signal at a sensor may be modeled as a plane wave. Let θ be the angle of incidence of the plane wave at any sensor, measured with respect to the line perpendicular to the array (Fig. 5.4.1). Because of the spatial displacement between the sensors, the plane wave impinges on the successive elements in the array with a time delay τ. Hence there is a corresponding phase difference ϕ between the signals received by successive elements of the sensor. For example, if the transmitted signal is a sine wave of the form $\alpha \cos(\omega_0 t + \psi)$, it readily follows that the received signal at the kth sensor is equal to $z_t(k) = \alpha_k \cos(\omega_0 t + \psi + k\phi)$, where $k = 0$ refers to the first sensor in the array, and $\phi = (\omega_0/c)\, d \sin\theta$, with c being the speed of propagation of the incident wave.

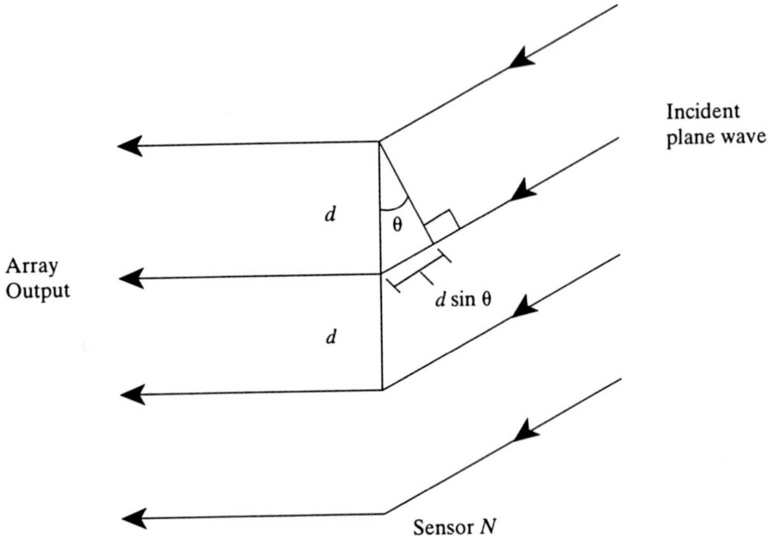

Figure 5.4.1 Linear array of uniformly spaced sensors and incident plane wave.

If there are M targets present, the received signal at the kth element will be the sum of all the incident plane waves due to the different targets and can be modeled as

$$z_t(k) = \sum_{l=1}^{M} \alpha_k \cos(\omega_l t + k\phi_l) + v_t(k) \qquad k = 0, 1, 2, \ldots, N-1$$

A set of measurements from the array at a fixed t (known as a *snapshot*) is then processed to determine the parameters associated with each target, such as the direction of arrival, θ, or center frequency, ω_0. More generally, a set of such snapshots may be processed together.

Let us now return to our original signal model and rewrite it as

$$z(n) = \sum_{m=1}^{M} A_m \cos(n\omega_m) + B_m \sin(n\omega_m) + v(n) \qquad n = 0, 1, \ldots, N-1$$

where $A_m = \alpha_m \cos\phi_m$ and $B_m = -\alpha_m \sin\phi_m$.

We will assume that ω_m, α_m, ϕ_m, and σ_v^2 are deterministic but unknown quantities and find an ML estimate of these parameters. Since the ML estimate commutes over nonlinear operations, we can first find the estimates $\hat{A}_m$ and $\hat{B}_m$ and then determine the estimates for α_m and ϕ_m as

$$\hat{\alpha}_m = \sqrt{\hat{A}_m^2 + \hat{B}_m^2} \qquad \hat{\phi}_m = \tan^{-1}\frac{-\hat{B}_m}{\hat{A}_m}$$

Now define the vectors $\mathbf{z}$, $\mathbf{v}$, and $\boldsymbol{\alpha}$ as

$$\mathbf{z} = [z(0) \quad z(1) \quad \cdots \quad z(N-1)]^T$$
$$\mathbf{v} = [v(0) \quad v(1) \quad \cdots \quad v(N-1)]^T$$
$$\boldsymbol{\alpha} = [A_1 \quad A_2 \quad \cdots \quad A_M \quad B_1 \quad B_2 \quad \cdots \quad B_M]^T$$

and let $\mathbf{S}(\omega)$ be the matrix

$$\mathbf{S}(\omega) = \begin{bmatrix} 1 & . & 1 & 0 & . & 0 \\ \cos \omega_1 & . & \cos \omega_M & \sin \omega_1 & . & \sin \omega_M \\ \cos 2\omega_1 & . & \cos 2\omega_M & \sin 2\omega_1 & . & \sin 2\omega_M \\ . & . & . & . & . & . \\ . & . & . & . & . & . \\ \cos(N-1)\omega_1 & . & \cos(N-1)\omega_M & \sin(N-1)\omega_1 & . & \sin(N-1)\omega_M \end{bmatrix}$$

We can then write

$$\mathbf{z} = \mathbf{S}(\omega)\boldsymbol{\alpha} + \mathbf{v}$$

Let

$$\theta = [\omega_1 \quad \cdots \quad \omega_M \quad A_1 \quad \cdots \quad A_M \quad B_1 \quad \cdots \quad B_M \quad \sigma_{\mathbf{v}}^2]^T$$

Since $\mathbf{z}$ given θ is multivariate Gaussian with mean $\mathbf{S}(\omega)\alpha$ and covariance $\sigma_{\mathbf{v}}^2\mathbf{I}$, the log-likelihood function easily follows as

$$\ln p(\mathbf{z}\,|\,\theta) = -\frac{N}{2}\ln 2\pi - \frac{N}{2}\ln \sigma_{\mathbf{v}}^2 - \frac{1}{2\sigma_{\mathbf{v}}^2}[\mathbf{z} - \mathbf{S}(\omega)\alpha]^T[\mathbf{z} - \mathbf{S}(\omega)\alpha]$$

The ML estimates of $\boldsymbol{\alpha}$ and $\sigma_{\mathbf{v}}^2$ obtained by setting the corresponding partial derivatives of the log-likelihood function to zero are

$$\hat{\boldsymbol{\alpha}} = [\mathbf{S}(\hat{\omega})^T\mathbf{S}(\hat{\omega})]^{-1}\mathbf{S}(\hat{\omega})^T\mathbf{z}$$

and

$$\hat{\sigma}_{\mathbf{v}}^2 = \frac{1}{N}[\mathbf{z} - \mathbf{S}(\hat{\omega})\hat{\boldsymbol{\alpha}}]^T[\mathbf{z} - \mathbf{S}(\hat{\omega})\hat{\boldsymbol{\alpha}}]$$
$$= \frac{1}{N}\mathbf{z}^T[\mathbf{I} - \mathbf{S}(\hat{\omega})[\mathbf{S}(\hat{\omega})\mathbf{S}(\hat{\omega})^T]^{-1}\mathbf{S}(\hat{\omega})]\mathbf{z}$$

Substituting in the expression for the log-likelihood function gives

$$\ln p(\mathbf{z}\,|\,\theta) = -\frac{N}{2}\ln 2\pi - \frac{N}{2}\ln \frac{1}{N}\mathbf{z}^T[\mathbf{I} - \mathbf{S}(\hat{\omega})[\mathbf{S}(\hat{\omega})\mathbf{S}(\hat{\omega})^T]^{-1}\mathbf{S}(\hat{\omega})]\mathbf{z} - \frac{N}{2}$$

Maximizing the log-likelihood function with respect to $\omega_1 \cdots \omega_M$ is therefore equivalent to maximization of the function

$$J = \mathbf{z}^T\mathbf{S}(\hat{\omega})[\mathbf{S}(\hat{\omega})\mathbf{S}(\hat{\omega})^T]^{-1}\mathbf{S}(\hat{\omega})\mathbf{z}$$

J is a nonlinear function of the frequencies ω_m, and typically has multiple maxima. The ML estimates of the frequencies are usually obtained using a suitable numerical technique. The choice of the initial values for the estimates is crucial to obtain the global maximum of J. It is shown in [9] that these estimates are strongly consistent, asymptotically efficient (attain the Cramér–Rao lower bound, see 5.5.1) and are asymptotically normal.

While exact determination of the ML estimates of the frequencies ω_m is difficult, it may be noted that computationally efficient methods have been proposed for determining these estimates approximately, based on an orthogonal decomposition of the observation space into a signal subspace and a noise subspace [8].

5.4.2 The EM Algorithm

In Section 5.4.1 we have seen that given a set of observations $\mathbf{z} = \{z_1, z_2, \ldots, z_n\}$, the ML estimate, $\hat{\theta}_{\text{ml}}(\mathbf{z})$, of a nonrandom parameter θ, is obtained by maximizing either the likelihood function $p(\mathbf{z}\,|\,\theta)$ or the log-likelihood function, $\ln p(\mathbf{z}\,|\,\theta)$. Let $\mathbf{y} = \{y_i, y_2, \ldots, y_m\}$ be a related set of observations depending on the parameter θ. If the transformation from the set $\{y_i\}$ to the set $\{z_i\}$ is many-to-one, it will be noninvertible. We can then consider the set of observations $\{z_i\}$ as being "incomplete" in relation to the "complete" set $\{y_i\}$. We will assume that the complete observations are not available and only the incomplete observations are.

By using Bayes' rule we can write the relation between the likelihood functions for the complete and incomplete data as

$$p(\mathbf{y}\,|\,\theta) = p(\mathbf{y}, \mathbf{z}\,|\,\theta) = p(\mathbf{y}\,|\,\mathbf{z}, \theta)\,p(\mathbf{z}\,|\,\theta) \tag{5.4.5}$$

Taking logarithms on both sides and rearranging terms yields

$$\ln p(\mathbf{z}\,|\,\theta) = \ln p(\mathbf{y}\,|\,\theta) - \ln p(\mathbf{y}\,|\,\mathbf{z}, \theta) \tag{5.4.6}$$

What we are looking for is the value of θ that maximizes the left-hand side of Eq. (5.4.6). In many cases, finding this maximum directly may not be an easy task. However, since the transformation from $\mathbf{y}$ to $\mathbf{z}$ is many-to-one, it is possible that there may be a "complete" data vector $\mathbf{y}$ associated with the "incomplete" observed vector $\mathbf{z}$, for which it is relatively straightforward to find the estimate $\hat{\theta}_{\text{ml}}(\mathbf{y})$. In such cases it is easier to find $\hat{\theta}_{\text{ml}}(\mathbf{z})$ by first finding $\hat{\theta}_{\text{ml}}(\mathbf{y})$. Since the vector $\mathbf{y}$ is not available, we find an iterative solution to this problem. Starting with an initial assumed value, $\hat{\theta}^0$, for the estimate $\hat{\theta}_{\text{ml}}(\mathbf{z})$, we seek successive estimates $\hat{\theta}^i$ such that at each iteration, the likelihood function $\ln p(\mathbf{z}\,|\,\theta)$ has a larger value [10].

Let us take the expectation on both sides of Eq. (5.4.6) with respect to the density function $p(\mathbf{y}\,|\,\mathbf{z}, \hat{\theta}^i)$. The expectation on the left-hand side is

$$\int_{\mathbf{y}} \ln p(\mathbf{z}\,|\,\theta)\,p(\mathbf{y}\,|\,\mathbf{z}, \hat{\theta}^i)\,d\mathbf{y} = \ln p(\mathbf{z}\,|\,\theta)$$

so that we get

$$\ln p(\mathbf{z}\,|\,\theta) = \mathrm{E}_{\mathbf{Y}|\mathbf{z},\hat{\theta}^i}\{\ln p(\mathbf{y}\,|\,\theta)\,|\,\mathbf{z}, \hat{\theta}^i\} - \mathrm{E}_{\mathbf{Y}|\mathbf{z},\hat{\theta}^i}\{\ln p(\mathbf{y}\,|\,\mathbf{z}, \theta)\,|\,\mathbf{z}, \hat{\theta}^i\} \tag{5.4.7}$$

We note that though both the terms on the right-hand side of Eq. (5.4.7) are functions of θ and $\hat{\theta}^i$, their difference is not.

It can be shown [10] that for any $\hat{\theta}^{i+1}$,

$$\mathrm{E}_{\mathbf{Y}|\mathbf{z},\hat{\theta}^i}\{\ln p(\mathbf{y}\,|\,\mathbf{z}, \hat{\theta}^{i+1})\,|\,\mathbf{z}, \hat{\theta}^i\} \le \mathrm{E}_{\mathbf{Y}|\mathbf{z},\hat{\theta}^i}\{\ln p(\mathbf{y}\,|\,\mathbf{z}, \hat{\theta}^i)\,|\,\mathbf{z}, \hat{\theta}^i\} \tag{5.4.8}$$

with equality holding only if $\hat{\theta}^{i+1} = \hat{\theta}^i$.

Let us denote the first term on the right-hand side of Eq. (5.4.7) by $\chi(\theta, \hat{\theta}^i)$. It follows that a sufficient condition for $\ln p(\mathbf{z} \mid \hat{\theta}^{i+1})$ to be larger than $\ln p(\mathbf{z} \mid \hat{\theta}^i)$ is that $\chi(\hat{\theta}^{i+1}, \hat{\theta}^i)$ is larger than $\chi(\hat{\theta}^i, \hat{\theta}^i)$. Thus at any iteration, the algorithm consists of the following two steps:

$$\begin{aligned} &\text{The E step:} \quad \text{Form } \chi(\theta, \hat{\theta}^i) = E_{\mathbf{Y}|\mathbf{z}}\{\ln p(\mathbf{y} \mid \theta) \mid \mathbf{z}, \hat{\theta}^i\} \\ &\text{The M step:} \quad \text{Find } \max_\theta \chi(\theta, \hat{\theta}^i) \end{aligned} \tag{5.4.9}$$

Because of the two steps involved, the resulting algorithm is known as the *expectation-maximization* (E-M) *algorithm*. If the likelihood function is bounded from above, and if $\chi(x_1, x_2)$ is continuous, the log-likelihood function $\ln p(z \mid \theta)$ converges to a stationary point (a local maximum) [11]. When the likelihood function has several maxima, the E-M algorithm is not guaranteed to converge to the global maximum. In general, it is preferable to start the algorithm at different initial values and compare the results.

Example 5.4.4

Let $Y_{ki}, 1 \le i \le N, 1 \le k \le M$, be a set of independent Gaussian variables with zero mean and variance θ_k and let $Z_i = \sum_{k=1}^M a_k Y_{ki}$ for $1 \le i \le N$, where the a_k's are known constants. Thus Z_i is Gaussian with mean zero and variance $\sum_{k=1}^M a_k^2 \theta_k$. We would like to find the ML estimates $\hat{\theta}_{k_{\text{ml}}}(\mathbf{z})$ where $\mathbf{z} = [z_1 \quad \cdots \quad z_N]^T$.

If we try to find these estimates directly by maximizing $\ln p(\mathbf{z} \mid \theta)$, we get the single equation

$$\sum_{k=1}^M a_m^2 \hat{\theta}_{k_{\text{ml}}} = \frac{1}{N}\sum_{i=1}^N \mathbf{z}_i^2$$

so that the estimates cannot be uniquely determined.

For this problem, the correct variances θ_k cannot be estimated uniquely. To see this, we note that

$$\text{var}(Z_i) = \sum_{k=1}^M a_m^2 \theta_k$$

Thus all θ_k that satisfy the foregoing constraint yield identical likelihood functions based on the Z_i. Even when N tends to infinity, although $\text{var}\{Z_i\}$ can be estimated exactly, the nonuniqueness of the estimates of θ_k remains.

We can find an estimate of the θ_k using the E-M algorithm as follows. We note from our assumptions that

$$\ln p(\mathbf{y} \mid \theta) = -\frac{MN}{2}\ln(2\pi) - \frac{N}{2}\sum_{m=1}^M \ln \theta_m - \sum_{j=1}^N \sum_{m=1}^M \frac{y_{mj}^2}{2\theta_m}$$

Also, from Eq. (5.4.5), we have

$$\begin{aligned} p(\mathbf{y} \mid \mathbf{z}, \hat{\theta}^i) &= \frac{p(\mathbf{y} \mid \hat{\theta}^i)}{p(\mathbf{z} \mid \hat{\theta}^i)} \\ &= \left(2\pi \sum_{k=1}^M a_k^2 \hat{\theta}_k^i\right)^{N/2} \exp\left[\frac{\sum_{j=1}^N z_j^2}{2\sum_{k=1}^M a_k^2 \hat{\theta}_k^i}\right] \prod_{j=1}^N \prod_{m=1}^M \frac{1}{(2\pi\hat{\theta}_m^i)^{1/2}} \exp\left(-\frac{y_{mj}^2}{2\hat{\theta}_m^i}\right) \end{aligned}$$

so that

$$\chi(\theta, \hat{\theta}^i) = -\frac{MN}{2}\ln(2\pi) - \frac{N}{2}\sum_{m=1}^{M}\ln\theta_m - \left(2\pi\sum_{k=1}^{M}a_k^2\hat{\theta}_k^i\right)^{N/2}\exp\left[\frac{\sum_{j=1}^{N}z_j^2}{2\sum_{k=1}^{M}a_k^2\hat{\theta}_k^i}\right]\sum_{m=1}^{M}\frac{N\hat{\theta}_m^i}{2\theta_m}$$

Maximizing $\chi(\theta, \hat{\theta}^i)$ with respect to θ_m yields

$$\hat{\theta}_m^{i+1} = \left(2\pi\sum_{k=1}^{M}a_k^2\hat{\theta}_k^i\right)^{N/2}\exp\left[\frac{\sum_{j=1}^{N}z_j^2}{2\sum_{k=1}^{M}a_k^2\hat{\theta}_k^i}\right]\hat{\theta}_m^i \qquad m = 1, 2, \ldots, M$$

We can therefore find $\hat{\theta}_m$ recursively, starting with arbitrary initial values for $\hat{\theta}_m^0$.

For this example, the E-M algorithm, if it converges, may pick any of the possible solutions, depending on the choice of our initial estimate $\hat{\theta}^0$. The difficulty in this example is that a statistical characterization of Y_{ki} cannot be uniquely determined from the available observations $Z_i, i = 1, \ldots, N$. Nevertheless, the example illustrates the steps involved in applying the E-M algorithm.

Example 5.4.5

Let $Z_1, Z_2, \ldots, Z_N$ be independent samples of a random variable drawn from the mixture density function

$$p(\mathbf{z}\,|\,\theta) = \prod_{k=1}^{N}[\rho p_1(z_k\,|\,\theta) + (1-\rho)p_2(z_k\,|\,\theta)]$$

That is, each sample z_k is drawn from either the density function $p_1(\cdot)$ or $p_2(\cdot)$ with probability ρ and $1-\rho$, respectively. We will assume that $p_i(\cdot), i = 1, 2$ is Gaussian with mean μ_i and variance σ_i^2, so that $\theta = (\rho, \mu_1, \mu_2, \sigma_1, \sigma_2)$.

The ML estimate of θ is obtained as a solution to the set of equations

$$\frac{\partial p(\mathbf{z}\,|\,\theta)}{\partial\rho} = 0, \quad \frac{\partial p(\mathbf{z}\,|\,\theta)}{\partial\mu_j} = 0, \quad \frac{\partial p(\mathbf{z}\,|\,\theta)}{\partial\sigma_j} = 0 \qquad j = 1, 2$$

Among several possible solutions, we seek one that maximizes $p(\mathbf{z}\,|\,\theta)$. This, in general, is not an easy task. Here we illustrate application of the E-M algorithm to find an estimate of θ.

Let $\mathbf{x}_k$ be a variable that indicates for each k whether z_k was drawn from $p_1(\cdot)$ or $p_2(\cdot)$. Thus $\mathbf{x}_k$ is either $[1 \quad 0]^T$ or $[0 \quad 1]^T$ with probabilities ρ and $1-\rho$, respectively. We consider the complete data to be $\mathbf{y} = (\mathbf{z}, \mathbf{x})$ with

$$p(\mathbf{y}\,|\,\theta) = \prod_{k=1}^{N}\mathbf{x}_k^T\begin{bmatrix}\rho p_1(z_k\,|\,\theta)\\(1-\rho)p_2(z_k\,|\,\theta)\end{bmatrix}$$

Thus the E-step is

$$\begin{aligned}\chi(\theta, \hat{\theta}^i) &= \mathrm{E}\{\ln p(\mathbf{Y}\,|\,\theta)\,|\,\mathbf{z}, \hat{\theta}^i\}\\ &= \mathrm{E}\left\{\sum_{k=1}^{N}\mathbf{X}_k^T\begin{bmatrix}\ln\rho p_1(z_k\,|\,\theta)\\ \ln(1-\rho)p_2(z_k\,|\,\theta)\end{bmatrix}\Bigg|\,\mathbf{z}, \hat{\theta}^i\right\}\\ &= \sum_{k=1}^{N}[\mathrm{E}\{X_{1k}\,|\,\mathbf{z}, \hat{\theta}^i\}\mathrm{E}\{X_{2k}\,|\,\mathbf{z}, \hat{\theta}^i\}]\begin{bmatrix}\ln\rho p_1(z_k\,|\,\theta)\\ \ln(1-\rho)p_2(z_k\,|\,\theta)\end{bmatrix}\end{aligned}$$

where

$$\mathbf{X}_k = [X_{1k} \quad X_{2k}]^T = [X_{1k} \quad (1 - X_{1k})]^T$$

$$\mathrm{E}\{X_{1k} \mid \mathbf{z}, \hat{\theta}^i\} = \frac{\hat{\rho}^i p_1(z_k \mid \hat{\theta}^i)}{p(z_k \mid \hat{\theta}^i)} \equiv w_k^i$$

$$p(z_k \mid \hat{\theta}^i) = \hat{\rho}^i p_1(z_k \mid \hat{\theta}^i) + (1 - \hat{\rho}^i) p_2(z_k \mid \hat{\theta}^i)$$

Maximizing $\chi(\theta, \hat{\theta}^i)$ with respect to θ yields

$$\frac{\partial \chi(\theta, \hat{\theta}^i)}{\partial \rho} = \sum_{k=1}^{N} \frac{w_k^i}{\rho} - \frac{1 - w_k^i}{1 - \rho} = 0$$

$$\frac{\partial \chi(\theta, \hat{\theta}^i)}{\partial \mu_j} = \hat{\rho}^i \sum_{k=1}^{N} \frac{p_j(z_k \mid \hat{\theta}^i)}{p(z_k \mid \hat{\theta}^i)} \frac{z_k - \mu_j}{\sigma_j^2} = 0$$

$$\frac{\partial \chi(\theta, \hat{\theta}^i)}{\partial \sigma_j} = \hat{\rho}^i \sum_{k=1}^{N} \frac{p_j(z_k \mid \hat{\theta}^i)}{p(z_k \mid \hat{\theta}^i)} \left[\frac{(z_k - \mu_j)^2}{\sigma_j^3} - \frac{1}{\sigma_j} \right] = 0$$

so that the M-step is

$$\hat{\rho}^{i+1} = \frac{1}{N} \sum_{k=1}^{N} w_k^i = \frac{1}{N} \sum_{k=1}^{N} \frac{\hat{\rho}^i p_1(z_k \mid \hat{\theta}^i)}{p(z_k \mid \hat{\theta}^i)}$$

$$\hat{\mu}_j^{i+1} = \frac{\displaystyle\sum_{k=1}^{N} \frac{p_j(z_k \mid \hat{\theta}^i) z_k}{p(z_k \mid \hat{\theta}^i)}}{\displaystyle\sum_{k=1}^{N} \frac{p_j(z_k \mid \hat{\theta}^i)}{p(z_k \mid \hat{\theta}^i)}} \qquad j = 1, 2$$

$$\hat{\sigma}_j^{i+1} = \left[\frac{\displaystyle\sum_{k=1}^{N} \frac{p_j(z_k \mid \hat{\theta}^i)(z_k - \hat{\mu}_j^i)^2}{p(z_k \mid \hat{\theta}^i)}}{\displaystyle\sum_{k=1}^{N} \frac{p_j(z_k \mid \hat{\theta}^i)}{p(z_k \mid \hat{\theta}^i)}} \right]^{1/2} \qquad j = 1, 2$$

A particular case of this example is the detection of a signal received in pulsed jamming noise [12]. The signal is received in thermal noise alone or thermal plus jammer noise, depending on whether the jammer is on or off. The E-M algorithm can be used to estimate the jammer parameters and a likelihood ratio test designed using these estimated parameters to detect the signal. With small sample sizes, the parameter estimates are quite poor. However, it has been shown in [12] that the detection performance of the algorithm is good over a certain range of parameters.

Note that in using the E-M algorithm, it is not necessary that the complete observations must actually exist. It is only required that we must be able to associate with the observed data, a set of complete observations such that the solution to the estimation problem is easier for the complete data than for the incomplete.

Finally, the E-M algorithm can be used for MAP estimation also by modifying the definition of $\chi(\theta, \hat{\theta}^i)$ as

$$\chi(\theta, \hat{\theta}^i) = \mathrm{E}\{\ln p(\mathbf{Y} \mid \theta) \mid \mathbf{z}, \hat{\theta}^i)\} + \ln p(\theta) \qquad (5.4.10)$$

5.4.3 Interval Estimates

We have discussed previously point estimates of a parameter θ, obtained by using a suitable criterion such as the Bayes criterion for random parameters or the maximum-likelihood criterion for nonrandom parameters. Quite often, instead of finding a point estimate, we may be interested in finding an estimate of an interval in which the parameter θ lies. Such an estimate is referred to as an *interval estimate*. We associate with the interval estimate a *confidence coefficient*, which represents a measure of how good our estimate is.

Example 5.4.6

Let $Z_i, i = 1, \ldots, n$ be a set of random samples drawn from a Gaussian distribution with unknown mean μ and known standard deviation σ. The maximum likelihood estimate of μ is given by the sample mean $\overline{Z}$. Let

$$Q = \frac{\overline{Z} - \mu}{\sigma/\sqrt{n}} \tag{5.4.11}$$

Then Q is Gaussian distributed with mean zero and unity variance. Since the density function of Q does not depend on the unknown parameter μ, we can determine the probability γ that Q lies between any two values a and b. We thus have

$$P(a < Q < b) = \int_a^b \frac{1}{\sqrt{2\pi}} e^{-(1/2)u^2}\, du \tag{5.4.12}$$

By substituting Eq. (5.4.11) in the inequality on the left side of Eq. (5.4.12), we can write this equation in the equivalent form

$$P\left(\overline{Z} - \frac{b\sigma}{\sqrt{n}} < \mu < \overline{Z} - \frac{a\sigma}{\sqrt{n}}\right) = \gamma \tag{5.4.13}$$

Equation (5.4.13) states that the probability that the unknown parameter μ lies in the interval $(\overline{Z} - b\sigma/\sqrt{n}, \overline{Z} - a\sigma/\sqrt{n})$ is equal to γ. This interval is referred to as a γ-*confidence interval* for μ. Clearly, the closer γ is to 1, the greater is our certainty that the true value μ lies in this interval. Since γ is a measure of how certain we are that the true value μ lies in this interval, it is referred to as the *confidence coefficient*.

The endpoints of the confidence interval as defined above are random variables. However, the specific interval obtained by substituting $\bar{z} = (1/n)\sum_{i=1}^{n} z_i$, where z_i are a specific set of samples, is also referred to as the γ-confidence interval for μ. Commonly used values for γ are 0.9, 0.95, 0.99, and 0.999.

In general, given random samples $Z_1, Z_2, \ldots, Z_n$ from a density $p(\cdot\,|\,\theta)$, let $g(\theta)$ be some function of θ. Let $T_1 = t_1(Z_1, Z_2, \ldots, Z_n)$ and $T_2 = t_2(Z_1, Z_2, \ldots, Z_n)$ be two statistics such that

$$P(T_1 < g(\theta) < T_2) = \gamma \tag{5.4.14}$$

where γ does not depend on θ. The random interval (T_1, T_2) is then a γ-confidence interval for $g(\theta)$ with confidence coefficient γ. The quantities T_1 and T_2 are called the lower and upper confidence limits.

In order to determine the interval estimate of $g(\theta)$, we can follow an approach similar to the one in Example 5.4.6. Let q be a function of $z_1, z_2, \ldots, z_n$ and θ such that the distribution of Q does not depend on θ. Then for any fixed $\gamma \in (0, 1)$, there will exist ζ_1 and ζ_2 such that $P(\zeta_1 < Q < \zeta_2) = \gamma$. If the inequality $\zeta_1 < q(z_1, z_2, \ldots, z_n; \theta) < \zeta_2$ is true if and only if $t_1(z_1, z_2, \ldots, z_n) < g(\theta) < t_2(z_1, z_2, \ldots, z_n)$ for each possible set of samples z_i, then (T_1, T_2) is a γ-confidence interval for $g(\theta)$. For obvious reasons, Q is called a pivotal quantity.

It is clear that for any fixed γ, we can find several values of ζ_1 and ζ_2 such that $P(\zeta_1 < Q < \zeta_2) = \gamma$. Each set of values corresponds to a different set of values for t_1 and t_2. Our aim is to select ζ_1 and ζ_2 so as to minimize the length of the interval (t_1, t_2), subject to the constraint that the confidence coefficient is γ.

Example 5.4.6 *(continued)*

The length of the confidence interval for a fixed γ for this example can be determined from Eq. (5.4.12) as $\sigma(b - a)/\sqrt{n}$. In addition, a and b must be chosen to satisfy the constraint of Eq. (5.4.12), that is,

$$P(a < Q < b) = \gamma$$

Thus we select a and b such that $b - a$ is minimum subject to the area under the density function $p(q)$ being fixed. Since $p(q)$ is symmetric about $q = 0$, the minimum occurs when $b = -a$. We can get an explicit expression for a or b in terms of the cumulative distribution function (CDF) $G(\cdot)$ of the standard Gaussian density function. From Eq. (5.4.11) we have $G(b) - G(a) = \gamma$. With $b = -a$, it follows that

$$G(b) = 1 - G(a) \tag{5.4.15}$$

so that

$$G(a) = \tfrac{1}{2}(1 - \gamma) \quad \text{and} \quad G(b) = \tfrac{1}{2}(1 + \gamma) \tag{5.4.16}$$

Given a random variable Z with CDF $F(\cdot)$, if $F(z_p) = p$, then z_p is the pth *percentile* of Z. For the standard Gaussian CDF $G(\cdot)$, these percentiles are available in the form of tabulated values. Thus given a specific value of γ, a and b can be determined as the $(1 - \gamma)/2$-percentile and $(1 + \gamma)/2$-percentile of $G(\cdot)$. For example, if $\gamma = 0.95$, $b = 1.96$.

Intuitively, we would expect that as the number of observations n increases, the closer the sample mean will be to the true mean. With $b = -a$, it follows that the length of the confidence interval is $2\sigma b/\sqrt{n}$. Thus the length of the confidence interval for a specified value of γ decreases as the number of observations increases. This fits in with our intuitive expectation. Finally, we note that by setting $(b\sigma)/\sqrt{n} = l$, Eq. (5.4.13) can be written as

$$P(\overline{Z} - l < \mu < \overline{Z} + l) = \gamma \tag{5.4.17}$$

The density function of Q in the foregoing example was unimodal and symmetric about the mode. For an arbitrary unimodal density $p(q)$, the length of the interval $b - a$ is minimum if $p(b) = p(a)$. Usually, a and b must then be determined by a trial-and-error procedure to satisfy the constraint of Eq. (5.4.12). Quite often, for simplicity, the confidence interval is chosen as in Eq. (5.4.15), that is, with

$F(b) = 1 - F(a)$, where $F(\cdot)$ denotes the corresponding CDF, although its length is no longer minimum.

A pivotal quantity Q can be chosen in many ways. For example, if θ is a location parameter of the density, the density of $[X_i - \theta]$ is by definition independent of θ and is hence a pivotal quantity; so are the quantities $[(1/n)\sum_{i=1}^{n} X_i - \theta]$, and $[Y_j - \theta]$ where Y_j is the jth-order statistic of $(X_1, X_2, \ldots, X_n)$. Similarly, if θ is a scale parameter of the density, then $[X_i/\theta]$, $[(\Sigma X_i)/\theta)]$, and $[Y_j/\theta]$ are all pivotal quantities.

In the preceding example, the samples Z_i were assumed to be Gaussian distributed, so that the variable Q was also Gaussian. Even in cases where the samples do not have a Gaussian distribution, if the number of samples are large, often we can use the central limit theorem to make the assumption that Q is Gaussian distributed. We can then use the results of the previous problem to determine confidence limits. However, if the mean and variance of this distribution are functionally related, the confidence limits cannot be so directly obtained and further manipulations may be required.

We illustrate this with the following example.

Example 5.4.7

Let the duration τ of telephone calls in an area be described by the exponential density

$$p(\tau) = \frac{1}{\lambda}\exp\left(-\frac{\tau}{\lambda}\right)u(\tau)$$

where λ represents the mean and $u(\cdot)$ is the unit step function. We use as an estimate of λ the observed average duration of calls, $\overline{T}$, obtained by measuring the durations $\tau_i, 1 \leq i \leq n$ of several calls. Under the large-sample condition, we can assume that $\overline{T}$ has a Gaussian distribution with mean λ and variance λ^2/n. From Eq. (5.4.17), it follows that the confidence interval satisfies

$$P(\overline{T} - l < \lambda < \overline{T} + l) = \gamma \tag{5.4.18}$$

Let $\alpha = (1 + \gamma)/2$ and let x_α denote the αth percentile of the standard Gaussian CDF $G(\cdot)$. That is,

$$\frac{1}{\sqrt{2\pi}}\int_{-\infty}^{x_\alpha} \exp\left(-\frac{y^2}{2}\right)dy = \alpha \tag{5.4.19}$$

It easily follows that

$$l = \frac{x_\alpha \lambda}{\sqrt{n}} \tag{5.4.20}$$

Substituting in Eq. (5.4.18) and rearranging the inequality yields, for $x_\alpha < \sqrt{n}$,

$$P\left(\frac{\overline{T}}{1 + x_\alpha/\sqrt{n}} < \lambda < \frac{\overline{T}}{1 - x_\alpha/\sqrt{n}}\right) = \gamma \tag{5.4.21}$$

Thus the confidence interval is given by

$$\left(\frac{\overline{T}}{1 + x_\alpha/\sqrt{n}}, \frac{\overline{T}}{1 - x_\alpha/\sqrt{n}}\right); \qquad G(x_\alpha) = \alpha = \frac{1+\gamma}{2} \tag{5.4.22}$$

As an example, if $\gamma = 0.95$, then $x_\alpha = 1.967$. Suppose that the average duration of 100 calls is 2.5 minutes. Substitution in Eq. (5.4.22) yields the confidence interval for λ as

(2.09, 3.11). If the average duration had been obtained from 250 calls, the confidence interval would have been (2.22, 2.85).

Example 5.4.8

As a final example, let us consider the estimation of the probability π of an event occurring in an experiment. We repeat the experiment n times and note the number k of successes. We use the ratio k/n as a point estimate $\hat{\Pi}$ of the probability π. For large n, the distribution of $\hat{\Pi}$ approaches a Gaussian distribution with mean π and variance $\pi(1 - \pi)/n$.

For a given value of γ, we can determine l as in Example 5.4.7 [see Eq. (5.4.20)] as

$$l = x_\alpha \sqrt{\frac{\pi(1 - \pi)}{n}}$$

where x_α is as defined earlier. It therefore follows that

$$P\left(\hat{\Pi} - x_\alpha \sqrt{\frac{\pi(1 - \pi)}{n}} < \pi < \hat{\Pi} + x_\alpha \sqrt{\frac{\pi(1 - \pi)}{n}}\right) = \gamma$$

To obtain the confidence interval, we rewrite this equation in the equivalent form

$$P\left[(\pi - \hat{\Pi})^2 < \frac{x_\alpha^2}{n}\pi(1 - \pi)\right] = \gamma$$

The values of π that satisfy this inequality are in the interior of the ellipse

$$(\pi - \hat{\Pi})^2 = \frac{x_\alpha^2}{n}\pi(1 - \pi)$$

Let π_1 and π_2 be the roots of this equation. Then the confidence interval is given by (π_1, π_2).

For example, suppose that a poll of 250 possible voters yielded 200 votes for candidate A and the remainder preferred candidate B. We want to estimate the probability p of the number of voters preferring candidate A. We use as our estimate $\hat{\Pi}$ the ratio $200/250 = 0.8$. The confidence interval associated with this estimate is obtained by solving the equation

$$(\pi - 0.8)^2 = \frac{x_\alpha^2}{n}\pi(1 - \pi)$$

where $x_\alpha = 1.967$ for $\gamma = 0.95$. The resulting confidence interval is (0.745, 0.845). Since the difference between the estimate $\hat{\Pi} = 0.8$ and the endpoints of the confidence interval is approximately ± 0.05, this result is usually stated as "80% of all voters favored candidate A. The margin of error is $\pm 5\%$." Clearly, to interpret this statement properly, the corresponding confidence coefficient γ must also be stated.

5.5 PROPERTIES OF ESTIMATORS

We discuss in this section some desirable properties of estimators. As noted earlier, the estimate is a function of the observations. As the observations change, the value of the estimate changes. Since the observations are random variables, so is the estimate. It is obviously desirable that given a large number of observations, the estimates group near the true value. For nonrandom parameters, we say that an estimate is *unbiased* if the mean value of the estimate is equal to the true value

$$\mathrm{E}\{\hat{\Theta}\} = \int_{-\infty}^{\infty} \hat{\theta} p(z \mid \theta)\, dz = \theta \tag{5.5.1}$$

The difference between the mean $\mathrm{E}\{\hat{\theta}\}$ and the true value is called the *bias* in the estimate. For random parameters, we say that the estimate is *unbiased* if the mean of the estimate is equal to the mean of the parameter being estimated:

$$\mathrm{E}\{\hat{\Theta}\} = \int\int_{-\infty}^{\infty} \hat{\theta} p(z, \theta)\, dz\, d\theta = \int_{-\infty}^{\infty} \theta p(\theta) d\theta = \mathrm{E}\{\Theta\} \tag{5.5.2}$$

In this case, the difference between the two means represents the bias in the estimate.

As the number of observations used in forming an estimate increases, we would like the estimate to tend toward its true value. Thus if we let $\hat{\Theta}_N$ denote the estimate based on N observations, we say that the estimate is consistent if

$$\lim_{N \to \infty} P(|\theta - \hat{\Theta}_N| < \epsilon) = 1 \tag{5.5.3}$$

for arbitrarily small $\epsilon > 0$.

Finally, an estimator is said to be a minimum variance estimator if the variance of the estimation error is less than that of any other estimator.

5.5.1 Bounds on Estimates: Nonrandom Parameters

In determining the effectiveness of an unbiased estimator $\hat{\theta}(z)$, we can compute its variance. Then by determining a lower bound for the variance of *any* unbiased estimator we can compare the variance of $\hat{\theta}(z)$ with this bound. While there are several such bounds available in the literature, we concern ourselves with the Cramér–Rao bound [1,3,5,13,14]. We derive the bound for the nonrandom parameter case. The derivation for the case of random parameters follows along essentially the same lines.

Since $\hat{\theta}$ is assumed to be an unbiased estimate, we have

$$\mathrm{E}[\hat{\Theta} - \theta] = \int_{-\infty}^{\infty} [\hat{\theta} - \theta] p(z \mid \theta)\, dz = 0 \tag{5.5.4}$$

Differentiate both sides with respect to θ to get

$$\begin{aligned} &\frac{d}{d\theta} \int_{-\infty}^{\infty} [\hat{\theta} - \theta] p(z \mid \theta)\, dz \\ &\qquad = \int_{-\infty}^{\infty} \frac{\partial}{\partial\theta} \{[\hat{\theta} - \theta] p(z \mid \theta)\}\, dz \qquad\qquad (5.5.5) \\ &\qquad = -\int_{-\infty}^{\infty} p(z \mid \theta)\, dz + \int_{-\infty}^{\infty} \frac{\partial p(z \mid \theta)}{\partial \theta} [\hat{\theta} - \theta]\, dz = 0 \end{aligned}$$

assuming that the conditions permitting an interchange of the orders of integration and differentiation have been satisfied. The first integral equals -1. We also note that for any function $f(x)$, $df(x)/dx$ may be expressed as $(d \ln f(x)/dx)f(x)$. Thus Eq. (5.5.5) may be written as

$$\int_{-\infty}^{\infty} \frac{\partial \ln p(z \mid \theta)}{\partial \theta} p(z \mid \theta)[\hat{\theta} - \theta]\, dz = 1 \tag{5.5.6}$$

Rewriting, we have

$$\int_{-\infty}^{\infty} \left[\frac{\partial \ln p(z \mid \theta)}{\partial \theta} \sqrt{p(z \mid \theta)}\right]\left[\sqrt{p(z \mid \theta)}(\hat{\theta} - \theta)\right] dz = 1 \tag{5.5.7}$$

Using Schwarz's inequality,[†] we find that

$$\int_{-\infty}^{\infty} \left[\frac{\partial \ln p(z \mid \theta)}{\partial \theta}\right]^2 p(z \mid \theta)\, dz \int_{-\infty}^{\infty} (\hat{\theta} - \theta)^2 p(z \mid \theta)\, dz \geq 1 \tag{5.5.8}$$

with equality holding if and only if

$$\frac{\partial \ln p(z \mid \theta)}{\partial \theta} = [\hat{\theta} - \theta]k(\theta) \tag{5.5.9}$$

for all z and θ. Here $k(\cdot)$ is not a function of z or $\hat{\theta}$ but may be a function of θ. From Eq. (5.5.8) we have the inequality

$$\operatorname{var}[\hat{\Theta} - \theta] = \mathrm{E}\{(\hat{\Theta} - \theta)^2 \mid \theta\} \geq \left\{\mathrm{E}\left\{\left[\frac{\partial \ln p(z \mid \theta)}{\partial \theta}\right]^2 \middle| \theta\right\}\right\}^{-1} \tag{5.5.10}$$

The right side of Eq. (5.5.10) represents a lower bound on the variance of any unbiased estimate of θ and is the Cramér–Rao bound referred to earlier. Any estimate that satisfies this bound with an equality is called an *efficient* estimate. We can derive an alternative expression for the right side of Eq. (5.5.10) by noting that

$$\int_{-\infty}^{\infty} p(z \mid \theta)\, dz = 1 \tag{5.5.11}$$

Differentiating both sides with respect to θ, we have

$$\int_{-\infty}^{\infty} \frac{\partial p(z \mid \theta)}{\partial \theta} dz = \int_{-\infty}^{\infty} \frac{\partial \ln p(z \mid \theta)}{\partial \theta} p(z \mid \theta)\, dz = 0 \tag{5.5.12}$$

Differentiating again with respect to θ yields

$$\int_{-\infty}^{\infty} \frac{\partial^2 \ln p(z \mid \theta)}{\partial \theta^2} p(z \mid \theta)\, dz + \int_{-\infty}^{\infty} \left[\frac{\partial \ln p(z \mid \theta)}{\partial \theta}\right]^2 p(z \mid \theta)\, dz = 0 \tag{5.5.13}$$

[†] The form of the Schwarz inequality used here is

$$\left[\int g(x)h(x)\, dx\right]^2 \leq \int g^2(x)\, dx \int h^2(x)\, dx$$

with equality holding when $g(x)$ and $h(x)$ are linearly related.

or

$$E\left[\frac{\partial^2 \ln p(z \mid \theta)}{\partial \theta^2} \middle| \theta\right] = -E\left[\left(\frac{\partial \ln p(z \mid \theta)}{\partial \theta}\right)^2 \middle| \theta\right] \tag{5.5.14}$$

As noted earlier, the maximum likelihood technique is used most often for estimation of nonrandom parameters. We can derive some useful properties of these estimators by using the Cramér–Rao bound.

Let us assume that Eq. (5.5.9) is satisfied and therefore an efficient estimate $\hat{\theta}$ exists. Suppose that we set $\theta = \hat{\theta}_{ml}$ on both sides of Eq. (5.5.9). Setting $\theta = \hat{\theta}_{ml}$ in the left side gives us the likelihood equation

$$\left.\frac{\partial \ln p(z \mid \theta)}{\partial \theta}\right|_{\theta=\hat{\theta}_{ml}} = 0 \tag{5.5.15}$$

Since the right side must also be zero, we get

$$[\hat{\theta} - \hat{\theta}_{ml}]k(\theta) = 0 \tag{5.5.16}$$

This is satisfied if

$$\hat{\theta} = \hat{\theta}_{ml} \tag{5.5.17}$$

Thus if an efficient estimate exists, it is $\hat{\theta}_{ml}$ and can be obtained as a solution to the likelihood equation.

If an efficient estimate does not exist, we will not know how good $\hat{\theta}_{ml}$ is. In fact, we will not know how closely the variance of any estimate will approach the bound.

Under reasonably general conditions it may be established that the ML estimate is consistent, asymptotically efficient, and asymptotically Gaussian [5].

Example 5.5.1

We consider the problem of Example 5.4.1, where

$$\frac{\partial \ln p(\mathbf{z} \mid \theta)}{\partial \theta} = \frac{N}{\sigma_v^2}\left(\frac{1}{N}\sum_{i=1}^{N} z_i - \theta\right)$$

so that

$$\hat{\theta}_{ml} = \frac{1}{N}\sum_{i=1}^{N} z_i$$

and

$$E[\hat{\Theta}_{ml}] = \frac{1}{N}\sum_{i=1}^{N} E\{Z_i\} = \frac{1}{N}\sum_{i=1}^{N} \theta = \theta$$

Thus $\hat{\theta}_{ml}$ is unbiased.

Since $\partial \ln p(\mathbf{z} \mid \theta)/\partial\theta$ has the form required by Eq. (5.5.9) an efficient estimate exists and is given by $\hat{\theta}_{ml}$. The variance of the estimate is given by the Cramér–Rao bound. We have

$$\frac{\partial^2 \ln p(\mathbf{z} \mid \theta)}{\partial \theta^2} = -\frac{N}{\sigma_v^2}$$

Using Eq. (5.5.14) and the fact that the estimate is efficient yields

$$\mathrm{var}[\hat{\theta}_{\mathrm{ml}} - \theta] = \frac{\sigma_v^2}{N}$$

Example 5.5.2

We consider the problem of estimating certain unknown (nonrandom) parameters of a signal of known form observed in noise [15]. The observations are assumed to be of the form

$$z(t) = y(t, \theta) + v(t) \qquad t_0 \leqslant t \leqslant t_f \tag{5.5.18}$$

where θ is an unknown parameter. In the first instance we assume that the noise is zero mean, Gaussian, white with spectral height $N_0/2$ and subsequently extend our results to the colored noise case.

The first step in determining the ML estimate of θ is to obtain the likelihood function (conditional density) $p(z \mid \theta)$. To do so, we use an argument similar to that in Section 4.2. If we use the sampling approach discussed there, we sample $z(t)$ so that we have N samples in the interval (t_0, t_f).

Let Z_k, y_k, and V_k denote the sample values at instant t_k. That is, $Z_k = z(t_k)$, $y_k = y(t_k, \Theta)$, and $V_k = v(t_k)$. As in Section 4.2, we initially assume that $v(t)$ is lowpass with (double-sided) bandwidth Ω and later let Ω tend to infinity. If we let $t_k = k\pi/\Omega$ so that $N = \Omega(t_f - t_0)/\pi$, the noise samples will be independent. The received samples can be written as

$$Z_k = y_k + V_k = y(t_k, \theta) + V_k \tag{5.5.19}$$

It follows that

$$\mathrm{E}\{Z_k \mid \theta\} = y_k$$

and

$$\mathrm{var}\{Z_k \mid \theta\} = \mathrm{var}\{V_k\} = \frac{N_0 \Omega}{2\pi} = \sigma^2 \tag{5.5.20}$$

Let $\mathbf{z}_N = [z_1 \quad \cdots \quad z_N]^T$.

Since the samples are independent, we have

$$\begin{aligned} p(z_N \mid \theta) &= \prod_{k=1}^{N} p(z_k \mid \theta) \\ &= \prod_{k=1}^{N} \frac{1}{\sqrt{2\pi}\,\sigma} \exp\left[-\frac{(z_k - y_k)^2}{2\sigma^2}\right] \\ &= \left(\frac{1}{\sqrt{2\pi}\,\sigma}\right)^N \exp\left[-\frac{1}{2\sigma^2}\sum_{k=1}^{N}(z_k - y_k)^2\right] \end{aligned} \tag{5.5.21}$$

It easily follows that

$$\frac{\partial \ln p(z_N \mid \theta)}{\partial \theta} = \frac{1}{\sigma^2}\sum_{k=1}^{N}(z_k - y_k)\frac{\partial y_k}{\partial \theta} \tag{5.5.22}$$

Letting $N \to \infty$, we obtain the maximum likelihood estimate of θ as the solution to the equation

$$\int_{t_0}^{t_f} [z(t) - y(t, \theta)] \frac{\partial y(t, \theta)}{\partial \theta} dt \bigg|_{\theta = \hat{\theta}_{ml}} = 0 \tag{5.5.23}$$

We can also use an expansion in a set of orthonormal basis functions $\{g_k(t)\}$ to derive this result. If we write the K-term approximation of $z(t)$ over the interval $[t_0, t_f]$ as

$$z_K(t) = \sum_{k=1}^{K} Z_k g_k(t) \tag{5.5.24}$$

we have

$$Z_k = y_k(\theta) + V_k$$

where now y_k and V_k are the coefficients of the expansion of $y(t, \theta)$ and $v(t)$, respectively.

It follows from the assumptions that

$$\mathrm{E}\{Z_k\} = y_k(\theta)$$

and

$$\mathrm{var}\{Z_k\} = \mathrm{var}\{V_k\} = \frac{N_0}{2} \tag{5.5.25}$$

Since the coefficients are independent, we have

$$\begin{aligned} p(z_K(t) \mid \theta) &= \prod_{k=1}^{K} p(z_k) \\ &= \prod_{k=1}^{K} \frac{1}{\sqrt{\pi N_0}} \exp\left[-\frac{(z_k - y_k(\theta))^2}{N_0}\right] \\ &= \left(\frac{1}{\sqrt{\pi N_0}}\right)^K \exp\left[-\frac{1}{N_0}\sum_{k=1}^{K} (z_k - y_k(\theta))^2\right] \end{aligned} \tag{5.5.26}$$

We now obtain the estimate of θ based on the K-term approximation by setting the partial derivative of $\ln p(z_k(t) \mid \theta)$ with respect to θ equal to zero to get

$$\sum_{k=1}^{K} (z_k - y_k(\theta)) \frac{\partial y_k(\theta)}{\partial \theta} = 0 \tag{5.5.27}$$

In the limit as $K \to \infty$, it can be shown that Eq. (5.5.27) yields the same result as Eq. (5.5.23). We note that Eq. (5.5.23) represents a necessary but not sufficient condition on $\hat{\theta}_{ml}$.

In particular, if θ is an amplitude parameter so that

$$y(t, \theta) = \theta y(t)$$

Eq. (5.5.23) reduces to

$$\int_{t_0}^{t_f} [z(t) - \theta y(t)] y(t)\, dt \bigg|_{\theta = \hat{\theta}_{ml}} = 0 \tag{5.5.28}$$

and the ML estimate is

$$\hat{\theta}_{ml} = \frac{\int_{t_0}^{t_f} z(t)y(t)\,dt}{\int_{t_0}^{t_f} y^2(t)\,dt} \tag{5.5.29}$$

If we normalize the signal $y(t)$ so that

$$\int_{t_0}^{t_f} y^2(t)\,dt = 1$$

we obtain for the estimate

$$\hat{\theta}_{ml} = \int_{t_0}^{t_f} z(t)y(t)\,dt \tag{5.5.30}$$

which corresponds to the correlation receiver operation discussed earlier. Since

$$\begin{aligned} E\{\hat{\theta}_{ml}\} &= E\left\{\int_{t_0}^{t_f} [\theta y(t) + v(t)]y(t)\,dt\right\} \\ &= \theta \end{aligned} \tag{5.5.31}$$

the estimate is unbiased. From Eq. (5.5.22) it follows easily that

$$\begin{aligned} \frac{\partial \ln p(z\,|\,\theta)}{\partial \theta} &= \frac{2}{N_0}\int_{t_0}^{t_f} [z(t) - \theta y(t)]y(t)\,dt \\ &= \frac{2}{N_0}[\hat{\theta}_{ml} - \theta] \end{aligned} \tag{5.5.32}$$

and is of the form of Eq. (5.5.9). Thus $\hat{\theta}_{ml}$ is an efficient estimate and its variance is given by Eq. (5.5.10) as $N_0/2$.

If the unknown parameter is a phase parameter, then

$$y(t, \theta) = A \sin(\omega_0 t + \theta)$$

With A and ω_0 known, the estimate is obtained from Eq. (5.5.23) by solving

$$\int_{t_0}^{t_f} [z(t) - A \sin(\omega_0 t + \hat{\theta})] \cos(\omega_0 t + \hat{\theta})\,dt = 0 \tag{5.5.33}$$

If we assume that $\omega_0(t_f - t_0) = k\pi$ for some integer k, the second integral is zero and the estimate is obtained as a solution of

$$\int_{t_0}^{t_f} z(t) \cos(\omega_0 t + \hat{\theta})\,dt = 0 \tag{5.5.34}$$

Use of trigonometric identities then enables us to write

$$\hat{\theta}_{ml} = \tan^{-1}\left\{\frac{\int_{t_0}^{t_f} z(t) \cos \omega_0 t\,dt}{\int_{t_0}^{t_f} z(t) \sin \omega_0 t\,dt}\right\} \tag{5.5.35}$$

If the measurement noise is colored, we can use the results of Section 4.4 to derive the estimator equations. We first pass the observations through a whitening filter with impulse response $h_w(\cdot,\cdot)$. The output of the whitening filter is then given by

$$\begin{aligned} z_w(t) &= \int_{t_0}^{t_f} h_w(t,u)z(u)\,du \\ &= \int_{t_0}^{t_f} h_w(t,u)y(u,\theta)\,du + \int_{t_0}^{t_f} h_w(t,u)v(u)\,du \\ &= y_w(t,\theta) + v_w(t) \qquad t_0 \leq t \leq t_f \end{aligned} \tag{5.5.36}$$

where $v_w(t)$ is white with spectral height arbitrarily chosen to be unity.

The estimator equations follow from Eq. (5.5.23) as

$$\int_{t_0}^{t_f} [z_w(t) - y_w(t,\theta)] \frac{\partial y_w(t,\theta)}{\partial \theta}\,dt = 0 \tag{5.5.37}$$

In terms of the original waveforms, Eq. (5.5.37) can be written as

$$\int_{t_0}^{t_f}\left[\int_{t_0}^{t_f} h_w(t,u)z(u)\,du - \int_{t_0}^{t_f} h_w(t,u)y(u,\theta)\,du\right] \times \left[\frac{\partial \int_{t_0}^{t_f} h_w(t,v)y(v,\theta)\,dv}{\partial \theta}\,dt\right] = 0 \tag{5.5.38}$$

Equation (5.5.38) can be written as

$$\int_{t_0}^{t_f}\left\{\int_{t_0}^{t_f} h_w(t,u)z(u)\,du - \int_{t_0}^{t_f} h_w(t,u)y(u,\theta)\,du\right\} \times \left\{\int_{t_0}^{t_f} h_w(t,v)\frac{\partial y(v,\theta)}{\partial \theta}\,dv\right\}dt = 0 \tag{5.5.39}$$

The ML estimate of θ can be obtained by solving Eq. (5.5.39). The determination of $h_w(\cdot,\cdot)$ was discussed in Section 4.5.

5.5.2 Uniformly Minimum Variance Unbiased Estimators

We have seen that given a set of samples drawn from a density function characterized by an unknown parameter θ, it is desirable to find an estimate of θ that is unbiased and has minimum variance. If the minimum variance condition holds for any value of θ in the parameter space, the estimate is known as a uniformly minimum variance unbiased estimate (UMVUE). Thus one should look for a UMVUE if such an estimate exists.

In the case of nonrandom parameters, if the density function of the observations satisfies certain restrictions, the maximum likelihood estimate is asymptotically unbiased with minimum possible variance. For large sample sizes, the ML estimate possesses an optimal property. In many situations the sample size is fixed. However, even with small sample sizes, it is still desirable to find an estimator that has minimum variance. Recall that an unbiased estimator is efficient if it attains the Cramér–Rao (CR) lower bound and that if an efficient estimator exists, it is the ML estimator.

However, it has been shown [14] that if $T^* = t^*(X_1, \ldots, X_n)$ is an unbiased estimator of some $\tau^*(\theta)$ whose variance coincides with the CR lower bound, then $p(\mathbf{x}\,|\,\theta)$ is a member of the exponential class; conversely, if $p(\cdot)$ is a member of the exponential class, there exists an unbiased estimator, say T^* of some function, say $\tau^*(\theta)$, whose variance coincides with the CR lower bound. Also, there is essentially one function (or any linear function of the one function) of the parameter for which there exists an unbiased estimator whose variance coincides with the CR lower bound.

Thus the CR bound is of limited use in finding a UMVUE and is useful only if the samples are from a member of the one-parameter exponential family. Even then, it is useful in finding the UMVUE of only one function of the parameter. Therefore, we need some other method that allows us to find a UMVUE. The following theorem due to Lehmann–Scheffe allows us to find a UMVUE under more general conditions.

Let $X_1, X_2, \ldots, X_n$ be a random sample from a density $p(\mathbf{x}\,|\,\theta)$. If $S = S(X_1, \ldots, X_n)$ is a complete sufficient statistic, and if $T^* = t^*(S)$, a function of S, is an unbiased estimator of $\tau(\theta)$, then T^* is a UMVUE of $\tau(\theta)$. The proof of this result can be found in [1,14].

Example 5.5.3

Consider a random sample of size n from the density

$$p(x\,|\,\theta) = \exp[-(x - \theta)]u(x - \theta),$$

$\theta > 0$, where $u(\cdot)$ is the unit step function. Let $(y_1, y_2, \ldots, y_n)$ be the ordered values obtained by arranging the samples $(x_1, x_2, \ldots, x_n)$ in increasing order of magnitude. We will show that Y_1, the minimum order statistic, is a complete and sufficient statistic for θ.

To show sufficiency, we write $p(\mathbf{x}\,|\,\theta)$ as

$$\begin{aligned} p(\mathbf{x}\,|\,\theta) &= \prod_{i=1}^{n} p(x_i\,|\,\theta) \\ &= \prod_{i=1}^{n} e^{-(x_i-\theta)} u(x_i - \theta) \\ &= e^{n\theta} e^{-\sum_{i=1}^{n} x_i} \prod_{k=2}^{n-1} I_{(y_{k-1}, y_{k+1})}(y_k) I_{(\theta,\infty)}(y_1) \end{aligned}$$

It follows from the factorization theorem, Eq. (2.6.12), that Y_1 is a sufficient statistic.

To show completeness, consider some function $t(Y_1)$ of Y_1. Then Y_1 is complete if $\mathrm{E}(t(Y_1)) = 0$ implies that $t(Y_1)$ is zero with probability 1. Let $p_{Y_1}(y)$ be the density of the minimum order statistic. Then

$$\begin{aligned} p_{Y_1}(y) &= \frac{d}{dy} F_{Y_1}(y) = \frac{d}{dy}[1 - P(Y_1 > y)] \\ &= \frac{d}{dy}[1 - P(X_1, X_2, \ldots, X_n > y)] \\ &= \frac{d}{dy}[1 - (1 - F_X(y))^n] \\ &= n(1 - F_X(y))^{n-1} p_X(y) \end{aligned}$$

Now

$$F_X(y) = \begin{cases} \int_\theta^y e^{-(x-\theta)}\,dx & y \geq \theta \\ 0 & y < \theta \end{cases}$$
$$= e^\theta(e^{-\theta} - e^{-y})u(y - \theta)$$

Hence

$$E(t(Y_1)) = \int_\theta^\infty t(y)n(1 - 1 + e^{-(y-\theta)})^{n-1}e^{-(y-\theta)}u(y - \theta)\,dy$$

Therefore, $E(t(Y_1)) = 0$ implies that

$$\int_\theta^\infty t(y)e^{-ny}\,dy = 0 \qquad \text{for all } \theta$$

Differentiating the equation above with respect to θ gives $t(\theta)e^{-n\theta} = 0$ for all θ so that $t(\theta) = 0$ for all θ. Hence $t(Y_1) = 0$ with probability 1, so that Y_1 is complete. Now

$$E(Y_1) = \int_\theta^\infty yne^{-n(y-\theta)}\,dy = ne^{n\theta}\int_\theta^\infty ye^{-ny}\,dy$$
$$= \theta + \frac{1}{n}$$

The statistic $Z = Y_1 - 1/n$ is an unbiased estimator of θ since $E(Z) = \theta$. Therefore, according to the Lehmann–Scheffé theorem, $(Y_1 - 1/n)$ is the UMVUE of θ.

Example 5.5.4

Consider the Poisson distribution

$$P(x \mid \theta) = \begin{cases} \dfrac{\theta^x e^{-\theta}}{x!} & x = 0, 1, 2, \ldots \\ 0 & \text{otherwise} \end{cases}$$

which belongs to the one-parameter exponential family. By writing the probability mass as $P(x \mid \theta) = e^{-\theta}(1/x!)e^{x \ln \theta}$, it follows that $d(x)$ in Eq. (2.6.21) is equal to x. Thus given a set of observations $X_1, X_2, \ldots, X_n$ from this distribution, the quantity $\sum_{i=1}^n X_i$ is a complete, minimal, sufficient statistic. Since $E\{(1/n)\sum_{i=1}^n X_i\} = \theta$, it follows that $(1/n)\sum_{i=1}^n X_i$ is the UMVUE of θ.

Example 5.5.5

Let $X_1, X_2, \ldots, X_n, n > 2$ be a random sample from the continuous density $p(x \mid \theta) = \theta e^{-\theta x}u(x), \theta > 0$.

Since $p(x \mid \theta)$ belongs to the single-parameter exponential family, as in Example 5.5.4, $\sum_{i=1}^n X_i$ is a complete and sufficient statistic for θ. Let $Y = \sum_{i=1}^n X_i$ and let $T = (n - 1)/Y$. Then, since the samples X_i are independent, identically distributed exponential random variables, Y has a gamma distribution with parameters n and θ. Hence

$$E(T) = \int_0^\infty \frac{n-1}{y}\frac{y^{n-1}e^{-\theta y}\theta^n}{(n-1)!}\,dy$$
$$= \theta\left[\int_0^\infty \frac{y^{n-2}\theta^{n-1}}{(n-2)!}e^{-\theta y}\,dy\right]$$
$$= \theta$$

so that T is unbiased. From the Lehmann–Scheffé theorem, it follows that T is the UMVUE of θ.

5.5.3 Random Parameter Estimation: Bounds and MAP Estimates

Analogous to our derivation for the nonrandom parameter case, we can establish bounds on the variance of any estimate of a random parameter. Here, in addition to requiring that $\partial p(z,\theta)/\partial\theta$ and $\partial^2 p(z,\theta)/\partial\theta^2$ both be absolutely integrable with respect to z and θ, we will also assume that the following condition holds:

$$\lim_{\theta\to\pm\infty} p(\theta)\int_{-\infty}^{\infty}[\hat{\theta}-\theta]p(z\,|\,\theta)\,dz = 0 \tag{5.5.40}$$

We can then establish along the same lines as in the nonrandom parameter case that if $\hat{\theta}$ is any unbiased estimate of θ, then

$$\begin{aligned} \mathrm{E}\{(\Theta-\hat{\Theta})^2\} &\geq \left(\mathrm{E}\left\{\left[\frac{\partial \ln p(z,\theta)}{\partial\theta}\right]^2\right\}\right)^{-1} \\ &= \left\{-\mathrm{E}\left[\frac{\partial^2 \ln p(z,\theta)}{\partial\theta}\right]\right\}^{-1} \end{aligned} \tag{5.5.41}$$

with equality if and only if

$$\frac{\partial \ln p(z,\theta)}{\partial\theta} = k[\hat{\theta}-\theta] \tag{5.5.42}$$

for all $\mathbf{z}$ and θ. We note that in Eq. (5.5.42) k cannot be a function of $\hat{\theta}$, $\mathbf{z}$, or θ.

Arguing as in Eqs. (5.5.15)–(5.5.17), we see that if Eq. (5.5.42) is satisfied, the MAP estimate will be efficient. Because the MSE estimate cannot have a larger variance, this implies that $\hat{\theta}_{ms} = \hat{\theta}_{MAP}$ whenever an efficient estimate exists. In such a case it is easier to solve the MAP equation to determine the estimate than to compute the conditional mean. The variance of the estimate can be determined from the bound given in Eq. (5.5.41).

Example 5.5.6

We consider the same problem as in Example 5.5.2, but assume that Θ is a Gaussian random variable with zero mean and variance σ_θ^2. We consider only the case when the observation noise is Gaussian, white. The extension to the colored noise problem is obvious.

To find the MAP estimate of Θ, we note that

$$p(\mathbf{z},\theta) = p(z\,|\,\theta)p(\theta)$$

so that to obtain the MAP estimate we need to add $d \ln p(\theta)/d\theta$ to the left side of Eq. (5.5.23). We then get

$$\hat{\theta}_{MAP} = \frac{2}{N_0}\sigma_\theta^2\int_{t_0}^{t_f}[z(t)-y(t,\theta)]\frac{\partial y(t,\theta)}{\partial\theta}\,dt\,\bigg|_{\theta=\hat{\theta}_{MAP}} \tag{5.5.43}$$

Again, Eq. (5.5.43) is necessary but not a sufficient condition on $\hat{\theta}_{MAP}$.

Simultaneous estimation of multiple parameters. The discussion so far has been concerned with the estimation of a single parameter. We can easily extend our results to the case of estimation of several parameters. For the parameter vector

$$\theta = [\theta_1, \theta_2, \dots, \theta_n]^T$$

we define the parameter estimation error as

$$\tilde{\theta}(\mathbf{z}) = \begin{bmatrix} \theta_1 - \hat{\theta}_1(\mathbf{z}) \\ \theta_2 - \hat{\theta}_2(\mathbf{z}) \\ \vdots \\ \theta_n - \hat{\theta}_n(\mathbf{z}) \end{bmatrix} \tag{5.5.44}$$

The mean-square estimate is then determined by minimizing the average risk

$$\overline{C}_{\text{ms}} = \int_{-\infty}^{\infty}\int_{-\infty}^{\infty} C(\tilde{\theta}(\mathbf{z}))p(\mathbf{z}, \theta)\, d\mathbf{z}\, d\theta \tag{5.5.45}$$

where

$$C(\tilde{\theta}(\mathbf{z})) = \sum_{i=1}^{n} \tilde{\theta}_i^2(\mathbf{z}) = \tilde{\theta}^T(\mathbf{z})\tilde{\theta}(\mathbf{z}) \tag{5.5.46}$$

Rewriting Eq. (5.5.45) in terms of the conditional density $p(\theta \mid \mathbf{z})$ gives

$$\overline{C}_{\text{ms}} = \int_{-\infty}^{\infty} p(\mathbf{z})\, d\mathbf{z} \int_{-\infty}^{\infty} \left[\sum_{i=1}^{n} (\theta_i - \hat{\theta}_i(\mathbf{z}))^2\right] p(\theta \mid \mathbf{z})\, d\theta \tag{5.5.47}$$

As before we can minimize the inner integral. Since the terms are positive, we minimize them separately to get the estimate of the ith element as

$$(\hat{\theta}_i)_{\text{ms}}(\mathbf{z}) = \int_{-\infty}^{\infty} \theta_i p(\theta \mid \mathbf{z})\, d\theta \tag{5.5.48}$$

or in vector notation,

$$\hat{\theta}_{\text{ms}} = \int_{-\infty}^{\infty} \theta p(\theta \mid \mathbf{z})\, d\theta \tag{5.5.49}$$

For MAP estimation, we must solve the set of simultaneous equations

$$\left.\frac{\partial \ln p(\theta \mid \mathbf{z})}{\partial \theta_i}\right|_{\theta = \hat{\theta}_{\text{MAP}}} = 0 \qquad i = 1, 2, \dots, n \tag{5.5.50}$$

We can write Eq. (5.5.50) compactly in terms of the partial derivative operator ∇_θ as

$$\nabla_\theta[\ln p(\theta \mid \mathbf{z})]_{\theta = \hat{\theta}_{\text{MAP}}} = 0 \tag{5.5.51}$$

For ML estimates we solve the set of equations

$$\nabla_\theta[\ln p(\mathbf{z} \mid \theta)]_{\theta = \hat{\theta}_{\text{ml}}} = 0 \tag{5.5.52}$$

We can obtain lower bounds on the variance of the estimation error for the multiple-parameter case similar to the ones derived earlier. For nonrandom variables this bound is given by

$$\operatorname{var} \tilde{\theta}_i(\mathbf{Z}) \geq \psi_{ii} \tag{5.5.53}$$

where ψ_{ii} is the iith element of the $n \times n$ matrix $\boldsymbol{\psi} = \mathbf{J}^{-1}$, with $\mathbf{J}$ given by

$$\begin{aligned} J_{ij} &= \mathrm{E}\left\{\frac{\partial \ln p(\mathbf{z}\,|\,\theta)}{\partial \theta_i}\frac{\partial \ln p(\mathbf{z}\,|\,\theta)}{\partial \theta_j}\right\} \\ &= -\mathrm{E}\left\{\frac{\partial^2 \ln p(\mathbf{z}\,|\,\theta)}{\partial \theta_i\,\partial \theta_j}\right\} \end{aligned} \tag{5.5.54}$$

The matrix $\mathbf{J}$ is commonly referred to as the *Fisher information matrix*. Equality in Eq. (5.5.53) holds if and only if

$$\hat{\theta}_i(\mathbf{z}) - \theta_i = \sum_{j=1}^{n} k_{ij}(\theta)\frac{\partial \ln p(\mathbf{z}\,|\,\theta)}{\partial \theta_j} \tag{5.5.55}$$

for all θ_i and $\mathbf{z}$.

For random parameters, we replace the conditional density function in the right sides of Eqs. (5.5.54) to (5.5.55) by the joint density function $p(\theta, \mathbf{z})$.

We conclude this section by considering an example.

Example 5.5.7

We consider the problem of estimating a random vector parameter Θ observed in white Gaussian noise. The observations are of the form

$$Z_i = \sum_{j=1}^{n} h_{ij}\Theta_j + V_i \qquad i = 1, \ldots, m \tag{5.5.56}$$

with V_i being a set of independent variables. In vector notation, the above can be written as

$$\mathbf{Z} = \mathbf{H}\boldsymbol{\Theta} + \mathbf{V} \tag{5.5.57}$$

where $\mathbf{Z}$ and $\mathbf{V}$ are m-dimensional column vectors, $\boldsymbol{\Theta}$ is n-dimensional, and the observation matrix $\mathbf{H}$ is $m \times n$. Let us assume that the prior density for $\boldsymbol{\Theta}$ is Gaussian, $N(\mu_\theta, \mathbf{V}_\theta)$ and that for $\mathbf{V}$ is $N(0, \mathbf{V}_v)$. Since the variables V_i are independent, the matrix $\mathbf{V}_v$ will be diagonal. The relevant densities for this problem will then be Gaussian with

$$p(\mathbf{z}|\theta) \sim N(\mathbf{H}\theta, \mathbf{V}_v) \tag{5.5.58a}$$

$$p(\mathbf{z}) \sim N(\mathbf{H}\boldsymbol{\mu}_\theta, \mathbf{H}\mathbf{V}_\theta\mathbf{H}^T + \mathbf{V}_v) \tag{5.5.58b}$$

Explicitly these densities can be written as

$$p(\mathbf{z}\,|\,\theta) = \frac{1}{(2\pi)^{m/2}|\mathbf{V}_v|^{1/2}}\exp\left[-\frac{1}{2}(\mathbf{z} - \mathbf{H}\theta)^T\mathbf{V}_v^{-1}(\mathbf{z} - \mathbf{H}\theta)\right] \tag{5.5.59a}$$

$$\begin{aligned} p(\mathbf{z}) &= \frac{1}{(2\pi)^{m/2}|\mathbf{H}\mathbf{V}_\theta\mathbf{H}^T + \mathbf{V}_v|^{1/2}} \\ &\quad \times \exp\left\{-\frac{1}{2}(\mathbf{z} - \mathbf{H}\boldsymbol{\mu}_\theta)^T[\mathbf{H}\mathbf{V}_\theta\mathbf{H}^T + \mathbf{V}_v]^{-1}(\mathbf{z} - \mathbf{H}\boldsymbol{\mu}_\theta)\right\} \end{aligned} \tag{5.5.59b}$$

$$p(\theta) = \frac{1}{(2\pi)^{n/2}|\mathbf{V}_\theta|^{1/2}}\exp\left[-\frac{1}{2}(\theta - \boldsymbol{\mu}_\theta)^T\mathbf{V}_\theta^{-1}(\theta - \boldsymbol{\mu}_\theta)\right] \tag{5.5.59c}$$

Use of the matrix lemma and some algebraic manipulations enables us to write the posterior density as

$$p(\theta|\mathbf{z}) = \frac{p(\mathbf{z}|\theta)p(\theta)}{p(\mathbf{z})} = \frac{|\mathbf{H}\mathbf{V}_\theta\mathbf{H}^T + \mathbf{V}_v|^{1/2}}{(2\pi)^{n/2}|\mathbf{V}_\theta|^{1/2}|\mathbf{V}_v|^{1/2}} \exp\left[-\frac{1}{2}(\theta - \xi)^T\Sigma^{-1}(\theta - \xi)\right] \tag{5.5.60}$$

where

$$\Sigma^{-1} = \mathbf{V}_\theta^{-1} + \mathbf{H}^T\mathbf{V}_v^{-1}\mathbf{H} \tag{5.5.61}$$

and

$$\xi = \Sigma(\mathbf{H}^T\mathbf{V}_v^{-1}\mathbf{z} + \mathbf{V}_\theta^{-1}\boldsymbol{\mu}_\theta) \tag{5.5.62}$$

Since this density is in a Gaussian form, we have

$$\begin{aligned}\hat{\theta}_{\mathrm{MAP}} = \hat{\theta}_{\mathrm{ms}} = \xi &= \Sigma(\mathbf{H}^T\mathbf{V}_v^{-1}\mathbf{z} + \mathbf{V}_\theta^{-1}\boldsymbol{\mu}_\theta) \\ &= (\mathbf{V}_\theta^{-1} + \mathbf{H}^T\mathbf{V}_v^{-1}\mathbf{H})^{-1}(\mathbf{H}^T\mathbf{V}_v^{-1}\mathbf{z} + \mathbf{V}_\theta^{-1}\boldsymbol{\mu}_\theta)\end{aligned} \tag{5.5.63}$$

The error variance is given by

$$\mathbf{V}_{\tilde{\theta}} = \Sigma = (\mathbf{V}_\theta^{-1} + \mathbf{H}^T\mathbf{V}_v^{-1}\mathbf{H})^{-1}$$

which can be written using matrix identities (see Appendix C) as

$$\mathbf{V}_{\tilde{\theta}} = \mathbf{V}_\theta - \mathbf{V}_\theta\mathbf{H}^T(\mathbf{V}_v + \mathbf{H}\mathbf{V}_\theta\mathbf{H}^T)^{-1}\mathbf{H}\mathbf{V}_\theta \tag{5.5.64}$$

The estimates above were obtained using a measurement vector of dimension n. If more measurements are available, we may construct a new estimate by repeating the procedure above and using the new observation vector. However, it seems reasonable to obtain the updated estimate as a function of the new observation and the old estimate since the first n measurements have already been used in an optimal fashion.

To see how this can be done, let us denote the estimate based on n measurements as $\hat{\theta}_n$. For simplicity, we will assume that $\boldsymbol{\mu}_\theta = 0$. Let $\mathbf{z}(n)$ denote the measurement vector and $\mathbf{H}(n)$ the corresponding observation matrix. The estimate based on $(n + 1)$ measurements is, from our previous derivation, given by

$$\hat{\theta}_{n+1} = \Sigma_{n+1}[\mathbf{H}^T(n + 1)\mathbf{V}_v^{-1}(n + 1)\mathbf{z}(n + 1)] \tag{5.5.65}$$

where

$$\Sigma_{n+1}^{-1} = \mathbf{V}_\theta^{-1} + \mathbf{H}^T(n + 1)\mathbf{V}_v^{-1}(n + 1)\mathbf{H}(n + 1) \tag{5.5.66}$$

$$\begin{aligned}\mathbf{H}(n + 1) &= \begin{bmatrix}\mathbf{H}(n) \\ \mathbf{h}_{n+1}\end{bmatrix}, \mathbf{z}(n + 1) = \begin{bmatrix}\mathbf{z}(n) \\ z_{n+1}\end{bmatrix} \\ \mathbf{V}_v(n + 1) &= \begin{bmatrix}\mathbf{V}_v(n) & \mathbf{0} \\ \mathbf{0} & \operatorname{cov} V_{n+1}\end{bmatrix}\end{aligned} \tag{5.5.67}$$

and $\mathbf{h}_{n+1} = [h_{1,n+1} \quad h_{2,n+1} \quad \cdots \quad h_{n+1,n+1}]$. Substituting in the expression for Σ_{n+1} yields

$$\begin{aligned}\Sigma_{n+1}^{-1} &= \mathbf{V}_\theta^{-1} + \mathbf{H}^T(n)\mathbf{V}_v^{-1}(n)\mathbf{H}(n) + \mathbf{h}_{n+1}^T(\operatorname{cov} V_{n+1})^{-1}\mathbf{h}_{n+1} \\ &= \Sigma_n^{-1} + \mathbf{h}_{n+1}^T(\operatorname{cov} V_{n+1})^{-1}\mathbf{h}_{n+1}\end{aligned} \tag{5.5.68}$$

Again using matrix identities, the preceding can be written as

$$\Sigma_{n+1} = \Sigma_n - \Sigma_n \mathbf{h}_{n+1}^T[\mathbf{h}_{n+1}\Sigma_n \mathbf{h}_{n+1}^T + \text{cov } V_{n+1}]^{-1}\mathbf{h}_{n+1}\Sigma_n \tag{5.5.69}$$

Thus the estimate $\hat{\theta}_{n+1}$ can be written as

$$\begin{aligned}
\hat{\theta}_{n+1} &= \Sigma_{n+1}[\mathbf{H}^T(n)\mathbf{V}_v^{-1}(n)\mathbf{z}(n) + \mathbf{h}_{n+1}^T(\text{cov } V_{n+1})^{-1} z_{n+1}] \\
&= \Sigma_n \mathbf{H}^T(n)\mathbf{V}_v^{-1}(n)\mathbf{z}(n) + \Sigma_n \mathbf{h}_{n+1}^T(\text{cov } V_{n+1})^{-1} z_{n+1} \\
&\quad - \Sigma_n \mathbf{h}_{n+1}^T(\mathbf{h}_{n+1}\Sigma_n \mathbf{h}_{n+1}^T + \text{cov } V_{n+1})^{-1}\mathbf{h}_{n+1} \\
&\quad \times \Sigma_n[\mathbf{H}^T(n)\mathbf{V}_v^{-1}(n)\mathbf{z}(n) + \mathbf{h}_{n+1}^T(\text{cov } V_{n+1})^{-1} z_{n+1}] \\
&= \hat{\theta}_n + \Sigma_n \mathbf{h}_{n+1}^T[\mathbf{h}_{n+1}\Sigma_n \mathbf{h}_{n+1}^T + \text{cov } V_{n+1}]^{-1}[z_{n+1} - \mathbf{h}_{n+1}\hat{\theta}_n]
\end{aligned} \tag{5.5.70}$$

Thus the present estimate is equal to the old estimate plus a linear correction term based on the new measurement. The computations involved in processing the data sequentially as in the preceding are much less than in obtaining the estimate directly from processing the measurements together in a batch.

5.6 LINEAR MEAN-SQUARE ESTIMATION

In our discussion of the estimation of a random parameter Θ from observations of a related random variable Z, we saw that the minimum-mean-square estimate was given by the mean of the posterior density $p(\theta \mid z)$:

$$\hat{\theta}_{\text{ms}} = \mathrm{E}\{\Theta \mid z\} = \int_{-\infty}^{\infty} \theta p(\theta \mid z)\, d\theta \tag{5.6.1}$$

As discussed earlier, the computation of this conditional mean becomes quite involved in many problems. In general, the conditional mean will be a nonlinear function of the observations z. Thus rather than seeking the mean-square estimate of θ, we look for the best linear mean-square estimate of θ, that is, the best estimate of θ expressible as a linear function of the observation z. If we write the linear minimum-mean-square estimate of θ as

$$\hat{\theta}_{\text{lms}} = az + b \tag{5.6.2}$$

the problem reduces to one of finding the best constants a and b so as to minimize

$$\begin{aligned}
\overline{C}_{\text{lms}} &= \mathrm{E}\{(\Theta - \hat{\Theta}_{\text{lms}})^2\} \\
&= \int\int_{-\infty}^{\infty} (\theta - \hat{\theta}_{\text{lms}})^2 p(z, \theta)\, dz\, d\theta
\end{aligned} \tag{5.6.3}$$

As in Section 5.3, we can easily show that $\hat{\theta}_{\text{lms}}$ can be obtained by minimizing

$$\int_{-\infty}^{\infty} (\theta - \hat{\theta}_{\text{lms}})^2 p(\theta \mid z)\, d\theta \tag{5.6.4}$$

If the conditional mean were expressible as a linear function of the observations z, the linear mean square estimate would also be the minimum mean square estimate, $\hat{\theta}_{\text{lms}} = \hat{\theta}_{\text{ms}}$.

It is instructive to determine under what conditions this can happen [16]. Assume that

$$E\{\Theta \mid z\} = az + b \tag{5.6.5}$$

We start with the characteristic equation associated with the density function $p(z, \theta)$

$$C_{z\theta}(\omega_1, \omega_2) = \iint_{-\infty}^{\infty} p(z, \theta) e^{j(\omega_1 z + \omega_2 \theta)}\, dz\, d\theta \tag{5.6.6}$$

Differentiate both sides with respect to $j\omega_2$ and set $\omega_2 = 0$ to get

$$\begin{aligned} \frac{\partial}{\partial j\omega_2} C_{z\theta}(\omega_1, \omega_2)\bigg|_{\omega_2=0} &= \iint_{-\infty}^{\infty} \theta p(z, \theta) e^{j\omega_1 z}\, dz\, d\theta \\ &= \int_{-\infty}^{\infty} \left[\int_{-\infty}^{\infty} \theta p(\theta \mid z)\, d\theta\right] e^{j\omega_1 z} p(z)\, dz \end{aligned} \tag{5.6.7}$$

Substituting for the inner integral on the right-hand side of Eq. (5.6.7) from Eq. (5.6.5) yields

$$\begin{aligned} \frac{\partial}{\partial j\omega_2} C_{z\theta}(\omega_1, \omega_2)\bigg|_{\omega_2=0} &= \int_{-\infty}^{\infty} (az + b) e^{j\omega_1 z} p(z)\, dz \\ &= a \frac{\partial}{\partial j\omega_1} \int_{-\infty}^{\infty} e^{j\omega_1 z} p(z)\, dz + b \int_{-\infty}^{\infty} e^{j\omega_1 z} p(z)\, dz \\ &= \left(a \frac{\partial}{\partial j\omega_1} + b\right) C_z(\omega_1) \end{aligned} \tag{5.6.8}$$

It can also be shown by retracing our steps that Eq. (5.6.8) is a sufficient condition for the conditional mean $E\{\Theta \mid z\}$ to be of the form of Eq. (5.6.5). It can easily be verified that Eq. (5.6.8) is satisfied if $p(z, \theta)$ is Gaussian. It follows that if the joint density function is Gaussian, the conditional mean is linear in the observations. Thus the minimum mean square estimate is linear.

To return to our original problem, we will now seek the best linear estimate $\hat{\theta}_{\text{lms}}$ of the parameter Θ by choosing the constants a and b of Eq. (5.6.2) to minimize the risk function of Eq. (5.6.3). Thus substituting for $\hat{\theta}_{\text{lms}}$ in Eq. (5.6.3) yields

$$\overline{C}_{\text{lms}} = \iint_{-\infty}^{\infty} [\theta - az - b]^2 p(z, \theta)\, dz\, d\theta \tag{5.6.9}$$

Setting the partial derivatives with respect to a and b equal to zero yields the two equations

$$\iint_{-\infty}^{\infty} z[\theta - az - b] p(z, \theta)\, dz\, d\theta = 0 \tag{5.6.10}$$

and

$$\iint_{-\infty}^{\infty} [\theta - az - b] p(z, \theta)\, dz\, d\theta = 0 \tag{5.6.11}$$

or, equivalently,

$$a\mathrm{E}\{Z^2\} + b\mathrm{E}\{Z\} = \mathrm{E}\{\Theta Z\} \tag{5.6.12}$$

and

$$a\mathrm{E}\{Z\} + b = \mathrm{E}\{\Theta\} \tag{5.6.13}$$

Solving for a and b then gives

$$a = \frac{\mathrm{E}\{Z\Theta\} - \mathrm{E}\{Z\}\mathrm{E}\{\Theta\}}{\mathrm{E}\{Z^2\} - \mathrm{E}^2\{Z\}} \tag{5.6.14}$$

$$b = \mathrm{E}\{\Theta\} - a\mathrm{E}\{Z\} \tag{5.6.15}$$

In terms of the correlation coefficient ρ,

$$\rho = \frac{\mathrm{E}\{[Z - \mathrm{E}\{Z\}][\Theta - \mathrm{E}\{\Theta\}]\}}{\sqrt{\mathrm{E}\{[Z - \mathrm{E}\{Z\}]^2\}\mathrm{E}\{[\Theta - \mathrm{E}\{\Theta\}]^2\}}} \tag{5.6.16}$$

we can write

$$a = \frac{\rho\sigma_\theta}{\sigma_z} \tag{5.6.17}$$

The optimal cost obtained by substituting these values of a and b in Eq. (5.6.9) can easily be shown to be

$$\overline{C}_{\text{lms}} = \sigma_\theta^2(1 - \rho^2) \tag{5.6.18}$$

If θ is a vector parameter, the optimal estimate can easily be shown to be

$$\hat{\boldsymbol{\theta}}_{\text{lms}} = \mathbf{Az} + \mathbf{b} \tag{5.6.19a}$$

where the matrix $\mathbf{A}$ and the vector $\mathbf{b}$ are given by

$$\begin{aligned}\mathbf{A} &= [\mathrm{E}\{\mathbf{ZZ}^T\} - \mathrm{E}\{\mathbf{Z}\}\mathrm{E}\{\mathbf{Z}^T\}]^{-1}[\mathrm{E}\{\boldsymbol{\Theta}\mathbf{Z}^T\} - \mathrm{E}\{\boldsymbol{\Theta}\}\mathbf{E}\{\mathbf{Z}^T\}] \\ &= \mathbf{V}_z^{-1}\mathbf{V}_{\theta z}\end{aligned} \tag{5.6.19b}$$

and

$$\mathbf{b} = \mathrm{E}\{\boldsymbol{\Theta}\} - \mathbf{A}\mathrm{E}\{\mathbf{Z}\} \tag{5.6.19c}$$

Example 5.6.1

We consider the problem of estimating a parameter Θ on the basis of a single observation Z. Let Z and Θ be jointly normal with means μ_z and μ_θ respectively. Let the elements of the covariance matrix

$$\mathrm{E}\left\{\begin{bmatrix} Z - \mu_z \\ \Theta - \mu_\theta \end{bmatrix}[Z - \mu_z \quad \Theta - \mu_\theta]\right\}$$

be given by $v_{ij}, i, j = 1, 2$.

The joint characteristic function of z and θ is given by

$$C_{z\theta}(\omega_1, \omega_2) = \exp\{j(\mu_z\omega_1 + \mu_\theta\omega_2) - \tfrac{1}{2}(v_{11}\omega_1^2 + 2v_{12}\omega_1\omega_2 + v_{22}\omega_2^2)\}$$

Thus

$$\frac{\partial}{\partial j\omega_2} C_{z\theta}(\omega_1, \omega_2)\bigg|_{\omega_2=0} = (\mu_\theta + \nu_{12} j\omega_1) \exp(j\mu_z \omega_1 - \tfrac{1}{2}\nu_{11}\omega_1^2)$$

Also since

$$C_z(\omega_1) = \exp(j\omega_1 \mu_z - \tfrac{1}{2}\nu_{11}\omega_1^2)$$

we have

$$\frac{\partial}{\partial j\omega_1} C_z(\omega_1) = (\mu_z + \nu_{11} j\omega_1) \exp(j\omega_1 \mu_z - \tfrac{1}{2}\nu_{11}\omega_1^2)$$

and

$$\left(a\frac{\partial}{\partial j\omega_1} + b\right) C_z(\omega_1) = (a\mu_z + b + a\nu_{11} j\omega_1) \exp(j\omega_1 \mu_z - \tfrac{1}{2}\nu_{11}\omega_1^2)$$

Thus Eq. (5.6.8) will be satisfied with

$$a\mu_z + b = \mu_\theta$$

and

$$a\nu_{11} = \nu_{12}$$

Solving for a and b yields

$$a = \frac{\nu_{12}}{\nu_{11}}$$

and

$$b = \mu_\theta - a\mu_z$$

so that

$$\hat{\theta}_{\text{ms}} = \text{E}\{\Theta \mid z\} = az + b$$

Thus the linear mean square estimate is also the minimum mean square estimate in this case. We note that a and b obtained previously are the same as in Eqs. (5.6.14) and (5.6.15).

It can easily be shown by direct calculation that

$$\text{E}\{\tilde{\Theta}Z\} = \text{E}\{\Theta Z\} - a\text{E}\{Z^2\} - b\text{E}\{Z\} = 0$$

so that the error in the estimate is orthogonal to the observation z.

Example 5.6.2

We can use our result on mean-square estimation of jointly Gaussian random variables of Example 5.6.1 to derive an expression for the conditional density $p(\theta \mid z)$.

Let us assume for simplicity that $\mu_z = \mu_\theta = 0$. Since $p(\theta \mid z)$ will be Gaussian, we need determine only its mean and variance. From Example 5.6.1 we know that

$$\text{E}\{\Theta \mid z\} = az + b$$

with

$$a = \frac{\rho\sigma_\theta}{\sigma_z} \quad \text{and} \quad b = 0$$

The variance is given by

$$\begin{aligned}\operatorname{var}\{\Theta \mid z\} &= \mathrm{E}\{(\Theta - az - b)^2 \mid z) = \overline{C}_{\text{ms}} \\ &= \sigma_\theta^2(1 - \rho^2)\end{aligned}$$

so that

$$p(\theta \mid z) = \frac{1}{\sqrt{2\pi\sigma_\theta^2(1-\rho^2)}} \exp\left[-\frac{(\theta - az)^2}{2\sigma_\theta^2(1-\rho^2)}\right]$$

5.6.1 Orthogonality Principle

It was shown in Example 5.6.1 that the error in the estimate was orthogonal to the observation z. This result is valid for general linear mean-square estimation also and is of significance in setting up the equations needed to obtain the estimate. We will now derive this result for the more general problem of estimating a vector parameter $\boldsymbol{\Theta}$ from the observation vector $\mathbf{z}$.

Let us assume that $\mathrm{E}\{\mathbf{Z}\} = \mathrm{E}\{\boldsymbol{\Theta}\} = 0$. Then we can write the linear mean-square estimate as

$$\hat{\boldsymbol{\theta}}_{\text{lms}} = \mathbf{A}\mathbf{z} \tag{5.6.20}$$

We will now show that the constant matrix $\mathbf{A}$ that minimizes the mean square error

$$\begin{aligned}\overline{C}_{\text{lms}} &= \mathrm{E}\{[\boldsymbol{\Theta} - \mathbf{AZ}]^T[\boldsymbol{\Theta} - \mathbf{AZ}]\} \\ &= \mathrm{E}\{\|\boldsymbol{\Theta} - \mathbf{AZ}\|^2\}\end{aligned} \tag{5.6.21}$$

is such that the error

$$\tilde{\boldsymbol{\theta}} = \boldsymbol{\theta} - \mathbf{Az} \tag{5.6.22}$$

is orthogonal to $\mathbf{z}$. That is,

$$\mathrm{E}\{(\boldsymbol{\Theta} - \mathbf{AZ})^T\mathbf{Z}\} = 0 \tag{5.6.23}$$

To see this, suppose that $\hat{\boldsymbol{\theta}} = \mathbf{A}_1\mathbf{z}$, represents any other linear estimate. Then

$$\begin{aligned}\mathrm{E}\{\|\boldsymbol{\Theta} - \mathbf{A}_1\mathbf{Z}\|^2 &= \mathrm{E}\{\|\boldsymbol{\Theta} - \mathbf{AZ} + (\mathbf{A} - \mathbf{A}_1)\mathbf{Z}\|^2\} \\ &= \mathrm{E}\{\|\boldsymbol{\Theta} - \mathbf{AZ}\|^2\} + (\mathbf{A} - \mathbf{A}_1)^T\mathrm{E}\{\mathbf{Z}^T(\boldsymbol{\Theta} - \mathbf{AZ})\} \\ &\quad + \mathrm{E}\{(\boldsymbol{\Theta} - \mathbf{AZ})^T\mathbf{Z}\}(\mathbf{A} - \mathbf{A}_1) \\ &\quad + \mathrm{E}\{\|(\mathbf{A} - \mathbf{A}_1)\mathbf{Z}\|^2\} \\ &\geq \mathrm{E}\{\|\boldsymbol{\Theta} - \mathbf{AZ}\|^2\}\end{aligned} \tag{5.6.24}$$

since the last term is nonnegative and the second and third terms are zero from Eq. (5.6.23).

The orthogonality principle can be interpreted in terms of projections on a linear space defined as follows [17]. Let $\mathcal{X}$ be the linear space formed from a finite collection of random vectors $\mathbf{z}_1, \mathbf{z}_2, \ldots, \mathbf{z}_n$ such that $\mathrm{E}\{\mathbf{z}_k \mathbf{z}_k^T\} < \infty$ for each k. Thus $\mathfrak{X}$ is the space of random vectors with finite second moments. We define the inner product of two vectors $\mathbf{x}$ and $\mathbf{y}$ on this space as

$$\langle \mathbf{x}, \mathbf{y} \rangle = \mathrm{E}\{\mathbf{x}^T\mathbf{y}\} \tag{5.6.25}$$

with associated norm

$$\|\mathbf{x}\| = [\mathrm{E}\{\mathbf{x}^T\mathbf{x}\}]^{1/2} \tag{5.6.26}$$

We say that the two vectors $\mathbf{x}$ and $\mathbf{y}$ are orthogonal if

$$\mathrm{E}\{\mathbf{x}^T\mathbf{y}\} = 0$$

If $\mathfrak{X}_1$ and $\mathfrak{X}_2$ are two complementary and orthogonal subspaces of $\mathfrak{X}$, then every vector $\mathbf{x} \in \mathfrak{X}$ can be decomposed as $\mathbf{x} = \mathbf{x}_1 + \mathbf{x}_2$ with $\mathbf{x}_1 \in \mathfrak{X}_1$ and $\mathbf{x}_2 \in \mathfrak{X}_2$ so that $\mathbf{x}_1$ and $\mathbf{x}_2$ are orthogonal. The operator P such that

$$P\mathbf{x} = \mathbf{x}_1$$

is then the projection operator onto the subspace $\mathfrak{X}_1$. The projection theorem for linear spaces says that if $\mathbf{x}$ is any vector in $\mathfrak{X}$ and $\mathbf{x}'$ is a vector in a subspace $\mathfrak{X}'$ of $\mathfrak{X}$, then the norm of the vector $\mathbf{x} - \mathbf{x}'$ has a minimum value if and only if $\mathbf{x}'$ is the projection of $\mathbf{x}$ on $\mathfrak{X}'$ [18].

Let us now return to the problem of estimating the parameter vector $\boldsymbol{\Theta}$. Suppose that we have N measurements of the vector $\mathbf{z}$. Then the best mean square estimate of $\boldsymbol{\Theta}$ is

$$\hat{\boldsymbol{\theta}}_{\mathrm{ms}} = \mathrm{E}\{\boldsymbol{\Theta} \mid \mathbf{z}_1, \ldots, \mathbf{z}_N\} \tag{5.6.27}$$

If we restrict ourselves to linear estimates, it follows from Eq. (5.6.23) and the discussions above that the best linear estimate $\hat{\boldsymbol{\theta}}_{\mathrm{lms}}$ is the projection of the vector θ onto the space generated by the vectors $\mathbf{z}_1, \mathbf{z}_2, \ldots, \mathbf{z}_N$. If we write

$$\hat{\boldsymbol{\theta}}_{\mathrm{lms}} = \sum_{i=1}^{N} \alpha_i \mathbf{z}_i \tag{5.6.28}$$

then the weight factors α_i are obtained from the orthogonality principle by solving the set of equations

$$\mathrm{E}\left\{\left(\Theta - \sum_{i=1}^{N} \alpha_i \mathbf{z}_i\right)\mathbf{z}_j^T\right\} = 0 \qquad j = 1, 2, \ldots, N \tag{5.6.29}$$

If the condition of Eq. (5.6.8) is satisfied, then $\hat{\theta}_{\mathrm{ms}} = \mathrm{E}\{\boldsymbol{\Theta} \mid \mathbf{z}\}$ will be an element of the space generated by $\mathbf{z}_1 \cdots \mathbf{z}_N$, and in this case $\hat{\boldsymbol{\theta}}_{\mathrm{ms}} = \hat{\boldsymbol{\theta}}_{\mathrm{lms}}$. When the conditional expectation is not an element of this space, the projection of θ on this space gives the best linear mean-square estimate. We denote this by

$$\hat{\boldsymbol{\theta}}_{\mathrm{lms}} = \hat{\mathrm{E}}\{\boldsymbol{\Theta} \mid \mathbf{z}_1, \ldots, \mathbf{z}_N\} \tag{5.6.30}$$

where $\hat{\mathrm{E}}$ is called the wide-sense conditional expectation of Θ relative to the set $\{z_1 \cdots z_N\}$. Since the wide-sense conditional expectation is equivalent to a projection operator [17], we can show that this expectation follows the same combinatorial rules as conventional conditional expectations. For example, for any scalars c_1 and c_2

$$\hat{\mathrm{E}}\{c_1\Theta_1 + c_2\Theta_2 | \mathbf{z}\} = c_1\hat{\mathrm{E}}\{\Theta_1 | \mathbf{z}\} + c_2\hat{\mathrm{E}}\{\Theta_2 | \mathbf{z}\} \tag{5.6.31}$$

Other relations may be stated similarly. These rules must be understood as equations valid with probability 1 and not as identities.

We conclude this chapter on parameter estimation with a discussion of reproducing densities, which is especially of use in pattern recognition problems.

5.7 REPRODUCING DENSITIES

We have seen in our discussion of the estimation of parameters that we need to specify the prior probability density function $p(\theta)$. The estimation procedure involves determining the posterior density function

$$p(\theta | z) = \frac{p(z | \theta)p(\theta)}{p(z)} \tag{5.7.1}$$

The estimation problem becomes particularly simple if the functional forms for $p(\theta)$ and $p(\theta | z)$ are the same, with only the associated parameters being different. We call such a density a reproducing density or a conjugate prior density [with respect to the transition density $p(z | \theta)$] [19,20]. Since the choice of the prior density is somewhat arbitrary in most cases, it is frequently convenient to choose a reproducing density.

We use the concept of reproducing densities to obtain estimates of parameters associated with probability density functions. We are given the functional form of the probability density function underlying a certain experiment so that the density functions are completely known except for some parameters. On the basis of measurements of the outcomes of the experiment, we are required to estimate the parameters by obtaining a reproducing density for the parameter. We generate this reproducing density [19] by choosing an a priori set $\mathbf{Z}_{n_0}$ of n_0 (fictitious) observations and a nonnegative function $g(\theta)$ and setting

$$p(\theta) = \frac{p(\mathbf{Z}_{n_0} | \theta)g(\theta)}{\int p(\mathbf{Z}_{n_0} | \theta)g(\theta)\, d\theta} \tag{5.7.2}$$

where $g(\theta)$ is any nonnegative function of θ such that the denominator in Eq. (5.7.2) exists. Usually, $g(\theta)$ can be chosen as unity. We illustrate the technique by considering two examples.

Example 5.7.1

Let Z be a scalar observation with

$$Z = \begin{cases} 1 & \text{with probability } \theta \\ 0 & \text{with probability } 1 - \theta \end{cases}$$

We make n observations of the random variable Z. Based on our observations we are required to estimate θ. To find a simple reproducing density, we consider a specific a priori sequence Z_{n_0} consisting of α_0 ones and $n_0 - \alpha_0$ zeros, for n_0 and α_0 any two positive integers, $\alpha_0 \leqslant n_0$. Then by Eq. (5.7.2) with $g(\theta) = 1$,

$$p(\theta) = \frac{p(\mathbf{Z}_{n_0}|\theta)}{\int p(\mathbf{Z}_{n_0}|\theta)\,d\theta} = \begin{cases} \dfrac{\theta^{\alpha_0}(1-\theta)^{n_0-\alpha_0}}{\int_0^1 \theta^{\alpha_0}(1-\theta)^{n_0-\alpha_0}\,d\theta} & 0<\theta<1 \\ 0 & \text{elsewhere} \end{cases}$$

$$= \begin{cases} \dfrac{\Gamma(n_0+2)}{\Gamma(n_0+1)\Gamma(n_0-\alpha_0+1)}\,\theta^{\alpha_0}(1-\theta)^{n_0-\alpha_0} & 0<\theta<1 \\ 0 & \text{elsewhere} \end{cases}$$

which is the beta density with parameters $\alpha_0 + 1$ and $n_0 - \alpha_0 + 1$. Let the sequence Z_n consisting of n observations have α ones and $n - \alpha$ zeros. Then

$$p(\theta|\mathbf{Z}_n) = \frac{p(\mathbf{Z}_n|\theta)p(\theta)}{p(\mathbf{Z}_n)} = \frac{p(\mathbf{Z}_n|\theta)p(\theta)}{\int p(\mathbf{Z}_n|\theta)p(\theta)\,d\theta}$$

$$= \begin{cases} \dfrac{\theta^{\alpha}(1-\theta)^{n-\alpha}\,\theta^{\alpha_0}(1-\theta)^{n_0-\alpha_0}}{\int_0^1 \theta^{(\alpha+\alpha_0)}(1-\theta)^{(n+n_0)-(\alpha+\alpha_0)}\,d\theta} & 0<\theta<1 \\ 0 & \text{elsewhere} \end{cases}$$

$$= \begin{cases} \dfrac{\Gamma(n+n_0+2)\theta^{(\alpha+\alpha_0)}(1-\theta)^{(n+n_0)-(\alpha+\alpha_0)}}{\Gamma(\alpha+\alpha_0+1)\Gamma(n+n_0-(\alpha+\alpha_0)+1)} & 0<\theta<1 \\ 0 & \text{elsewhere} \end{cases}$$

which is also a beta density. The mean-square estimate of Θ is given by

$$\hat{\theta}_{\text{ms}}(n) = \mathrm{E}\{\Theta|\mathbf{Z}_n\} = \frac{\alpha+\alpha_0+1}{n+n_0+2}$$

To put the estimate in a sequential form, we note that if $z_n = 1$, then with $(n - 1)$ observations, there would have been $\alpha - 1$ ones and $n - \alpha$ zeros. Thus

$$\hat{\theta}_{\text{ms}}(n-1) = \frac{(\alpha-1)+\alpha_0+1}{(n-1)+n_0+2} = \frac{\alpha+\alpha_0}{n+n_0+1}$$

If $z_n = 0$, then the sequence $\mathbf{Z}_{n-1}$ would consist of α ones and $n - 1 - \alpha$ zeros and

$$\hat{\theta}_{\text{ms}}(n-1) = \frac{\alpha+\alpha_0+1}{n+n_0+1}$$

We can combine the two to write

$$\hat{\theta}_{\text{ms}}(n-1) = \frac{\alpha+\alpha_0+1-z_n}{n+n_0+1}$$

Thus

$$\hat{\theta}_{ms}(n) = \frac{z_n}{n + n_0 + 2} + \frac{\alpha + \alpha_0 + 1 - z_n}{n + n_0 + 2}$$

$$= \frac{z_n}{n + n_0 + 2} + \frac{n + n_0 + 1}{n + n_0 + 2}\hat{\theta}_{ms}(n - 1)$$

with

$$\hat{\theta}_{ms}(0) = \frac{\alpha_0 + 1}{n_0 + 2}$$

Example 5.7.2

The number of events over an interval τ in an experiment obey a Poisson law with mean value θ per unit interval

$$\Pr(n \text{ events} \mid \theta) = \begin{cases} \dfrac{(\theta\tau)^n e^{-\theta\tau}}{n!} & \theta \geq 0 \\ 0 & \text{otherwise} \end{cases}$$

We want to observe the number of events and estimate the parameter θ. To find a reproducing density, we assume a specific a priori sample $\mathbf{Z}_{n_0}$ of n_0 events in time τ_0. Then

$$p(\theta) = \frac{p(\mathbf{Z}_{n_0} \mid \theta)}{\int p(\mathbf{Z}_{n_0} \mid \theta)\, d\theta} = \begin{cases} \dfrac{\dfrac{(\theta\tau_0)^{n_0} e^{-\theta\tau_0}}{n_0!}}{\displaystyle\int_0^\infty \frac{(\theta\tau_0)^{n_0} e^{-\theta\tau_0}}{n_0!}\, d\theta} & \theta \geq 0 \\ 0 & \text{otherwise} \end{cases}$$

But

$$\int_0^\infty (\theta\tau_0)^{n_0} e^{-\theta\tau_0}\, d\theta = \frac{1}{\tau_0}\Gamma(n_0 + 1)$$

so that a reproducing density is

$$p(\theta) = \begin{cases} \dfrac{\tau_0^{n_0+1}}{\Gamma(n_0 + 1)}\theta^{n_0} e^{-\theta\tau_0} & \theta \geq 0 \\ 0 & \text{otherwise} \end{cases}$$

which is the gamma density with parameters τ_0, $n_0 + 1$. Now if we observed n_1 events in time τ_1

$$p(\theta \mid \mathbf{Z}_{n_1}) = \frac{p(\mathbf{Z}_{n_1} \mid \theta) p(\theta)}{\int [\text{num}]\, d\theta}$$

$$= \begin{cases} \dfrac{\dfrac{(\theta\tau_1)^{n_1} e^{-\theta\tau_1}}{n_1!} \dfrac{\tau_0^{n_0+1}}{\Gamma(n_0 + 1)}\theta^{n_0} e^{-\theta\tau_0}}{\int [\text{num}]\, d\theta} & \theta \geq 0 \\ 0 & \text{otherwise} \end{cases}$$

$$= \begin{cases} \dfrac{(\tau_0 + \tau_1)^{n_0+n_1+1}}{\Gamma(n_0 + n_1 + 1)}\theta^{(n_0+n_1)} e^{-\theta(\tau_0+\tau_1)} & \theta \geq 0 \\ 0 & \text{otherwise} \end{cases}$$

where [num] indicates the expression in the numerator in the corresponding term. This is a gamma density with parameters $\tau^1 = \tau_0 + \tau_1$ and $n^1 = n_0 + n_1 + 1$. The mean-square estimate $\hat{\theta}_{ms}$ is given by

$$\hat{\theta}_{ms} = E\{\Theta \mid Z_{n_1}\} = \frac{n^1}{\tau^1} = \frac{1 + n_0 + n_1}{\tau_0 + \tau_1}$$

If we conduct N successive trials of the experiment and observe n_1 events in τ_1, n_2 events in $\tau_2 \cdots n_N$ events in τ_N, we can easily show using the preceding procedure that $p(\theta \mid Z_{n1}, Z_{n2}, \ldots, Z_{nN})$ will also be a gamma density with parameters

$$\tau^N = \sum_{i=0}^{N} \tau_1, \; n^N = \sum_{i=0}^{N} n_i + 1$$

and

$$\hat{\theta}_{ms} = \frac{\sum_{i=1}^{N} n_i + 1}{\sum_{i=1}^{N} \tau_i}$$

We list here a few other commonly encountered distributions, the reproducing densities for estimating unknown parameters associated with these, and the mean square estimates of the parameters:

1. *N-variate Gaussian density, mean* **m** *and covariance matrix* **V**:
   (a) If the unknown parameter is the mean **m**, the appropriate prior density is Gaussian with mean $\mathbf{m}_0$ and covariance matrix $\mathbf{V}_0$:

$$p(\mathbf{m}) = \frac{1}{(2\pi)^{N/2} |\mathbf{V}_0|^{1/2}} \exp\left[-\frac{1}{2}(\mathbf{m} - \mathbf{m}_0)^T \mathbf{V}_0^{-1} (\mathbf{m} - \mathbf{m}_0)\right]$$

   The estimate after n observations is, if we let $\mathbf{V}_0^{-1} = n_0 \mathbf{V}^{-1}$

$$\hat{\mathbf{m}}(n) = \frac{n_0}{n + n_0}\mathbf{m}_0 + \frac{n}{n + n_0}\left(\frac{1}{n}\sum_{j=1}^{n} \mathbf{z}_j\right),$$

   where n_0 is an arbitrary number.
   (b) If the unknown parameter is **V**, we assume for the prior density of $\mathbf{V}^{-1}$ the Wishart density of the form

$$p(\mathbf{V}^{-1}) = \begin{cases} k(n_0)\left|\dfrac{n_0}{2}\mathbf{V}_0\right|^{(n_0-1)/2} |\mathbf{V}^{-1}|^{(n_0-N-2)/2} \exp\left(-\dfrac{1}{2}\operatorname{tr}(n_0 \mathbf{V}_0 \mathbf{V}^{-1})\right) \\ \qquad \text{if } \mathbf{V}^{-1} \text{ is positive definite} \\ 0 \qquad \text{otherwise} \end{cases}$$

   where n_0 is a scalar and

$$k(n_0) = \left[(\pi)^{N(N-1)/4} \prod_{\alpha=1}^{N} \Gamma\left(\frac{n_0 - \alpha}{2}\right)\right]^{-1}$$

The mean-square estimate of $\mathbf{V}$ is

$$\hat{\mathbf{V}}(n) = \frac{n_0}{n + n_0}\mathbf{V}_0 + \frac{n}{n + n_0}\left[\frac{1}{n}\sum_{j=1}^{n}(\mathbf{z}_j - \mathbf{m})(\mathbf{z}_j - \mathbf{m})^T\right]$$

(c) If both $\mathbf{m}$ and $\mathbf{V}$ are unknown, we choose a Gaussian form for the prior density of $\mathbf{m}$ as in (a) preceding and a Wishart form as in (b) for the prior density of $\mathbf{V}$. The mean square estimate of $\mathbf{m}$ is the same as in (a). The estimate of $\mathbf{V}$ is now given by

$$\hat{\mathbf{V}}(n) = \frac{1}{n + n_0}\left[(n - 1)\mathbf{S}_n + \frac{1}{n}\left(\sum_{j=1}^{n}\mathbf{z}_j\right)\left(\sum_{j=1}^{n}\mathbf{z}_j\right)^T\right]$$
$$+ (n_0\mathbf{V}_0 + n_0\mathbf{m}_0\mathbf{m}_0^T) - (n + n_0)\hat{\mathbf{m}}(n)\hat{\mathbf{m}}^T(n)$$

where

$$\mathbf{S}_n = \frac{1}{n - 1}\sum_{j=1}^{n}\left(\mathbf{z}_j - \frac{1}{n}\sum_{i=1}^{n}\mathbf{z}_i\right)\left(\mathbf{z}_j - \frac{1}{n}\sum_{i=1}^{n}\mathbf{z}_i\right)^T$$

2. *Rayleigh density, parameter* σ^2: (see Exercise 3.10 for the definition of Rayleigh density). We assume a gamma density with parameters τ_0 and $n_0 + 1$ as the prior density for $\theta = 1/\sigma^2$:

$$p(\theta) = \begin{cases} \dfrac{(\tau_0)^{n_0+1}}{\Gamma(n_0 + 1)}\theta^{n_0}e^{-\theta\tau_0} & \theta \geq 0 \\ 0 & \text{otherwise} \end{cases}$$

The mean-square estimate of θ is

$$\hat{\theta}(n) = \frac{n_0 + n + 1}{\tau_0 + \sum^{n} z_j^2/2}$$

3. *Exponential density, parameter* θ: The prior density for θ is again a gamma density with parameters τ_0 and $n_0 + 1$. The estimate is

$$\hat{\theta}(n) = \frac{n_0 + n + 1}{\tau_0 + \sum_{j=1}^{n} z_j}$$

5.8 SUMMARY

In this chapter we have considered the estimation of parameters from noisy observations. We started in Section 5.3 with the problem of Bayesian estimation of random parameters. The density function of interest in the estimation problem is the posterior density $p(\theta \mid z)$. Two criteria of particular interest are the minimum-mean-square error (MMSE) criterion and the maximum posterior (MAP) criterion. The

MMSE estimate is the mean of the posterior density (conditional mean), while the MAP estimate is the value of θ at which the posterior density is a maximum. It was also shown that for a large class of cost functions, the conditional mean is the optimum estimate whenever $p(\theta \mid z)$ is unimodal and symmetric about its mean.

We discussed the estimation of nonrandom parameters in Section 5.4. The maximum likelihood estimate (MLE) of a parameter θ is a widely used point estimate and is obtained as the value of θ that maximizes the likelihood function $p(z \mid \theta)$. The MLE is also applicable when θ is a random parameter with unknown prior density. Next, the EM algorithm was introduced as a way of finding an estimate of a parameter based on the likelihood function. In general, the estimate found using the EM algorithm may not correspond to the ML estimate. We concluded this section with a brief discussion of an interval estimate which defines an interval in which the true value of the parameter is expected to lie. The confidence coefficient associated with the interval represents a measure of how good the estimate is.

In Section 5.5 we discussed the properties of estimators. Two measures of the performance of an estimator are the bias and the variance. In general, the bias should be zero and the variance as small as possible. Since the variance is usually difficult to determine, we established the Cramér–Rao inequality to provide a bound on the variance of any unbiased estimate. If any estimate exists that meets this bound, it will be equal to the ML estimate. Similar bounds were also stated for random parameter estimation. In this case, if an efficient estimate exists, it can be determined using the MAP technique. We also discussed uniformly minimum variance unbiased estimates (UMVUE) and the application of the Lehmann–Scheffé theorem in finding the UMVUE. The extension to multiple parameters is straightforward and involves extensions to multiple dimensions of the results obtained for the scalar case.

The problem of MMSE estimation involves determining the conditional density $p(\theta \mid z)$. In many applications of interest, the determination of this density is not an easy task. It then becomes appropriate to use a suboptimal estimate, which can easily be determined. Section 5.6 considers the problem of determining the best estimate in the class of linear estimates for this problem. The conditions under which the linear mean-square (LMS) estimate is also the MMSE estimate are presented in terms of the characteristic function associated with the joint density function $p(z, \theta)$. Since this condition is satisfied if $p(z, \theta)$ is Gaussian, it follows that for this case the LMS estimate is the optimum estimate. By defining a linear space formed from a collection of random vectors of finite variance, the LMS estimate can be interpreted as the projection on the subspace of observation vectors. An important consequence of this result is the orthogonality principle, which states that the LMS estimate is such that the estimation error is orthogonal to the observations.

We concluded the chapter with a brief discussion of reproducing densities. These are particularly useful in estimation of the parameters of probability density functions.

The discussions in this chapter also bring out the similarities between the detection and estimation problems. Both are based on the determination of the likelihood function or posterior density and in manipulations of these functions.

EXERCISES

5.1. A random parameter λ is observed through another random variable Z with

$$p(z \mid \lambda) = \lambda e^{-\lambda z} \qquad z \geqslant 0, \lambda > 0$$
$$= 0 \qquad \lambda < 0$$

Assuming the prior density of λ to be

$$p(\lambda) = \frac{l^n}{\Gamma(n)} e^{-\lambda l} \lambda^{n-1} \qquad \lambda \geqslant 0$$
$$= 0 \qquad \lambda < 0$$

obtain $\hat{\lambda}_{\text{MAP}}$ and $\hat{\lambda}_{\text{ms}}$. Find $\mathrm{E}\{[\lambda - \hat{\lambda}_{\text{ms}}]^2\}$.

5.2. In Example 5.3.1, denote the estimate of θ with N observations as $\hat{\theta}(N)$, and the corresponding variance as $\sigma^2(N)$. Now another sample z_{N+1} is observed. Express $\hat{\theta}(N+1)$ in terms of $\hat{\theta}(N)$, z_{N+1}, and $\sigma^2(N)$. This provides a sequential estimation algorithm for $\hat{\theta}(N)$.

5.3. Consider the extension of Example 5.3.3 to the case of multiple observations. Show that

$$\hat{y}_{\text{MAP}} = \operatorname{sgn}\left(\sum_{i=1}^{N} z_i\right)$$

$$\hat{y}_{\text{ms}} = \tanh\left(\frac{1}{\sigma_v^2}\sum_{i=1}^{N} z_i\right)$$

Show that both estimates are asymptotically consistent as N tends to infinity.

5.4. We are given a sequence of independent observations $Z_1, Z_2, \ldots, Z_N$ with mean m and variance σ^2.

(a) Is the sample mean

$$\overline{Z} = \frac{1}{N}\sum_{i=1}^{N} z_i$$

an unbiased estimate of m? What is its variance?

(b) We may seek to estimate the variance as

$$V = \frac{1}{N}\sum_{i=1}^{N} (z_i - \overline{Z})^2$$

Is this an unbiased estimate of σ^2? Can you find its variance?

5.5. We have N independent observations, $z_i, i = 1, \ldots, N$ of a Gaussian variable with mean m and variance σ^2, which is unknown. Obtain $\hat{\sigma}^2_{\text{ml}}$. Is this estimate unbiased? Is this efficient? What is the variance of this estimate?

5.6. Consider Exercise 5.5 with both mean m and variance σ^2 unknown. Obtain the maximum likelihood estimates.

5.7. We observe a random parameter X, Gaussian distributed with zero mean and variance σ_x^2, through two independent observation channels:

$$Z_1 = X + V_1$$

$$Z_2 = X + V_2$$

where

$$p(v_i) = \frac{1}{\sqrt{2\pi}\,\sigma_i} \exp\left(-\frac{v_i^2}{2\sigma_i^2}\right)$$

Obtain $\hat{x}_{ms}$ and $\hat{x}_{MAP}$ as a function of z_1 and z_2. What is the mean square error?

5.8. In Exercise 5.7, consider x to be nonrandom real. Obtain $\hat{x}_{ml}$. Is this an efficient estimate? What is the variance of the estimate?

5.9. Obtain a MAP and MMSE estimate of x_1 from the observation

$$Z = X_1 + X_2$$

where X_1 and X_2 are independent and are Rayleigh distributed with parameters σ_1^2 and σ_2^2.

5.10. Consider the problem of pulse amplitude modulation communication. A signal $x(t)$ is sampled at regular intervals to give a sequence of amplitudes $x_1, x_2, \ldots, x_n$. The x_i's modulate a carrier and are received in a T-second interval in additive Gaussian white noise with zero mean and spectral height $N_0/2$. The received signal in the jth T-second interval is

$$z(t) = x_j \sin\omega_c t + v(t) \qquad jT \leq t \leq (j+1)T$$

How can you estimate x_j in the sense of maximum likelihood?

5.11. The log-normal distribution is useful in characterizing the density functions of the radar cross section (RCS) of large metal objects of irregular shape, sea clutter, and so on. This density function is given by

$$p(z \mid m, \sigma) = \frac{1}{\sqrt{2\pi}\, z\sigma} \exp\left[-\frac{1}{2\sigma^2} \ln^2\left(\frac{z}{m}\right)\right] \qquad z > 0, m > 0, \sigma > 0$$

where m is the median of Z and σ is the standard deviation of $\ln(Z/m)$. An important quantity of interest is the ratio of the mean to median of the variable Z:

$$\rho = \frac{E\{Z\}}{m} = \exp\left(\frac{\sigma^2}{2}\right)$$

Assume that we have N independent observations of the variable z. Show that the maximum likelihood estimates of the parameters m and ρ are

$$\hat{m} = \left(\prod_{i=1}^{N} z_i\right)^{1/N}$$

and

$$\hat{\rho} = \left[\prod_{i=1}^{N} \left(\frac{z_i}{\hat{m}}\right)^{\ln(z_i/\hat{m})}\right]^{1/2N}$$

Obtain the variance of the estimates as the number of observations becomes very large. (Note that the ML estimate commutes over nonlinear operations. See [21] for details and some numerical experiments.)

5.12. We consider the problem of estimating the parameters of signals of known form observed in zero-mean additive Gaussian noise, $\mathbf{v}(n)$, with covariance matrix $E\{\mathbf{v}(n)\mathbf{v}^{*T}(m)\} = \mathbf{V}_v\,\delta(n,m)$. Let the observation vector be given by

$$\mathbf{z}(n) = \sum_{k=1}^{K} \mathbf{s}_k(n, \theta_k) + \mathbf{v}(n) \qquad n = 1, 2, \ldots, N$$

where θ_k is a vector of unknown parameters associated with signal $\mathbf{s}_k(\cdot)$. We wish to find the ML estimates of these parameters based on the observations. This is a generalization of the problem of estimating the parameters of sinusoidal signals observed in noise that we considered in Example 5.4.3. As in that problem, direct determination of the ML estimates requires solving a complicated multiparameter optimization problem. Instead, we will use the E-M algorithm to determine these parameters.

Let us decompose the observation noise $\mathbf{v}(n)$ into components $\mathbf{v}_k(n)$ such that $\sum_{k=1}^{K} \mathbf{v}_k(n) = \mathbf{v}(n)$. We can then define the complete observations $\mathbf{y}(n)$ as

$$\mathbf{y}(n) = \begin{bmatrix} \mathbf{y}_1(n) \\ \mathbf{y}_2(n) \\ \vdots \\ \mathbf{y}_K(n) \end{bmatrix}$$

where

$$\mathbf{y}_k(n) = \mathbf{s}_k(n_1 \theta_k) + \mathbf{v}_k(n)$$

so that

$$\mathbf{z}(n) = \sum_{k=1}^{K} \mathbf{y}_k(n) = \mathbf{H}\mathbf{y}(n)$$

with

$$\mathbf{H} = [\mathbf{I} \;\; \mathbf{I} \;\; \cdots \;\; \mathbf{I}]$$

Assume that $\mathbf{v}_k(n)$ are statistically independent, zero-mean Gaussian with covariance matrix $\mathbf{V}_k$ satisfying $\sum_{k=1}^{K} \mathbf{V}_k = \mathbf{V}_v$. To avoid having to estimate $\mathbf{V}_k$, we can set $\mathbf{V}_k = \beta_k \mathbf{V}_v$ with β_k being chosen arbitrarily to satisfy the constraint $\sum_{k=1}^{K} \beta_k = 1$. The convergence of the algorithm will depend on the choice of the β_k. Set up the E-M algorithm for determining the parameters θ_k [22].

5.13. Let $X_1, \ldots, X_n$ be a random sample from a density that is uniform on $(\theta - \frac{1}{2}, \theta + \frac{1}{2})$. Note that θ is a location parameter and that $X_i - \theta$ is uniformly distributed on $(-\frac{1}{2}, \frac{1}{2})$. Let $Y_1, \ldots, Y_n$ denote the corresponding ordered sample. Show that (Y_1, Y_n) is a confidence interval for θ. Find the confidence coefficient for this interval.

5.14. Let $X_1, \ldots, X_n$ be independent, identically distributed random variables with a Poisson distribution with parameter λ. Determine a two-sided confidence interval for λ at confidence coefficient $(1 - \alpha)$. [*Hint:* Let $T_n = \sum_{i=1}^{n} X_i$. Then $P(T_n \le t)$ is equal to the probability that a gamma $(t + 1, 1)$ density exceeds $n\lambda$.]

5.15. Let X_1, X_2 denote a random sample of size 2 from a $N(\theta, 1)$ distribution. Let Y_1, Y_2 be the corresponding ordered sample. Determine γ where $P(Y_1 < \theta < Y_2) = \gamma$. Find the expected length of the interval (Y_1, Y_2).

5.16. Let $X_1, X_2, \ldots, X_n$ be a random sample from the Poisson density

$$p(x \mid \lambda) = \frac{e^{-\lambda} \lambda^x}{x!} \qquad x = 0, 1, \ldots$$

Show that the UMVUE of $e^{-\lambda}$ for $n > 1$ is $(n - 1/n)^{\sum_{i=1}^{n} X_1}$.

5.17. Let $X_1, X_2, \ldots, X_n$ be a random sample from the density function $p(x \mid \theta) = \theta x^{\theta - 1} I_{(0,1)}(x), \theta > 0$. Find the UMVUE of each of the following: θ, $1/\theta$, $\theta/(1 + \theta)$.

5.18. Consider a set of N observations $X_1, X_2, \ldots, X_N$ given by

$$X_k = \theta s_k + V_k$$

where $\mathbf{s} = [s_1 \cdots s_N]^T$ is known, and $\mathbf{V} = [V_1 \ldots V_N]^T$ is Gaussian, zero mean, with covariance matrix $\mathbf{V}_v$. We want to estimate θ assuming that it is an unknown constant.

(a) With $\mathbf{V}_v$ known, show that $p(\mathbf{x})$ belongs to a one-parameter exponential family and that the statistic $\mathbf{X}^T\mathbf{V}_v^{-1}\mathbf{s}$ is sufficient and complete.

(b) Show that $[\mathbf{X}^T\mathbf{V}_v^{-1}\mathbf{s}]/[\mathbf{s}^T\mathbf{V}_v^{-1}\mathbf{s}]$ is an unbiased estimator of θ and therefore it is UMVUE. (Note that the estimator is a linear combination of X_i.)

(c) Show that $[\mathbf{X}^T\mathbf{V}_v^{-1}\mathbf{s}]/[\mathbf{s}^T\mathbf{V}_v^{-1}\mathbf{s}]$ is also the ML estimate.

(d) Find the variance of $\hat{\theta}_{\text{ml}}$.

5.19. We observe a random parameter X in multiplicative and additive noise as

$$Z = V_1 X + V_2$$

where

$$p(x) = \frac{1}{\sqrt{2\pi}\,\sigma_x} \exp\left[-\frac{(x - m_x)^2}{2\sigma_x^2}\right]$$

$$p(v_1) = \frac{1}{\sqrt{2\pi}\,\sigma_1} \exp\left[-\frac{(v_1 - m_1)^2}{2\sigma_1^2}\right]$$

and

$$p(v_2) = \frac{1}{\sqrt{2\pi}\,\sigma_2} \exp\left[-\frac{(v_2 - m_2)^2}{2\sigma_2^2}\right]$$

Assume X, V_1 and V_2 are independent. Obtain the linear MS estimate of x based on the observation z. Extend your results to N independent observations.

5.20. Repeat Exercise 5.10 when the observation model is

$$z(t) = x_j \sin(\omega_c t + \theta) + v(t) \qquad jT < t \leq (j + 1)T$$

when θ is unknown and is uniformly distributed between 0 and 2π radians.

5.21. In a pulse frequency modulation (PFM) system, the samples $x_1, x_2, \ldots, x_N$ of a message $x(t)$ change the frequency of the carrier and are observed with additive noise. The observed signal in a T-second interval is

$$z(t) = \sqrt{\frac{2\epsilon}{T}} \sin(\omega_c t + \beta x_j) + v(t)$$

where β is the *modulation index* and $v(t)$ is Gaussian white with zero mean and spectral height $N_0/2$. Assume that x_j is uniformly distributed between $-M$ and M. Obtain the estimator equations and suggest a practical implementation.

5.22. If $\mathbf{X}$, $\mathbf{Y}$, and $\mathbf{Z}$ are zero-mean Gaussian vectors such that $E\{\mathbf{Y}^T\mathbf{Z}\} = 0$, show that

$$E\{\mathbf{X}\,|\,\mathbf{Y}, \mathbf{Z}\} = E\{\mathbf{X}\,|\,\mathbf{Y}\} + E\{\mathbf{X}\,|\,\mathbf{Z}\}$$

5.23. Consider independent samples $X_1, X_2, \ldots, X_n$ from a Gaussian distribution

$$p(x) = \frac{1}{\sqrt{2\pi}\,\sigma} \exp\left[-\frac{(x - m)^2}{2\sigma^2}\right]$$

We have to estimate the unknown mean m, and variance σ^2 assuming a proper reproducing density function for m and σ^2. Set up the estimation algorithms in a sequential manner.

REFERENCES

1. A. M. Mood, F. A. Graybill, and D. C. Boes, *Introduction to the Theory of Statistics*, McGraw-Hill, New York, 1974.
2. D. Middleton, *Introduction to Statistical Communication Theory*, McGraw-Hill, New York, 1963.
3. H. L. Van Trees, *Detection, Estimation and Modulation Theory*, Part I, Wiley, New York, 1968.
4. R. Deutsch, *Estimation Theory*, Prentice Hall, Englewood Cliffs, NJ, 1965.
5. H. Cramer, *Mathematical Methods of Statistics*, Princeton University Press, Princeton, NJ, 1946.
6. P. W. Zehna, Invariance of maximum likelihood estimation, *Ann. Math. Stat.*, **37**, No. 744 (1966).
7. B. Porat, *Digital Processing of Random Signals*, Prentice Hall, Englewood Cliffs, NJ, 1994.
8. S. Haykin, *Adaptive Filter Theory*, 2nd ed., Prentice Hall, Englewood Cliffs, NJ, 1991.
9. C. R. Rao and L. C. Zhao, Asymptotic behavior of the maximum likelihood estimates of superimposed exponential signals, *IEEE Trans. Signal Process.*, **41**, 1461–1463 (1993).
10. A. P. Dempster, N. M. Laird, and D. B. Rubin, Maximum likelihood from incomplete data via the EM algorithm, *J. Roy. Stat. Soc. (B)*, **39**, 1–22 (1977).
11. C. F. J. Wu, On the convergence properties of the EM algorithm, *Ann. Stat.*, No. 11, 95–103 (1983).
12. A. Ansari and R. Viswanathan, Application of expectation-maximization algorithm to the detection of direct sequence signal in pulsed noise jamming, *IEEE Trans. Commun.*, August, 1151–1154 (1993).
13. C. R. Rao, Information and the accuracy attainable in the estimation of statistical parameters, *Bull. Calcutta Math. Soc.*, **37**, 81–91 (1945).
14. E. L. Lehmann, *Theory of Point Estimation*, Wiley, New York, 1983.
15. P. Swerling, Parameter estimation for waveforms in additive Gaussian noise, *J. SIAM*, **7**, No. 2, 152–166 (1959).
16. A. V. Balakrishnan, On a characterization of processes for which optimal mean-square systems are of a specified form, *IRE Trans. Inf. Theory*, **IT-6**, September, 491–500 (1960).
17. J. L. Doob, *Stochastic Processes*, Wiley, New York, 1950.
18. K. Hoffman and R. Kunze, *Linear Algebra*, Prentice Hall, Englewood Cliffs, NJ, 1961.
19. E. A. Patrick, *Fundamentals of Pattern Recognition*, Prentice Hall, Englewood Cliffs, NJ, 1972.
20. J. D. Spragins, Reproducing distributions for machine learning, *Technical Report 6103-7*, Stanford Electronics Laboratories, Stanford, CA, November 1963.
21. W. J. Sajnowski, Estimators of log normal distribution parameters, *IEEE Trans. Aerosp. Electron. Syst.*, **AES-13**, No. 5, September, 533–536 (1977).
22. M. Feder and E. Weinstein, Parameter estimation of superimposed signals using the EM algorithm, *IEEE Trans. Acoustics Speech Signal Process.*, **36**, No. 4, April, 477–489 (1988).

6

Estimation of Waveforms

6.1 INTRODUCTION

In Chapter 5 we developed estimation techniques for both random and nonrandom parameters. Many physical situations, for example in analog communications, radar tracking, and recognition of patterns with time-varying characteristics, involve the estimation of waveforms. These waveforms to be estimated can be modeled as realizations of stochastic processes and are usually observed in noise. When unknown deterministic signals are observed in noise, essentially the same estimation techniques can be applied. If a signal that is known except for certain nonrandom parameters is observed in a noisy background, the MLE techniques of Chapter 5 can be used.

In this chapter we consider the mean square estimation of stochastic processes observed in additive noise. We define the three basic estimation problems: prediction, filtering, and smoothing. We first develop the orthogonal projection property of linear estimators of waveforms. The classical theory of Wiener filtering for both continuous-time and discrete-time stationary stochastic processes is presented with examples in Section 6.3. The FIR Wiener filter and its application in the forward and backward prediction of discrete-time processes is considered. The Levinson–Durbin algorithm for computation of the partial correlation coefficients and the lattice structure implementation of forward-backward recursion are given. In Section 6.4 we consider recursive estimation schemes for the estimation of nonstationary processes. The discrete- and continuous-time Kalman filters are derived using system theory concepts. For nonlinear signal and observation models, approximate estimation schemes are presented in Section 6.5. We conclude with a consideration of MAP estimation in certain nonlinear observation models.

6.2 LINEAR MMSE ESTIMATION OF WAVEFORMS: PRELIMINARIES

We are given two related stochastic processes $y(t)$ and $z(t)$. Our problem is to estimate various parameters of the first process (signal) in terms of certain values $z(\xi)$ of the second process. We will assume without loss of generality that the processes $y(t)$ and $z(t)$ have zero mean. We shall refer to the quantity to be estimated as $g(t)$. For example,

$$g(t) = y(t) \quad \text{or} \quad \dot{y}(t) \quad \text{or} \quad y(t + \alpha)$$

The estimation is to be based on the values of $z(t)$ at certain instants $t = \xi$, forming a set I on the time axis. This set could consist of a number of discrete points or an interval on the time axis. Our problem is to find a suitable transformation of the data that can be used as a best estimate $\hat{g}(t)$ of $g(t)$ [1–12].

Next we consider several specific problems.

1. If $g(t) = y(t + \alpha)$ with $\alpha > 0$, we have the *prediction* problem. If $z(t) = y(t)$ and if the observation interval I consists of the single point t, we are required to predict the value of $y(\cdot)$, α units in the future, in terms of its value at the present time. If the observation interval is $(-\infty, t)$, our data consist of the entire past history of $y(\cdot)$.
2. If $g(t) = y(t)$, we have the *filtering* problem.
3. If the observation interval is $[t_0, t_f]$ and we are required to estimate $y(t)$ for any $t \in [t_0, t_f]$, the problem is one of *smoothing* or interpolation. Of particular interest is the *fixed-lag* smoothing problem, where we are required to estimate $y(t + \alpha)$ for $\alpha < 0$ based on observations over the interval $[t_0, t]$.

If we want to use the mean square cost function in our estimation problem, we form the estimation error $\tilde{g}(t)$, defined as

$$\tilde{g}(t) = g(t) - \hat{g}(t) \tag{6.2.1}$$

The risk is

$$\overline{C} = \mathrm{E}\{[g(t) - \hat{g}(t)]^2\} \tag{6.2.2}$$

If I is a set of discrete points, we are required to find the estimate $\hat{g}(t)$ based on the set of random variables $z(\xi), \xi \in I$. From our discussions in Chapter 5 it follows readily that for fixed t, the estimate $\hat{g}(t)$ is given by the conditional mean

$$\hat{g}(t) = \mathrm{E}\{g(t) \mid z(\xi), \xi \in I\} \tag{6.2.3}$$

However, if I is an interval, say $[t_0, t_f]$, $z(\xi)$ represents a sample function of a stochastic process and the conditioning in Eq. (6.2.3) has to be interpreted a little more carefully. In this case the conditioning is really on the σ-algebra generated by $z(\cdot)$. For a detailed discussion, see, for example, [5]. We can overcome this problem by using an approach similar to the one used in Chapter 4. Thus if we use a K-term approximation of $z(\xi)$ in terms of a complete orthonormal set, the conditioning can be expressed in terms of a set of random variables, the coefficients of the expansion. We can then use a suitable limiting process to obtain $\hat{g}(t)$. We leave the details to the interested reader.

The determination of the conditional mean in Eq. (6.2.3) is not straightforward, since it requires the joint distribution of two variables $g(t)$ and $z(\xi)$ for each

t and ξ of interest. Rather, just as we did in the case of random parameters, we confine ourselves to linear estimates. In this case our estimate is of the form

$$\hat{g}(t) = L[z(\xi), \xi \in I] \tag{6.2.4}$$

where L is a linear operator.

For example, if we want to predict $y(t + \alpha)$ in terms of $y(t)$ and its derivative $\dot{y}(t)$, we write for the estimate

$$\hat{y}(t + \alpha) = ay(t) + b\dot{y}(t) \tag{6.2.5}$$

The problem then will be to choose the constants a and b so as to minimize the mean-square error

$$\mathrm{E}\{[y(t + \alpha) - ay(t) - b\dot{y}(t)]^2\} \tag{6.2.6}$$

In general, our problem is to find a linear operator

$$L[z(\xi), \xi \in I]$$

operating on the data $z(\xi)$ so as to minimize the mean-square error

$$\overline{C}_{\text{lms}} = \mathrm{E}\{[g(t) - L(z(\xi))]^2\} \tag{6.2.7}$$

The principle of orthogonality applied to this problem then implies that if an optimum solution exists, this solution is found by choosing L such that the estimation error

$$\tilde{g}(t) = g(t) - L[z(\xi)]$$

is orthogonal to $z(\xi)$ for every $\xi \in I$.

That is, if

$$\mathrm{E}\{[g(t) - L(z(\xi))]z(\xi_i)\} = 0 \qquad \text{for each } \xi_i \in I \tag{6.2.8}$$

then $\overline{C}_{\text{lms}}$ is minimum and is given by

$$\overline{C}_{\text{lms}} = \mathrm{E}\{[g(t) - L(z(\xi))]g(t)\} \tag{6.2.9}$$

The proof is similar to the one used in Chapter 5. Suppose that L is such that it satisfies Eq. (6.2.8). Let L_1 be any other linear operator operating on the set of data $z(\xi), \xi \in I$. We can write

$$\tilde{g}(t) = g(t) - L_1[z(\xi)] = g(t) - L[z(\xi)] + L[z(\xi)] - L_1[z(\xi)] \tag{6.2.10}$$

Since the difference of two linear operators is linear, we write

$$L[z(\xi)] - L_1[z(\xi)] = L_2[z(\xi)] \tag{6.2.11}$$

Then

$$\begin{aligned}\mathrm{E}\{[g(t) - L_1[z(\xi)]]^2 &= \mathrm{E}\{[g(t) - L[z(\xi)] + L_2[z(\xi)]]^2\} \\ &= \mathrm{E}\{[g(t) - L[z(\xi)]]^2\} + 2\mathrm{E}\{[g(t) \\ &\quad - L[z(\xi)]]L_2[z(\xi)]\} + \mathrm{E}\{L_2[z(\xi)]^2\}\end{aligned} \tag{6.2.12}$$

The difference $g(t) - L[z(\xi)]$ is by assumption orthogonal to the data. It must therefore be orthogonal to a linear combination of $z(\xi)$. Thus the second term on the right side of Eq. (6.2.12) is zero. Since the last term is nonnegative, we have

$$\mathrm{E}\{[g(t) - L_1[z(\xi)]]^2\} \geqslant \mathrm{E}\{[g(t) - L[z(\xi)]]^2\} \tag{6.2.13}$$

thus establishing the orthogonality principle. We note that equality holds in Eq. (6.2.13) only when $\mathrm{E}\{[L_2[z(\xi)]]^2\} = 0$ implying that $L_2[z(\xi)] = 0$ (that is, $L_1 = L$ with probability 1). The expression for $\overline{C}_{\text{lms}}$ follows from the definition and Eq. (6.2.8).

As in the case of random parameters, if $g(t)$ and $z(t)$ are jointly normal (with zero mean), then

$$\mathrm{E}\{g(t) \,|\, z(\xi), \xi \in I\}$$

is a linear combination of the values of $z(\xi)$. In this case the linear mean-square estimate of $g(t)$ is also the minimum-mean-square estimate.

We now consider a few examples.

Example 6.2.1

Consider the problem of predicting a signal based on present measured values. Thus we want to estimate $y(t + \alpha)$ in terms of $y(t)$ as

$$\hat{y}(t + \alpha) = ay(t)$$

We assume that $y(t)$ is a stationary process.

From the orthogonality principle, it follows that the constant a must be such that the error $y(t + \alpha) - ay(t)$ is orthogonal to the data $y(t)$:

$$\mathrm{E}\{[y(t + \alpha) - ay(t)]y(t)\} = 0$$

This yields

$$a = \frac{\phi_y(\alpha)}{\phi_y(0)}$$

where $\phi_y(\alpha)$ denotes the autocorrelation function of $y(t)$. The mean-square error is given by Eq. (6.2.9) as

$$\begin{aligned}\overline{C}_{\text{lms}} &= \mathrm{E}\{[y(t + \alpha) - ay(t)]y(t + \alpha)\} \\ &= \phi_y(0) - a\phi_y(\alpha) \\ &= \phi_y(0) - \frac{\phi_y^2(\alpha)}{\phi_y(0)}\end{aligned}$$

Example 6.2.2

Suppose that we want to estimate $y(t + \alpha)$ in terms of $y(t)$ and its derivative $\dot{y}(t)$:

$$\hat{y}(t + \alpha) = ay(t) + b\dot{y}(t)$$

The error must be orthogonal to $y(t)$ and $\dot{y}(t)$, so that

$$\mathrm{E}\{[y(t+\alpha) - ay(t) - b\dot{y}(t)]y(t)\} = 0$$

and

$$\mathrm{E}\{[y(t+\alpha) - ay(t) - b\dot{y}(t)]\dot{y}(t)\} = 0$$

We note that

$$\phi_{y\dot{y}}(\tau) = -\dot{\phi}_y(\tau), \phi_{\dot{y}\dot{y}}(\tau) = -\ddot{\phi}_y(\tau)$$

and that $\dot{\phi}_y(0) = 0$ for $\dot{y}(t)$ to exist.

We thus have

$$a = \frac{\phi_y(\alpha)}{\phi_y(0)} \qquad b = \frac{\dot{\phi}_y(\alpha)}{\ddot{\phi}_y(0)}$$

The mean-square error is

$$\begin{aligned}\overline{C}_{\text{lms}} &= \mathrm{E}\{[y(t+\alpha) - ay(t) - b\dot{y}(t)]y(t+\alpha)\} \\ &= \phi_y(0) - a\phi_y(\alpha) + b\dot{\phi}_y(\alpha)\end{aligned}$$

Example 6.2.3

As a final example we consider an interpolation problem in which we are given the values of a signal at the endpoints of an interval and are required to estimate the signal at any instant in the interval. Thus given $y(0)$ and $y(T)$ we write for the estimate

$$\hat{y}(t) = ay(0) + by(T) \qquad \text{for } t \in [0, T]$$

Since the error is orthogonal to the data, we have

$$\mathrm{E}\{[y(t) - ay(0) - by(T)]y(0)\} = 0$$

and

$$\mathrm{E}\{[y(t) - ay(0) - by(T)]y(T)\} = 0$$

Hence

$$a\phi_y(0) + b\phi_y(T) = \phi_y(t)$$

and

$$a\phi_y(T) + b\phi_y(0) = \phi_y(T - t)$$

Solving these two equations for a and b yields

$$a = \frac{\phi_y(0)\phi_y(t) - \phi_y(T)\phi_y(T-t)}{\phi_y^2(0) - \phi_y^2(T)}$$

and

$$b = \frac{\phi_y(0)\phi_y(T-t) - \phi_y(t)\phi_y(T)}{\phi_y^2(0) - \phi_y^2(T)}$$

We note that for $t = 0$, $a = 1$, $b = 0$; for $t = T$, $a = 0$, and $b = 1$.

6.3 ESTIMATION OF STATIONARY PROCESSES: THE WIENER FILTER

In the examples in Section 6.2, the data consisted of a finite number of random variables. We now consider the problem of estimating a stochastic process $g(t)$ based on observations of a related process $z(t)$ by a linear combination of known values

$$\{z(\xi) : t_0 \leq \phi \leq t_f\}$$

of $z(t)$, so that the observation interval consists of the closed interval $[t_0, t_f]$. We assume that the processes $g(t)$ and $z(t)$ have zero mean. We write for our estimate

$$\begin{aligned}\hat{g}(t) &= \lim_{\substack{\Delta\xi \to 0 \\ N \to \infty}} \sum_{i=1}^{N} h(t, \xi_i) z(\xi_i)\, \Delta\xi \\ &= \int_{t_0}^{t_f} h(t, \xi) z(\xi)\, d\xi\end{aligned} \tag{6.3.1}$$

We note that $g(t)$ represents the output of a linear time-varying system whose input is $z(t)$.

Since the integral in Eq. (6.3.1) is the limit of a sum, the problem is one of estimating $g(t)$ by a linear combination of the random variables $z(\xi_i)$. The weights $h(t, \xi_i)\, \Delta\xi$ associated with these variables are to be chosen so as to minimize the mean square error. From the orthogonality principle, it follows that these weights have to be chosen such that the error in the estimate is orthogonal to the data; that is,

$$\mathrm{E}\left\{\left[g(t) - \lim_{\substack{\Delta\xi \to 0 \\ N \to \infty}} \sum_{i=1}^{N} h(t, \xi_i) z(\xi_i)\, \Delta\xi\right] z(\tau)\right\} = 0 \qquad t_0 \leq \tau \leq t_f \tag{6.3.2}$$

In terms of the integral, the preceding can be written as

$$\mathrm{E}\left\{\left[g(t) - \int_{t_0}^{t_f} h(t, \xi) z(\xi)\, d\xi\right] z(\tau)\right\} = 0 \qquad t_0 \leq \tau \leq t_f \tag{6.3.3}$$

In terms of correlation functions, this becomes

$$\phi_{gz}(t, \tau) = \int_{t_0}^{t_f} h(t, \xi) \phi_z(\xi, \tau)\, d\xi \qquad t_0 \leq \tau \leq t_f \tag{6.3.4}$$

which is an integral equation of the type we encountered in Chapter 4. This is referred to in the literature as the Wiener–Hopf integral equation [1–11]. The mean-square error can be written from Eq. (6.2.9) as

$$\begin{aligned}\overline{C}_{\text{lms}} &= \mathrm{E}\left\{\left[g(t) - \int_{t_0}^{t_f} h(t, \xi) z(\xi)\, d\xi\right] g(t)\right\} \\ &= \phi_g(t, t) - \int_{t_0}^{t_f} h(t, \xi) \phi_{zg}(\xi, t)\, d\xi\end{aligned} \tag{6.3.5}$$

We now consider the solution of Eq. (6.3.4) for some special cases.

6.3.1 Estimation of Signals Observed in Additive Noise

Of particular importance is the case where the observations $z(t)$ consist of a signal observed in additive noise:

$$z(t) = y(t) + v(t) \qquad t_0 \leq t \leq t_f \tag{6.3.6}$$

We desire to estimate $g(t)$ in the interval $t_0 \leq t \leq t_f$. This was first considered by Wiener in connection with problems in tracking and control [10]. As discussed in Section 6.2, if we choose $g(t) = y(t + \alpha), \alpha > 0$, we have a prediction problem. If $g(t) = y(t)$, we have a filtering problem, and if $g(t) = y(t + \alpha), \alpha < 0$, the problem is one of interpolation or smoothing.

Let us now consider the problem of solving Eq. (6.3.4). From our discussions in Chapter 4, we know that the solution is rather difficult except in the special case when the observations $z(t)$ and the desired estimate $\hat{g}(t)$ are jointly stationary. For this to happen, the observation interval must be semi-infinite, that is, $t_0 = -\infty$ and the system (filter) must be time invariant. If we restrict ourselves to causal systems, the filter is permitted to use only past data in constructing the estimate. Under these conditions Eq. (6.3.1) becomes

$$\hat{g}(t) = \int_{-\infty}^{t} h(t - \xi)z(\xi)\, d\xi \tag{6.3.7}$$

Application of the orthogonality principle yields the equation that the optimal filter must satisfy:

$$\phi_{gz}(t - \tau) = \int_{-\infty}^{t} h(t - \xi)\phi_z(\xi - \tau)\, d\xi \qquad -\infty < \tau \leq t \tag{6.3.8}$$

We solve this equation in a manner similar to our solution in Chapter 4 [11]. Thus we note that if $z(t)$ were a white process with

$$\phi_z(\tau) = \delta(\tau) \tag{6.3.9}$$

the equation that the optimum filter satisfies reduces to

$$\phi_{gz}(t - \tau) = \int_{-\infty}^{t} h(t - \xi)\delta(\xi - \tau)\, d\xi = h(t - \tau) \qquad \text{for} \quad -\infty < \tau \leq t \tag{6.3.10}$$

or in terms of transfer functions

$$H(s) = [\Phi_{gz}(s)]_+ \tag{6.3.11}$$

We see that the solution of the integral equation is straightforward.

When the observation $z(t)$ is not white, we can consider passing it through a whitening filter $H_w(s)$. The input to the optimum filter is now the whitened observation $v(t)$ and the filter is designed so that its output is the desired mean square estimate $\hat{g}(t)$. This is illustrated in Fig. 6.3.1.

The structure of the filter is similar to the one we encountered in our solution to the matched filter problem (see Fig. 4.3.1). The filter $H_w(s)$ in the estimation

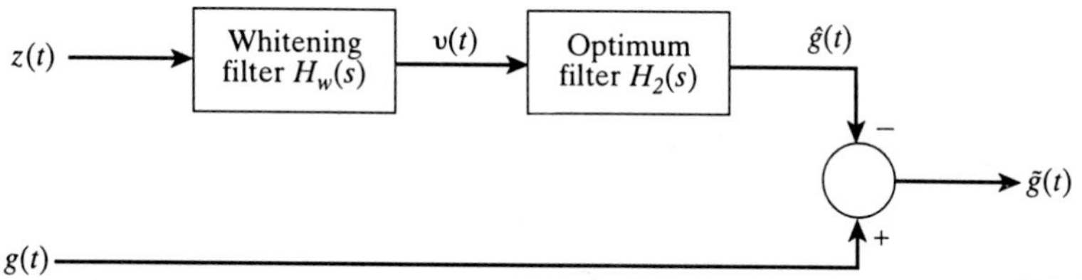

Figure 6.3.1 Wiener filter for colored observations.

problem is, however, designed to whiten the observation $z(t)$, whereas in the detection problem, only the observation noise was whitened.

The optimum filter $H_2(s)$ operates on the whitened observations $v(t)$ and is obtained from Eq. (6.3.11) by replacing $z(t)$ with $v(t)$. We thus get

$$H_2(s) = [\Phi_{gv}(s)]_+ \tag{6.3.12}$$

The whitening filter is given by

$$H_w(s) = \frac{1}{\Phi_z^+(s)} \tag{6.3.13}$$

It only remains to determine $\Phi_{gv}(s)$. We note that

$$\begin{aligned} \Phi_{gv}(s) &= H_w(-s)\Phi_{gz}(s) \\ &= \frac{\Phi_{gz}(s)}{\Phi_z^-(s)} \end{aligned} \tag{6.3.14}$$

so that the overall filter is

$$H(s) = H_w(s)H_2(s) = \frac{1}{\Phi_z^+(s)}\left[\frac{\Phi_{gz}(s)}{\Phi_z^-(s)}\right]_+ \tag{6.3.15}$$

The filter so obtained is usually referred to as the *Wiener filter* [10,11]. The error associated with the Wiener filter can be determined as follows. From Eq. (6.3.5), we have, with $t_0 = -\infty$, and for stationary signals

$$\begin{aligned} \overline{C}_{\text{lms}} &= \phi_g(0) - \int_{-\infty}^{t} h(t-\xi)\phi_{gz}(t-\xi)\,d\xi \\ &= \phi_g(0) - \int_0^{\infty} h(\lambda)\phi_{gz}(\lambda)\,d\lambda \end{aligned} \tag{6.3.16}$$

A problem of special interest is the estimation of $y(t+\alpha)$. In this case

$$\begin{aligned} \phi_{gz}(\tau) &= \mathrm{E}\{g(t+\tau)z(t)\} \\ &= \mathrm{E}\{y(t+\tau+\alpha)z(t)\} \\ &= \phi_{yz}(\tau+\alpha) \end{aligned} \tag{6.3.17}$$

Thus

$$\Phi_{gz}(s) = \Phi_{yz}(s)e^{\alpha s} \tag{6.3.18}$$

and the optimal filter obtained from Eq. (6.3.15) is

$$H(s) = \frac{1}{\Phi_z^+(s)}\left[\frac{e^{\alpha s}\Phi_{yz}(s)}{\Phi_z^-(s)}\right]_+ \tag{6.3.19}$$

The mean-square error in this case is, from Eq. (6.3.16),

$$\overline{C}_{\text{lms}} = \phi_y(0) - \int_0^\infty h(\lambda)\phi_{yz}(\lambda + \alpha)\,d\lambda \tag{6.3.20}$$

Let us denote the integral on the right side of Eq. (6.3.20) as

$$F(\alpha) = \int_0^\infty h(\lambda)\phi_{yz}(\lambda + \alpha)\,d\lambda \tag{6.3.21}$$

Since $h(\lambda) = 0$ for $\lambda < 0$, we can replace the lower limit by $-\infty$. Let

$$k(t) = \mathscr{L}^{-1}\left[\frac{\Phi_{yz}(s)}{\Phi_z^-(s)}\right] \tag{6.3.22}$$

where $\mathscr{L}^{-1}$ denotes the inverse Laplace transform. Then

$$\mathscr{L}^{-1}\left\{\left[\frac{\Phi_{yz}(s)e^{\alpha s}}{\Phi_z^-(s)}\right]_+\right\} = \begin{cases} k(t + \alpha) & t \geqslant 0 \\ 0 & t < 0 \end{cases} \tag{6.3.23}$$

We can substitute Eq. (6.3.23) in Eq. (6.3.19) and write

$$H(s) = \frac{1}{\Phi_z^+(s)}\int_0^\infty k(t + \alpha)e^{-st}\,dt \tag{6.3.24}$$

so that

$$h(\lambda) = \frac{1}{2\pi j}\int_{c-j\infty}^{c+j\infty} \frac{e^{s\lambda}\,ds}{\Phi_z^+(s)}\int_0^\infty k(t + \alpha)e^{-st}\,dt \tag{6.3.25}$$

Substitution of Eq. (6.3.25) in Eq. (6.3.21) yields

$$\begin{aligned} F(\alpha) = \int_0^\infty k(t + \alpha)\frac{1}{2\pi j}\int_{c-j\infty}^{c+j\infty} \frac{e^{-s(t+\alpha)}}{\Phi_z^+(s)} \\ \times \int_{-\infty}^{\infty} \phi_{yz}(\lambda + \alpha)e^{s(\lambda+\alpha)}\,d\lambda\,ds\,dt \end{aligned} \tag{6.3.26}$$

We note that the innermost integral is $\Phi_{yz}(-s)$, so that Eq. (6.3.26) becomes

$$F(\alpha) = \int_0^\infty k(t + \alpha)\left[\frac{1}{2\pi j}\int_{c-j\infty}^{c+j\infty} K(-s)e^{-s(t+\alpha)}\,ds\right]dt \tag{6.3.27}$$

where

$$K(s) = \mathscr{L}[k(t)] = \frac{\Phi_{yz}(s)}{\Phi_z^-(s)} \tag{6.3.28}$$

Finally, we have

$$F(\alpha) = \int_0^\infty [k(t+\alpha)]^2\,dt$$
$$= \int_\alpha^\infty k^2(t)\,dt \tag{6.3.29}$$

We see that $F(\alpha)$ is a monotone decreasing function of α.

Substituting for $F(\alpha)$ in Eq. (6.3.20) yields the mean-square error

$$\overline{C}_{\text{lms}} = \phi_y(0) - \int_\alpha^\infty k^2(t)\,dt \tag{6.3.30}$$

The error is maximum for $\alpha = \infty$ (prediction time is infinite) and is minimum for $\alpha = -\infty$ (smoothing time or lag is infinite). In fact, for the case of $\alpha = -\infty$, the mean-square error obtained is called the *irreducible error*. This irreducible error is also obtained for the filtering problem ($\alpha = 0$), but with the observation interval being infinity, $(-\infty, \infty)$. In such a case, Eq. (6.3.8) has the upper limit ∞ instead of t. The solution for $H(s)$ can be obtained by taking the bilateral Laplace transform, but the filter will be physically unrealizable (noncausal).

We illustrate the material in this section by considering four examples.

Example 6.3.1

Let the input signal be stationary, zero mean, with autocorrelation function

$$\phi_y(\tau) = \frac{7}{12}e^{-|\tau|/2}$$

and power spectral density

$$\Phi_y(s) = \frac{7/3}{-4s^2+1}$$

Assume that the noise is zero mean with spectral density

$$\Phi_v(s) = \frac{5/3}{-s^2+1}$$

and is uncorrelated with the signal. We seek to estimate $y(t+\alpha)$.

In this case we note that since the signal and noise are uncorrelated,

$$\Phi_{yz}(s) = \Phi_y(s) = \frac{7/3}{-4s^2+1}$$

and

$$\Phi_z(s) = \Phi_y(s) + \Phi_v(s) = \frac{7/3}{-4s^2+1} + \frac{5/3}{-s^2+1} = \frac{-9s^2+4}{(-4s^2+1)(-s^2+1)}$$
$$= \frac{(3s+2)}{(2s+1)(s+1)}\cdot\frac{(-3s+2)}{(-2s+1)(-s+1)}$$

The optimal filter as given by Eq. (6.3.15) is, therefore,

$$H(s) = \frac{(2s+1)(s+1)}{3s+2}\left[e^{\alpha s}\frac{7/3}{-4s^2+1}\frac{(-2s+1)(-s+1)}{(-3s+2)}\right]_+$$
$$= \frac{(2s+1)(s+1)}{3s+2}\left[e^{\alpha s}\left\{\frac{1}{2s+1}+\frac{1/3}{-3s+2}\right\}\right]_+$$

Let

$$K(s) = \frac{1}{2s+1} + \frac{1/3}{-3s+2}$$

Then

$$k(t) = \begin{cases} \frac{1}{2}e^{-t/2} & t \geqslant 0 \\ \frac{1}{9}e^{2t/3} & t < 0 \end{cases}$$

The causal part of $k(t+\alpha)$ is given by

$$k(t+\alpha) = \tfrac{1}{2}e^{-(t+\alpha)/2} \qquad \alpha > 0$$
$$= \begin{cases} \frac{1}{9}e^{2(t+\alpha)/3} & 0 \leqslant t < |\alpha| \\ \frac{1}{2}e^{-(t+\alpha)/2} & t \geqslant |\alpha| \end{cases} \qquad \alpha < 0$$

so that

$$[K(s)e^{\alpha s}]_+ = \begin{cases} \dfrac{e^{-\alpha/2}}{2s+1} & \alpha > 0 \\ \dfrac{1}{2s+1} & \alpha = 0 \\ \dfrac{1}{3}\dfrac{e^{2\alpha/3}-e^{s\alpha}}{3s-2} + \dfrac{1}{2}\dfrac{e^{s\alpha}}{s+\frac{1}{2}} & \alpha < 0 \end{cases}$$

The optimum filter in each case is given by

$$H(s) = \frac{(2s+1)(s+1)}{3s+2}[K(s)e^{\alpha s}]_+$$

To determine the mean-square error, we compute

$$F(\alpha) = \int_\alpha^\infty k^2(t)\,dt$$

We have

$$F(\alpha) = \begin{cases} \frac{1}{4}e^{-\alpha} & \alpha > 0 \\ \frac{1}{4} & \alpha = 0 \\ \frac{1}{108}(1-e^{4\alpha/3}) + \frac{1}{4} & \alpha < 0 \end{cases}$$

and the mean-square error is

$$\overline{C}_{\text{lms}} = \phi_y(0) - F(\alpha) = \tfrac{7}{12} - F(\alpha)$$

These various functions are sketched in Fig. 6.3.2. The minimum error that can be obtained when $\alpha = -\infty$ is 35/108.

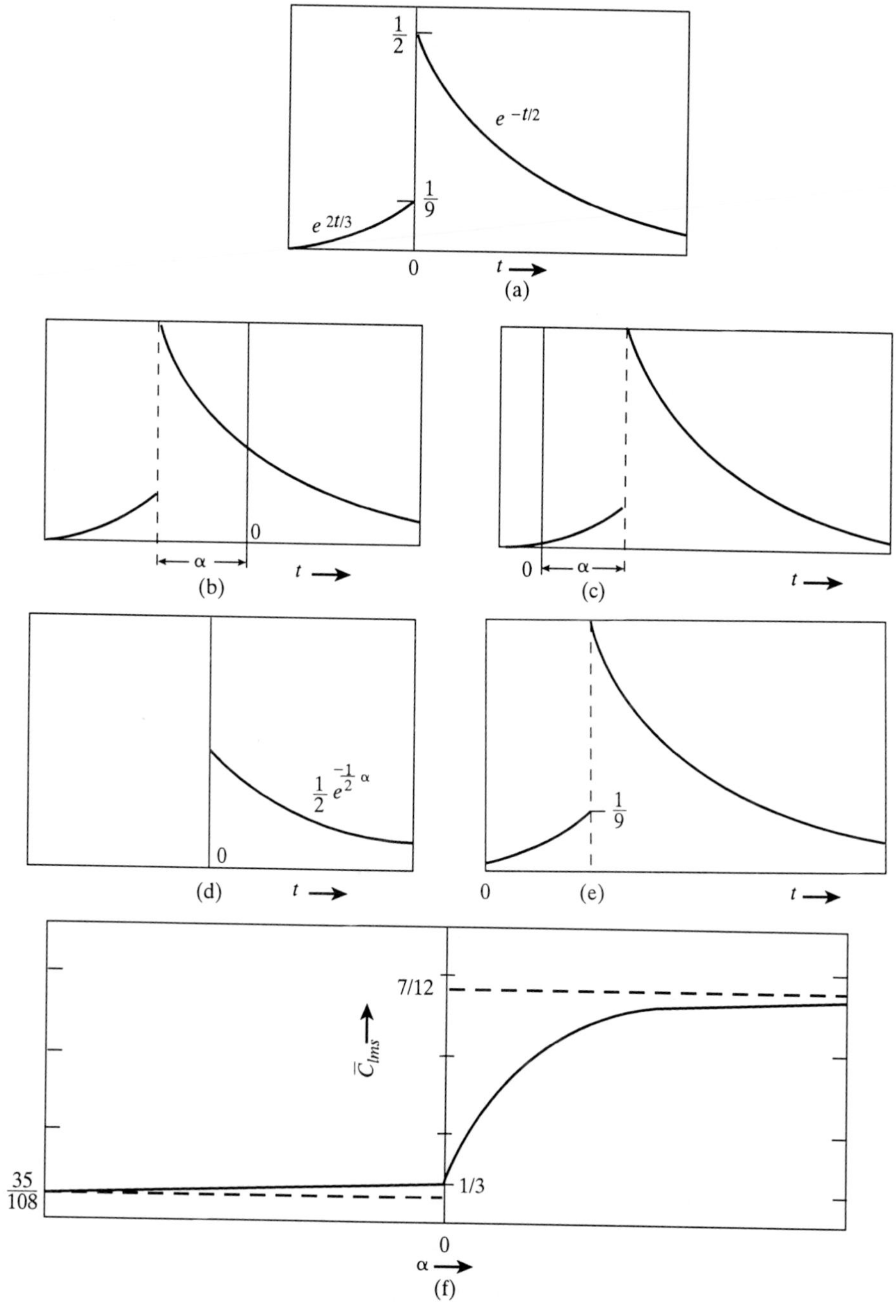

Figure 6.3.2 Computation of error for Example 6.3.1: (a) $k(t)$; (b) $k(t + \alpha)$ for $\alpha > 0$; (c) $k(t + \alpha)$ for $\alpha < 0$; (d) causal part of (b); (e) causal part of (c); (f) cost versus α.

Example 6.3.2

Consider the problem of estimating a stationary signal $y(t)$ with rational power spectrum, in the presence of white noise of spectral height $N_0/2$. If the signal and noise are uncorrelated, we can write

$$\Phi_z(s) = \Phi_y(s) + \Phi_v(s) = \Phi_y(s) + \frac{N_0}{2}$$

From Eq. (6.3.15) the optimum filter is given by

$$H(s) = \frac{1}{\Phi_z^+(s)}\left[\frac{\Phi_{gz}(s)}{\Phi_z^-(s)}\right]_+$$

Because $g(t) = y(t)$ in this case, we have

$$\Phi_{gz}(s) = \Phi_{yz}(s) = \Phi_y(s)$$

By adding and subtracting $\Phi_v(s)$ to the numerator of the term in brackets, we can write the expression for $H(s)$ as

$$H(s) = \frac{1}{\Phi_z^+(s)}\left[\Phi_z^+(s) - \frac{\Phi_v(s)}{\Phi_z^-(s)}\right]_+$$

$$= 1 - \frac{1}{\Phi_z^+(s)}\left[\frac{N_0/2}{\Phi_z^-(s)}\right]_+$$

From the form of $\Phi_z(s)$ we note that it is a rational function with equal-order numerator and denominator. The leading coefficient in the numerator will be $N_0/2$. Thus $\Phi_z^-(s)$ will also be a rational function with equal-order numerator and denominator with leading coefficient $\sqrt{N_0/2}$. Correspondingly, $(N_0/2)/\Phi_z^-(s)$ will have leading coefficient $\sqrt{N_0/2}$ and can be written as

$$\frac{N_0/2}{\Phi_z^-(s)} = \sqrt{\frac{N_0}{2}} + \Phi_1(s)$$

where $\Phi_1(s)$ has numerator degree at least one less than denominator. Since $\Phi_1(s)$ has all its poles in the right half-plane,

$$\left[\frac{N_0/2}{\Phi_z^-(s)}\right]_+ = \sqrt{\frac{N_0}{2}}$$

and the optimum filter is

$$H(s) = 1 - \frac{\sqrt{N_0/2}}{\Phi_z^+(s)}$$

For example, if $\Phi_y(s) = 1/(1 - s^2)$,

$$\Phi_z(s) = \frac{1}{1 - s^2} + \frac{N_0}{2} = \frac{1 + N_0/2 - N_0/2s^2}{1 - s^2}$$

and

$$\Phi_z^+(s) = \frac{\sqrt{1 + N_0/2} + \sqrt{N_0/2}s}{1 + s}$$

so that the optimum filter is

$$H(s) = 1 - \frac{\sqrt{N_0/2}(1 + s)}{\sqrt{1 + N_0/2} + \sqrt{N_0/2}\,s} = \frac{\sqrt{1 + 2/N_0} - 1}{s + 1 + 2/N_0}$$

Example 6.3.3

We consider the problem of Example 6.3.2 but with $N_0/2 = 1$. The estimate is then given by

$$\hat{Y}(s) = H(s)Z(s) = \left[1 - \frac{1}{\Phi_z^+(s)}\right]Z(s)$$

Since $[1/\Phi_z^+(s)]Z(s)$ represents the output of the whitening filter, we have

$$\hat{Y}(s) = Z(s) - \eta(s)$$

so that the whitened observation is given by

$$\eta(t) = z(t) - \hat{y}(t)$$

By rewriting the equation as

$$z(t) = \hat{y}(t) + \eta(t)$$

we see that the observation has essentially been decomposed into two components, one of which is white. Furthermore, our discussions in this section show that the solution to the estimation problem can be obtained completely in terms of this white process.

Example 6.3.4

Consider the feedback control system shown in Fig. 6.3.3, which consists of a fixed plant $G_p(s)$ and a compensator $G_c(s)$. The input $z(t)$ to the plant consists of a stationary signal $y(t)$ corrupted by additive noise $v(t)$. It is desired to design the compensator $G_c(s)$ so that the system output $c(t)$ follows the signal input $y(t)$ in a mean-square sense. It can easily be verified that the input–output transfer function $T(s)$ is given by

$$T(s) = \frac{C(s)}{Z(s)} = \frac{G_c(s)G_p(s)}{1 + G_c(s)G_p(s)}$$

If $G_p(s)$ has no poles or zeros in the right half-plane, we can solve the problem by posing it as a Wiener filtering problem. We then design the system (filter) $T(s)$ such that $c(t)$ is the mean-square estimate of $y(t)$ by using Eq. (6.3.15). Once $T(s)$ is determined, $G_c(s)$ can easily be obtained.

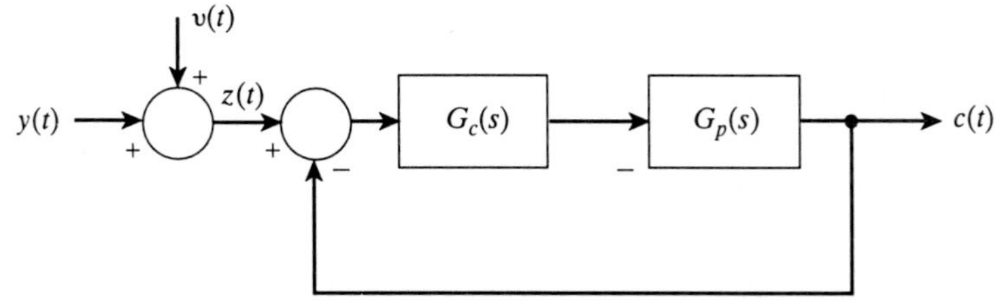

Figure 6.3.3 Block diagram for control example.

For example, let the signal $y(t)$ be obtained by passing white noise of spectral height unity through an integrator. Let $v(t)$ be white with spectral height $N_0/2$. Let $G_p(s) = 1/s(s+1)$.

We note that the two-sided Laplace transform of $\phi_y(\tau)$ does not exist in this case. We overcome this problem by replacing the integrator by a system with transfer function $1/(s+\epsilon)$ and let ϵ tend to zero later. We can then write for the power spectrum of $y(t)$

$$\Phi_y(s) = \frac{1}{\epsilon^2 - s^2}$$

Since $v(t)$ is white, we can use the results of Example 6.3.2 to obtain the optimum system $T(s)$. We have

$$\Phi_z(s) = \Phi_y(s) + \Phi_v(s) = \frac{1}{\epsilon^2 - s^2} + \frac{N_0}{2}$$

$$= \frac{N_0}{2}\left[\frac{\sqrt{(2/N_0) + \epsilon^2} + s}{\epsilon + s}\right]\left[\frac{\sqrt{(2/N_0) + \epsilon^2} - s}{\epsilon - s}\right]$$

and

$$\Phi_z^+(s) = \frac{s + \sqrt{\epsilon^2 + (2/N_0)}}{s + \epsilon}\sqrt{\frac{N_0}{2}}$$

so that

$$T(s) = 1 - \frac{s + \epsilon}{s + \sqrt{\epsilon^2 + (2/N_0)}} = \frac{\sqrt{\epsilon^2 + (2/N_0)} - \epsilon}{s + \sqrt{\epsilon^2 + (2/N_0)}}$$

Now if we let $\epsilon \to 0$, we obtain the optimal system as

$$T(s) = \frac{\sqrt{2/N_0}}{s + \sqrt{2/N_0}}$$

The compensator $G_c(s)$ is obtained as

$$G_c(s) = \frac{1}{G_p(s)}\frac{T(s)}{1 - T(s)}$$

$$= \sqrt{\frac{2}{N_0}}(s+1)$$

To obtain a practical realization of this compensator, it is usual to add poles in the interior of the left-half s-plane so that the degree of denominator is at least as large as that of the numerator. The poles are chosen so that they will not significantly alter the response of the system. In case the plant has poles or zeros in the right half-plane, the design of the overall system must be modified to take into account certain practical constraints. This is because the system may become unstable due to the presence of right half-plane poles. The problem must be solved as a constrained optimization problem. The reader is referred to [13] for further details.

6.3.2 Estimation of Vector Processes

So far we have restricted ourselves to the estimation of scalar processes. The extension to vector processes is quite straightforward. In this case we are given values of the observed process $\mathbf{z}(t)$ which is assumed to be an n-vector, and are required to estimate the related m-vector process $\mathbf{g}(t)$ by a linear combination of the data. Thus if $g_i(t)$ is the ith component of the vector $\mathbf{g}(t)$, we write for its estimate

$$\hat{g}_i(t) = \sum_{j=1}^{n} \int_{t_0}^{t_f} H_{ij}(t, \xi) z_j(\xi)\, d\xi \qquad i = 1, \ldots, m \tag{6.3.31}$$

In vector-matrix notation, this can be written as

$$\hat{\mathbf{g}}(t) = \int_{t_0}^{t_i} \mathbf{H}(t, \xi)\mathbf{z}(\xi)\, d\xi \tag{6.3.32}$$

We could choose as our estimate either the minimum-mean-square estimate or the minimum variance estimate. In the former case our risk function is

$$\begin{aligned} \overline{C}_{\text{lms}} &= \mathrm{E}\{[\mathbf{g}(t) - \hat{\mathbf{g}}(t)]^T[\mathbf{g}(t) - \hat{\mathbf{g}}(t)]\} \\ &= \mathrm{E}\left\{\sum_{i=1}^{m} [\tilde{g}_i(t)]^2\right\} \end{aligned} \tag{6.3.33}$$

In the latter case we seek to minimize the error covariance. Since this is now a matrix, this implies that we minimize the associated quadratic form

$$\begin{aligned} \overline{C}_{\text{lmv}} &= \mathrm{E}\{\mathbf{a}^T[\mathbf{g}(t) - \hat{\mathbf{g}}(t)][\mathbf{g}(t) - \hat{\mathbf{g}}(t)]^T\mathbf{a}\} \\ &= \mathbf{a}^T \operatorname{cov} \tilde{\mathbf{g}}(t)\mathbf{a} \end{aligned} \tag{6.3.34}$$

where $\mathbf{a}$ is any arbitrary vector.

In either case, it can be shown by using the orthogonality principle that the matrix $\mathbf{H}(t, \xi)$ must be chosen such that

$$\phi_{gz}(t, \tau) = \int_{t_0}^{t_f} \mathbf{H}(t, \xi)\phi_z(\xi, \tau)\, d\tau \qquad t_0 \leqslant \tau \leqslant t_f \tag{6.3.35}$$

Again the solution of this equation is not an easy task. We might expect that if we restrict ourselves to stationary processes and semi-infinite observation intervals, we may be able to get a solution as in the scalar case. In this case Eq. (6.3.35) becomes

$$\phi_{gz}(t - \tau) = \int_{-\infty}^{t} \mathbf{H}(t - \xi)\phi_z(\xi - \tau)\, d\xi \qquad t_0 \leqslant \tau \leqslant t_f \tag{6.3.36}$$

We may again seek to obtain the overall filter as a cascade of a whitening filter and an optimal filter that operates on the whitened observations to yield an estimate of the desired signal. We recall that in the scalar problem, this involved a spectral factorization of $\Phi_z(s)$ into terms containing poles and zeros only in the left half-plane. We might seek a similar factorization for the present problem also. Unfortunately, in this case $\Phi_z(s)$ is a matrix and the spectral factorization is not trivial. Here we are seeking a matrix $\Phi^+(s)$ such that

$$\Phi_z(s) = \Phi^+(s)[\Phi^+(-s)]^T \tag{6.3.37}$$

such that $\Phi^+(s)$ is analytic (has no poles) in the right half-plane and $[\Phi^+(s)]^{-1}$ is also analytic in the right half-plane. We will denote $[\Phi^+(-s)]^T$ as $\Phi^-(s)$ for convenience.

Assuming such a matrix has been found, the transfer matrix of the whitening filter is given by

$$H_w(s) = [\Phi^+(s)]^{-1} \tag{6.3.38}$$

The transfer function of the optimal filter follows analogous to our derivation in the scalar case as

$$H_2(s) = \{\Phi_{gz}(s)[\Phi^-(s)]^{-1}\}_+ \tag{6.3.39}$$

and the overall filter obtained as a cascade of $H_w(s)$ and $H_2(s)$ is

$$\mathbf{H}(s) = \{\Phi_{gz}(s)[\Phi^-(s)]^{-1}\}_+[\Phi^+(s)]^{-1} \tag{6.3.40}$$

As explained earlier, the critical part of this solution is in the spectral factorization of Eq. (6.3.37). While several techniques are available for accomplishing this, the techniques are usually cumbersome [14,15]. In Section 6.5 we discuss one such technique which involves the solution to a nonlinear algebraic equation.

6.3.3 Linear Estimation of Discrete-Time Processes

We can extend our results of Section 6.3.2 to the estimation of discrete-time processes. The problem is to estimate a discrete-time process $g(k)$ given observations of a related process $z(k)$ over an interval $[k_0, k_f]$. We will again assume that $g(k)$ and $z(k)$ are zero mean.

As in Section 6.3.2, the problem is a filtering problem when $k = k_f$, it is a prediction problem when $k > k_f$, and it is a smoothing problem when $k_0 < k < k_f$. As we will see later, of particular interest is the *one-stage prediction* problem, in which we are required to find the estimate of $g(k)$ given observations in the range $[k_0, k-1]$ [16,17].

If we are seeking a linear minimum-mean-square estimate, we can write

$$\hat{g}(k) = \sum_{j=k_0}^{k_f} h(k,j)z(j) \tag{6.3.41}$$

where $h(k,j)$ represents the impulse response of the system. The coefficients (weights) $h(k,j)$ are determined by the orthogonality principle so as to make the estimation error orthogonal to the observations. Thus we have

$$\mathrm{E}\left\{\left[g(k) - \sum_{j=k_0}^{k_f} h(k,j)z(j)\right]z(l)\right\} = 0 \qquad \text{for} \quad l \in [k_0, k_f] \tag{6.3.42}$$

so that

$$\sum_{j=k_0}^{k_f} h(k,j)\phi_z(j,l) = \phi_{gz}(k,l) \qquad \text{for} \quad l \in [k_0, k_f] \tag{6.3.43}$$

Equation (6.3.43) is the discrete analog of Eq. (6.3.4). The solution of this equation is again difficult for the general case of nonstationary signals. We defer until the next section, the solution of this equation for a particular class of nonstationary signals.

Here we derive a solution under the assumption that the processes $g(k)$ and $z(k)$ are jointly stationary and that the filter is time invariant and causal. Then Eq. (6.3.41) becomes

$$\begin{aligned}\hat{g}(k) &= \sum_{j=k_0}^{k} h(k-j)z(j) \\ &= \sum_{j=0}^{k-k_0} h(j)z(k-j)\end{aligned} \tag{6.3.44}$$

When $k_0 = -\infty$, we can determine the optimum filter by following a procedure similar to that for the continuous-time case in Section 6.3.2. In this case it follows from Eq. (6.3.43) that

$$\sum_{j=0}^{\infty} h(j)\phi_z(k-j-l) = \phi_{gz}(k-l) \tag{6.3.45}$$

Setting $k - l = n$ yields

$$\sum_{j=0}^{\infty} h(j)\phi_z(n-j) = \phi_{gz}(n) \qquad n \in [0, \infty) \tag{6.3.46}$$

which is the discrete analog of Eq. (6.3.8). We can solve Eq. (6.3.46) in a fashion similar to the solution of Eq. (6.3.8). We thus again note that if $z(\cdot)$ were a white sequence such that

$$\phi_z(n-j) = \delta_{nj} \tag{6.3.47}$$

the solution to Eq. (6.3.46) is easily obtained as

$$h(n) = \phi_{gz}(n) \qquad n \in [0, \infty) \tag{6.3.48}$$

with corresponding transfer function

$$H(z) = [\Phi_{gz}(z)]_+ \tag{6.3.49}$$

where for any function $X(z)$, $[X(z)]_+$ denotes the z-transform of the causal part of the corresponding time sequence $x(n)$. For rational $X(z)$, $[X(z)]_+$ can be obtained by making a partial fraction expansion over the poles and collecting together all terms corresponding to poles within the unit circle in the z-plane.

If $z(\cdot)$ is not a white sequence, we can again convert it into a white sequence by passing $z(\cdot)$ through a whitening filter given by

$$H_w(z) = \frac{1}{\Phi_z^+(z)} \tag{6.3.50}$$

The filter $H_o(z)$ operates on the whitened signal $\eta(\cdot)$ to yield the desired estimate $\hat{g}(\cdot)$. We can obtain $H_o(z)$ from Eq. (6.3.49) by replacing $z(\cdot)$ by $\eta(\cdot)$ so that

$$H_o(z) = [\Phi_{g\eta}(z)]_+ \tag{6.3.51}$$

Finally, since

$$\Phi_{g\eta}(z) = H_w(z^{-1})\Phi_{gz}(z) \\ = \frac{\Phi_{gz}(z)}{\Phi_z^-(z)} \tag{6.3.52}$$

the overall filter obtained by cascading $H_w(z)$ and $H_o(z)$ is

$$H(z) = \frac{1}{\Phi_z^+(z)}\left[\frac{\Phi_{gz}(z)}{\Phi_z^-(z)}\right]_+ \tag{6.3.53}$$

The minimum error associated with the filter can be shown to be

$$\overline{C}_{lms} = \phi_g(0) - \sum_{j=0}^{\infty} h(j)\phi_{gz}(j) \tag{6.3.54}$$

Example 6.3.5

Consider the problem of estimating a signal $g(k)$ with correlation function $\phi_g(k) = (0.5)^{|k|}$ from noise-corrupted observations of the form

$$z(k) = g(k) + v(k)$$

with $v(k)$ being a zero-mean, white process with variance 0.5, uncorrelated with $g(k)$.

For this problem we have

$$\Phi_g(z) = \Phi_{gz}(z) = \frac{-1.5z}{(z-0.5)(z-2)}$$

$$\Phi_z(z) = \Phi_g(z) + \Phi_v(z) = \frac{0.5(z-5.312)(z-0.1883)}{(z-0.5)(z-2)}$$

so that

$$\left[\frac{\Phi_{gz}(z)}{\Phi_z^-(z)}\right]_+ = \left[\frac{-1.5z}{\sqrt{0.5}(z-0.5)(z-5.312)}\right]_+ = \frac{0.4408z}{z-0.5}$$

From Eq. (6.3.53), the optimal filter is therefore given by

$$H(z) = \frac{0.6234z}{z-0.1883}$$

The impulse response of the optimal filter is

$$h(k) = 0.6234(0.1883)^k \qquad k \geq 0$$

so that the minimum error as given by Eq. (6.3.54) is

$$\overline{C}_{lms} = 1 - \sum_{j=0}^{\infty}(0.6234)(0.1883)^j(0.5)^j = 0.3118$$

6.3.4 Optimum FIR Filter

Let us now consider the determination of the optimum filter when the estimate $\hat{g}(k)$ is based on a finite number, m, of past observations, $\{z_k \quad z_{k-1} \quad \cdots \quad z_{k-m+1}\}$. The optimum filter is a *finite impulse response* (FIR) *filter*. Let us denote the weights of this filter by h_j^m, where we use the superscript to denote explicitly the order of the filter. The estimate can then be written as

$$\hat{g}(k) = \sum_{j=1}^{m} h_j^m z(k - j + 1) \tag{6.3.55}$$

with the weights h_j^m satisfying

$$\sum_{j=1}^{m} h_j^m \phi_z(k - j + 1) = \phi_{gz}(k) \qquad \text{for} \quad 0 \leq k \leq m - 1 \tag{6.3.56}$$

This constitutes a set of m simultaneous linear algebraic equations in the weights $h_j^m, 1 \leq j \leq m$. We can write these equations in vector-matrix notation as

$$\begin{bmatrix} \phi_z(0) & \phi_z(1) & \cdots & \phi_z(m-1) \\ \phi_z(1) & \phi_z(0) & \cdots & \phi_z(m-2) \\ \phi_z(2) & \phi_z(1) & \cdots & \phi_z(m-3) \\ \vdots & \vdots & \vdots & \vdots \\ \phi_z(m-1) & \phi_z(m-2) & \cdots & \phi_z(0) \end{bmatrix} \begin{bmatrix} h_1^m \\ h_2^m \\ h_3^m \\ \vdots \\ h_m^m \end{bmatrix} = \begin{bmatrix} \phi_{gz}(0) \\ \phi_{gz}(1) \\ \phi_{gz}(2) \\ \vdots \\ \phi_{gz}(m-1) \end{bmatrix} \tag{6.3.57}$$

Let us define the m-vectors

$$\mathbf{h}^m = [h_1^m \quad h_2^m \quad \cdots \quad h_m^m]^T$$
$$\boldsymbol{\phi}_{\mathbf{gz}}^{\mathbf{m}} = [\phi_{gz}(0) \quad \phi_{gz}(1) \quad \cdots \quad \phi_{gz}(m-1)]^T$$
$$\mathbf{z}^m(k) = [z(k) \quad z(k-1) \quad \cdots \quad z(k-m+1)]^T$$

Let $\boldsymbol{\Phi}^m$ be the matrix whose (k, j)th element is given by $[\boldsymbol{\Phi}^m]_{kj} = \phi_z(k - j)$. Then Eq. (6.3.56) can be written compactly as

$$\boldsymbol{\Phi}^m \mathbf{h}^m = \boldsymbol{\phi}_{\mathbf{gz}}^{\mathbf{m}} \tag{6.3.58}$$

Clearly, $\boldsymbol{\Phi}^m$ is the correlation matrix of the vector $\mathbf{z}^m(k)$,

$$\boldsymbol{\Phi}^m = \mathrm{E}\{\mathbf{z}^m(k)[\mathbf{z}^m(k)]^T\}$$

Thus $\boldsymbol{\Phi}^m$ is a positive-definite matrix and is hence invertible. We can therefore find the optimum filter as

$$\mathbf{h}^m = [\boldsymbol{\Phi}^m]^{-1} \boldsymbol{\phi}_{gz}^m \tag{6.3.59}$$

The transfer function of the filter is given by

$$H^m(z) = \sum_{j=0}^{m-1} h_{j+1}^m z^{-j} \tag{6.3.60}$$

The filter is usually referred to as the *Wiener filter for discrete signals*. The structure of the filter is shown in Fig. 6.3.4. The filter is a *transversal filter* or a *tapped delay line* (TDL) *filter*.

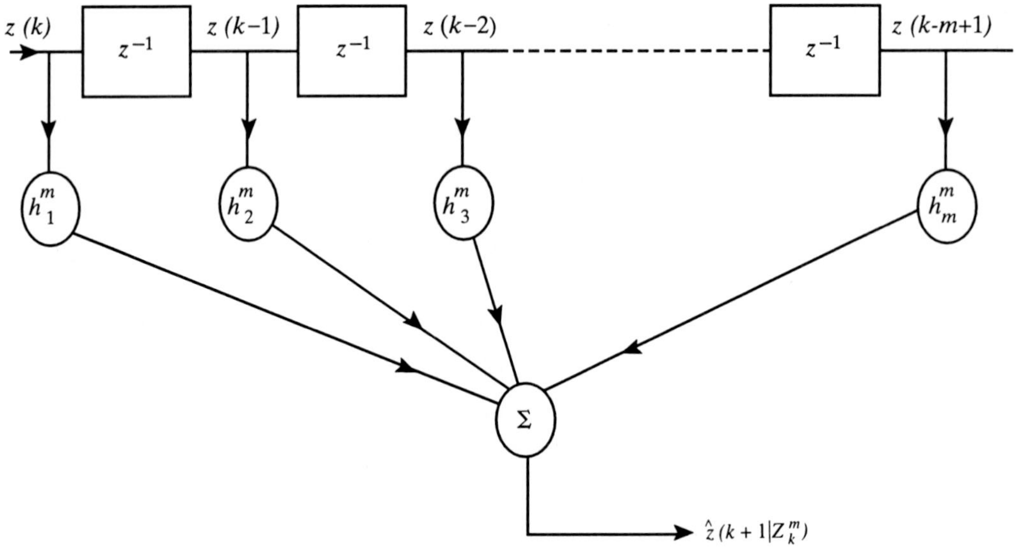

Figure 6.3.4 Transversal Wiener filter.

We can find the minimum mean-square error associated with the filter as

$$\begin{aligned}\overline{C}_{\text{lms}} &= \text{E}\{[g(k) - \hat{g}(k)]^2\} \\ &= \text{E}\left\{\left[g(k) - \sum_{j=1}^{m} h_j^m z(k - j + 1)\right]g(k)\right\} \\ &= \phi_g(0) - \sum_{j=1}^{m} h_j^m \phi_{gz}(j - 1)\end{aligned} \tag{6.3.61}$$

where $\phi_g(0) = \text{E}\{g^2(k)\}$ and the second step follows from the principle of orthogonality, Eq. (6.3.42).

In terms of the vectors $\mathbf{h}^m$ and $\boldsymbol{\phi}_{gz}^m$ defined earlier, it follows that the minimum error is

$$\begin{aligned}\overline{C}_{\text{lms}} &= \phi_g(0) - [\boldsymbol{\phi}_{\mathbf{gz}}^{\mathbf{m}}]^T \mathbf{h}^m \\ &= \phi_g(0) - [\boldsymbol{\phi}_{\mathbf{gz}}^{\mathbf{m}}]^T [\boldsymbol{\Phi}^m]^{-1} \boldsymbol{\phi}_{\mathbf{gz}}^{\mathbf{m}}\end{aligned} \tag{6.3.62}$$

where we have used Eq. (6.3.59).

Example 6.3.6

We consider the design of a third-order transversal filter for the problem of Example 6.3.5. From Eq. (6.3.57) it follows that the weights of the filter satisfy

$$\begin{bmatrix} 1.5 & 0.5 & 0.25 \\ 0.5 & 1.5 & 0.5 \\ 0.25 & 0.5 & 1.5 \end{bmatrix} \begin{bmatrix} h_1^3 \\ h_2^3 \\ h_3^3 \end{bmatrix} = \begin{bmatrix} 1.0 \\ 0.5 \\ 0.25 \end{bmatrix}$$

so that

$$h_1^3 = 0.6235 \qquad h_2^3 = 0.1176 \qquad h_3^3 = 0.0235$$

The error corresponding to this filter can be determined from Eq. (6.3.61) to be

$$\overline{C}_{\text{lms}} = 0.311825$$

As can be expected, the error for the optimal transversal filter is somewhat larger than for the IIR Wiener filter.

We note that $\mathbf{\Phi}^m$ is a special kind of matrix known as a *Toeplitz matrix*. As such, it has some properties that can be exploited to yield other filter structures. We discuss such structures in the next section.

While Eq. (6.3.59) can be used to find the optimum impulse response, we can use an alternative approach which does not require the inversion of matrix $\mathbf{\Phi}^m$. This approach is usually referred to as a *U-L factorization* or *Cholesky decomposition* and follows from the fact that the positive definite matrix $\mathbf{\Phi}^m$ can be expressed as the product of an upper triangular matrix $\mathbf{U}$ and a lower triangular matrix $\mathbf{L}$, so that $\mathbf{\Phi} = \mathbf{UL}$ [18]. The triangular decomposition can be obtained using Gauss reduction. Equation (6.3.58) therefore becomes

$$\mathbf{ULh}^m = \boldsymbol{\phi}_{\mathbf{gz}}^{\mathbf{m}} \tag{6.3.63}$$

Now define the vector $\boldsymbol{\psi}$ as

$$\mathbf{Lh}^m = \boldsymbol{\psi} \tag{6.3.64}$$

Substituting in Eq. (6.3.63) yields

$$\mathbf{U}\boldsymbol{\psi} = \boldsymbol{\phi}_{\mathbf{gz}}^{\mathbf{m}} \tag{6.3.65}$$

Since $\mathbf{U}$ is an upper triangular matrix, the vector $\boldsymbol{\psi}$ can easily be determined by solving for the elements of $\boldsymbol{\psi}$ in reverse order starting with the last element and substituting recursively in the preceding equation. These values of $\boldsymbol{\psi}$ can now be used in Eq. (6.3.64) to find $\mathbf{h}^m$. The solution is again straightforward since $\mathbf{L}$ is a lower triangular matrix and the equations can be solved recursively starting with h_1^m.

Finally, we note that if $z(\cdot)$ is a white sequence with variance σ_z^2, then $\mathbf{\Phi}^m = \sigma_z^2\mathbf{I}$ and the solution to Eq. (6.3.58) is easily obtained as

$$\mathbf{h}^m = \frac{1}{\sigma_z^2}\boldsymbol{\phi}_{\mathbf{gz}}^{\mathbf{m}} \tag{6.3.66}$$

6.3.5 Forward–Backward Prediction

Let us now consider a special case of the discrete Wiener filter problem in which we are required to predict the value $z(k+1)$ given the set of m observations $Z_k^m = \{z(k), z(k-1), \ldots, z(k-m+1)\}$. Let us denote this predicted value (estimate) as $\hat{z}(k+1 \mid Z_k^m)$. This problem is usually referred to as one of *forward linear prediction*.

We can use the results of our previous discussions to obtain the optimal estimate by setting $g(k) = z(k + 1)$. It then follows from Eq. (6.3.55) that

$$\hat{z}(k+1 \mid Z_k^m) = \sum_{j=1}^{m} h_j^m z(k-j+1) \tag{6.3.67}$$

where $h_j^m = 0$ if $j > m$. From Eq. (6.3.56), the weights $h_j^m, 1 \le j \le m$, satisfy the relation

$$\sum_{j=1}^{m} h_j^m \phi_z(k-j+1) = \phi_z(k+1) \tag{6.3.68}$$

The vector-matrix form of this equation follows from Eq. (6.3.57) as

$$\begin{bmatrix} \phi_z(0) & \phi_z(1) & \cdots & \phi_z(m-1) \\ \phi_z(1) & \phi_z(0) & \cdots & \phi_z(m-2) \\ \phi_z(2) & \phi_z(1) & \cdots & \phi_z(m-3) \\ \vdots & \vdots & \vdots & \vdots \\ \phi_z(m-1) & \phi_z(m-2) & \cdots & \phi_z(0) \end{bmatrix} \begin{bmatrix} h_1^m \\ h_2^m \\ h_3^m \\ \vdots \\ h_m^m \end{bmatrix} = \begin{bmatrix} \phi_z(1) \\ \phi_z(2) \\ \phi_z(3) \\ \vdots \\ \phi_z(m) \end{bmatrix} \tag{6.3.69}$$

By reversing the order of these equations, this equation can also be written as

$$\begin{bmatrix} \phi_z(0) & \phi_z(1) & \cdots & \phi_z(m-1) \\ \phi_z(1) & \phi_z(0) & \cdots & \phi_z(m-2) \\ \phi_z(2) & \phi_z(1) & \cdots & \phi_z(m-3) \\ \vdots & \vdots & \vdots & \vdots \\ \phi_z(m-1) & \phi_z(m-2) & \cdots & \phi_z(0) \end{bmatrix} \begin{bmatrix} h_m^m \\ h_{m-1}^m \\ h_{m-2}^m \\ \vdots \\ h_1^m \end{bmatrix} = \begin{bmatrix} \phi_z(m) \\ \phi_z(m-1) \\ \phi_z(m-2) \\ \vdots \\ \phi_z(1) \end{bmatrix} \tag{6.3.70}$$

Let us now consider a related problem, the *backward linear prediction problem*, in which, given the set of observations $\{z(k), z(k-1), \ldots, z(k-m+1)\}$, we want to find the estimate $\hat{z}(k-m \mid Z_k^m)$. As was the case with the forward prediction problem, we can write this estimate as

$$\hat{z}(k-m \mid Z_k^m) = \sum_{j=1}^{m} g_j^m z(k-j+1) \tag{6.3.71}$$

From the principle of orthogonality, we can solve for the weights g_j^m from the set of equations

$$\sum_{j=1}^{m} g_j^m \phi_z(k-j+1) = \phi_z(m-k) \qquad \text{for} \quad 0 \le k \le m-1 \tag{6.3.72}$$

By explicitly writing out this set of equations and comparing it with Eq. (6.3.70), it can easily be verified that

$$g_{m-j+1}^m = h_j^m \qquad \text{for} \quad 1 \le j \le m \tag{6.3.73}$$

From Eqs. (6.3.67) and (6.3.73) it follows that the tap weights of the backward prediction filter are the same as those of the forward prediction filter, but in reverse order. The backward prediction estimate is therefore given by

$$\hat{z}(k-m \mid Z_k^m) = \sum_{j=1}^{m} h_{m-j+1}^m z(k-j+1) \tag{6.3.74}$$

The minimum cost associated with these two estimates can be obtained from Eq. (6.3.61). These are

$$\overline{C}_f^m = \phi_z(0) - \sum_{j=1}^{m} h_j^m \phi_z(j), \tag{6.3.75}$$

$$\overline{C}_b^m = \phi_z(0) - \sum_{j=1}^{m} h_{m-j+1}^m \phi_z(m-j+1). \tag{6.3.76}$$

from which it follows that

$$\overline{C}_f^m = \overline{C}_b^m \tag{6.3.77}$$

The Levinson–Durbin algorithm. Clearly, the previous results show that the estimates $\hat{z}(k+1 \mid Z_k^m)$ and $\hat{z}(k-m \mid Z_k^m)$ are closely related. We can use these and other interesting relationships to obtain computationally efficient methods for determining these estimates. One such is the Levinson–Durbin recursion [19].

Suppose that we increase the estimator order by one and write the estimate of $z(k+1)$ as

$$\hat{z}(k+1 \mid Z_k^{m+1}) = \sum_{j=1}^{m+1} h_j^{m+1} z(k-j+1) \tag{6.3.78}$$

We can write the equation for the weights of this $(m+1)$st-order filter as in Eq. (6.3.70). In partitioned form we get

$$\left[\begin{array}{cccc|c} \phi_z(0) & \phi_z(1) & \cdots & \phi_z(m-1) & \phi_z(m) \\ \phi_z(1) & \phi_z(0) & \cdots & \phi_z(m-2) & \phi_z(m-1) \\ \phi_z(2) & \phi_z(1) & \cdots & \phi_z(m-3) & \phi_z(m-2) \\ \vdots & \vdots & \vdots & \vdots & \vdots \\ \phi_z(m-1) & \phi_z(m-2) & \cdots & \phi_z(0) & \phi_z(1) \\ \hline \phi_z(m) & \phi_z(m-1) & \cdots & \phi_z(1) & \phi_z(0) \end{array}\right] \left[\begin{array}{c} h_1^{m+1} \\ h_2^{m+1} \\ h_3^{m+1} \\ \vdots \\ h_m^{m+1} \\ \hline h_{m+1}^{m+1} \end{array}\right] = \left[\begin{array}{c} \phi_z(1) \\ \phi_z(2) \\ \phi_z(3) \\ \vdots \\ \phi_z(m) \\ \hline \phi_z(m+1) \end{array}\right] \tag{6.3.79}$$

or, equivalently,

$$\begin{bmatrix} \phi_z(0) & \cdots & \phi_z(m-1) \\ \phi_z(1) & \cdots & \phi_z(m-2) \\ \phi_z(2) & \cdots & \phi_z(m-3) \\ \vdots & \vdots & \vdots \\ \phi_z(m-1) & \cdots & \phi_z(0) \end{bmatrix} \begin{bmatrix} h_1^{m+1} \\ h_2^{m+1} \\ h_3^{m+1} \\ \vdots \\ h_m^{m+1} \end{bmatrix} + h_{m+1}^{m+1} \begin{bmatrix} \phi_z(m) \\ \phi_z(m-1) \\ \phi_z(m-2) \\ \vdots \\ \phi_z(1) \end{bmatrix} = \begin{bmatrix} \phi_z(1) \\ \phi_z(2) \\ \phi_z(3) \\ \vdots \\ \phi_z(m) \end{bmatrix} \tag{6.3.80}$$

and

$$\sum_{j=1}^{m} h_j^{m+1} \phi_z(m - j + 1) + h_{m+1}^{m+1} \phi_z(0) = \phi_z(m + 1) \tag{6.3.81}$$

Premultiplying by the inverse of the correlation matrix Φ^m and using Eqs. (6.3.69) and (6.3.80), we get

$$\begin{bmatrix} h_1^{m+1} \\ h_2^{m+2} \\ h_3^{m+3} \\ \vdots \\ h_{m+1}^{m} \end{bmatrix} + h_{m+1}^{m+1} \begin{bmatrix} h_m^m \\ h_{m-1}^m \\ h_{m-2}^m \\ \vdots \\ h_1^m \end{bmatrix} = \begin{bmatrix} h_1^m \\ h_2^m \\ h_3^m \\ \vdots \\ h_m^m \end{bmatrix} \tag{6.3.82}$$

Let us now define

$$h_0^m = h_0^{m+1} = -1 \tag{6.3.83}$$

Equations (6.3.82) and (6.3.83) constitute a set of equations for recursively computing the weights for an estimator of order $m + 1$ given the weights for an estimator of order m. This can be seen by explicitly writing these equations as follows. Let us denote $-h_{m+1}^{m+1}$ by p_{m+1}. We can then write

$$h_0^{m+1} = h_0^m = -1 \tag{6.3.84a}$$

$$h_j^{m+1} = h_j^m + p_{m+1} h_{m-j+1}^m \qquad j = 1, \ldots, m \tag{6.3.84b}$$

where

$$p_{m+1} = -h_{m+1}^{m+1} \tag{6.3.85}$$

The parameters p_m are known as the *partial correlation coefficients* or *reflection coefficients*. It is clear from the equations above that the weights h_j^m for any order m can be computed if we know $p_1, p_2, \ldots, p_m$.

To find p_m, we substitute Eqs. (6.3.84) and (6.3.85) in Eq. (6.3.81) and rearrange terms to get

$$p_{m+1}\left[\sum_{j=1}^{m} h_{m-j+1}^m \phi_z(m - j + 1) - \phi_z(0)\right] = \phi_z(m + 1) - \sum_{j=1}^{m} h_j^m \phi_z(m - j + 1) \tag{6.3.86}$$

Use of Eqs. (6.3.76), (6.3.77) and (6.3.84a) then yields

$$p_{m+1} = \frac{1}{\overline{C}_f^m}\left[\sum_{j=1}^{m+1} h_{m-j+1}^m \phi_z(j)\right] \tag{6.3.87}$$

Finally, we note that by using Eqs. (6.3.84), (6.3.85), and (6.3.87) in Eq. (6.3.75), we can obtain a recursion relation for the minimum mean-square error as

$$\begin{aligned} \overline{C}_f^{m+1} &= \overline{C}_f^m(1 - p_{m+1}^2) \\ \overline{C}_f^0 &= \phi_z(0) \end{aligned} \tag{6.3.88}$$

Equation (6.3.88) shows that the minimum error decreases as the order of the filter increases.

Equations (6.3.84), (6.3.85), (6.3.87), and (6.3.88) together constitute the recursions for the Levinson–Durbin algorithm. Given the set of autocorrelation coefficients $\phi_z(j)$ for $j \geq 0$, we can find the coefficients for any order filter as follows:

1. Initialize the algorithm by setting $h_0^0 = -1$ and $\overline{C}_f^0 = \phi_z(0)$.
2. At any stage m, compute the values of h_j^m, p_m, and $\overline{C}_f^m$ from the corresponding values from the previous stage as

$$h_0^m = h_0^{m-1} \tag{6.3.89a}$$

$$p_m = \frac{1}{\overline{C}_f^{m-1}}\left[\sum_{j=1}^{m} h_{m-j}^{m-1}\,\phi_z(j)\right] \tag{6.3.89b}$$

$$h_j^m = h_j^{m-1} + p_m h_{m-j}^{m-1} \qquad j = 1, \ldots, m-1 \tag{6.3.89c}$$

$$h_m^m = -p_m \tag{6.3.89d}$$

$$\overline{C}_f^m = \overline{C}_f^{m-1}(1 - p_m^2) \tag{6.3.89e}$$

Note that $h_j^m = 0$ if $j > m$.

The estimates can then be found from Eqs. (6.3.67) and (6.3.74).

Example 6.3.7

We will find a third-order predictor for the observed signal $z(k)$ of Example 6.3.5. The first four correlation coefficients of this signal are given by

$$\phi_z(0) = 1.5 \qquad \phi_z(1) = 0.5 \qquad \phi_z(2) = 0.25 \qquad \phi_z(3) = 0.125$$

The coefficients of the third-order forward predictor can be found by substituting these values in Eq. (6.3.69) as

$$\begin{bmatrix} 1.5 & 0.5 & 0.25 \\ 0.5 & 1.5 & 0.5 \\ 0.25 & 0.5 & 1.5 \end{bmatrix}\begin{bmatrix} h_1^3 \\ h_2^3 \\ h_3^3 \end{bmatrix} = \begin{bmatrix} 0.5 \\ 0.25 \\ 0.125 \end{bmatrix}$$

which gives

$$h_1^3 = 0.31176 \qquad h_2^3 = 0.05882 \qquad h_3^3 = 0.01176$$

To use the Levinson–Durbin recursion, we set

$$h_0^0 = -1 \qquad \overline{C}_f^0 = \phi_z(0) = 1.5$$

From Eq. (6.3.89b)

$$p_1 = \frac{h_0^0\,\phi_2(1)}{\overline{C}_f^0} = -0.33333$$

so that

$$h_1^1 = -p_1 = 0.33333$$

From Eq. (6.3.89e) we get

$$\overline{C}_f^1 = \overline{C}_f^0(1 - p_1^2) = 1.33333$$

Continuing the recursions gives

$$p_2 = \frac{h_1^1 \phi_z(1) + h_0^1 \phi_z(2)}{\overline{C}_f^1} = -0.625$$

$$h_2^2 = -p_2 = 0.0625$$

$$h_1^2 = h_1^1 + p_2 h_1^1 = 0.3125$$

$$\overline{C}_f^2 = \overline{C}_f^1(1 - p_2^2) = 1.32185$$

$$p_3 = \frac{h_2^2 \phi_z(1) + h_1^2 \phi_z(2) + h_0^2 \phi_z(3)}{\overline{C}_f^2} = -0.011765$$

$$h_1^3 = h_1^2 + p_3 h_2^2 = 0.311764$$

$$h_2^3 = h_2^2 + p_3 h_1^2 = 0.058823$$

$$h_3^3 = -p_3 = 0.011764$$

which checks out with our previous results.

Lattice structures. We can exploit the relationships between the forward and backward prediction problems to derive an alternative structure for the estimator. To do so, let us define the *forward estimation error* $\epsilon_f^m(k+1)$ as

$$\epsilon_f^m(k+1) = z(k+1) - \hat{z}(k+1 \mid Z_k^m) \tag{6.3.90}$$

and the *backward estimation error* $\epsilon_b^m(k)$ as

$$\epsilon_b^m(k) = z(k-m) - \hat{z}(k-m \mid Z_k^m) \tag{6.3.91}$$

It then follows that

$$\begin{aligned} \epsilon_f^{m+1}(k+1) &= z(k+1) - \hat{z}(k+1 \mid Z_k^{m+1}) \\ &= z(k+1) - \sum_{j=1}^{m+1} h_j^{m+1} z(k-j+1) \end{aligned} \tag{6.3.92}$$

Use of Eqs. (6.3.84) and (6.3.85) then yields

$$\begin{aligned} \epsilon_f^{m+1}(k+1) &= z(k+1) - \sum_{j=1}^{m} [(h_j^m + p_{m+1} h_{m-j+1}^m) z(k-j+1)] - h_{m+1}^{m+1} z(k-m) \\ &= \epsilon_f^m(k+1) + p_{m+1}\left[z(k-m) - \sum_{j=1}^{m} h_{m-j+1}^m z(k-j+1)\right] \\ &= \epsilon_f^m(k+1) + p_{m+1}\epsilon_b^m(k) \end{aligned} \tag{6.3.93}$$

where the last step follows from Eq. (6.3.74).

We can find a similar relationship for the backward prediction error. From Eq. (6.3.91) we can write

$$\epsilon_b^{m+1}(k+1) = z(k-m) - \sum_{j=1}^{m+1} h_{m-j+2}^{m+1} z(k-j+2) \tag{6.3.94}$$

It can then be shown that (see Exercise 6.12)

$$\epsilon_b^{m+1}(k+1) = \epsilon_b^m(k) + p_{m+1}\epsilon_f^m(k+1) \tag{6.3.95}$$

Equations (6.3.93) and (6.3.95) constitute a set of recursion relations for computing the forward and backward prediction errors for filter order $m + 1$ in terms of the errors for filter order m. As such, they are usually referred to as *order-update* equations. The structure of this recursion is shown in Fig. 6.3.5. The initial conditions for this recursion can be obtained by noting that for $m = 0$, both $\hat{z}(k+1 \mid Z_k^m)$ and $\hat{z}(k-m \mid Z_k^m)$ are zero. It therefore follows that

$$\epsilon_f^0(k) = z(k) = \epsilon_b^0(k) \tag{6.3.96}$$

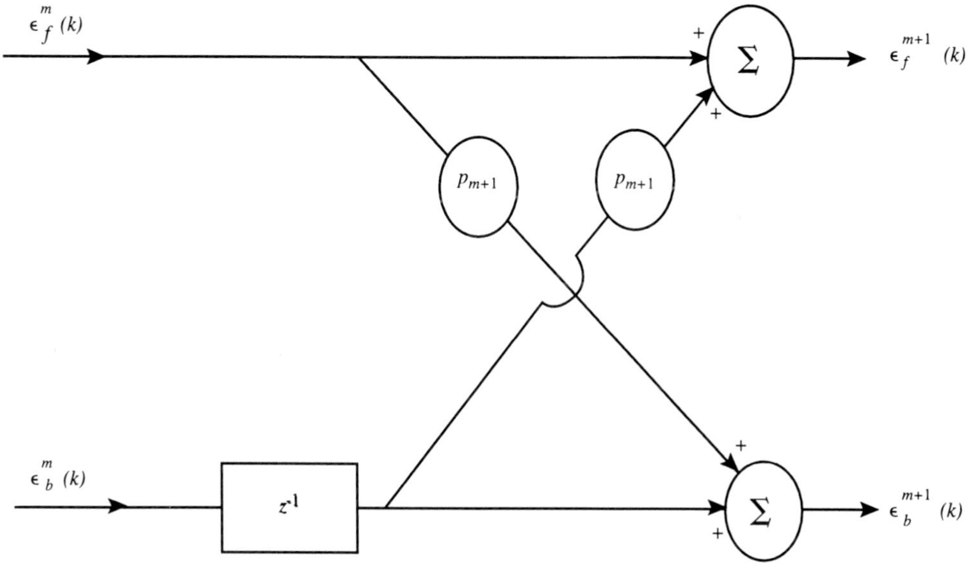

Figure 6.3.5 Recursive computation of prediction errors.

The complete structure for generating the errors $\epsilon_f^m(k)$ and $\epsilon_b^m(k)$ for any order filter is therefore as shown in Fig. 6.3.6. Because of its form, the filter is known as a *lattice filter*. The lattice filter has been widely used in a variety of applications.

While we can determine the reflection coefficients from Eq. (6.3.87), we can derive alternative expressions for p_m which are useful. To do so, we note that our goal is to find the estimates $\hat{z}(k+1 \mid Z_k^m)$ and $\hat{z}(k-m \mid Z_k^m)$ so as to minimize the mean-square error in the estimates. If we minimize $E\{\epsilon_f^{m+1}(k+1)^2\}$, from Eq. (6.3.93) we get the following expression for p_{m+1}:

$$p_{m+1} = -\left\{\frac{E\{\epsilon_f^m(k+1)\epsilon_b^m(k)\}}{E\{[\epsilon_b^m(k)]^2\}}\right\} \tag{6.3.97}$$

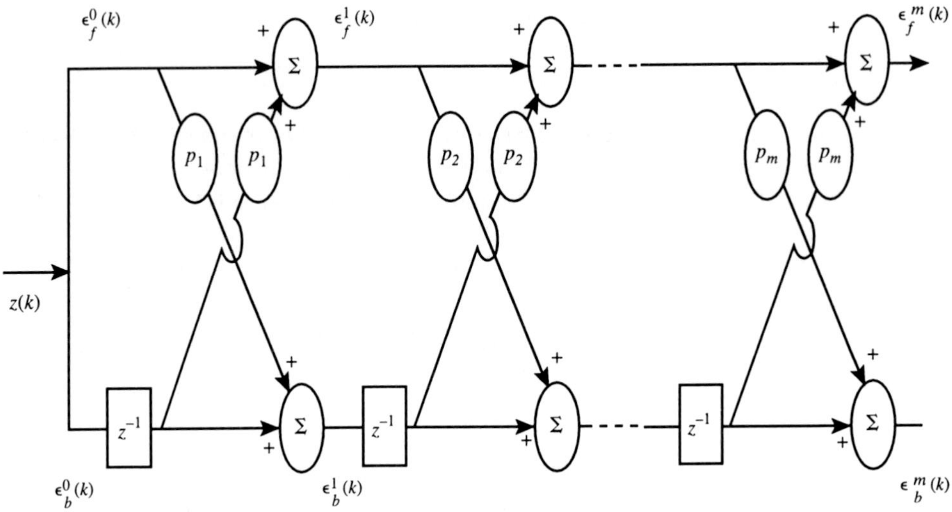

Figure 6.3.6 Lattice filter for forward–backward prediction.

Similarly minimizing $\mathrm{E}\{\epsilon_b^{m+1}(k+1)^2\}$ yields, from Eq. (6.3.95),

$$p_{m+1} = -\left\{\frac{\mathrm{E}\{\epsilon_f^m(k+1)\epsilon_b^m(k)\}}{\mathrm{E}\{[\epsilon_f^m(k+1)]^2\}}\right\} \tag{6.3.98}$$

Since $\mathrm{E}\{[\epsilon_f^m(k+1)]^2\} = \overline{C}_f^m$ and $\mathrm{E}\{[\epsilon_b^m(k)]^2\} = \overline{C}_b^m$, it follows from Eq. (6.3.77) that the two expressions for p_{m+1} are identical. It can also be easily verified that these expressions are equivalent to the expression for p_{m+1} in Eq. (6.3.87).

We conclude this section by noting that for fixed k, the backward prediction errors $\epsilon_b^0(k), \epsilon_b^1(k), \epsilon_b^2(k), \ldots, \epsilon_b^m(k)$ are orthogonal to each other. Specifically,

$$\mathrm{E}\{\epsilon_b^i(k)\epsilon_b^j(k)\} = \overline{C}_f^i\,\delta_{ij} \tag{6.3.99}$$

This result can be easily established as follows. Suppose that $j > i$. Then since the error $\epsilon_b^j(k)$ is orthogonal to $z(k), z(k-1), \ldots, z(k-j+1)$, it is certainly orthogonal to $\epsilon_b^i(k)$, which is a linear combination of $z(k), z(k-1), \ldots, z(k-i)$ [see Eq. (6.3.91)]. It can be seen by reversing the roles of i and j that the result holds for $i > j$. For $i = j$, the result follows by definition and Eq. (6.3.77).

The previous result shows that the lattice filter can be used as a whitening filter for the observation sequence $z(k)$ since the backward errors are orthogonal to one another. We can use this result to find an mth-order linear estimate of a process $g(k)$ from observations of a related process $z(k)$. We first generate the backward errors corresponding to forward–backward estimation of $z(k)$ and write the required estimate as

$$\hat{g}(k \mid Z_k^m) = \sum_{j=0}^{m} \Gamma_j\,\epsilon_b^j(k) \tag{6.3.100}$$

It easily follows that

$$\Gamma_j = \frac{1}{\overline{C}_f^j} E\{g(k)\epsilon_b^j(k)\} = -\frac{1}{\overline{C}_f^j}\sum_{l=0}^{j} h_l^j \phi_{gz}(j-l) \tag{6.3.101}$$

Example 6.3.8

Consider the problem of Example 6.3.6 in which we designed a third-order filter to estimate the signal $g(k)$ from observations $z(k)$. We will now obtain a second order filter for this problem by first designing a whitening filter for the observations using the lattice predictor and use the backward residuals (errors) as in Eq. (6.3.100) to find the estimate $\hat{g}(k)$. To find $\Gamma_i^2, 0 \le i \le 2$, we use Eq. (6.3.101), with the coefficients of the lattice recursion as determined in Example 6.3.7. We therefore have

$$\Gamma_0 = -\frac{1}{\overline{C}_f^0 h_0^0}\phi_{gz}(0) = 0.66666$$

$$\Gamma_1 = -\frac{1}{\overline{C}_f^1}[h_0^1\phi_{gz}(1) + h_1^1\phi_{gz}(0)] = 0.125$$

and

$$\Gamma_2 = -\frac{1}{\overline{C}_f^2}[h_0^2\phi_{gz}(2) + h_1^2\phi_{gz}(1) + h_2^2\phi_{gz}(0)] = 0.023529$$

The structure of the filter is shown in Fig. 6.3.7. It can easily be verified that the transfer function of this filter is the same as that of a second-order tapped delay line filter determined as in Example 6.3.5.

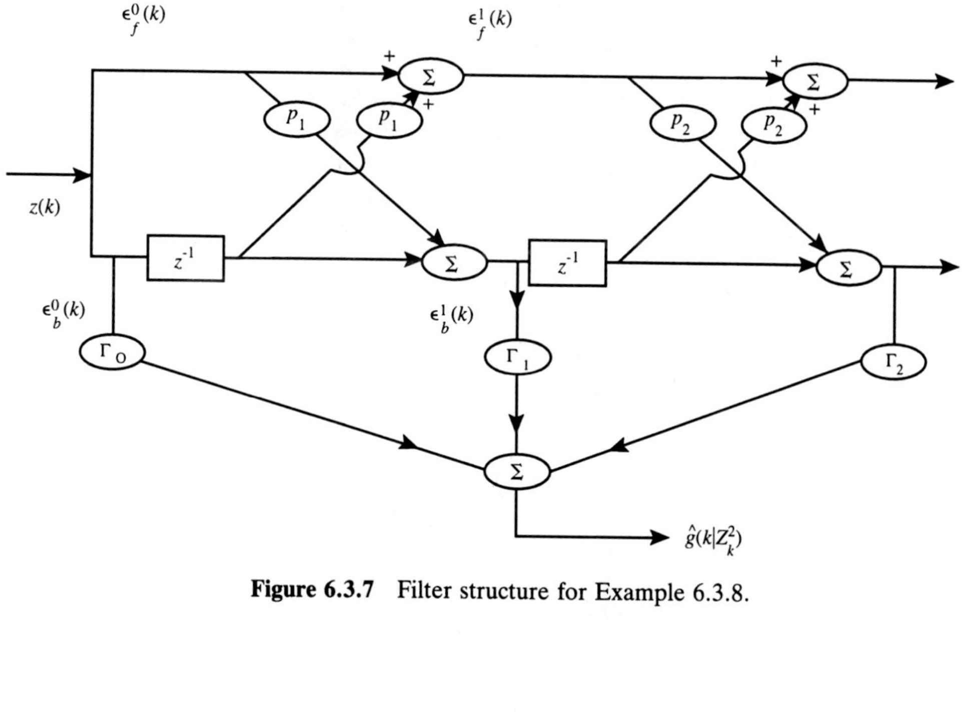

Figure 6.3.7 Filter structure for Example 6.3.8.

6.4 ESTIMATION OF NONSTATIONARY PROCESSES: THE KALMAN FILTER

We now turn to the solution of the integral Eq. (6.3.4) or (6.3.43) for nonstationary processes. We consider measurement models of the type of Eq. (6.3.6) and assume that the observation noise is white. Since the approach that we take for the solution of the estimation problem is no more complicated for vector processes than for scalar processes, we assume that our processes are in general vectors. The observations are thus given by

$$\mathbf{z}(t) = \mathbf{y}(t) + \mathbf{v}(t) \qquad t_0 \leqslant t \leqslant t_f \tag{6.4.1}$$

where $\mathbf{z}(t)$, $\mathbf{y}(t)$, and $\mathbf{v}(t)$ are $m \times 1$ vectors and $\mathbf{v}(t)$ has covariance

$$\mathrm{E}\{\mathbf{v}(t)\mathbf{v}^T(s)\} = \mathbf{V}_\mathbf{v}(t)\delta(t - s) \tag{6.4.2}$$

where $\mathbf{V}_\mathbf{v}(t)$ is now an $m \times m$ matrix.

The integral equation for the estimation problem is then given by Eq. (6.3.35) as [20–22]

$$\boldsymbol{\phi}_{\mathbf{gz}}(t, \tau) = \int_{t_0}^{t_f} \mathbf{H}(t, \xi)\boldsymbol{\phi}_\mathbf{z}(\xi, \tau)\, d\xi \qquad t_0 \leqslant \tau \leqslant t_f \tag{6.4.3}$$

We have seen earlier that the solution to the integral equation is straightforward if $\mathbf{z}(\cdot)$ is a white process. In the event that $\mathbf{z}(t)$ is not white, we have seen in our earlier discussions that for the stationary problem, $\mathbf{z}(t)$ can be decomposed into two components, one of which is white and has mean zero. In the general case under discussion here also, it can be shown that $\mathbf{z}(t)$ can be decomposed into two components, one of which can be determined from past values of $\mathbf{z}(\cdot)$ and the other, a white process $\boldsymbol{\eta}(t)$, containing the new information in $\mathbf{z}(t)$. This result is known as the *Wold decomposition* [1,5]. The process $\boldsymbol{\eta}(t)$ is referred to as the *innovations* process associated with the observations $\mathbf{z}(t)$ [22] and is obtained by a suitable causal linear transformation of the process $\mathbf{z}(t)$. It follows that the estimate $\hat{\mathbf{g}}(t)$ can be obtained completely in terms of $\boldsymbol{\eta}(t)$ as

$$\hat{\mathbf{g}}(t) = \int_{t_0}^{t_f} \mathbf{H}(t, \xi)\boldsymbol{\eta}(\xi)\, d\xi \tag{6.4.4}$$

where, by the principle of orthogonality, the matrix $\mathbf{H}(t, \xi)$ satisfies

$$\mathrm{E}\left\{\left[\mathbf{g}(t) - \int_{t_0}^{t_f} \mathbf{H}(t, \xi)\boldsymbol{\eta}(\xi)\, d\xi\right]\boldsymbol{\eta}^T(\tau)\right\} = 0 \qquad t_0 \leqslant \tau \leqslant t_f \tag{6.4.5}$$

or

$$\mathrm{E}\{\mathbf{g}(t)\boldsymbol{\eta}^T(\tau)\} = \int_{t_0}^{t_f} \mathbf{H}(t, \xi)\mathrm{E}\{\boldsymbol{\eta}(\xi)\boldsymbol{\eta}^T(\tau)\}\, d\xi \qquad t_0 \leqslant \tau \leqslant t_f \tag{6.4.6}$$

Let the covariance matrix of $\boldsymbol{\eta}(t)$ be given by

$$\operatorname{cov} \boldsymbol{\eta}(t) = \mathrm{E}\{\boldsymbol{\eta}(t)\boldsymbol{\eta}^T(\tau)\} = \mathbf{V}_\eta(t)\delta(t - \tau) \tag{6.4.7}$$

Substitution in Eq. (6.4.6) then yields

$$\mathrm{E}\{\mathbf{g}(t)\boldsymbol{\eta}^T(\tau)\} = \mathbf{H}(t,\tau)\mathbf{V}_\eta(\tau) \tag{6.4.8}$$

Solving for $\mathbf{H}(t,\tau)$, we obtain

$$\mathbf{H}(t,\tau) = \mathrm{E}\{\mathbf{g}(t)\boldsymbol{\eta}^T(\tau)\}\mathbf{V}_\eta^{-1}(\tau) \tag{6.4.9}$$

Using the preceding in Eq. (6.4.4) yields

$$\hat{\mathbf{g}}(t) = \int_{t_0}^{t_f} \mathrm{E}\{\mathbf{g}(t)\boldsymbol{\eta}^T(\tau)\}\mathbf{V}_\eta^{-1}(\tau)\boldsymbol{\eta}(\tau)\,d\tau \tag{6.4.10}$$

It is worthwhile at this point to derive analogous equations for the corresponding discrete-time problem also. We seek an estimate of the process $\mathbf{g}(k)$ based on observations of the form

$$\mathbf{z}(k) = \mathbf{y}(k) + \mathbf{v}(k) \qquad k_0 \leq k \leq k_f \tag{6.4.11}$$

where $\mathbf{v}(k)$ is an independent random process with covariance matrix

$$\mathrm{E}\{\mathbf{v}(k)\mathbf{v}^T(j)\} = \mathbf{V}_v(k)\delta_{kj} \tag{6.4.12}$$

If we convert the observations $\mathbf{z}(k)$ into an independent (white) process $\boldsymbol{\eta}(k)$, the estimate can be written as

$$\hat{\mathbf{g}}(k) = \sum_{j=k_0}^{k_f} \mathbf{H}(k,j)\boldsymbol{\eta}(j) \tag{6.4.13}$$

The weights $\mathbf{H}(k,j)$ are such that the estimation error is orthogonal to the data set $\{\boldsymbol{\eta}(j)\}$. We thus have

$$\mathrm{E}\left\{\left[\mathbf{g}(k) - \sum_{j=k_0}^{k_f} \mathbf{H}(k,j)\boldsymbol{\eta}(j)\right]\boldsymbol{\eta}^T(l)\right\} = 0 \qquad k_0 \leq l \leq k_f \tag{6.4.14}$$

or, equivalently,

$$\mathrm{E}\{\mathbf{g}(k)\boldsymbol{\eta}^T(l)\} = \sum_{j=k_0}^{k_f} \mathbf{H}(k,j)\mathrm{E}\{\boldsymbol{\eta}(j)\boldsymbol{\eta}^T(l)\} \qquad k_0 \leq l \leq k_f \tag{6.4.15}$$

Since $\boldsymbol{\eta}(k)$ is white, its covariance is

$$\mathrm{E}\{\boldsymbol{\eta}(k)\boldsymbol{\eta}^T(l)\} = \mathbf{V}_\eta(k)\delta_{kl} \tag{6.4.16}$$

Substituting Eq. (6.4.16) into Eq. (6.4.15) and solving for $\mathbf{H}(\cdot,\cdot)$ yields

$$\mathbf{H}(k,l) = \mathrm{E}\{\mathbf{g}(k)\boldsymbol{\eta}^T(l)\}\mathbf{V}_\eta^{-1}(l) \tag{6.4.17}$$

and the required estimate is

$$\hat{\mathbf{g}}(k) = \sum_{j=k_0}^{k_f} \mathrm{E}\{\mathbf{g}(k)\boldsymbol{\eta}^T(j)\}\mathbf{V}_\eta^{-1}(j)\boldsymbol{\eta}(j) \tag{6.4.18}$$

It only remains to find the transformation that yields the innovations process $\boldsymbol{\eta}(t)$[$\boldsymbol{\eta}(k)$ in the discrete case] and its associated covariance. For stationary processes, at least for the scalar case, this transformation is easy to find. For nonstationary processes, however, this is not so easy. For such processes, even if we find the innovations process and obtain a solution as in Eq. (6.4.10) or (6.4.18), a major difficulty remains in the practical implementation of the estimator. The solution is specified in terms of the impulse response matrix of a time-varying system. To translate the impulse response into physical hardware is again not an easy task. We therefore seek an alternative solution to the problem which can easily be implemented using either analog or digital hardware. We do this by deriving a differential equation (a difference equation for the discrete-time problem) which the estimate must satisfy. The solution to the estimation problem is thus given not in explicit analytical terms but rather in an algorithmic form; that is, we specify a set of algorithms which when applied to the data give us the desired estimate. This approach to the problem was first suggested by Kalman and the resulting estimator is usually referred to as the *Kalman filter* [20,21]. For reasons that will soon become apparent, we start our derivation of the Kalman filter algorithms with the discrete-time problem.

6.4.1 The Optimum Linear Discrete Filter

We now consider the determination of the estimate in Eq. (6.4.18). Since the estimate is based on measurements up to k_f, we denote the estimate as $\hat{\mathbf{g}}(k \mid k_f)$. To proceed with our discussion, at this point we need to make some specific assumptions about the signal $\mathbf{y}(k)$. In particular, we assume that the signal $\mathbf{y}(k)$ is obtained as the output of a linear discrete system driven by white noise. We use the state-variable model†

$$\begin{aligned} \mathbf{x}(k+1) &= \boldsymbol{\phi}(k+1, k)\mathbf{x}(k) + \boldsymbol{\Gamma}(k)\mathbf{w}(k) \\ \mathbf{y}(k) &= \mathbf{C}(k)\mathbf{x}(k) \end{aligned} \tag{6.4.19}$$

where the input $\mathbf{w}(k)$ is zero-mean, white with covariance matrix

$$\operatorname{cov} \mathbf{w}(k) = \mathrm{E}\{\mathbf{w}(k)\mathbf{w}^T(j)\} = \mathbf{V}_{\mathbf{w}}(k)\delta_{kj} \tag{6.4.20}$$

The observation equation is

$$\mathbf{z}(k) = \mathbf{y}(k) + \mathbf{v}(k) \qquad k_0 \leqslant k \leqslant k_f \tag{6.4.21}$$

where $\mathbf{v}(k)$ is zero mean, white with covariance matrix

$$\operatorname{cov} \mathbf{v}(k) = \mathrm{E}\{\mathbf{v}(k)\mathbf{v}^T(j)\} = \mathbf{V}_{\mathbf{v}}(k)\delta_{kj} \tag{6.4.22}$$

and is uncorrelated with $\mathbf{w}(k)$.

†The matrix $\boldsymbol{\phi}(\cdot,\cdot)$ represents the state-transition matrix of the discrete-time system and should not be confused with the autocorrelation matrix.

We further assume that $\mathbf{x}(k_0)$ is a random variable with known mean $\boldsymbol{\mu}_x(k_0)$ and covariance $\mathbf{V}_x(k_0)$ uncorrelated with $\mathbf{w}(k)$ and $\mathbf{v}(k)$ for all k. Thus

$$E\{\mathbf{w}(k)\mathbf{v}^T(j)\} = 0 \qquad \text{all } k, j$$

$$E\{\mathbf{x}(k_0)\mathbf{w}^T(j)\} = 0 \qquad \text{all } j$$

and

$$E\{\mathbf{x}(k_0)\mathbf{v}^T(j)\} = 0 \qquad \text{all } j \tag{6.4.23}$$

The problem is hence slightly different from the one considered in the preceding discussions in that $\mathbf{x}(k)$ [and hence $\mathbf{y}(k)$] is no longer zero mean.

We will seek to obtain an estimate of the state of the model $\mathbf{x}(k)$ based on the noisy observations. Let us denote by $\mathbf{Z}(j)$ the set of observations $\{\mathbf{z}(k_0), \mathbf{z}(k_0+1), \ldots, \mathbf{z}(j)\}$:

$$\mathbf{Z}(j) = \{\mathbf{z}(i), i \in [k_0, j]\} \tag{6.4.24}$$

The estimate we are seeking is

$$\hat{\mathbf{x}}(k \mid k_f) = \hat{E}\{\mathbf{x}(k) \mid \mathbf{Z}(k_f)\} \tag{6.4.25}$$

where $\hat{E}$ denotes the wide-sense conditional expectation, as explained earlier.

The estimate of the signal $\mathbf{y}(k)$ can be obtained as

$$\begin{aligned}\hat{\mathbf{y}}(k \mid k_f) &= \hat{E}\{\mathbf{y}(k) \mid \mathbf{Z}(k_f)\} \\ &= \hat{E}\{\mathbf{C}(k)\mathbf{x}(k) \mid \mathbf{Z}(k_f)\} \\ &= \mathbf{C}(k)\hat{E}\{\mathbf{x}(k) \mid \mathbf{Z}(k_f)\}\end{aligned} \tag{6.4.26}$$

where the last expression follows since $\hat{E}$ commutes over linear operations. We may therefore write

$$\hat{\mathbf{y}}(k \mid k_f) = \mathbf{C}(k)\hat{\mathbf{x}}(k \mid k_f) \tag{6.4.27}$$

In terms of the innovations sequence $\boldsymbol{\eta}(k)$, the estimate $\hat{\mathbf{x}}(k \mid k_f)$ can be written from Eq. (6.4.18) as

$$\hat{\mathbf{x}}(k \mid k_f) = \sum_{j=k_0}^{k_f} E\{\mathbf{x}(k)\boldsymbol{\eta}^T(j)\}\mathbf{V}_{\boldsymbol{\eta}}^{-1}(j)\boldsymbol{\eta}(j) \tag{6.4.28}$$

If $k_f = k$, the problem corresponds to one of filtering a signal in the presence of noise, and the estimate $\hat{\mathbf{x}}(k \mid k)$ is referred to as the *filtered estimate*. For $k > k_f$, we have the prediction problem, and the corresponding estimate is the predicted estimate. Of particular importance is the one-stage predicted estimate $\hat{\mathbf{x}}(k \mid k-1)$. If $k < k_f$, we have the interpolation problem and the estimate is called a *smoothed estimate*.

We now determine the innovations sequence $\boldsymbol{\eta}(k)$ associated with the observation model Eq. (6.4.21). We have seen in Example 6.3.3 that for the continuous-time stationary problem in which the signal and observation noise were uncorrelated, the innovations were given by

$$\eta(t) = z(t) - \hat{y}(t)$$

so that $\hat{y}(t)$ corresponds to that component of $z(t)$ which is determined by past values of $z(\cdot)$. For the discrete-time problem under consideration, it follows that the innovations sequence should be

$$\begin{aligned} \boldsymbol{\eta}(k) &= \mathbf{z}(k) - \hat{\mathbf{y}}(k \mid k-1) \\ &= \tilde{\mathbf{y}}(k \mid k-1) + \mathbf{v}(k) \end{aligned} \tag{6.4.29}$$

where $\tilde{\mathbf{y}}(k \mid k-1)$ represents the estimation error. For the estimate $\tilde{\mathbf{y}}(k \mid k-1)$ to be unbiased, $\tilde{\mathbf{y}}(k \mid k-1)$ must have mean zero. We will see later, how this can be achieved. It then follows that since $\mathbf{v}(k)$ is a zero-mean sequence, $\boldsymbol{\eta}(k)$ will also be a zero-mean sequence. We now establish that $\boldsymbol{\eta}(k)$ as defined previously is a white sequence.

Let $k > l$. Then

$$\begin{aligned} \mathrm{E}\{\boldsymbol{\eta}(k)\boldsymbol{\eta}^T(l)\} &= \mathrm{E}\{[\tilde{\mathbf{y}}(k \mid k-1) + \mathbf{v}(k)]\boldsymbol{\eta}^T(l)\} \\ &= \mathrm{E}\{\mathbf{v}(k)\boldsymbol{\eta}^T(l)\} \end{aligned}$$

[since $\tilde{\mathbf{y}}(k \mid k-1)$ is orthogonal to $\boldsymbol{\eta}(l)$ for $l \leqslant k-1$]

$$\begin{aligned} &= \mathrm{E}\{\mathbf{v}(k)[\tilde{\mathbf{y}}(l \mid l-1) + \mathbf{v}(l)]^T\} \\ &= 0 \end{aligned} \tag{6.4.30}$$

The last result follows by causality [$\mathbf{v}(k)$ is in the future as far as $\tilde{\mathbf{y}}(l \mid l-1)$ is concerned and hence cannot be correlated with it] and the fact that $\mathbf{v}(\cdot)$ is a white sequence. The result holds for $k < l$, by interchanging k and l in the preceding.

For $k = l$ we have

$$\begin{aligned} \mathrm{E}\{\boldsymbol{\eta}(k)\boldsymbol{\eta}^T(k)\} &= \mathrm{E}\{[\tilde{\mathbf{y}}(k \mid k-1) + \mathbf{v}(k)][\tilde{\mathbf{y}}(k \mid k-1) + \mathbf{v}(k)]^T\} \\ &= \mathbf{V}_{\tilde{y}}(k \mid k-1) + \mathbf{V}_v(k) \end{aligned} \tag{6.4.31}$$

where

$$\begin{aligned} \mathbf{V}_{\tilde{y}}(k \mid k-1) &= \mathrm{E}\{\tilde{\mathbf{y}}(k \mid k-1)\tilde{\mathbf{y}}^T(k \mid k-1)\} \\ &= \mathbf{C}(k)\mathbf{V}_{\tilde{x}}(k \mid k-1)\mathbf{C}^T(k) \end{aligned} \tag{6.4.32}$$

Thus $\boldsymbol{\eta}(k)$ is a zero-mean white sequence with covariance kernel

$$\mathbf{V}_\eta(k) = \mathbf{C}(k)\mathbf{V}_{\tilde{x}}(k \mid k-1)\mathbf{C}^T(k) + \mathbf{V}_v(k) \tag{6.4.33}$$

6.4.2 One-Stage Predictor Algorithms

Since the computation of the innovations sequence involves determining the estimate $\hat{\mathbf{y}}(k \mid k-1)$, we first set up algorithms for obtaining the one-stage predicted estimate $\hat{\mathbf{x}}(k \mid k-1)$. If we set $\hat{\mathbf{g}}(k) = \hat{\mathbf{x}}(k \mid k-1)$ in Eq. (6.4.18), we obtain

$$\hat{\mathbf{x}}(k \mid k-1) = \sum_{j=k_0}^{k-1} \mathrm{E}\{\mathbf{x}(k)\boldsymbol{\eta}^T(j)\}\mathbf{V}_\eta^{-1}(j)\boldsymbol{\eta}(j) \tag{6.4.34}$$

Indexing upward by one gives

$$\begin{aligned}\hat{\mathbf{x}}(k+1|k) &= \sum_{j=k_0}^{k} \mathrm{E}\{\mathbf{x}(k+1)\boldsymbol{\eta}^T(j)\}\mathbf{V}_\eta^{-1}(j)\boldsymbol{\eta}(j) \\ &= \sum_{j=k_0}^{k} \mathrm{E}\{[\boldsymbol{\phi}(k+1,k)\mathbf{x}(k) + \boldsymbol{\Gamma}(k)\mathbf{w}(k)]\boldsymbol{\eta}^T(j)\}\mathbf{V}_\eta^{-1}(j)\boldsymbol{\eta}(j)\end{aligned} \tag{6.4.35}$$

We note that in Eq. (6.4.35), $\boldsymbol{\eta}(j)$ depends on values of $\mathbf{x}(l)$ only for $l = k_0$, $k_0 + 1, \ldots, j - 1$. Thus since $\mathbf{w}(k)$ is white,

$$\mathrm{E}\{\mathbf{w}(k)\boldsymbol{\eta}^T(j)\} = 0 \qquad j \leq k \tag{6.4.36}$$

We can therefore write Eq. (6.4.35) as

$$\begin{aligned}\hat{\mathbf{x}}(k+1|k) &= \boldsymbol{\phi}(k+1,k) \sum_{j=k_0}^{k} \mathrm{E}\{\mathbf{x}(k)\boldsymbol{\eta}^T(j)\}\mathbf{V}_\eta^{-1}(j)\boldsymbol{\eta}(j) \\ &= \boldsymbol{\phi}(k+1,k)\mathrm{E}\{\mathbf{x}(k)\boldsymbol{\eta}^T(k)\}\mathbf{V}_\eta^{-1}(k)\boldsymbol{\eta}(k) \\ &\quad + \boldsymbol{\phi}(k+1,k) \sum_{j=k_0}^{k-1} \mathrm{E}\{\mathbf{x}(k)\boldsymbol{\eta}^T(j)\}\mathbf{V}_\eta^{-1}(j)\boldsymbol{\eta}(j)\end{aligned} \tag{6.4.37}$$

Define $\mathbf{K}(k+1,k)$ by

$$\mathbf{K}(k+1,k) \triangleq \boldsymbol{\phi}(k+1,k)\mathrm{E}\{\mathbf{x}(k)\boldsymbol{\eta}^T(k)\}\mathbf{V}_\eta^{-1}(k) \tag{6.4.38}$$

Substitution of Eqs. (6.4.38) and (6.4.34) in Eq. (6.4.37) then yields

$$\hat{\mathbf{x}}(k+1|k) = \boldsymbol{\phi}(k+1,k)\hat{\mathbf{x}}(k|k-1) + \mathbf{K}(k+1,k)\boldsymbol{\eta}(k) \tag{6.4.39}$$

or

$$\begin{aligned}\hat{\mathbf{x}}(k+1|k) &= \boldsymbol{\phi}(k+1,k)\hat{\mathbf{x}}(k|k-1) \\ &\quad + \mathbf{K}(k+1,k)[\mathbf{z}(k) - \mathbf{C}(k)\hat{\mathbf{x}}(k|k-1)]\end{aligned} \tag{6.4.40}$$

where we have used Eq. (6.4.29) and the fact that

$$\hat{\mathbf{y}}(k|k-1) = \mathbf{C}(k)\hat{\mathbf{x}}(k|k-1)$$

The matrix $\mathbf{K}(k+1,k)$ is referred to as the *Kalman gain for one-stage prediction*.

The form of the solution in Eq. (6.4.40) is computationally very useful. The result is a sequential algorithm for determining $\hat{\mathbf{x}}(k+1|k)$ based on $\hat{\mathbf{x}}(k|k-1)$ and the new observation $\mathbf{z}(k)$. The new estimate is formed by predicting forward from the old estimate and then correcting it by the innovation term that contains the new information. A block diagram of the estimator, including the message model, is shown in Fig. 6.4.1.

All that is needed to complete the specification of the estimator are the Kalman gain $\mathbf{K}(k+1,k)$ and the error variance $\mathbf{V}_{\tilde{y}}(k|k-1)$ needed to specify the covariance kernel $\mathbf{V}_\eta(k)$. Substitution of Eq. (6.4.29) in Eq. (6.4.38) yields

$$\begin{aligned}\mathbf{K}(k+1,k) &= \boldsymbol{\phi}(k+1,k)\mathrm{E}\{\mathbf{x}(k)[\mathbf{C}(k)\tilde{\mathbf{x}}(k|k-1) + \mathbf{v}(k)]^T\}\mathbf{V}_\eta^{-1}(k) \\ &= \boldsymbol{\phi}(k+1,k)\mathrm{E}\{\mathbf{x}(k)\tilde{\mathbf{x}}^T(k|k-1)\}\mathbf{C}^T(k)\mathbf{V}_\eta^{-1}(k)\end{aligned} \tag{6.4.41}$$

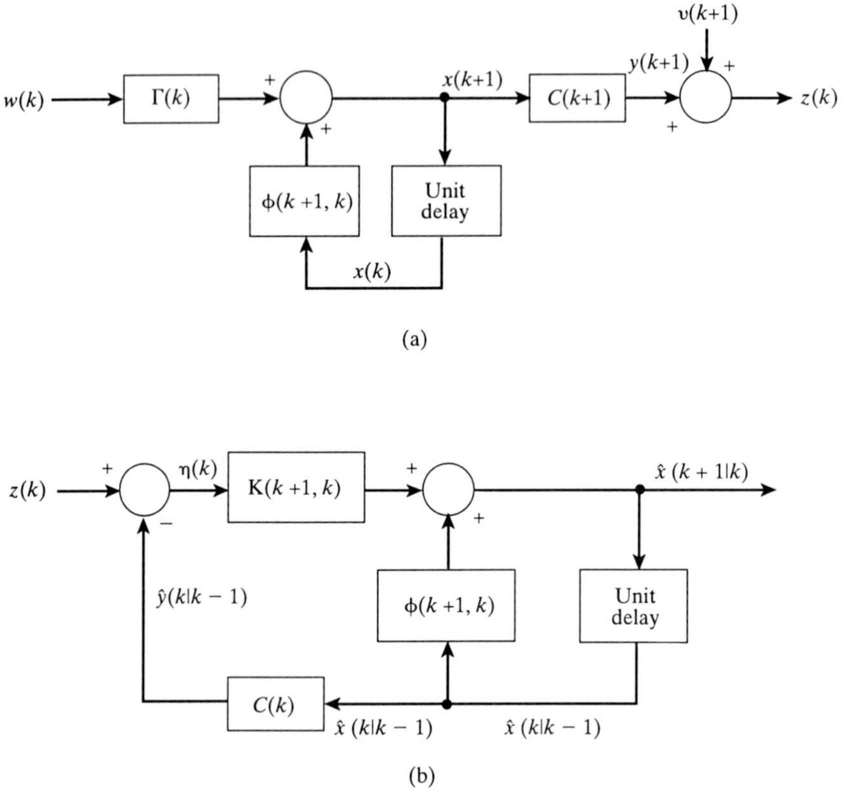

Figure 6.4.1 Block diagram for one-stage prediction for the discrete Kalman filter: (a) message model; (b) one-stage predictor.

since $\mathbf{x}(k)$ and $\mathbf{v}(k)$ are uncorrelated. We can write

$$\mathbf{x}(k) = \hat{\mathbf{x}}(k \mid k-1) + \tilde{\mathbf{x}}(k \mid k-1) \tag{6.4.42}$$

Since $\tilde{\mathbf{x}}(k \mid k-1)$ is orthogonal to $\boldsymbol{\eta}(j)$ for $j \leqslant k-1$ it is also orthogonal to $\hat{\mathbf{x}}(k \mid k-1)$ [since $\hat{\mathbf{x}}(k \mid k-1)$ is a linear combination of $\boldsymbol{\eta}(j)$ for $j \leqslant k-1$]. Equation (6.4.41) can therefore be written as

$$\mathbf{K}(k+1, k) = \boldsymbol{\phi}(k+1, k)\mathbf{V}_{\tilde{\mathbf{x}}}(k \mid k-1)\mathbf{C}^T(k)\mathbf{V}_{\boldsymbol{\eta}}^{-1}(k) \tag{6.4.43}$$

To obtain an expression for the error variance $\mathbf{V}_{\tilde{\mathbf{x}}}(k \mid k-1)$, we subtract Eq. (6.4.40) from Eq. (6.4.19) and use Eq. (6.4.29) to get

$$\begin{aligned}\tilde{\mathbf{x}}(k+1 \mid k) &= \boldsymbol{\phi}(k+1, k)\tilde{\mathbf{x}}(k \mid k-1) + \boldsymbol{\Gamma}(k)\mathbf{w}(k) \\ &\quad - \mathbf{K}(k+1, k)[\mathbf{C}(k)\tilde{\mathbf{x}}(k \mid k-1) + \mathbf{v}(k)]\end{aligned} \tag{6.4.44}$$

Thus we have

$$\begin{aligned}\tilde{\mathbf{x}}(k+1|k) = {} & [\boldsymbol{\phi}(k+1,k) - \mathbf{K}(k+1,k)\mathbf{C}(k)]\tilde{\mathbf{x}}(k|k-1) \\ & + \boldsymbol{\Gamma}(k)\mathbf{w}(k) - \mathbf{K}(k+1,k)\mathbf{v}(k)\end{aligned} \tag{6.4.45}$$

Since $\mathbf{w}(\cdot)$ and $\mathbf{v}(\cdot)$ are white and uncorrelated with each other, the difference equation for the error variance may be written from Eq. (2.4.22) as

$$\begin{aligned}\mathbf{V}_{\tilde{\mathbf{x}}}(k+1|k) = {} & [\boldsymbol{\phi}(k+1,k) - \mathbf{K}(k+1,k)\mathbf{C}(k)] \\ & \times \mathbf{V}_{\tilde{\mathbf{x}}}(k|k-1)[\boldsymbol{\phi}(k+1,k) - \mathbf{K}(k+1,k)\mathbf{C}(k)]^T \\ & + \boldsymbol{\Gamma}(k)\mathbf{V}_{\mathbf{w}}(k)\boldsymbol{\Gamma}^T(k) \\ & + \mathbf{K}(k+1,k)\mathbf{V}_{\mathbf{v}}(k)\mathbf{K}^T(k+1,k)\end{aligned} \tag{6.4.46}$$

Substituting for $\mathbf{K}(k+1,k)$ from Eq. (6.4.43) and simplifying yields

$$\begin{aligned}\mathbf{V}_{\tilde{\mathbf{x}}}(k+1|k) = {} & \boldsymbol{\phi}(k+1,k)\mathbf{V}_{\tilde{\mathbf{x}}}(k|k-1)\boldsymbol{\phi}^T(k+1,k) + \boldsymbol{\Gamma}(k)\mathbf{V}_{\mathbf{w}}(k)\boldsymbol{\Gamma}^T(k) \\ & - \boldsymbol{\phi}(k+1,k)\mathbf{V}_{\tilde{\mathbf{x}}}(k|k-1)\mathbf{C}^T(k) \\ & \times [\mathbf{C}(k)\mathbf{V}_{\tilde{\mathbf{x}}}(k|k-1)\mathbf{C}^T(k) + \mathbf{V}_{\mathbf{v}}(k)]^{-1} \\ & \times \mathbf{C}(k)\mathbf{V}_{\tilde{\mathbf{x}}}(k|k-1)\boldsymbol{\phi}^T(k+1,k)\end{aligned} \tag{6.4.47}$$

To specify the initial conditions for Eqs. (6.4.40) and (6.4.47), we note from Eq. (6.4.44) that

$$\mathrm{E}\{\tilde{\mathbf{x}}(k+1|k)\} = [\boldsymbol{\phi}(k+1,k) - \mathbf{K}(k+1,k)\mathbf{C}(k)]\mathrm{E}\{\tilde{\mathbf{x}}(k|k-1)\} \tag{6.4.48}$$

For an unbiased estimate, we must have

$$\mathrm{E}\{\mathbf{x}(k) - \hat{\mathbf{x}}(k|k-1)\} = \mathrm{E}\{\tilde{\mathbf{x}}(k|k-1)\} = 0 \tag{6.4.49}$$

Since Eq. (6.4.48) is a homogeneous difference equation, if we choose $\mathrm{E}\{\tilde{\mathbf{x}}(k_0)\} = 0$, then $\mathrm{E}\{\tilde{\mathbf{x}}(k+1|k)\} = 0$ for all $k \geqslant k_0$. We therefore have

$$\hat{\mathbf{x}}(k_0) = \mathrm{E}\{\mathbf{x}(k_0)\} = \boldsymbol{\mu}_{\mathbf{x}}(k_0) \tag{6.4.50}$$

In this case

$$\begin{aligned}\mathbf{V}_{\tilde{\mathbf{x}}}(k_0) &= \mathrm{E}[[\mathbf{x}(k_0) - \hat{\mathbf{x}}(k_0)][\mathbf{x}(k_0) - \hat{\mathbf{x}}(k_0)]^T] \\ &= \mathbf{V}_{\mathbf{x}}(k_0)\end{aligned} \tag{6.4.51}$$

The algorithms are summarized in Table 6.4.1a. As can be seen from Eq. (6.4.47), the error variance $\mathbf{V}_{\tilde{\mathbf{x}}}(k|k-1)$ and hence the gain $\mathbf{K}(k+1,k)$ do not depend on the observations and may be precomputed and stored. From a practical viewpoint it is easier to process these equations sequentially along with the estimator equation. The error variance is directly available and can be used to measure the performance of the filter.

TABLE 6.4.1 ONE-STAGE PREDICTION AND FILTERING ALGORITHMS FOR THE DISCRETE KALMAN FILTER

Message model:

$$\begin{aligned}\mathbf{x}(k+1) &= \boldsymbol{\phi}(k+1,k)\mathbf{x}(k) + \boldsymbol{\Gamma}(k)\mathbf{w}(k)\\ \mathbf{y}(k) &= \mathbf{C}(k)\mathbf{x}(k)\end{aligned} \tag{6.4.19}$$

Observation model:

$$\mathbf{z}(k) = \mathbf{y}(k) + \mathbf{v}(k) \tag{6.4.21}$$

Prior statistics:

$$\begin{aligned}\mathrm{E}\{\mathbf{w}(k)\} &= \mathrm{E}\{\mathbf{v}(k)\} = \mathbf{0}, \mathrm{E}\{\mathbf{x}(k_0)\} = \boldsymbol{\mu}_{\mathbf{x}}(k_0)\\ \mathrm{E}\{\mathbf{w}(k)\mathbf{w}^T(l)\} &= \mathbf{V}_{\mathbf{w}}(k)\delta_{kl}, \mathrm{E}\{\mathbf{v}(k)\mathbf{v}^T(l)\} = \mathbf{V}_{\mathbf{v}}(k)\delta_{kl}\\ \mathrm{E}\{\mathbf{w}(k)\mathbf{v}^T(l)\} &= \mathrm{E}\{\mathbf{w}(k)\mathbf{x}^T(k_0)\} = \mathrm{E}\{\mathbf{v}(l)\mathbf{x}^T(k_0)\} = \mathbf{0}\\ \text{covar } \mathbf{x}(k_0) &= \mathbf{V}_{\mathbf{x}}(k_0)\end{aligned}$$

(a) Predictor Algorithms

Predictor algorithm:

$$\hat{\mathbf{x}}(k+1|k) = \boldsymbol{\phi}(k+1,k)\hat{\mathbf{x}}(k|k-1) + \mathbf{K}(k+1,k)\boldsymbol{\eta}(k) \tag{6.4.39}$$

$$\boldsymbol{\eta}(k) = \mathbf{z}(k) - \mathbf{C}(k)\hat{\mathbf{x}}(k|k-1) \tag{6.4.29}$$

$$\hat{\mathbf{x}}(k_0) = \boldsymbol{\mu}_{\mathbf{x}}(k_0) \tag{6.4.50}$$

Gain algorithm:

$$\mathbf{K}(k+1,k) = \boldsymbol{\phi}(k+1,k)\mathbf{V}_{\tilde{\mathbf{x}}}(k|k-1)\mathbf{C}^T(k)\mathbf{V}_{\boldsymbol{\eta}}^{-1}(k) \tag{6.4.43}$$

Variance equations:

$$\mathbf{V}_{\boldsymbol{\eta}}(k) = \mathbf{C}(k)\mathbf{V}_{\tilde{\mathbf{x}}}(k|k-1)\mathbf{C}^T(k) + \mathbf{V}_{\mathbf{v}}(k) \tag{6.4.33}$$

$$\begin{aligned}\mathbf{V}_{\tilde{\mathbf{x}}}(k+1|k) &= \boldsymbol{\phi}(k+1,k)\mathbf{V}_{\tilde{\mathbf{x}}}(k|k-1)\boldsymbol{\phi}^T(k+1,k) + \boldsymbol{\Gamma}(k)\mathbf{V}_{\mathbf{w}}(k)\boldsymbol{\Gamma}^T(k)\\ &\quad - \boldsymbol{\phi}(k+1,k)\mathbf{V}_{\tilde{\mathbf{x}}}(k|k-1)\mathbf{C}^T(k)\mathbf{V}_{\boldsymbol{\eta}}^{-1}(k)\mathbf{C}(k)\\ &\quad \times \mathbf{V}_{\tilde{\mathbf{x}}}(k|k-1)\boldsymbol{\phi}^T(k+1,k)\end{aligned} \tag{6.4.47}$$

$$\mathbf{V}_{\tilde{\mathbf{x}}}(k_0) = \mathbf{V}_{\mathbf{x}}(k_0) \tag{6.4.51}$$

(b) Filtering Algorithms

Filtering algorithm:

$$\hat{\mathbf{x}}(k|k) = \boldsymbol{\phi}(k,k-1)\hat{\mathbf{x}}(k-1|k-1) + \mathbf{K}(k)\boldsymbol{\eta}(k) \tag{6.4.57}$$

$$\hat{\mathbf{x}}(k+1|k) = \boldsymbol{\phi}(k+1,k)\hat{\mathbf{x}}(k|k) \tag{6.4.56}$$

Gain algorithm:

$$\mathbf{K}(k) = \mathbf{V}_{\tilde{\mathbf{x}}}(k|k-1)\mathbf{C}^T(k)\mathbf{V}_{\boldsymbol{\eta}}^{-1}(k) \tag{6.4.61}$$

Variance equations:

$$\mathbf{V}_{\tilde{\mathbf{x}}}(k|k) = [\mathbf{I} - \mathbf{K}(k)\mathbf{C}(k)]\mathbf{V}_{\tilde{\mathbf{x}}}(k|k-1) \tag{6.4.62}$$

$$\mathbf{V}_{\tilde{\mathbf{x}}}(k+1|k) = \boldsymbol{\phi}(k+1,k)\mathbf{V}_{\tilde{\mathbf{x}}}(k|k)\boldsymbol{\phi}^T(k+1,k) + \boldsymbol{\Gamma}(k)\mathbf{V}_{\mathbf{w}}(k)\boldsymbol{\Gamma}^T(k) \tag{6.4.63}$$

6.4.3 Discrete Filtering Algorithms

We now consider the problem of obtaining the filtered estimate $\hat{\mathbf{x}}(k|k)$. If we set $k_f = k$ in Eq. (6.4.28) we have

$$\begin{aligned}\hat{\mathbf{x}}(k|k) &= \sum_{j=k_0}^{k} \mathrm{E}\{\mathbf{x}(k)\boldsymbol{\eta}^T(j)\}\mathbf{V}_{\eta}^{-1}(j)\boldsymbol{\eta}(j) \\ &= \sum_{j=k_0}^{k-1} \mathrm{E}\{\mathbf{x}(k)\boldsymbol{\eta}^T(j)\}\mathbf{V}_{\eta}^{-1}(j)\boldsymbol{\eta}(j) + \mathrm{E}\{\mathbf{x}(k)\boldsymbol{\eta}^T(k)\}\mathbf{V}_{\eta}^{-1}(k)\boldsymbol{\eta}(k)\end{aligned} \tag{6.4.52}$$

We recognize from Eq. (6.4.34) that the first term on the right-hand side in the equation preceding is the one-stage predicted estimate $\hat{\mathbf{x}}(k|k-1)$. We can therefore write for the filtered estimate

$$\hat{\mathbf{x}}(k|k) = \hat{\mathbf{x}}(k|k-1) + \mathbf{K}(k)\boldsymbol{\eta}(k) \tag{6.4.53}$$

where we have defined the gain $\mathbf{K}(k)$ as

$$\mathbf{K}(k) = \mathrm{E}\{\mathbf{x}(k)\boldsymbol{\eta}^T(k)\}\mathbf{V}_{\eta}^{-1}(k) \tag{6.4.54}$$

Comparison with Eq. (6.4.38) shows that

$$\mathbf{K}(k+1,k) = \boldsymbol{\phi}(k+1,k)\mathbf{K}(k) \tag{6.4.55}$$

Since $\hat{\mathbf{x}}(k+1|k)$ is the wide-sense conditional expectation of $\mathbf{x}(k)$ given the data sequence $\mathbf{Z}(k)$, we have

$$\begin{aligned}\hat{\mathbf{x}}(k|k-1) &= \hat{\mathrm{E}}[\mathbf{x}(k)|\mathbf{Z}(k)] \\ &= \hat{\mathrm{E}}[\boldsymbol{\phi}(k,k-1)\mathbf{x}(k-1) + \boldsymbol{\Gamma}(k-1)\mathbf{w}(k-1)|\mathbf{Z}(k-1)] \\ &= \boldsymbol{\phi}(k,k-1)\hat{\mathrm{E}}\{\mathbf{x}(k-1)|\mathbf{Z}(k-1)\} \\ &= \boldsymbol{\phi}(k,k-1)\hat{\mathbf{x}}(k-1|k-1)\end{aligned} \tag{6.4.56}$$

Substitution in Eq. (6.4.53) yields a direct sequential algorithm for $\hat{\mathbf{x}}(k|k)$ as

$$\hat{\mathbf{x}}(k|k) = \boldsymbol{\phi}(k,k-1)\hat{\mathbf{x}}(k-1|k-1) + \mathbf{K}(k)\boldsymbol{\eta}(k) \tag{6.4.57}$$

where

$$\begin{aligned}\boldsymbol{\eta}(k) &= \mathbf{z}(k) - \mathbf{C}(k)\hat{\mathbf{x}}(k|k-1) \\ &= \mathbf{z}(k) - \mathbf{C}(k)\boldsymbol{\phi}(k,k-1)\hat{\mathbf{x}}(k-1|k-1)\end{aligned} \tag{6.4.58}$$

To obtain an expression for the filtered estimate error variance we proceed as follows. From Eq. (6.4.57) we have

$$\begin{aligned}\tilde{\mathbf{x}}(k|k) &= \mathbf{x}(k) - \hat{\mathbf{x}}(k|k) \\ &= \mathbf{x}(k) - \hat{\mathbf{x}}(k|k-1) - \mathbf{K}(k)\boldsymbol{\eta}(k) \\ &= \tilde{\mathbf{x}}(k|k-1) - \mathbf{K}(k)[\mathbf{C}(k)\tilde{\mathbf{x}}(k|k-1) + \mathbf{v}(k)] \\ &= [\mathbf{I} - \mathbf{K}(k)\mathbf{C}(k)]\tilde{\mathbf{x}}(k|k-1) - \mathbf{K}(k)\mathbf{v}(k)\end{aligned} \tag{6.4.59}$$

Thus the error variance is given by

$$\mathbf{V}_{\tilde{\mathbf{x}}}(k\,|\,k) = [\mathbf{I} - \mathbf{K}(k)\mathbf{C}(k)]\mathbf{V}_{\tilde{\mathbf{x}}}(k\,|\,k-1)[\mathbf{I} - \mathbf{K}(k)\mathbf{C}(k)]^T + \mathbf{K}(k)\mathbf{V}_{\mathbf{v}}(k)\mathbf{K}^T(k) \tag{6.4.60}$$

which upon substituting for $\mathbf{K}(k)$ as (see 6.4.43, 6.4.55)

$$\mathbf{K}(k) = \mathbf{V}_{\tilde{\mathbf{x}}}(k\,|\,k-1)\mathbf{C}^T(k)\mathbf{V}_{\boldsymbol{\eta}}^{-1}(k) \tag{6.4.61}$$

can be simplified to yield

$$\mathbf{V}_{\tilde{\mathbf{x}}}(k\,|\,k) = [\mathbf{I} - \mathbf{K}(k)\mathbf{C}(k)]\mathbf{V}_{\tilde{\mathbf{x}}}(k\,|\,k-1) \tag{6.4.62}$$

We can simplify the equation for the predicted error variance by using $\mathbf{V}_{\tilde{\mathbf{x}}}(k\,|\,k)$. Substitution of Eqs. (6.4.61) and (6.4.62) in Eq. (6.4.47) yields

$$\mathbf{V}_{\tilde{\mathbf{x}}}(k+1\,|\,k) = \boldsymbol{\phi}(k+1,k)\mathbf{V}_{\tilde{\mathbf{x}}}(k\,|\,k)\boldsymbol{\phi}^T(k+1,k) + \boldsymbol{\Gamma}(k)\mathbf{V}_{\mathbf{w}}(k)\boldsymbol{\Gamma}^T(k) \tag{6.4.63}$$

These algorithms are summarized in Table 6.4.1b.

Example 6.4.1

Consider the scalar message model given by

$$x(k+1) = 0.8x(k) + w(k)$$
$$z(k) = h(k)x(k) + v(k)$$

where $w(k)$ and $v(k)$ are stationary, zero mean, white with unity covariance. $h(k)$ is a periodic sequence as follows

$$h(k) = \begin{cases} 0.1 & k \text{ odd} \\ 1 & k \text{ even} \end{cases}$$

The one-stage predictor equations as obtained from Table 6.4.1 are given by

$$\hat{x}(k+1\,|\,k) = 0.8\hat{x}(k\,|\,k-1) + K(k+1,k)[z(k) - h(k)\hat{x}(k\,|\,k-1)]$$
$$K(k+1,k) = 0.8V_{\tilde{x}}(k\,|\,k-1)h(k)[h^2(k)V_{\tilde{x}}(k\,|\,k-1) + 1]^{-1}$$
$$V_{\tilde{x}}(k+1\,|\,k) = 0.64V_{\tilde{x}}(k\,|\,k-1)[h^2(k)V_{\tilde{x}}(k\,|\,k-1) + 1]^{-1} + 1$$

Note that the equations for the error variance $V_{\tilde{x}}(k+1\,|\,k)$ and the filter gain $K(k+1,k)$ can be processed independently of the observations. Figure 6.4.2a shows the true state and the estimate for a single realization of the input and observation noise processes as obtained by simulation on a digital computer. Since the odd-numbered measurements have a relatively larger noise component, we should expect that these measurements do not play as significant a role in updating our estimate as the even-numbered measurements do: the value of the gain $K(k+1,k)$ should be smaller for odd values of k than for even k. In fact, as can be seen from Fig. 6.4.2b, the gain becomes periodic after a fairly short transient period.

We note that the estimate $\hat{x}(k)$ follows the true value $x(k)$ fairly reasonably for this particular simulation. We may not obtain similar results with other realizations of the input $w(k)$ or the observation noise $v(k)$ since $x(k)$ and $\hat{x}(k)$ are close only in the mean-square sense. If, however, we average the results of a large number of simulations using a different realization of $w(k)$ and $v(k)$ in each simulation, we may expect the average values of true state and the estimate to be close to each other.

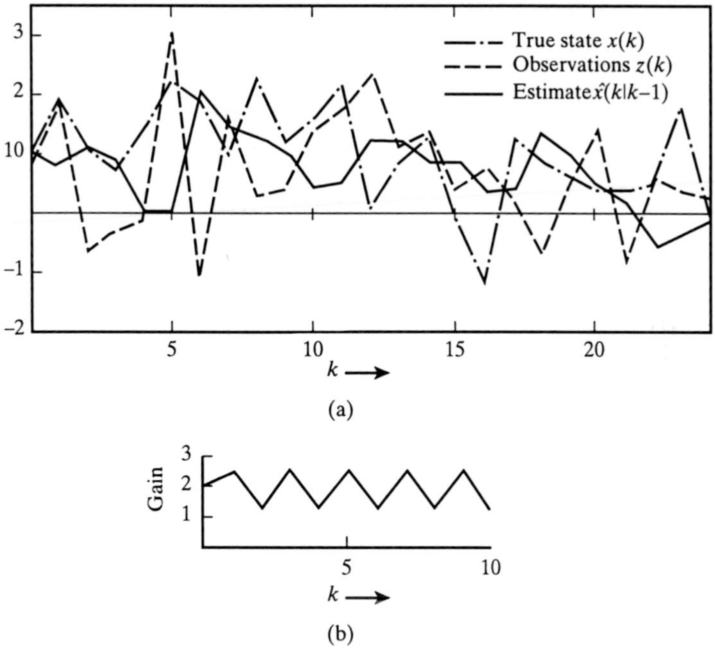

Figure 6.4.2 Simulation results for discrete Kalman filter example: (a) true state and estimated values; (b) filter gain.

6.4.4 Continuous-Time Filter

We now derive the analogous estimator equations for the continuous-time problem. If we define the set of observations over the interval $t_0 \leqslant \tau < t$ by

$$\mathbf{Z}(t) = \{\mathbf{z}(\tau) : t_0 \leqslant \tau < t\} \tag{6.4.64}$$

we see from Eq. (6.4.10) that the estimate $\hat{\mathbf{g}}(t \mid t_f)$ is given by

$$\hat{\mathbf{g}}(t \mid t_f) = \hat{\mathrm{E}}\{\mathbf{g}(t) \mid \mathbf{Z}(t_f)\} = \int_{t_0}^{t_f} \mathrm{E}\{\mathbf{g}(t)\boldsymbol{\eta}^T(\tau)\}\mathbf{V}_{\eta}^{-1}(\tau)\boldsymbol{\eta}(\tau)\, d\tau \tag{6.4.65}$$

Analogous to our derivation in the discrete-time case, we will show that the innovations $\boldsymbol{\eta}(t)$ for the observation model Eq. (6.4.1) is given by

$$\boldsymbol{\eta}(t) = \mathbf{z}(t) - \hat{\mathbf{y}}(t \mid t) \tag{6.4.66}$$

where

$$\hat{\mathbf{y}}(t \mid t) = \hat{\mathrm{E}}\{\mathbf{y}(t) \mid \mathbf{Z}(t)\} \tag{6.4.67}$$

We note that Eq. (6.4.66) can also be written as

$$\boldsymbol{\eta}(t) = \tilde{\mathbf{y}}(t \mid t) + \mathbf{v}(t) \tag{6.4.68}$$

To establish that $\boldsymbol{\eta}(t)$ is white, we note that for $t > s$,

$$\begin{aligned} \mathrm{E}\{\boldsymbol{\eta}(t)\boldsymbol{\eta}^T(s)\} &= \mathrm{E}\{[\tilde{\mathbf{y}}(t\,|\,t) + \mathbf{v}(t)]\boldsymbol{\eta}^T(s)\} \\ &= \mathrm{E}\{\mathbf{v}(t)\boldsymbol{\eta}^T(s)\} \qquad \text{by orthogonality} \\ &= \mathrm{E}\{\mathbf{v}(t)[\tilde{\mathbf{y}}(s\,|\,s) + \mathbf{v}(s)]^T\} \\ &= \mathrm{E}\{\mathbf{v}(t)\mathbf{v}^T(s)\} \qquad \text{by causality} \\ &= 0 \end{aligned} \tag{6.4.69}$$

It can be seen that this result holds for $s > t$ also. For $t = s$ we have

$$\begin{aligned} \mathrm{E}\{\boldsymbol{\eta}(t)\boldsymbol{\eta}^T(t)\} &= \mathrm{E}\{[\tilde{\mathbf{y}}(t\,|\,t) + \mathbf{v}(t)][\tilde{\mathbf{y}}(t\,|\,t) + \mathbf{v}(t)]^T\} \\ &= \mathrm{E}\{\tilde{\mathbf{y}}(t\,|\,t)\tilde{\mathbf{y}}^T(t\,|\,t)\} + \mathrm{E}\{\mathbf{v}(t)\mathbf{v}^T(t)\} \end{aligned}$$

We note that in the preceding expression, $\mathrm{E}\{\mathbf{v}(t)\mathbf{v}^T(t)\}$ is an impulse of strength $\mathbf{V}_\mathbf{v}(t)$. The variance of the error, $\mathrm{E}\{\tilde{\mathbf{y}}(t\,|\,t)\tilde{\mathbf{y}}^T(t\,|\,t)\}$ is a finite quantity (as otherwise there would be no solution to the minimization problem) and can be neglected in comparison with the other term. We conclude, therefore, that the variance of $\mathbf{v}(t)$ is also an impulse of strength $\mathbf{V}_\mathbf{v}(t)$. Thus we have

$$\mathrm{E}\{\boldsymbol{\eta}(t)\boldsymbol{\eta}^T(s)\} = \mathbf{V}_\mathbf{v}(t)\delta(t - s) \tag{6.4.70}$$

We point out that in the discrete-time problem, $\mathbf{V}_\mathbf{v}(k)$ is a finite quantity and hence $\mathbf{V}_{\tilde{\mathbf{y}}}(k\,|\,k-1)$ cannot be neglected.

In order to derive a differential equation for the estimate, we will need to assume a specific model for the signal process. We assume again that the signal $\mathbf{y}(t)$ is the output of a linear system driven by white noise:

$$\begin{aligned} \dot{\mathbf{x}}(t) &= \mathbf{A}(t)\mathbf{x}(t) + \mathbf{B}(t)\mathbf{w}(t) \\ \mathbf{y}(t) &= \mathbf{C}(t)\mathbf{x}(t) \end{aligned} \tag{6.4.71}$$

where $\mathbf{w}(t)$ is zero mean, white with covariance

$$\mathrm{E}\{\mathbf{w}(t)\mathbf{w}^T(s)\} = \mathbf{V}_\mathbf{w}(t)\delta(t - s) \tag{6.4.72}$$

and is uncorrelated with the observation noise $\mathbf{v}(t)$. The initial mean and variance of the state are assumed known

$$\mathrm{E}\{\mathbf{x}(t_0)\} = \boldsymbol{\mu}_\mathbf{x}(t_0) \quad \text{and} \quad \operatorname{cov}\mathbf{x}(t_0) = \mathbf{V}_\mathbf{x}(t_0)$$

The function we are seeking to estimate is the state vector $\mathbf{x}(t)$ so that $\mathbf{g}(t) = \mathbf{x}(t)$. We restrict our discussion to filtered estimates.

From Eq. (6.4.65), for $t_f = t$, we have

$$\hat{\mathbf{x}}(t\,|\,t) = \int_{t_0}^{t} \mathrm{E}\{\mathbf{x}(t)\boldsymbol{\eta}^T(\tau)\}\mathbf{V}_\eta^{-1}(\tau)\boldsymbol{\eta}(\tau)\,d\tau \tag{6.4.73}$$

Differentiating both sides with respect to t yields

$$\begin{aligned}\dot{\hat{\mathbf{x}}}(t\,|\,t) &= \frac{d}{dt}\int_{t_0}^{t} \mathrm{E}\{\mathbf{x}(t)\boldsymbol{\eta}^T(\tau)\}\mathbf{V}_\eta^{-1}(\tau)\boldsymbol{\eta}(\tau)\,d\tau \\ &= \int_{t_0}^{t} \mathrm{E}\{\dot{\mathbf{x}}(t)\boldsymbol{\eta}^T(\tau)\}\mathbf{V}_\eta^{-1}(\tau)\boldsymbol{\eta}(\tau)\,d\tau + \mathrm{E}\{\mathbf{x}(t)\boldsymbol{\eta}^T(t)\}\mathbf{V}_\eta^{-1}(t)\boldsymbol{\eta}(t)\end{aligned} \tag{6.4.74}$$

Substituting for $\dot{\mathbf{x}}(t)$ from Eq. (6.4.71) and noting that by causality

$$\dot{\hat{\mathbf{x}}}(t\,|\,t) = \mathbf{A}(t)\int_{t_0}^{t_f} \mathrm{E}\{\mathbf{x}(t)\boldsymbol{\eta}^T(\tau)\}\mathbf{V}_\eta^{-1}(\tau)\boldsymbol{\eta}(\tau)\,d\tau + \mathrm{E}\{\mathbf{x}(t)\boldsymbol{\eta}^T(t)\}\mathbf{V}_\eta^{-1}(t)\boldsymbol{\eta}(t) \tag{6.4.75}$$

Using Eq. (6.4.73) then yields

$$\dot{\hat{\mathbf{x}}}(t\,|\,t) = \mathbf{A}(t)\hat{\mathbf{x}}(t\,|\,t) + \mathbf{K}(t)\boldsymbol{\eta}(t) \tag{6.4.76}$$

where we have defined

$$\mathbf{K}(t) = \mathrm{E}\{\mathbf{x}(t)\boldsymbol{\eta}^T(t)\}\mathbf{V}_\eta^{-1}(t) \tag{6.4.77}$$

Now from Eq. (6.4.68) and the assumptions on $\mathbf{v}(t)$, we can write

$$\begin{aligned}\mathrm{E}\{\mathbf{x}(t)\boldsymbol{\eta}^T(t)\} &= \mathrm{E}\{\mathbf{x}(t)[\mathbf{C}(t)\tilde{\mathbf{x}}(t\,|\,t) + \mathbf{v}(t)]^T\} \\ &= \mathrm{E}\{\mathbf{x}(t)\tilde{\mathbf{x}}^T(t\,|\,t)\}\mathbf{C}^T(t) \\ &= \mathrm{E}\{[\hat{\mathbf{x}}(t\,|\,t) + \tilde{\mathbf{x}}(t\,|\,t)]\tilde{\mathbf{x}}^T(t\,|\,t)\}\mathbf{C}^T(t) \\ &= \mathbf{V}_{\tilde{\mathbf{x}}}(t\,|\,t)\mathbf{C}^T(t)\end{aligned} \tag{6.4.78}$$

The last equality follows since $\tilde{\mathbf{x}}(t\,|\,t)$ is orthogonal to $\hat{\mathbf{x}}(t\,|\,t)$. Substitution of Eq. (6.4.78) in Eq. (6.4.77) then yields

$$\mathbf{K}(t) = \mathbf{V}_{\tilde{\mathbf{x}}}(t\,|\,t)\mathbf{C}^T(t)\mathbf{V}_\eta^{-1}(t) \tag{6.4.79}$$

where the last equality follows from Eq. (6.4.70). $\mathbf{K}(t)$ is referred to as the Kalman gain for the continuous-filter case. All that remains is to determine $\mathbf{V}_{\tilde{\mathbf{x}}}(t\,|\,t)$. To this end, we subtract Eq. (6.4.76) from Eq. (6.4.71) to get

$$\begin{aligned}\dot{\tilde{\mathbf{x}}}(t\,|\,t) &= \dot{\mathbf{x}}(t) - \dot{\hat{\mathbf{x}}}(t\,|\,t) \\ &= \mathbf{A}(t)\tilde{\mathbf{x}}(t\,|\,t) + \mathbf{B}(t)\mathbf{w}(t) - \mathbf{K}(t)\boldsymbol{\eta}(t) \\ &= [\mathbf{A}(t) - \mathbf{K}(t)\mathbf{C}(t)]\tilde{\mathbf{x}}(t\,|\,t) + \mathbf{B}(t)\mathbf{w}(t) - \mathbf{K}(t)\mathbf{v}(t)\end{aligned} \tag{6.4.80}$$

We thus have, in view of the fact that $\mathbf{w}(\cdot)$ and $\mathbf{v}(\cdot)$ are white and uncorrelated, from Eq. (2.4.44),

$$\begin{aligned}\dot{\mathbf{V}}_{\tilde{\mathbf{x}}}(t\,|\,t) &= [\mathbf{A}(t) - \mathbf{K}(t)\mathbf{C}(t)]\mathbf{V}_{\tilde{\mathbf{x}}}(t\,|\,t) \\ &\quad + \mathbf{V}_{\tilde{\mathbf{x}}}(t\,|\,t)[\mathbf{A}(t) - \mathbf{K}(t)\mathbf{C}(t)]^T \\ &\quad + \mathbf{B}(t)\mathbf{V}_{\mathbf{w}}(t)\mathbf{B}^T(t) + \mathbf{K}(t)\mathbf{V}_{\mathbf{v}}(t)\mathbf{K}^T(t)\end{aligned} \tag{6.4.81}$$

Substitution of Eq. (6.4.79) into Eq. (6.4.81) and simplification yields

$$\begin{aligned}\dot{\mathbf{V}}_{\tilde{\mathbf{x}}}(t\,|\,t) = \mathbf{A}(t)\mathbf{V}_{\tilde{\mathbf{x}}}(t\,|\,t) + \mathbf{V}_{\tilde{\mathbf{x}}}(t\,|\,t)\mathbf{A}^T(t) + \mathbf{B}(t)\mathbf{V}_{\mathbf{w}}(t)\mathbf{B}^T(t)\\ - \mathbf{V}_{\tilde{\mathbf{x}}}(t\,|\,t)\mathbf{C}^T(t)\mathbf{V}_{\mathbf{v}}^{-1}(t)\mathbf{C}(t)\mathbf{V}_{\tilde{\mathbf{x}}}(t\,|\,t)\end{aligned} \tag{6.4.82}$$

Equation (6.4.82) is usually referred to as the *matrix Riccati differential equation*. It can easily be shown that for unbiased estimates, we must choose the initial conditions for Eqs. (6.4.75) and (6.4.81) as

$$\hat{\mathbf{x}}(t_0) = \boldsymbol{\mu}_{\mathbf{x}}(t_0) \quad \text{and} \quad \mathbf{V}_{\tilde{\mathbf{x}}}(t_0) = \mathbf{V}_{\mathbf{x}}(t_0) \tag{6.4.83}$$

The algorithms are summarized in Table 6.4.2. Figure 6.4.3 shows the structure of the Kalman filter.

TABLE 6.4.2 CONTINUOUS-TIME KALMAN FILTER ALGORITHMS

Message model:

$$\begin{aligned}\dot{\mathbf{x}}(t) &= \mathbf{A}(t)\mathbf{x}(t) + \mathbf{B}(t)\mathbf{w}(t)\\ \mathbf{y}(t) &= \mathbf{C}(t)\mathbf{x}(t)\end{aligned} \tag{6.4.71}$$

Observation model:

$$\mathbf{z}(t) = \mathbf{y}(t) + \mathbf{v}(t) \tag{6.4.1}$$

Prior statistics:

$$\begin{aligned}\mathrm{E}\{\mathbf{w}(t)\} &= \mathrm{E}\{\mathbf{v}(t)\} = \mathbf{0}, \mathrm{E}\{\mathbf{x}(t_0)\} = \boldsymbol{\mu}_{\mathbf{x}}(t_0)\\ \mathrm{E}\{\mathbf{w}(t)\mathbf{w}^T(s)\} &= \mathbf{V}_{\mathbf{w}}(t)\delta(t-s), \mathrm{E}\{\mathbf{v}(t)\mathbf{v}^T(s)\} = \mathbf{V}_{\mathbf{v}}(t)\delta(t-s)\\ \mathrm{E}\{\mathbf{w}(t)\mathbf{v}^T(s)\} &= \mathrm{E}\{\mathbf{w}(t)\mathbf{x}^T(t_0)\} = \mathrm{E}\{\mathbf{v}(t)\mathbf{x}^T(t_0)\} = \mathbf{0}\\ \operatorname{cov}\mathbf{x}(t_0) &= \mathbf{V}_{\mathbf{x}}(t_0)\end{aligned}$$

Filter algorithm:

$$\begin{aligned}\dot{\hat{\mathbf{x}}}(t\,|\,t) &= \mathbf{A}(t)\hat{\mathbf{x}}(t\,|\,t) + \mathbf{K}(t)\boldsymbol{\eta}(t)\\ \boldsymbol{\eta}(t) &= \mathbf{z}(t) - \mathbf{C}(t)\hat{\mathbf{x}}(t\,|\,t)\end{aligned} \tag{6.4.76}$$

Gain algorithm:

$$\mathbf{K}(t) = \mathbf{V}_{\tilde{\mathbf{x}}}(t\,|\,t)\mathbf{C}^T(t)\mathbf{V}_{\mathbf{v}}^{-1}(t) \tag{6.4.79}$$

Variance algorithm:

$$\begin{aligned}\dot{\mathbf{V}}_{\tilde{\mathbf{x}}}(t\,|\,t) = \mathbf{A}(t)\mathbf{V}_{\tilde{\mathbf{x}}}(t\,|\,t) + \mathbf{V}_{\tilde{\mathbf{x}}}(t\,|\,t)\mathbf{A}^T(t)\\ + \mathbf{B}(t)\mathbf{V}_{\mathbf{w}}(t)\mathbf{B}^T(t) - \mathbf{V}_{\tilde{\mathbf{x}}}(t\,|\,t)\mathbf{C}^T(t)\mathbf{V}_{\mathbf{v}}^{-1}(t)\mathbf{C}(t)\mathbf{V}_{\tilde{\mathbf{x}}}(t\,|\,t)\end{aligned} \tag{6.4.82}$$

Initial conditions:

$$\hat{\mathbf{x}}(t_0) = \boldsymbol{\mu}_{\mathbf{x}}(t_0), \mathbf{V}_{\tilde{\mathbf{x}}}(t_0) = \mathbf{V}_{\mathbf{x}}(t_0) \tag{6.4.83}$$

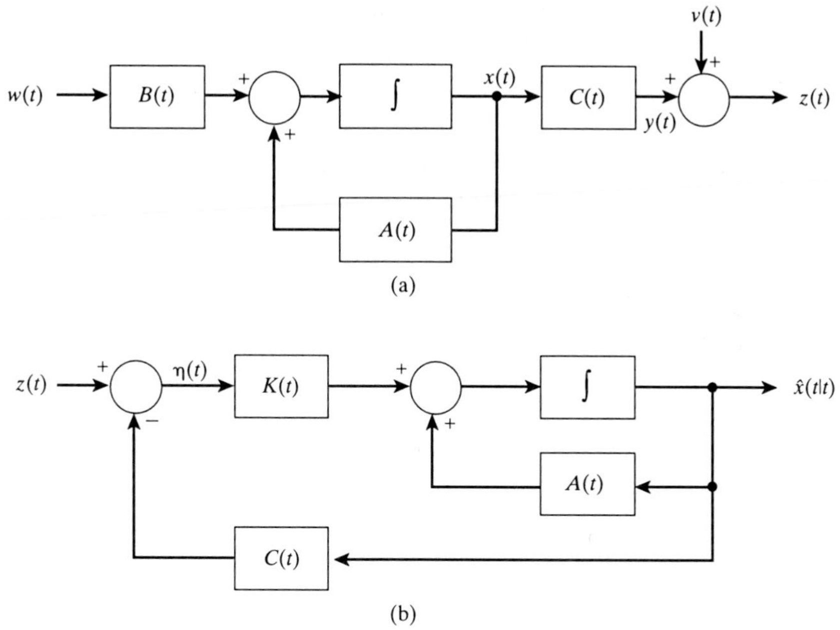

Figure 6.4.3 Structure of continuous-time Kalman filter: (a) message model; (b) filter.

Example 6.4.2

We consider the estimation of the state in a continuous-time model. The message and observation models are assumed to be as follows:

$$\dot{x}(t) = -x(t) + w(t) \qquad x(0) = 1$$

$$z(t) = \tfrac{1}{2}x(t) + v(t)$$

$$\mathrm{E}\{w(t)w(\tau)\} = 6\delta(t-\tau), \mathrm{E}\{v(t)v(\tau)\} = \tfrac{1}{2}\delta(t-\tau)$$

$$\mathrm{E}\{w(t)\} = E\{v(t)\} = \mathrm{E}\{w(t)v(\tau)\} = 0$$

The estimator equations for this problem are easily seen to be

$$\dot{\hat{x}}(t\,|\,t) = -\hat{x}(t\,|\,t) + K(t)[z(t) - \tfrac{1}{2}\hat{x}(t\,|\,t)]$$

with

$$K(t) = V_{\tilde{x}}(t\,|\,t)$$

The error variance equation is given by

$$\dot{V}_{\tilde{x}}(t\,|\,t) = -2V_{\tilde{x}}(t\,|\,t) - \tfrac{1}{2}V_{\tilde{x}}^2(t\,|\,t) + 6$$

While this equation can be solved using numerical techniques, it is instructive to obtain an analytical solution in this simple case. Since $x(0)$ is known (deterministic), we must choose $\hat{x}(0) = x(0)$ to obtain an unbiased estimate. It follows that $\tilde{x}(0)$ and $V_{\tilde{x}}(0)$ should be zero. Let us assume a solution of the form

$$V_{\tilde{x}}(t\,|\,t) = \alpha + \beta \tanh(\gamma t + \delta) = \alpha + \beta \tanh\theta$$

where α, β, γ, δ, and θ are unknown constants. Then

$$\dot{V}_{\tilde{x}}(t\,|\,t) = \beta\gamma(1 - \tanh^2\theta)$$

Substituting in the variance equation yields

$$\beta\gamma(1 - \tanh^2\theta) = -2(\alpha + \beta\tanh\theta) - \tfrac{1}{2}(\alpha^2 + \beta^2\tanh^2\theta + 2\alpha\beta\tanh\theta) + 6$$

Collecting like terms together and equating to zero, we have

$$\alpha = -2 \qquad \beta = 2\gamma \qquad \gamma = \pm 2$$

Choosing $\gamma = +2$ yields

$$\begin{aligned} V_{\tilde{x}}(t\,|\,t) &= -2 + 4\tanh(2t + \delta) \\ &= -2 + 4\tanh(2t + \tanh^{-1}\tfrac{1}{2}) \end{aligned}$$

where the last step is obtained by making use of the fact that $V_{\tilde{x}}(0) = 0$. The steady-state value of the error variance is obtained by letting $t \to \infty$ and is given by

$$V_{\tilde{x}\,ss} = -2 + 4 = 2$$

It follows therefore that the filter gain $K(t)$ also approaches a constant value as $t \to \infty$, given by

$$K_{ss} = 2$$

The estimator equation then becomes

$$\begin{aligned} \dot{\hat{x}}(t\,|\,t) &= -\hat{x}(t\,|\,t) + 2[z(t) - \tfrac{1}{2}\hat{x}(t\,|\,t)] \\ &= -2\hat{x}(t\,|\,t) + 2z(t) \end{aligned}$$

Laplace transformation of this equation yields

$$\hat{X}(s) = \frac{2}{s + 2} Z(s)$$

so that the estimator is asymptotically equivalent to a time-invariant linear system with transfer function $2/(s + 2)$.

The system was also simulated on a digital computer for different values of the initial condition $\hat{x}(0)$. The corresponding error variances are shown in Fig. 6.4.4a. As can be seen, the error variances approach the steady-state value fairly rapidly. Since the filter gain tends to a constant value after an initial transient period, we might expect that the estimates for various initial values $x(0)$ tend to converge asymptotically. This is borne out by the plots of the estimate $\hat{x}(t)$ in Fig. 6.4.4b. We therefore expect that the estimate (at least after an initial transient period) is not significantly dependent on our choice of initial conditions. This is very desirable since the initial conditions in many cases represent no more than an educated guess on the part of the designer. Estimators whose performance does not depend on the choice of initial conditions (or more generally on the prior statistics) are called *robust estimators*.

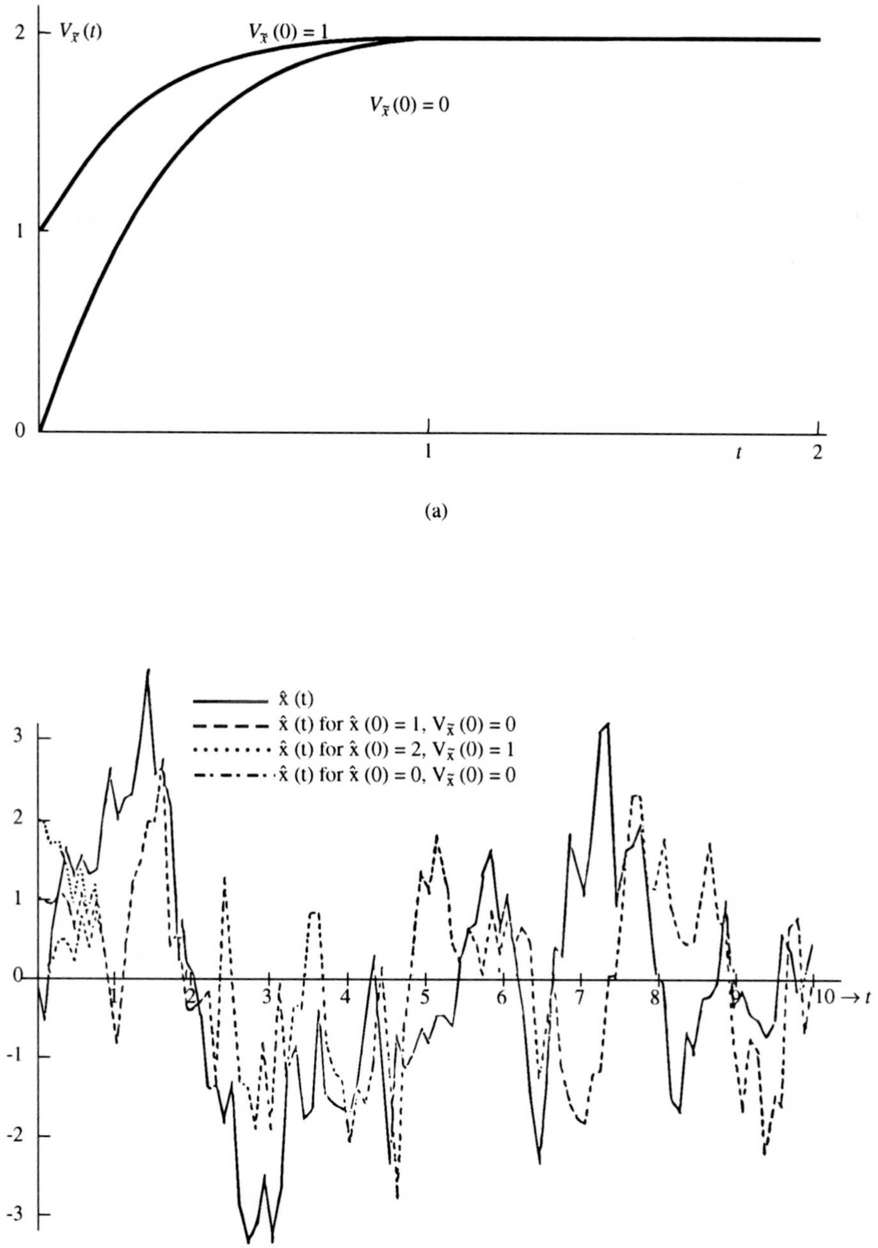

Figure 6.4.4 Simulation results for continuous-time Kalman filter example: (a) error variance; (b) true state and estimate for various initial conditions.

6.4.5 Continuous-Time Estimation with Discrete Observations

In many instances, even though the signal model is continuous, the observations are available only at discrete intervals of time. The relevant filtering algorithms follow easily from our previous discussions. We assume the message model of Eq. (6.4.71):

$$\dot{\mathbf{x}}(t) = \mathbf{A}(t)\mathbf{x}(t) + \mathbf{B}(t)\mathbf{w}(t)$$
$$\mathbf{y}(t) = \mathbf{C}(t)\mathbf{x}(t) \tag{6.4.84}$$

for $t \geqslant t_0$.

The prior statistics of the noise input $\mathbf{w}(t)$ and the initial vector $\mathbf{x}(t_0)$ are as before. If we assume that noisy measurements of $\mathbf{y}(t)$ are available at discrete instants kT, the observation model becomes

$$\mathbf{z}(kT) = \mathbf{y}(kT) + \mathbf{v}(kT) \tag{6.4.85}$$

The solution to the differential equation of Eq. (6.4.84) over any interval $[\tau, t]$ is

$$\mathbf{x}(t) = \boldsymbol{\psi}(t,\tau)\mathbf{x}(\tau) + \int_\tau^t \boldsymbol{\psi}(t,\sigma)\mathbf{B}(\sigma)\mathbf{w}(\sigma)\,d\sigma \tag{6.4.86}$$

where $\boldsymbol{\psi}(\cdot,\cdot)$ is the state transition matrix corresponding to the matrix $\mathbf{A}(t)$. By setting $\tau = kT$ and $t = (k+1)T$ in Eq. (6.4.86), we obtain

$$\mathbf{x}[(k+1)T] = \boldsymbol{\psi}[(k+1)T, kT]\mathbf{x}(kT) + \int_{kT}^{(k+1)T} \boldsymbol{\psi}[(k+1)T,\sigma]\mathbf{B}(\sigma)\mathbf{w}(\sigma)\,d\sigma \tag{6.4.87}$$

We now make the following definitions:

$$\mathbf{w}_1(kT) \triangleq \int_{kT}^{(k+1)T} \boldsymbol{\psi}[(k+1)T,\sigma]\mathbf{B}(\sigma)\mathbf{w}(\sigma)\,d\sigma \tag{6.4.88a}$$

$$\mathbf{V}_{\mathbf{w}_1}(kT) \triangleq \int_{kT}^{(k+1)T} \boldsymbol{\psi}[(k+1)T,\sigma]\mathbf{B}(\sigma)\mathbf{V}_{\mathbf{w}}(\sigma)\mathbf{B}^T(\sigma) \times \boldsymbol{\psi}^T[(k+1)T,\sigma]\,d\sigma \tag{6.4.88b}$$

Substitution in Eq. (6.4.87) then yields

$$\mathbf{x}[(k+1)T] = \boldsymbol{\psi}[(k+1)T, kT]\mathbf{x}(kT) + \mathbf{w}_1(kT) \tag{6.4.89a}$$

Also,

$$\mathbf{y}(kT) = \mathbf{C}(kT)\mathbf{x}(kT) \tag{6.4.89b}$$

It can easily be verified that $\mathbf{w}_1(kT)$ is white, zero mean with

$$\operatorname{cov}[\mathbf{w}_1(kT), \mathbf{w}_1(jT)] = \mathbf{V}_{\mathbf{w}_1}(kT)\delta_{kj} \tag{6.4.90}$$

Comparison of Eqs. (6.4.89a,b) and (6.4.85) with Eqs. (6.4.19) and (6.4.21) shows that the algorithms of Table 6.4.1 can now be directly applied to this problem to obtain the estimates $\hat{\mathbf{x}}(kT\,|\,kT)$ and $\hat{\mathbf{y}}(kT\,|\,kT)$.

To obtain the estimate of the state in between observations, we note from Eq. (6.4.86) that in the interval $kT \leqslant t < (k+1)T$, we can write

$$\mathbf{x}(t) = \boldsymbol{\psi}(t, kT)\mathbf{x}(kT) + \int_{kT}^{t} \boldsymbol{\psi}(t, \sigma)\mathbf{B}(\sigma)\mathbf{w}(\sigma)\, d\sigma \tag{6.4.91}$$

Let us denote the observations up to time kT by $\mathbf{Z}(kT)$. Then it follows from Eq. (6.4.91) that

$$\begin{aligned} \hat{\mathrm{E}}[\mathbf{x}(t) \mid \mathbf{Z}(kT)] &= \boldsymbol{\psi}(t, kT)\hat{\mathrm{E}}[\mathbf{x}(kT) \mid \mathbf{Z}(kT)] \\ &\quad + \hat{\mathrm{E}}\left[\int_{kT}^{t} \boldsymbol{\psi}(t, \sigma)\mathbf{B}(\sigma)\mathbf{w}(\sigma)\, d\sigma \mid \mathbf{Z}(kT)\right] \end{aligned} \tag{6.4.92}$$

where $\hat{\mathrm{E}}(\cdot)$ denotes the wide-sense conditional exception. The second term on the right-hand side of Eq. (6.4.92) is zero since $\mathbf{w}(\cdot)$ is a zero-mean, white process and $\mathbf{w}(\sigma)$ is independent of $\mathbf{Z}(kT)$ for $\sigma > kT$. We therefore have

$$\begin{aligned} \hat{\mathbf{x}}(t \mid kT) &= \hat{\mathrm{E}}[\mathbf{x}(t) \mid \mathbf{Z}(kT)] \\ &= \boldsymbol{\psi}(t, kT)\hat{\mathbf{x}}(kT \mid kT) \end{aligned} \tag{6.4.93}$$

The estimate of the signal between observations follows as

$$\hat{\mathbf{y}}(t \mid kT) = \mathbf{C}(t)\hat{\mathbf{x}}(t \mid kT) \tag{6.4.94}$$

Equation (6.4.93) can be used to obtain the initial condition for processing the observation $\mathbf{z}[(k+1)T]$. We have

$$\hat{\mathbf{x}}(kT + T \mid kT) = \boldsymbol{\psi}(kT + T, kT)\hat{\mathbf{x}}(kT \mid kT) \tag{6.4.95}$$

The predicted error covariance can easily be shown to be

$$\begin{aligned} \mathbf{V}_{\tilde{\mathbf{x}}}(kT + T \mid kT) &= \mathbf{x}(kT + T, kT)\mathbf{V}_{\tilde{\mathbf{x}}}(kT \mid kT)\boldsymbol{\psi}^{T}(kT + T, kT) \\ &\quad + \mathbf{V}_{\mathbf{w}_1}(kT) \end{aligned} \tag{6.4.96}$$

where $\mathbf{V}_{\mathbf{w}_1}(kT)$ is defined in Eq. (6.4.88b).

Various extensions of the algorithms presented here are available in the literature. These extensions consider noise processes with nonzero means, cases where the input $\mathbf{w}(t)$ and the observation noise $\mathbf{v}(t)$ are correlated and the case where the input to the message model is a known signal plus a random process. The colored noise problem has also been considered in the literature. We do not consider these extensions here but invite the interested reader to look through the appropriate references listed at the end of this chapter [3–6].

Next, we consider briefly the relation between the Wiener and Kalman filters.

6.5 RELATION BETWEEN THE KALMAN AND WIENER FILTERS

The Kalman filtering algorithms are rather general in nature. The Wiener filter, on the other hand, was derived for estimation of stationary processes. It is of interest to explore the relation between the two filters when the underlying assumptions are the same. Specifically, we establish in this section that the Kalman filter reduces to the classical Wiener filter when the processes involved are stationary and the observation interval is semi-infinite. Incidentally, we are also able to solve the spectral factorization problem for vector processes that we encountered in Section 6.3. We restrict our discussion to the continuous-time filter. Since we have considered only white observation noise in the Kalman filter problem, we assume that the noise in the Wiener formulation is also white and uncorrelated with the signal. The estimate that is desired is that of the signal $\mathbf{y}(t)$.

Let us consider the solution using the Kalman filter for the problem under consideration. Since we are assuming stationary processes, it is clear that the matrices $\mathbf{A}$, $\mathbf{B}$, and $\mathbf{C}$ and the covariance kernels $\mathbf{V}_{\mathbf{w}}$ and $\mathbf{V}_{\mathbf{v}}$ in Eqs. (6.4.70) and (6.4.72) are constants. Since the estimate $\hat{\mathbf{x}}(t \mid t)$ will now be asymptotically a stationary process, the error variance $\mathbf{V}_{\tilde{\mathbf{x}}}(t)$ must be asymptotically a constant. The error variance can thus be obtained as the steady-state solution to Eq. (6.4.82) by setting the derivative equal to zero. Thus the error variance satisfies the algebraic equation

$$\mathbf{A}\mathbf{V}_{\tilde{\mathbf{x}}} + \mathbf{V}_{\tilde{\mathbf{x}}}\mathbf{A}^T + \mathbf{B}\mathbf{V}_{\mathbf{w}}\mathbf{B}^T - \mathbf{V}_{\tilde{\mathbf{x}}}\mathbf{C}^T\mathbf{V}_{\mathbf{v}}^{-1}\mathbf{C}\mathbf{V}_{\tilde{\mathbf{x}}} = 0 \tag{6.5.1}$$

This is a quadratic algebraic equation and is a reduced Riccati equation. Since the equation is nonlinear, it has many solutions. However, since $\mathbf{V}_{\tilde{\mathbf{x}}}$ is a covariance matrix, it must be positive definite; thus we pick as our solution the (unique) positive definite solution to Eq. (6.5.1) [3].

The solution of the algebraic equation above is tedious and difficult. We may rather seek the solution to the algebraic equation as the steady-state solution to the corresponding Riccati differential equation, which is comparatively straightforward to solve on a computer.

We now discuss the relationship of the stationary Kalman filter with the Wiener filter. The appropriate message and observation models are

$$\begin{aligned} \dot{\mathbf{x}}(t) &= \mathbf{A}\mathbf{x}(t) + \mathbf{B}\mathbf{w}(t) \\ \mathbf{y}(t) &= \mathbf{C}\mathbf{x}(t) \end{aligned} \tag{6.5.2}$$

and

$$\mathbf{z}(t) = \mathbf{y}(t) + \mathbf{v}(t) \tag{6.5.3}$$

The transfer matrix relating the output $\mathbf{y}(t)$ to the input $\mathbf{w}(t)$ is given by

$$\mathbf{G}(s) = \mathbf{C}[s\mathbf{I} - \mathbf{A}]^{-1}\mathbf{B} = \mathbf{C}\boldsymbol{\Omega}(s)\mathbf{B} \tag{6.5.4}$$

where

$$\boldsymbol{\Omega}(s) = [s\mathbf{I} - \mathbf{A}]^{-1} \tag{6.5.5}$$

The filter equations are

$$\begin{aligned} \dot{\hat{\mathbf{x}}}(t \mid t) &= \mathbf{A}\hat{\mathbf{x}}(t \mid t) + \mathbf{K}[\mathbf{z}(t) - \mathbf{C}\hat{\mathbf{x}}(t \mid t)] \\ &= [\mathbf{A} - \mathbf{K}\mathbf{C}]\hat{\mathbf{x}}(t \mid t) + \mathbf{K}\mathbf{z}(t) \\ \mathbf{y}(t) &= \mathbf{C}\hat{\mathbf{x}}(t \mid t) \end{aligned} \tag{6.5.6}$$

where

$$\mathbf{K} = \mathbf{V}_{\tilde{\mathbf{x}}}\mathbf{C}^T\mathbf{V}_{\mathbf{v}}^{-1} \tag{6.5.7}$$

with $\mathbf{V}_{\tilde{\mathbf{x}}}$ the positive definite solution to the algebraic equation

$$\mathbf{A}\mathbf{V}_{\tilde{\mathbf{x}}} + \mathbf{V}_{\tilde{\mathbf{x}}}\mathbf{A}^T - \mathbf{V}_{\tilde{\mathbf{x}}}\mathbf{C}^T\mathbf{V}_{\mathbf{v}}^{-1}\mathbf{C}\mathbf{V}_{\tilde{\mathbf{x}}} + \mathbf{B}\mathbf{V}_{\mathbf{w}}\mathbf{B}^T = 0 \tag{6.5.8}$$

The transfer function of the filter as obtained from Eq. (6.5.6) is

$$\mathbf{G}_{fK}(s) = \mathbf{C}[s\mathbf{I} - \mathbf{A} + \mathbf{K}\mathbf{C})]^{-1}\mathbf{K} \tag{6.5.9}$$

From our discussion in Section 6.3 [Eq. (6.3.40)] we see that the Wiener filter is determined in terms of the spectrum of the observed signal $\mathbf{z}(t)$. For the model of Eq. (6.5.2), since the input signal is white with variance $\mathbf{V}_{\mathbf{w}}$, the power spectrum of the signal output is given by

$$\begin{aligned}\mathbf{\Phi}_{\mathbf{y}}(s) &= \mathbf{G}(s)\mathbf{V}_{\mathbf{w}}\mathbf{G}^T(-s)\\ &= \mathbf{C}\mathbf{\Omega}(s)\mathbf{B}\mathbf{V}_{\mathbf{w}}\mathbf{B}^T\mathbf{\Omega}^T(-s)\mathbf{C}^T\end{aligned} \tag{6.5.10}$$

The spectrum of the observed signal is

$$\begin{aligned}\mathbf{\Phi}_{\mathbf{z}}(s) &= \mathbf{\Phi}_{\mathbf{y}}(s) + \mathbf{\Phi}_{\mathbf{v}}(s)\\ &= \mathbf{C}\mathbf{\Omega}(s)\mathbf{B}\mathbf{V}_{\mathbf{w}}\mathbf{B}^T\mathbf{\Omega}^T(-s)\mathbf{C}^T + \mathbf{V}_{\mathbf{v}}\end{aligned} \tag{6.5.11}$$

Since the desired estimate is $\hat{\mathbf{y}}(t)$ and since $\mathbf{y}(\cdot)$ and $\mathbf{v}(\cdot)$ are uncorrelated, from Eq. (6.3.40), the Wiener filter for this problem is specified by

$$\mathbf{G}_{fW}(s) = \{\mathbf{\Phi}_{\mathbf{y}}(s)[\mathbf{\Phi}^-(s)]^{-1}\}_+[\mathbf{\Phi}^+(s)]^{-1} \tag{6.5.12}$$

where $\mathbf{\Phi}^+(s)$ is obtained from a spectral factorization of $\mathbf{\Phi}_{\mathbf{z}}(s)$

$$\mathbf{\Phi}_{\mathbf{z}}(s) = \mathbf{\Phi}^+(s)\mathbf{\Phi}^-(s) \tag{6.5.13}$$

Since $\mathbf{V}_{\mathbf{v}}$ is a positive definite matrix, we can write

$$\mathbf{V}_{\mathbf{v}} = (\mathbf{V}_{\mathbf{v}})^{1/2}(\mathbf{V}_{\mathbf{v}})^{1/2} \tag{6.5.14}$$

where $(\mathbf{V}_{\mathbf{v}})^{1/2}$ is a positive definite symmetric matrix. We claim that $\mathbf{\Phi}^+(s)$ is given by

$$\mathbf{\Phi}^+(s) = \mathbf{V}_{\mathbf{v}}^{1/2} + \mathbf{C}\mathbf{\Omega}(s)\mathbf{V}_{\tilde{\mathbf{x}}}\mathbf{C}^T\mathbf{V}_{\mathbf{v}}^{-1/2} \tag{6.5.15}$$

We verify this by direct computation. We have

$$\begin{aligned}\mathbf{\Phi}^+(s)\mathbf{\Phi}^-(s) &= [\mathbf{V}_{\mathbf{v}}^{1/2} + \mathbf{C}\mathbf{\Omega}(s)\mathbf{V}_{\tilde{\mathbf{x}}}\mathbf{C}^T\mathbf{V}_{\mathbf{v}}^{-1/2}]\\ &\quad\times[\mathbf{V}_{\mathbf{v}}^{1/2} + \mathbf{C}\mathbf{\Omega}(-s)\mathbf{V}_{\tilde{\mathbf{x}}}\mathbf{C}^T\mathbf{V}_{\mathbf{v}}^{-1/2}]^T\\ &= \mathbf{V}_{\mathbf{v}} + \{\mathbf{C}\mathbf{\Omega}(s)\mathbf{V}_{\tilde{\mathbf{x}}}\mathbf{C}^T + \mathbf{C}\mathbf{V}_{\tilde{\mathbf{x}}}\mathbf{\Omega}^T(-s)\mathbf{C}^T\\ &\quad+ \mathbf{C}\mathbf{\Omega}(s)\mathbf{V}_{\tilde{\mathbf{x}}}\mathbf{C}^T\mathbf{V}_{\mathbf{v}}^{-1}\mathbf{C}\mathbf{V}_{\tilde{\mathbf{x}}}\mathbf{\Omega}^T(-s)\mathbf{C}^T\}\end{aligned} \tag{6.5.16}$$

The second term on the right side may be written as

$$\mathbf{C}\boldsymbol{\Omega}(s)\{\mathbf{V}_{\tilde{\mathbf{x}}}[\boldsymbol{\Omega}^T(-s)]^{-1} + [\boldsymbol{\Omega}(s)]^{-1}\mathbf{V}_{\tilde{\mathbf{x}}} + \mathbf{V}_{\tilde{\mathbf{x}}}\mathbf{C}^T\mathbf{V}_{\mathbf{v}}^{-1}\mathbf{C}\mathbf{V}_{\tilde{\mathbf{x}}}\}\boldsymbol{\Omega}^T(-s)\mathbf{C}^T \tag{6.5.17}$$

By noting that $\boldsymbol{\Omega}^{-1}(s) = (s\mathbf{I} - \mathbf{A})$ and $\boldsymbol{\Omega}^{-T}(-s) = (-s\mathbf{I} - \mathbf{A})^T$ and canceling out the resulting $s\mathbf{V}_{\tilde{\mathbf{x}}}$ and $-s\mathbf{V}_{\tilde{\mathbf{x}}}$ terms, we can write the expression in Eq. (6.5.17) as

$$\mathbf{C}\boldsymbol{\Omega}(s)[-\mathbf{V}_{\tilde{\mathbf{x}}}\mathbf{A}^T - \mathbf{A}\mathbf{V}_{\tilde{\mathbf{x}}} + \mathbf{V}_{\tilde{\mathbf{x}}}\mathbf{C}^T\mathbf{V}_{\mathbf{v}}^{-1}\mathbf{C}\mathbf{V}_{\tilde{\mathbf{x}}}]\boldsymbol{\Omega}^T(-s)\mathbf{C}^T \tag{6.5.18}$$

Substitution of Eq. (6.5.8) in Eq. (6.5.18) then enables us to write this expression as

$$\mathbf{C}\boldsymbol{\Omega}(s)\mathbf{B}\mathbf{V}_{\mathbf{w}}\mathbf{B}^T\boldsymbol{\Omega}^T(-s)\mathbf{C}^T \tag{6.5.19}$$

We can thus write

$$\begin{aligned}\boldsymbol{\Phi}^+(s)\boldsymbol{\Phi}^-(s) &= \mathbf{V}_{\mathbf{v}} + \mathbf{C}\boldsymbol{\Omega}(s)\mathbf{B}\mathbf{V}_{\mathbf{w}}\mathbf{B}^T\boldsymbol{\Omega}^T(-s)\mathbf{C}^T \\ &= \boldsymbol{\Phi}_{\mathbf{z}}(s)\end{aligned} \tag{6.5.20}$$

It remains to show that $\boldsymbol{\Phi}^+(s)$ is analytic in the right half-plane. This will be true if $\det\boldsymbol{\Phi}^+(s)$ has no poles in the right half-plane. We see from Eq. (6.5.15) that

$$\begin{aligned}\det\boldsymbol{\Phi}^+(s) &= \det\{[\mathbf{I} + \mathbf{C}\boldsymbol{\Omega}(s)\mathbf{V}_{\tilde{\mathbf{x}}}\mathbf{C}^T\mathbf{V}_{\mathbf{v}}^{-1}]\mathbf{V}_{\mathbf{v}}^{1/2}\} \\ &= \det\{[\mathbf{I} + \mathbf{C}\boldsymbol{\Omega}(s)\mathbf{K}]\}\det(V_{\mathbf{v}}^{1/2})\end{aligned} \tag{6.5.21}$$

where we have used Eq. (6.5.7). By use of a matrix identity (see Appendix C), Eq. (6.5.21) can be written as

$$\begin{aligned}\det\boldsymbol{\Phi}^+(s) &= \det\boldsymbol{\Omega}(s)\det[(s\mathbf{I} - \mathbf{A}) + \mathbf{K}\mathbf{C}]\det\mathbf{V}_{\mathbf{v}}^{1/2} \\ &= \frac{\det[s\mathbf{I} - \mathbf{A} + \mathbf{K}\mathbf{C}]}{\det(s\mathbf{I} - \mathbf{A})}\det\mathbf{V}_{\mathbf{v}}^{1/2}\end{aligned} \tag{6.5.22}$$

Since we have assumed that our signal process has finite variance (implying that $\mathbf{A}$ is a stable matrix) and since the error variance is bounded, it can be shown that the estimator must also be a stable structure [20,23]. Thus the poles of $(s\mathbf{I} - \mathbf{A} + \mathbf{K}\mathbf{C})$ will all lie in the left half-plane, implying that roots of $\det\boldsymbol{\Phi}^+(s)$ will also be in the left half-plane, thus establishing our result. We note that the solution to the spectral factorization problem as given in Eq. (6.5.15) involves the solution of the nonlinear algebraic equation (6.5.8).

To obtain the Wiener filter, we now substitute Eqs. (6.5.10) and (6.5.13) in Eq. (6.5.12) to get

$$\mathbf{G}_{fW}(s) = \{\mathbf{C}\boldsymbol{\Omega}(s)\mathbf{B}\mathbf{V}_{\mathbf{w}}\mathbf{B}^T\boldsymbol{\Omega}^T(-s)\mathbf{C}^T[\boldsymbol{\Phi}^-(s)]^{-1}\}_+[\boldsymbol{\Phi}^+(s)]^{-1} \tag{6.5.23}$$

From Eq. (6.5.16), we can write

$$\mathbf{C}\boldsymbol{\Omega}(s)\mathbf{B}\mathbf{V}_{\mathbf{w}}\mathbf{B}^T\boldsymbol{\Omega}^T(-s)\mathbf{C}^T = \boldsymbol{\Phi}^+(s)\boldsymbol{\Phi}^-(s) - \mathbf{V}_{\mathbf{v}} \tag{6.5.24}$$

Thus we can write from Eq. (6.5.23)

$$\mathbf{G}_{fW}(s) = \{\boldsymbol{\Phi}^+(s) - \mathbf{V}_{\mathbf{v}}[\boldsymbol{\Phi}^-(s)]^{-1}\}_+[\boldsymbol{\Phi}^+(s)]^{-1} \tag{6.5.25}$$

By use of a matrix lemma (Appendix C) we can write

$$\begin{aligned}[\mathbf{\Phi}^-(s)]^{-1} &= [\mathbf{V}_v^{1/2} + \mathbf{V}_v^{-1/2}\mathbf{C}\mathbf{V}_{\tilde{x}}\mathbf{\Omega}^T(-s)\mathbf{C}^T]^{-1} \\ &= \mathbf{V}_v^{-1/2} - \mathbf{V}_v^{-1}\mathbf{C}\mathbf{V}_{\tilde{x}}(\mathbf{I} + \mathbf{\Omega}^T(-s)\mathbf{C}^T\mathbf{V}_v^{-1}\mathbf{C}\mathbf{V}_{\tilde{x}})^{-1}\mathbf{\Omega}^T(-s)\mathbf{C}^T\mathbf{V}_v^{-1/2} \\ &= \mathbf{V}_v^{-1/2} - \mathbf{K}^T(-s\mathbf{I} - \mathbf{A} + \mathbf{C}^T\mathbf{K}^T)^{-1}\mathbf{C}^T\mathbf{V}_v^{-1/2}\end{aligned} \tag{6.5.26}$$

where we have used Eq. (6.5.7).

Substituting in Eq. (6.5.25) then yields

$$\begin{aligned}\mathbf{G}_{fW}(s) = {} & [\mathbf{\Phi}^+(s) - \mathbf{V}_v^{1/2} + \mathbf{V}_v\mathbf{K}^T(-s\mathbf{I} - \mathbf{A} + \mathbf{C}^T\mathbf{K}^T)^{-1}\mathbf{C}^T\mathbf{V}_v^{-1/2}]_+ \\ & \times [\mathbf{\Phi}^+(s)]^{-1}\end{aligned} \tag{6.5.27}$$

From Eq. (6.5.15) we see that the first two terms in brackets on the right-hand side of Eq. (6.5.27) are together equal to $\mathbf{C\Omega}(s)\mathbf{K}\mathbf{V}_v^{1/2}$. Since the second term in the braces corresponds to a noncausal function (it has poles in the right half-plane), Eq. (6.5.27) reduces to

$$\mathbf{G}_{fW}(s) = \mathbf{C\Omega}(s)\mathbf{K}\mathbf{V}_v^{1/2}[\mathbf{V}_v^{1/2} + \mathbf{C\Omega}(s)\mathbf{K}\mathbf{V}_v^{1/2}]^{-1} \tag{6.5.28}$$

Use of the matrix lemma as in Eq. (6.5.26), then yields

$$\begin{aligned}\mathbf{G}_{fW}(s) &= \mathbf{C\Omega}(s)\mathbf{K}\mathbf{V}_v^{1/2}[\mathbf{V}_v^{-1/2} - \mathbf{V}_v^{-1/2}\mathbf{C}(s\mathbf{I} - \mathbf{A} + \mathbf{KC})^{-1}\mathbf{K}] \\ &= \mathbf{C\Omega}(s)[\mathbf{I} - \mathbf{KC}(s\mathbf{I} - \mathbf{A} + \mathbf{KC})^{-1}]\mathbf{K} \\ &= \mathbf{C}[s\mathbf{I} - \mathbf{A} + \mathbf{KC}]^{-1}\mathbf{K}\end{aligned} \tag{6.5.29}$$

Comparison with Eq. (6.5.9) shows that the Wiener filter is identical with the Kalman filter.

6.6 NONLINEAR ESTIMATION

We now present a brief discussion of estimation in certain nonlinear signal models. We consider two problems. First, we present algorithms for the approximate estimation of the signals in a nonlinear state-space model using extensions of the Kalman filter algorithms. In the second type of problem, the transmitted signal is a function of the waveform to be estimated. We discuss the use of the MAP estimator when the transmitted signal is observed in Gaussian noise.

6.6.1 Nonlinear State Models

We consider the following message and observation models:

$$\text{Message model:} \quad \begin{aligned}\dot{\mathbf{x}}(t) &= \mathbf{f}[\mathbf{x}(t), t] + \mathbf{g}[\mathbf{x}(t), t]\mathbf{w}(t) \\ \mathbf{y}(t) &= \mathbf{h}[\mathbf{x}(t), t]\end{aligned} \tag{6.6.1}$$

$$\text{Observation model:} \quad \mathbf{z}(t) = \mathbf{y}(t) + \mathbf{v}(t) \text{ for } t \geq t_0 \tag{6.6.2}$$

The statistics of the processes $\mathbf{w}(t)$ and $\mathbf{v}(t)$ and the initial state $\mathbf{x}(t_0)$ are assumed to be

$$\begin{gathered} \mathrm{E}\{\mathbf{x}(t_0)\} = \boldsymbol{\mu}_x(t_0) \qquad \mathrm{E}\{\mathbf{w}(t)\} = \mathrm{E}\{\mathbf{v}(t)\} = \mathbf{0} \\ \operatorname{cov}\{\mathbf{w}(t), \mathbf{w}(\tau)\} = \mathbf{V}_w(t)\delta(t-\tau), \operatorname{cov}\{\mathbf{v}(t), \mathbf{v}(\tau)\} = \mathbf{V}_v(t)\delta(t-\tau) \\ \operatorname{cov}\{\mathbf{w}(t), \mathbf{v}(\tau)\} = \operatorname{cov}\{\mathbf{w}(t), \mathbf{x}(t_0)\} = \operatorname{cov}\{\mathbf{v}(t), \mathbf{x}(t_0)\} = \mathbf{0} \end{gathered} \tag{6.6.3}$$

For mean square estimation, the estimate based on the set of observations over the interval $[t_0, t_f]$ can be obtained as the mean of the conditional density functions of $\mathbf{x}(t)$ or $\mathbf{y}(t)$. Unfortunately, in most cases, the determination of these density functions is a formidable task. The conditional density functions are usually obtained as the solutions to certain partial differential equations. These equations have been derived by Stratanovich [24] and by Kushner [25,26] for the case when all the processes involved are Gaussian and by Snyder [27] for Poisson processes. The solution of these equations, however, is not straightforward. While a representation of the estimator using the innovations approach has been obtained in [28], the determination of the actual filter is still difficult in most cases. In most applications we are therefore constrained to seek an approximate solution to our estimation problem. In this section we discuss two approximate schemes which have a simple structure and have been widely applied.

6.6.2 Linearized Kalman Filter

A common technique for solving nonlinear problems is to linearize around a known (nominal) trajectory. The perturbation equations so obtained will be linear and hence an approximate solution to the nonlinear problem can be easily obtained as the nominal value plus the perturbation term.

Let us denote by $\mathbf{x}_n(t)$ the solution to Eq. (6.6.1) with $\mathbf{w}(t) = 0$. That is,

$$\dot{\mathbf{x}}_n(t) = \mathbf{f}(\mathbf{x}_n(t), t) \qquad \mathbf{x}_n(t_0) = \boldsymbol{\mu}_x(t_0) \tag{6.6.4}$$

Let

$$\mathbf{y}_n(t) = \mathbf{h}[\mathbf{x}_n(t), t] \tag{6.6.5}$$

and

$$\delta\mathbf{z}(t) = \mathbf{z}(t) - \mathbf{y}_n(t) \tag{6.6.6}$$

If we write $\mathbf{x}(t)$ as $\mathbf{x}_n(t) + \delta\mathbf{x}(t)$, it can easily be verified that $\delta\mathbf{x}(t)$ satisfies the equation

$$\begin{aligned} \delta\dot{\mathbf{x}}(t) &= \mathbf{f}(\mathbf{x}_n(t) + \delta\mathbf{x}(t), t) - \mathbf{f}(\mathbf{x}_n(t), t) \\ &\quad + \mathbf{g}[\mathbf{x}_n(t) + \delta\mathbf{x}(t), t]\mathbf{w}(t) \end{aligned}$$

Assuming that $\mathbf{f}(\cdot)$ and $\mathbf{g}(\cdot)$ are differentiable in their arguments, we can expand $\mathbf{f}(\cdot,\cdot)$ and $\mathbf{g}(\cdot,\cdot)$ in Taylor's series about $\mathbf{x}_n(t)$. If we assume that $\delta\mathbf{x}(t)$ is small enough, we can neglect higher-order terms and write

$$\delta\dot{\mathbf{x}}(t) \approx \frac{\partial \mathbf{f}[\mathbf{x}_n(t), t]}{\partial \mathbf{x}_n(t)} \delta\mathbf{x}(t) + \mathbf{g}[\mathbf{x}_n(t), t]\mathbf{w}(t) \tag{6.6.7}$$

Similarly substituting Eqs. (6.6.2) and (6.6.5) in Eq. (6.6.6), expanding $\mathbf{h}(\mathbf{x}(t), t)$ in a Taylor's series about $\mathbf{x}_n(t)$ and neglecting higher-order terms yields

$$\delta\mathbf{z}(t) \simeq \frac{\partial \mathbf{h}[\mathbf{x}_n(t), t]}{\partial \mathbf{x}_n(t)} \delta\mathbf{x}(t) + \mathbf{v}(t) \tag{6.6.8}$$

Equations (6.6.7) and (6.6.8) are now in a form where the linear Kalman filter algorithms can be used to obtain an estimate of $\delta\mathbf{x}(t)$. From Table 6.4.2, we can write

$$\delta\dot{\hat{x}}(t \mid t) = \frac{\partial \mathbf{f}[\mathbf{x}_n(t), t]}{\partial \mathbf{x}_n(t)} \delta\hat{\mathbf{x}}(t \mid t) + \mathbf{K}(t)\left[\delta\mathbf{z}(t) - \frac{\partial \mathbf{h}[\mathbf{x}_n(t), t]}{\partial \mathbf{x}_n(t)} \delta\hat{\mathbf{x}}(t \mid t)\right] \tag{6.6.9}$$

where

$$\mathbf{K}(t) = \mathbf{V}_{\delta\tilde{\mathbf{x}}}(t \mid t) \frac{\partial \mathbf{h}^T[\mathbf{x}_n(t), t]}{\partial \mathbf{x}_n(t)} \mathbf{V}_{\mathbf{v}}^{-1}(t) \tag{6.6.10}$$

and

$$\begin{aligned}\frac{d}{dt}\mathbf{V}_{\delta\tilde{\mathbf{x}}}(t \mid t) = {} & \frac{\partial \mathbf{f}[\mathbf{x}_n(t), t]}{\partial \mathbf{x}_n(t)} \mathbf{V}_{\delta\tilde{\mathbf{x}}}(t \mid t) + \mathbf{V}_{\delta\tilde{\mathbf{x}}}(t \mid t) \frac{\partial \mathbf{f}^T(\mathbf{x}_n(t), t]}{\partial \mathbf{x}_n(t)} \\ & - \mathbf{V}_{\delta\tilde{\mathbf{x}}}(t \mid t) \frac{\partial \mathbf{h}^T[\mathbf{x}_n(t), t]}{\partial \mathbf{x}_n(t)} \mathbf{V}_{\mathbf{v}}^{-1} \frac{\partial \mathbf{h}[\mathbf{x}_n(t), t]}{\partial \mathbf{x}_n(t)} \mathbf{V}_{\delta\tilde{\mathbf{x}}}(t \mid t) \\ & + \mathbf{g}[\mathbf{x}_n(t), t]\mathbf{V}_{\mathbf{w}}(t)\mathbf{g}^T[\mathbf{x}_n(t), t]\end{aligned} \tag{6.6.11}$$

where

$$\delta\tilde{\mathbf{x}}(t \mid t) = \delta\mathbf{x}(t) - \delta\hat{\mathbf{x}}(t \mid t)$$

and

$$\mathbf{V}_{\delta\tilde{\mathbf{x}}}(t \mid t) = \text{cov}[\delta\tilde{\mathbf{x}}(t \mid t)]$$

These equations are known as the *linearized Kalman algorithms*. The estimate $\hat{\mathbf{x}}(t \mid t)$ can be determined as

$$\hat{\mathbf{x}}(t \mid t) = \mathbf{x}_n(t) + \delta\hat{\mathbf{x}}(t \mid t)$$

6.6.3 Extended Kalman Filter Algorithm

We can now write an expression for $\dot{\hat{\mathbf{x}}}(t \mid t)$ by adding Eqs. (6.6.4) and (6.6.9). We then have

$$\begin{aligned}\dot{\hat{\mathbf{x}}}(t \mid t) &= \dot{\mathbf{x}}_n(t) + \delta\dot{\hat{\mathbf{x}}}(t \mid t) \\ &= \mathbf{f}(\mathbf{x}_n(t), t) + \frac{\partial \mathbf{f}[\mathbf{x}_n(t), t]}{\partial \mathbf{x}_n(t)} \delta\hat{\mathbf{x}}(t \mid t) \\ &\quad + \mathbf{K}(t)\left[\mathbf{z}(t) - \mathbf{y}_n(t) - \frac{\partial \mathbf{h}[\mathbf{x}_n(t), t]}{\partial \mathbf{x}_n(t)} \delta\hat{\mathbf{x}}(t \mid t)\right]\end{aligned} \tag{6.6.12}$$

where we have used Eq. (6.6.6).

By noting that

$$\mathbf{f}(\hat{\mathbf{x}}(t\,|\,t),t) \simeq \mathbf{f}(\mathbf{x}_n(t),t) + \frac{\partial \mathbf{f}[\mathbf{x}_n(t),t]}{\partial \mathbf{x}_n(t)}\delta\hat{\mathbf{x}}(t\,|\,t)$$

and

$$\mathbf{h}(\hat{\mathbf{x}}(t\,|\,t),t)) \simeq \mathbf{h}(\mathbf{x}_n(t),t) + \frac{\partial \mathbf{h}[\mathbf{x}_n(t),t]}{\partial \mathbf{x}_n(t)}\delta\hat{\mathbf{x}}(t\,|\,t)$$

we can write Eq. (6.6.12) as

$$\dot{\hat{\mathbf{x}}}(t\,|\,t) = \mathbf{f}(\hat{\mathbf{x}}(t\,|\,t),t) + \mathbf{K}(t)[\mathbf{z}(t) - \mathbf{h}(\hat{\mathbf{x}}(t\,|\,t)] \tag{6.6.13}$$

Since

$$\begin{aligned}\tilde{\mathbf{x}}(t\,|\,t) &= \mathbf{x}(t) - \hat{\mathbf{x}}(t\,|\,t) = [\mathbf{x}_n(t) + \delta\mathbf{x}(t)] - [\mathbf{x}_n(t) + \delta\hat{\mathbf{x}}(t\,|\,t)] \\ &= \delta\tilde{\mathbf{x}}(t\,|\,t)\end{aligned} \tag{6.6.14}$$

the gain and variance equations (6.6.10) and (6.6.11) can also be written in terms of $\tilde{\mathbf{x}}(t\,|\,t)$ as follows:

$$\mathbf{K}(t) = \mathbf{V}_{\tilde{\mathbf{x}}}(t\,|\,t)\frac{\partial \mathbf{h}^T[\hat{\mathbf{x}}(t\,|\,t),t]}{\partial \hat{\mathbf{x}}(t\,|\,t)}\mathbf{V}_{\mathbf{v}}^{-1}(t) \tag{6.6.15}$$

and

$$\begin{aligned}\dot{\mathbf{V}}_{\tilde{\mathbf{x}}}(t\,|\,t) = &\frac{\partial \mathbf{f}[\hat{\mathbf{x}}(t\,|\,t),t]}{\partial \hat{\mathbf{x}}(t\,|\,t)}\mathbf{V}_{\tilde{\mathbf{x}}}(t\,|\,t) + \mathbf{V}_{\tilde{\mathbf{x}}}(t\,|\,t)\frac{\partial \mathbf{f}^T[\hat{\mathbf{x}}(t\,|\,t),t]}{\partial \hat{\mathbf{x}}(t\,|\,t)} \\ &- \mathbf{V}_{\tilde{\mathbf{x}}}(t\,|\,t)\frac{\partial \mathbf{h}^T[\hat{\mathbf{x}}(t\,|\,t),t]}{\partial \hat{\mathbf{x}}(t\,|\,t)}\mathbf{V}_{\mathbf{v}}^{-1}(t)\frac{\partial \mathbf{h}[\hat{\mathbf{x}}(t\,|\,t),t]}{\partial \hat{\mathbf{x}}(t\,|\,t)}\mathbf{V}_{\tilde{\mathbf{x}}}(t\,|\,t) \\ &+ \mathbf{g}[\hat{\mathbf{x}}(t\,|\,t),t]\mathbf{V}_{\mathbf{w}}(t)\mathbf{g}^T[\hat{\mathbf{x}}(t\,|\,t),t]\end{aligned} \tag{6.6.16}$$

where we have used the fact that to first order

$$\frac{\partial \mathbf{f}[\mathbf{x}_n(t),t]}{\partial \mathbf{x}_n(t)} \simeq \frac{\partial \mathbf{f}[\hat{\mathbf{x}}(t\,|\,t),t]}{\partial \hat{\mathbf{x}}(t\,|\,t)}$$

and similarly for

$$\frac{\partial \mathbf{h}(\hat{\mathbf{x}}(t\,|\,t),t)}{\partial \hat{\mathbf{x}}(t\,|\,t)}$$

Equations (6.6.13), (6.6.15), and (6.6.16) constitute a complete set of approximate algorithms for evaluating the estimate $\hat{\mathbf{x}}(t\,|\,t)$. These algorithms are known as the *extended Kalman algorithms* and are summarized in Table 6.6.1. The algorithms can be thought of as being obtained from the linearized algorithms by choosing $\hat{\mathbf{x}}(t\,|\,t)$ as the nominal trajectory. They also represent a first-order approximate solution to the partial differential equations for the conditional density corresponding to this estimation problem [3–5].

In our derivation of the linearized algorithms, we assumed that the perturba-

TABLE 6.6.1 EXTENDED KALMAN FILTER ALGORITHM: CONTINUOUS TIME

Message model:

$$\dot{\mathbf{x}}(t) = \mathbf{f}[\mathbf{x}(t), t] + \mathbf{g}[\mathbf{x}(t), t]\mathbf{w}(t)$$
$$\mathbf{y}(t) = \mathbf{h}[\mathbf{x}(t), t] \tag{6.6.1}$$

Observations:

$$\mathbf{z}(t) = \mathbf{y}(t) + \mathbf{v}(t) \tag{6.6.2}$$

Prior statistics:

$$\begin{aligned} E\{\mathbf{w}(t)\} &= E\{\mathbf{v}(t)\} = \mathbf{0} \\ \text{cov}\{\mathbf{w}(t), \mathbf{w}(\tau)\} &= \mathbf{V}_{\mathbf{w}}(t)\delta(t-\tau) \\ \text{cov}\{\mathbf{v}(t), \mathbf{v}(\tau)\} &= \mathbf{V}_{\mathbf{v}}(t)\delta(t-\tau) \\ \text{cov}\{\mathbf{w}(t), \mathbf{v}(\tau)\} &= \text{cov}\{\mathbf{w}(t), \mathbf{x}(t_0)\} = \text{cov}\{\mathbf{v}(t), \mathbf{x}(t_0)\} = \mathbf{0} \end{aligned} \tag{6.6.3}$$

Filter algorithm:

$$\dot{\hat{\mathbf{x}}}(t \mid t) = \mathbf{f}(\hat{\mathbf{x}}(t \mid t), t) + \mathbf{K}(t)[\mathbf{z}(t) - \mathbf{h}(\hat{\mathbf{x}}(t \mid t), t)] \tag{6.6.13}$$

Filter gain:

$$\mathbf{K}(t) = \mathbf{V}_{\tilde{\mathbf{x}}}(t \mid t) \frac{\partial \mathbf{h}^T[\hat{\mathbf{x}}(t \mid t), t]}{\partial \hat{\mathbf{x}}(t \mid t)} \mathbf{V}_{\mathbf{v}}^{-1}(t) \tag{6.6.15}$$

Error variance:

$$\begin{aligned} \dot{\mathbf{V}}_{\tilde{\mathbf{x}}}(t \mid t) = {} & \frac{\partial \mathbf{f}[\hat{\mathbf{x}}(t \mid t), t]}{\partial \hat{\mathbf{x}}(t \mid t)} \mathbf{V}_{\tilde{\mathbf{x}}}(t \mid t) + \mathbf{V}_{\tilde{\mathbf{x}}}(t \mid t) \frac{\partial \mathbf{f}^T[\mathbf{x}(t \mid t), t]}{\partial \hat{\mathbf{x}}(t \mid t)} \\ & - \mathbf{V}_{\tilde{\mathbf{x}}}(t \mid t) \frac{\partial \mathbf{h}^T[\hat{\mathbf{x}}(t \mid t), t]}{\partial \hat{\mathbf{x}}(t \mid t)} \mathbf{V}_{\mathbf{v}}^{-1}(t) \frac{\partial \mathbf{h}[\hat{\mathbf{x}}(t \mid t), t]}{\partial \hat{\mathbf{x}}(t \mid t)} \mathbf{V}_{\tilde{\mathbf{x}}}(t \mid t) \\ & + \mathbf{g}[\hat{\mathbf{x}}(t \mid t), t] \mathbf{V}_{\mathbf{w}}(t) \mathbf{g}^T[\hat{\mathbf{x}}(t \mid t), t] \end{aligned} \tag{6.6.16}$$

Initial conditions:

$$\hat{\mathbf{x}}(t_0) = E\{\mathbf{x}(t_0)\} = \boldsymbol{\mu}_{\mathbf{x}}(t_0) \text{ (known)}$$
$$\mathbf{V}_{\tilde{\mathbf{x}}}(t_0) = \mathbf{V}_{\tilde{\mathbf{x}}_0} \text{ (known)}$$

tion $\delta\mathbf{x}(t)$ around the nominal trajectory $\mathbf{x}_n(t)$ was small. In certain applications, this is not a valid assumption. The linearized algorithms perform poorly under these conditions especially when the variance of the input process $\mathbf{V}_{\mathbf{w}}(t)$ is high. The extended Kalman algorithms yield fairly good estimates when the signal-to-noise ratio is large, and are easy to implement. Higher-order approximate algorithms have been obtained in the literature which yield better estimates, but at the expense of added complexity in the estimator structure. The interested reader is referred to the various excellent texts on estimation theory [3–5] for these algorithms (see also [29]).

The extended Kalman filter algorithms for discrete-time models can be derived along the same lines. The algorithms are summarized in Table 6.6.2.

TABLE 6.6.2 EXTENDED KALMAN FILTER ALGORITHM: DISCRETE TIME

Message model:

$$\begin{aligned} \mathbf{x}(k+1) &= \boldsymbol{\phi}[\mathbf{x}(k), k] + \boldsymbol{\Gamma}[\mathbf{x}(k), k]\mathbf{w}(k) \\ \mathbf{y}(k) &= \mathbf{h}[\mathbf{x}(k), k] \end{aligned} \tag{6.6.17}$$

Observations:

$$\mathbf{z}(k) = \mathbf{h}[\mathbf{x}(k), k] + \mathbf{v}(k) \tag{6.6.18}$$

Prior statistics:

$$\begin{aligned} \mathrm{E}\{\mathbf{w}(k)\} &= \mathrm{E}\{\mathbf{v}(k)\} = \mathbf{0} \\ \mathrm{cov}\{\mathbf{w}(k), \mathbf{w}(j)\} &= \mathbf{V}_{\mathbf{w}}(k)\delta_{kj} \\ \mathrm{cov}\{\mathbf{v}(k), \mathbf{v}(j)\} &= \mathbf{V}_{\mathbf{v}}(k)\delta_{kj} \\ \mathrm{cov}\{\mathbf{w}(k), \mathbf{v}(j)\} &= \mathrm{cov}\{\mathbf{w}(k), \mathbf{x}(k_0)\} = \mathrm{cov}\{\mathbf{v}(k), \mathbf{x}(k_0)\} = 0 \end{aligned} \tag{6.6.19}$$

Filter algorithm:

$$\begin{aligned} \hat{\mathbf{x}}(k+1 \mid k+1) = \hat{\mathbf{x}}(k+1 \mid k) \\ + \mathbf{K}(k+1)\{\mathbf{z}(k+1) - \mathbf{h}[\hat{\mathbf{x}}(k+1 \mid k), k+1]\} \end{aligned} \tag{6.6.20}$$

One-stage prediction:

$$\hat{\mathbf{x}}(k+1 \mid k) = \boldsymbol{\phi}[\hat{\mathbf{x}}(k \mid k), k] \tag{6.6.21}$$

Filter gain:

$$\mathbf{K}(k+1) = \mathbf{V}_{\tilde{\mathbf{x}}}(k+1 \mid k+1)\frac{\partial \mathbf{h}^T[\hat{\mathbf{x}}(k+1 \mid k), k+1]}{\partial \hat{\mathbf{x}}(k+1 \mid k)}\mathbf{V}_{\boldsymbol{\eta}}^{-1}(k+1) \tag{6.6.22}$$

Error variance:

$$\begin{aligned} \mathbf{V}_{\boldsymbol{\eta}}(k+1) = \frac{\partial \mathbf{h}[\hat{\mathbf{x}}(k+1 \mid k), k+1]}{\partial \hat{\mathbf{x}}(k+1 \mid k)}\mathbf{V}_{\tilde{\mathbf{x}}}(k+1 \mid k)\frac{\partial \mathbf{h}^T[\hat{\mathbf{x}}(k+1 \mid k), k+1]}{\partial \hat{\mathbf{x}}(k+1 \mid k)} \\ + \mathbf{V}_{\mathbf{v}}(k+1) \end{aligned} \tag{6.6.23}$$

$$\begin{aligned} \mathbf{V}_{\tilde{\mathbf{x}}}(k+1 \mid k) = \frac{\partial \boldsymbol{\phi}[\hat{\mathbf{x}}(k \mid k), k]}{\partial \hat{\mathbf{x}}(k \mid k)}\mathbf{V}_{\tilde{\mathbf{x}}}(k \mid k)\frac{\partial \boldsymbol{\phi}^T[\hat{\mathbf{x}}(k \mid k), k]}{\partial \hat{\mathbf{x}}(k \mid k)} \\ + \boldsymbol{\Gamma}[\hat{\mathbf{x}}(k \mid k), k]\mathbf{V}_{\mathbf{w}}(k)\boldsymbol{\Gamma}^T[\hat{\mathbf{x}}(k \mid k), k] \end{aligned} \tag{6.6.24}$$

$$\begin{aligned} \mathbf{V}_{\tilde{\mathbf{x}}}(k+1 \mid k+1) = \mathbf{V}_{\tilde{\mathbf{x}}}(k+1 \mid k) - \mathbf{V}_{\tilde{\mathbf{x}}}(k+1 \mid k)\frac{\partial \mathbf{h}^T[\hat{\mathbf{x}}(k+1 \mid k), k+1]}{\partial \hat{\mathbf{x}}(k+1 \mid k)} \\ \times \mathbf{V}_{\boldsymbol{\eta}}^{-1}(k+1)\frac{\partial \mathbf{h}[\hat{\mathbf{x}}(k+1 \mid k), k+1]}{\partial \hat{\mathbf{x}}(k+1 \mid k)}\mathbf{V}_{\tilde{\mathbf{x}}}(k+1 \mid k) \end{aligned} \tag{6.6.25}$$

Initial conditions:

$$\begin{aligned} \hat{\mathbf{x}}(k_0) &= \mathrm{E}\{\mathbf{x}(k_0)\} = \boldsymbol{\mu}_{\mathbf{x}}(k_0) \text{ (known)} \\ \mathbf{V}_{\tilde{\mathbf{x}}}(k_0) &= \mathbf{V}_{\mathbf{x}_0} \text{ (known)} \end{aligned}$$

Example 6.6.1

We consider the estimation of a modulated signal observed in noise. The message and observation models are assumed to be

$$x(k+1) = 0.8x(k) + w(k)$$
$$y(k) = x(k)$$
$$z(k) = A\cos[\omega_0 k + 0.5y(k)] + v(k)$$

The problem, therefore, is one of phase modulation. The processes $w(k)$ and $v(k)$ are zero mean, stationary with unity variance. The estimator equations as obtained from Table 6.6.2 are given by

$$\hat{x}(k+1|k+1) = \hat{x}(k+1|k) + K(k+1)\{z(k+1) - A\cos[\omega_0(k+1) + 0.5\hat{x}(k+1|k)]\}$$
$$\hat{x}(k+1|k) = 0.8\hat{x}(k|k)$$
$$V_\eta(k+1) = 0.25A^2 V_{\tilde{x}}(k+1|k)\sin^2[\omega_0(k+1) + 0.5\hat{x}(k+1|k)] + 1$$
$$K(k+1) = 0.5AV_{\tilde{x}}(k+1|k+1)\sin[\omega_0(k+1) + 0.5\hat{x}(k+1|k)]V_\eta^{-1}(k+1)$$
$$V_{\tilde{x}}(k+1|k) = 0.64V_{\tilde{x}}(k|k) + 1$$
$$V_x(k+1|k+1) = V_{\tilde{x}}(k+1|k) - K^2(k+1)V_\eta^{-1}(k+1)$$

The system was simulated on a digital computer and the mean-squared error obtained by averaging the results of 5000 different runs of the example. The results are shown in Fig. 6.6.1, where the quantity

$$\frac{1}{5000}\sum_{i=1}^{5000}[\tilde{x}^2(k|k)]_i$$

is plotted for three values of A. This quantity is usually referred to as the *actual* error variance, whereas the solutions to Eqs. (6.6.23) or (6.6.24) are referred to as *theoretical* error variances. The actual error variance is a measure of how the estimator performs with actual data. This is especially important with nonlinear models, since the algorithms are only approximately optimal, and the theoretical error variances may be deceptively small. The curves in Fig. 6.6.1 show that actual error variance increases as A increases, thereby indicating that the performance of the estimator is better at the lower values of A. Figure 6.6.1 shows the error in a single run of the experiment for the three values of A. As can be seen from the figure, the error seems to be approximately the same in all three cases.

6.6.4 Nonlinear MAP Estimation

We consider the problem of estimating a random signal $x(t)$ when the observations are of the form

$$z(t) = y(x(t), t) + v(t) \qquad 0 \leq t \leq T \tag{6.6.26}$$

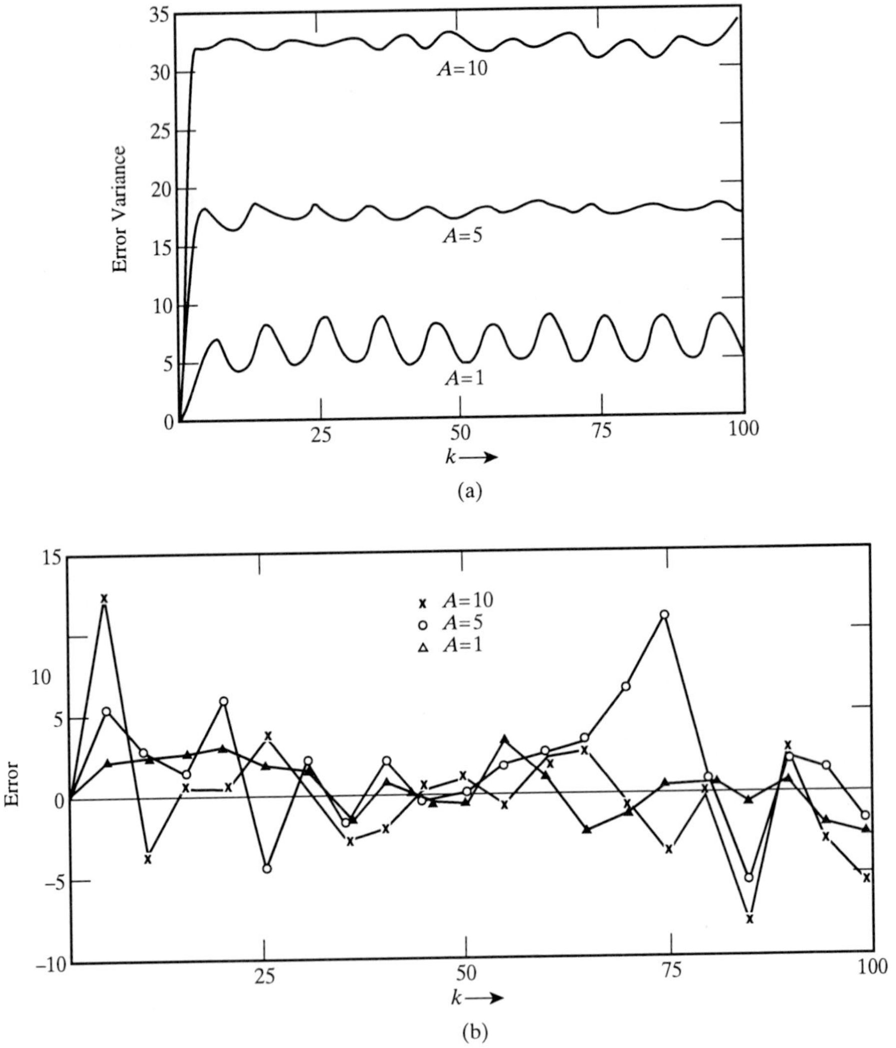

Figure 6.6.1 Extended Kalman filter example: (a) error variance for three values of A; (b) error corresponding to a single run.

where $y(\cdot,\cdot)$ represents the transmitted signal and the additive noise $v(t)$ is zero mean, Gaussian.

The observation model of Eq. (6.6.26) represents a generalization of the models considered in Examples 5.5.2 and 5.5.6 [see Eq. (5.5.18)]. Equation (6.6.26) arises in the study of certain modulation schemes in communication systems [2,30].

Our aim in this section is to obtain an estimate of the signal $x(t)$ based on the observations of $z(t)$ over the interval $[0, T]$. The technique that we use to obtain the estimate $\hat{x}(t)$ is a modification of the MAP procedure used in Example 5.5.6. We derive the estimator equations only for the case when the observation noise $v(t)$ is white, with spectral height $N_0/2$. We leave the extension to the case of colored observation noise to the interested reader.

As in Example 5.5.2, the first step in deriving the estimator equations is to sample the observation $z(t)$ so as to obtain N samples in the interval $[0, T]$. The sampled observations at instant t_k are given by

$$z_k = y_k + v_k \tag{6.6.27}$$

where

$$z_k = z(t_k), y_k = y(x_k, t_k), v_k = v(t_k)$$

and

$$x(t_k) = x_k \tag{6.6.28}$$

Let

$$\mathbf{z}_N = [z_1 z_2 \cdots z_N]^T, \mathbf{y}_N = [y_1 y_2 \cdots y_N]^T$$

and

$$\mathbf{x}_N = [x_1 x_2 \cdots x_N]^T$$

We will initially assume that $v(t)$ is lowpass with bandwidth 2Ω. Following an argument similar to the one used in Example 5.5.2, the conditional density $p(\mathbf{z}_N \mid \mathbf{x}_N)$ follows from Eq. (5.5.21) as

$$p(\mathbf{z}_N \mid \mathbf{x}_N) = \left(\frac{1}{2\pi\sigma^2}\right)^{N/2} \exp\left[-\frac{1}{2\sigma^2}\sum_{k=1}^{N}(z_k - y_k)^2\right] \tag{6.6.29}$$

where

$$\sigma^2 = \frac{N_0 \Omega}{2\pi}$$

To obtain the joint density function $p(\mathbf{z}_N, \mathbf{x}_N)$ required for deriving the MAP estimate of $\mathbf{x}_N$, we assume that the samples $\{x_i\}$ are zero-mean and Gaussian distributed so that

$$p(\mathbf{x}_N) = \frac{1}{(2\pi)^{N/2} \det \mathbf{V}_{\mathbf{x}_N}} \exp\left(-\frac{1}{2}\mathbf{x}_N^T \mathbf{V}_{\mathbf{x}_N}^{-1} \mathbf{x}_N\right) \tag{6.6.30}$$

where

$$\mathbf{V}_{\mathbf{x}_N} = \mathrm{E}\{\mathbf{x}_N \mathbf{x}_N^T\} \tag{6.6.31}$$

We note that the (i, j)th element of $\mathbf{V}_{\mathbf{x}_N}$ is given by

$$(\mathbf{V}_{\mathbf{x}_N})_{ij} = \mathrm{E}\{x_i x_j\} = \mathrm{E}\{x(t_i)x(t_j)\} = V_x(t_i, t_j) \tag{6.6.32}$$

Combining Eqs. (6.6.29) and (6.6.30) yields for the joint density function

$$p(\mathbf{z}_N, \mathbf{x}_N) = p(\mathbf{z}_N \mid \mathbf{x}_N)p(\mathbf{x}_N)$$
$$= \frac{1}{(2\pi\sigma)^N \det \mathbf{V}_{\mathbf{x}_N}} \exp\left\{-\frac{1}{2\sigma^2}[\mathbf{z}_N - \mathbf{y}_N]^T[\mathbf{z}_N - \mathbf{y}_N] - \frac{1}{2}\mathbf{x}_N^T \mathbf{V}_{\mathbf{x}_N}^{-1} \mathbf{x}_N\right\} \tag{6.6.33}$$

The MAP equation is obtained by partial differentiation of Eq. (6.6.33) with respect to $\mathbf{x}_N$ and by setting the result equal to zero. Noting that $\mathbf{y}_N$ is explicitly a function of $\mathbf{x}_N$, we obtain

$$\frac{1}{\sigma^2}\frac{\partial^T \mathbf{y}_N}{\partial \mathbf{x}_N}[\mathbf{z}_N - \mathbf{y}_N] - \mathbf{V}_{\mathbf{x}_N}^{-1}\mathbf{x}_N \bigg|_{\mathbf{x}_N = (\hat{\mathbf{x}}_N)_{\text{MAP}}} = 0 \tag{6.6.34}$$

Solving for $\mathbf{x}_N$ yields

$$(\hat{\mathbf{x}}_N)_{\text{MAP}} = \frac{1}{\sigma^2}\mathbf{V}_{\mathbf{x}_N}\frac{\partial^T \mathbf{y}_N}{\partial \mathbf{x}_N}[\mathbf{z}_N - \mathbf{y}_N]\bigg|_{\mathbf{x}_N = (\hat{\mathbf{x}}_N)_{\text{MAP}}} \tag{6.6.35}$$

The Jacobian matrix $\partial\mathbf{y}_N/\partial\mathbf{x}_N$ can easily be shown to be a diagonal matrix

$$\frac{\partial \mathbf{y}_N}{\partial \mathbf{x}_N} = \begin{bmatrix} \frac{\partial y_1}{\partial x_1} & & & 0 \\ & \frac{\partial y_2}{\partial x_2} & & \\ & & \ddots & \\ 0 & & & \frac{\partial y_N}{\partial x_N} \end{bmatrix} \tag{6.6.36}$$

so that Eq. (6.6.35) can be written in component form as

$$(\hat{x}_k)_{\text{MAP}} = \frac{1}{\sigma^2}\sum_{j=1}^{N} V_x(t_k, t_j)\frac{\partial y_j}{\partial x_j}(z_j - y_j)\bigg|_{x_k = (\hat{x}_k)_{\text{MAP}}} \tag{6.6.37}$$

We now let the number of samples, N, become infinity. In the limit the summation becomes an integral and the variance σ^2 becomes an impulse of strength $N_0/2$ (see Example 5.5.2). We therefore get the MAP equation as

$$\hat{x}_{\text{MAP}}(t) = \frac{2}{N_0}\int_0^T V_x(t, \tau)\frac{\partial y(x(\tau), \tau)}{\partial x(\tau)}[z(\tau) - y(\tau)]\,d\tau\bigg|_{\hat{x}_{\text{MAP}}(t)} \tag{6.6.38}$$

We point out that the estimate $\hat{x}_{\text{MAP}}(\cdot)$ in Eq. (6.6.38) refers to the smoothed estimate based on the record of observations over the entire interval $[0, T]$.

6.7 SUMMARY

In this chapter we have considered the estimation of stochastic processes observed in additive noise. We have for the most part restricted ourselves to linear mean-square estimates of these processes. If the processes involved are all Gaussian, the linear estimate is the optimum estimate.

In Section 6.2 the basic concepts of linear mean-square estimation of signal waveforms were presented. For continuous observations over an interval $[t_0, t_f]$, the optimum estimator is a linear time-varying system. The orthogonality principle was used in Section 6.3 to determine the optimum system. It was shown that the optimum system satisfies the Wiener–Hopf integral equation. As in Chapter 4, the solution to this equation is easily obtained if the observation noise is white. If the observation noise is colored, we can use a whitening filter first to convert the observation noise into white noise. The optimum filter operates on the whitened observations to yield the desired estimate. For stationary processes with rational spectra, the whitening filter can be determined by spectral factorization of the spectrum of the observations. These concepts extend to the case of discrete-time processes. For such processes we can also obtain a finite impulse response filter. We also considered the special case of forward–backward prediction. The Levinson–Durbin recursion provides a computationally efficient method of solving the resulting equations in terms of the partial correlation coefficients. Finally, we derived the lattice structure for the forward–backward prediction problem.

For stationary processes, the extension of the Wiener filter to multivariate signals is conceptually straightforward. However, the resulting spectral-factorization problem is not easy to solve.

In Section 6.4 we considered the important problem of estimation in linear state models using the Kalman filter. This algorithmic approach to the estimation problem was derived using the concept of innovations. Both discrete- and continuous-time models were considered. The relation between the Kalman filter and the Wiener filter was explored in Section 6.5.

Nonlinear state models were considered in Section 6.6. In this case the signal process will not be Gaussian and the optimum estimator is no longer linear. Since determination of the optimum estimator is not easy, two approximate methods for estimation were discussed. Essentially, these algorithms were obtained by linearizing about certain nominal trajectories and can be considered as first-order approximations to the optimal estimator.

The chapter concluded with a derivation of the MAP estimator for certain signal models which are of interest in the study of modulation schemes in communication systems.

EXERCISES

6.1. Find the linear mean-square estimate of $y(t + \lambda)$ in terms of $y(t)$ and its first two derivatives. That is,

$$\hat{y}(t + \lambda) = ay(t) + b\dot{y}(t) + c\ddot{y}(t)$$

6.2. Let $y(t)$ be a bandlimited process such that

$$\Phi_y(\omega) = 0 \qquad |\omega| > \omega_c, \quad \omega_c = \frac{\pi}{T}$$

We wish to obtain the mean-square estimate of $y(t)$ in terms of its sample values at instants nT as

$$\hat{y}(t) = \sum_{n=-\infty}^{\infty} a_n(t) y(nT)$$

Show that the mean-square error is minimized if

$$a_n(t) = \frac{\sin \omega_c(t - nT)}{\omega_c(t - nT)}$$

(This is the sampling theorem for random processes.)

6.3. Find the optimum filter to estimate $y(t + \alpha)$ in the presence of noise when the signal and noise spectrums are given by

$$\Phi_y(s) = \frac{-s^2}{-s^2 + 2a^2} \qquad \Phi_v(s) = \frac{a^2}{-s^2 + a^2}$$

Assume that $y(\cdot)$ and $v(\cdot)$ are uncorrelated and that $z(u) = y(u) + v(u), -\infty < u \leq t$. Let $a = 1$. Sketch the minimum mean-square error as a function of α and determine the irreducible error.

6.4. Determine the optimum filter to estimate $y(t)$ given that the signal and noise are uncorrelated and have spectral densities

$$\Phi_y(s) = \frac{-s^2}{(1 - s^2)^2} \qquad \Phi_v(s) = \frac{1}{2}$$

and

$$z(u) = y(u) + v(u) \qquad -\infty < u \leq t$$

6.5. Consider the control problem of Fig. E.6.1, in which the fixed plant has the transfer function

$$G_p(s) = \frac{1}{s(s + 1)}$$

The signal input $y(t)$ has the spectrum

$$\Phi_y(s) = \frac{A^2}{(-s^2 + a^2)(-s^2 + b^2)}$$

and is uncorrelated with the load disturbance $v(t)$, which has a spectrum

$$\Phi_v(s) = \frac{1}{-s^2 + d^2}$$

The gain K_T is assumed to be fixed.

(a) Find the optimum compensator $G_c(s)$ so that the output $c(t)$ follows the signal input in a mean-square sense.

(b) Find $G_c(s)$ if the signal input to the plant $y_p(t)$ is constrained such that

$$\mathrm{E}\{y_p^2(t)\} < B \qquad \text{for } B \text{ a positive constant}$$

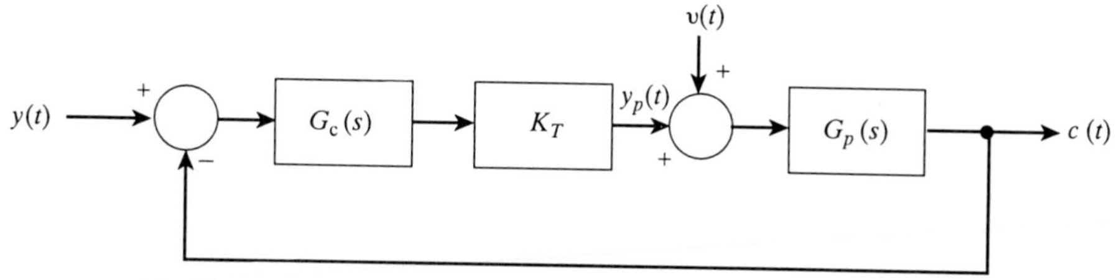

Figure E.6.1

6.6. Find the optimum predictor for estimating $y(t + \alpha)$ when the signal spectrum is

(a) $\Phi_y(s) = \dfrac{2a}{-s^2 + a^2}$ **(b)** $\Phi_y(s) = \dfrac{-s^2 + 1}{s^4 + 2}$

if $\Phi_v(s) = 1$ and $z(u) = y(u) + v(u)$ for $-\infty < u \leq t$.

6.7. The observed signal is

$$z(u) = y(u) + v(u) \qquad -\infty < u \leq t$$

(a) Find the optimum filter to estimate $(d/dt)y(t)$ when the signal and noise spectra are given by

$$\Phi_y(s) = \frac{1}{1 - s^2} \qquad \Phi_v(s) = \frac{1}{2}$$

Assume that the signal and noise are uncorrelated.

(b) Is

$$\frac{\hat{d}}{dt}y(t) = \frac{d}{dt}\hat{y}(t)?$$

6.8. Consider the autoregressive process described by the equation

$$y(k) - 0.8y(k-1) + 0.15y(k-2) = w(k)$$

where $w(k)$ is a zero-mean, white sequence with variance $\sigma_w^2 = 1.5$. Given observations of the form

$$z(k) = y(k) + v(k)$$

where $v(k)$ is a zero-mean, white sequence with variance $\sigma_v^2 = 0.5$ and which is independent of $w(k)$, find the optimal IIR Wiener filter to find the estimate $\hat{y}(k \mid k)$. Find the mean-square error obtainable with this filter.

6.9. Find the optimal third-order transversal filter for estimating $y(k)$ in Exercise 6.8. What is the mean-square error for this estimate?

6.10. Find the coefficients of a third-order forward–backward predictor for the signal $z(k)$ of Exercise 6.8. What are the coefficients of the lattice implementation of this filter?

6.11. Use the lattice filter of Exercise 6.10 as a whitening filter for the observation sequence $z(k)$ of Exercise 6.8 to find the optimal third-order estimator for the signal $y(k)$. How does this filter compare with the filter you obtained in Exercise 6.9?

6.12. Derive Eq. (6.3.95) for the order update of the backward-prediction error $\epsilon_b^m(k)$.

6.13. Show that the expressions for p_{m+1} in Eqs. (6.3.97) and (6.3.98) are equivalent to the one in Eq. (6.3.87).

6.14. We noted in the text that for fixed k, the backward errors $\epsilon_b^i(k), 0 \leq i \leq m$ were orthogonal to one another. Show that for the forward prediction errors

$$\mathrm{E}\{\epsilon_f^i(k)\,\epsilon_f^j(k)\} = \overline{C}_f^j \qquad \text{if } i \geq j$$

so that these errors are not orthogonal.

6.15. In Eqs. (6.3.97) and (6.3.98), we presented alternative but equivalent expressions for p_{m+1}. Since the denominators are equal, we can combine the two expressions to get

$$p_{m+1} = -\frac{\mathrm{E}\{\epsilon_f^m(k+1)\epsilon_b^m(k)\}}{\alpha\mathrm{E}\{[\epsilon_f^m(k+1)]^2\} + (1-\alpha)\mathrm{E}\{[\epsilon_b^m(k)]^2\}}$$

where α is a constant taking on values between zero and 1.

Show that this expression results from minimizing the cost function

$$(1-\alpha)\mathrm{E}\{[\epsilon_f^{m+1}(k+1)]^2\} + \alpha\mathrm{E}\{[\epsilon_b^{m+1}(k+1)]^2\}$$

This expression, for the case $\alpha = \frac{1}{2}$, is usually referred to as the *Burg formula* [31].

6.16. Consider the following message and observation models:

$$\dot{x}_1(t) = x_2(t)$$

$$\dot{x}_2(t) = -2x_1(t) - 3x_2(t) + w(t)$$

where $w(t)$ is zero-mean, white, with unit variance. Let

$$y(t) = x_1(t)$$

and

$$z(u) = 2y(u) + v(u) \qquad t_0 \leqslant u \leqslant t$$

where $v(t)$ is zero-mean, white, with variance 4, and uncorrelated with $w(t)$.

(a) Determine the Kalman filter equations for this problem.

(b) Find the estimator for $t_0 = -\infty$. Verify that this estimator is the same as the Wiener filter for this problem.

6.17. In our derivation of the Kalman filter equations, we assumed that the input $\mathbf{w}(t)$ and the observation noise $\mathbf{v}(t)$ were uncorrelated. Derive the corresponding equations for the continuous-time problem when

$$\mathrm{E}\{\mathbf{w}(t)\mathbf{v}^T(\tau)\} = \mathbf{V}_{wv}(t)\delta(1-\tau)$$

[Note that in this case $\mathbf{x}(t)$ and $\mathbf{v}(t)$ will not be uncorrelated.]

6.18. Consider the following signal and observation models:

$$\dot{x}_1(t) = x_2(t)$$

$$\dot{x}_2(t) = -x_1(t) - 3x_2(t) + w(t)$$

where $w(t)$ is zero mean, white, with unity variance and

$$z(t) = x_1(t) + v(t)$$

where $v(t)$ is nonwhite (colored) noise. Suppose that $v(t)$ can be modeled as the output of a linear system driven by white noise:

$$\dot{v}(t) = -v(t) + \xi(t)$$

where $\xi(t)$ is zero mean, white, with unity covariance and uncorrelated with $w(t)$. We can then form an augmented message model by letting $v(t) = x_3(t)$, so that we have

$$\dot{x}_1(t) = x_2(t)$$
$$\dot{x}_2(t) = -x_1(t) - 3x_2(t) + w(t)$$
$$\dot{x}_3(t) = -x_3(t) + \xi(t)$$

However, the message model now becomes

$$z(t) = x_1(t) + x_3(t)$$

and is hence noise-free. Since the continuous-time Kalman filter equations assume that all observations are corrupted by white noise (why?), we have to modify our observations to introduce a white component. We therefore use the derived observation $\dot{z}(t)$ since

$$\dot{z}(t) = \dot{x}_1(t) + \dot{x}_3(t)$$
$$= x_2(t) - x_3(t) + \xi(t)$$

and has a white component.

(a) Derive the filter equations for this problem using the augmented message model and the derived observations. Note that the input and observation noises are correlated in this formulation.

(b) Derive the equations for the general model of Eqs. (6.4.71) and (6.4.1). Assume that $\mathbf{v}(t)$ is colored and can be modeled as

$$\dot{\mathbf{v}}(t) = \mathbf{D}\mathbf{v}(t) + \xi(t)$$

where $\xi(t)$ is zero-mean, white, with

$$\mathrm{E}\{\xi(t)\xi^T(s)\} = \mathbf{V}_\xi(t)\delta(t - s), \mathrm{E}\{\xi(t)\mathbf{w}^T(s)\} = 0$$

6.19. Derive the estimation algorithms for the colored noise problem for discrete-time models. Is there a difference between the continuous- and discrete-time problems?

6.20. Consider the message model of Eq. (6.4.71) and the observation model

$$\mathbf{z}(u) = m(u)\mathbf{y}(u) + \mathbf{v}(u) \qquad t_0 \leq u \leq t$$

where $m(t)$ is a scalar white process with nonzero mean $\mu(t)$ and covariance kernel $V_m(t)$. Determine the linear minimum mean-square estimation algorithms for this problem. Show that they reduce to the Kalman filter algorithms when $\mu(t) = 1$ and $V_m(t) = 0$.

6.21. Consider the message and observation models for the system

$$\dot{x}(t) = -x(t) + w(t)$$
$$z(t) = A\cos[w_0 t + 0.5x(t)] + v(t)$$

The input $w(t)$ is zero mean, white, with unity variance as is the observation noise $v(t)$.

(a) Obtain the linearized Kalman filter and extended Kalman filter algorithms for this problem.

(b) Simulate the system on a computer and compare the performance of the algorithms for values of $A = 0.01$, 1, and 100.

6.22. Consider the following message and observation models in which α is an unknown constant parameter.

$$\dot{x}_1(t) = x_2(t)$$
$$\dot{x}_2(t) = -x_1(t) - \alpha x_2(t) + w(t)$$
$$z(t) = x_1(t) + v(t)$$

where $w(t)$ and $v(t)$ are zero-mean, white, with unity variance, and uncorrelated with each other. We can estimate the states and simultaneously identify the parameter α by considering $\alpha(t) = x_3(t)$ as a state and noting that

$$\dot{x}_3(t) = 0$$

(a) Set up the extended Kalman filter algorithms for this problem.
(b) Simulate the system on a computer to determine the effectiveness of the algorithm.

6.23. Set up the MAP equation for the estimation problem of Exercise 6.21.

6.24. Derive the MAP equation analogous to Eq. (6.6.38) for colored observation noise.

6.25. The output in a discrete-time, single-input/single-output system is modeled by the *autoregressive* (AR) equation

$$y(k) = \sum_{i=1}^{N} a_i y(k-i) + bw(k)$$

where $\{a_i\}$ and b are known constants. Assume that $w(k)$ is zero mean, Gaussian, with $E\{w(k)w(j)\} = \sigma_w^2 \delta_{kj}$. Let the observation model be

$$z(k) = y(k) + v(k)$$

where $v(\cdot)$ is zero mean, Gaussian, with $E\{v(k)v(j)\} = \sigma_v^2 \delta_{kj}$, uncorrelated with $w(\cdot)$. Find the estimate $\hat{y}(k \mid k-1)$ [32].

6.26. *Uncertain observations:* In some applications we are not sure whether the observation is signal plus noise or noise alone. We can formulate the problem in terms of two hypotheses as

$$H_1\colon z(k) = y(k) + v(k) \qquad k = 1, 2, \ldots, N$$
$$H_0\colon z(k) = v(k)$$

The posterior density function of $y(k)$ can then be written as

$$p(y(k) \mid \mathbf{z}) = p(y(k) \mid \mathbf{z}, H_1)P(H_1 \mid \mathbf{z}) + p(y(k) \mid \mathbf{z}, H_0)P(H_0 \mid \mathbf{z})$$

Find the conditional mean estimate of $y(k)$ if all the processes are Gaussian.

6.27. *Divergence:* The Kalman filter algorithms assume that the system model and priors are known exactly. If there is a mismatch between the assumed model and the actual system, the estimates may become biased and the actual mean-square error may be very large even though the theoretical error variance obtained by solving the error variance equation may be small. We illustrate this in this problem.

Assume that the actual message model is

$$x_a(k+1) = x_a(k) + m$$

where m is a constant. The observations are given by

$$z(k) = x_a(k) + v(k)$$

We design the estimator, however, by assuming that the message and observation models are

$$x_d(k + 1) = x_d(k)$$
$$z(k) = x_d(k) + v(k)$$

Let $\tilde{x}(k \mid k)$ denote the actual error in the estimate

$$\tilde{x}(k \mid k) = x_a(k) - \hat{x}_d(k \mid k)$$

Determine expressions for the propagation of the variance of $\tilde{x}(k \mid k)$. How does the actual error variance $V_{\tilde{x}}(k \mid k)$ compare with the theoretical error variance $V_{\tilde{x}_d}(k \mid k)$ at the end of N stages?

6.28. Consider the nonlinear message model of Eq. (6.6.1):

$$\dot{x}(t) = \mathbf{f}(\mathbf{x}(t), t) + \mathbf{g}(\mathbf{x}(t), t)\mathbf{w}(t)$$

but suppose that the observations are discrete, of the form

$$\mathbf{z}(k) = \mathbf{h}[\mathbf{x}(k), k] + \mathbf{v}(k)$$

where $x(k) = x(kT)$ for some T. Derive approximate algorithms for obtaining the estimate for this model.

6.29. Use the results of Exercise 6.28 to develop approximate estimation algorithms for the following problem. A radar located at a height RA from ground level observes the position of a freely falling object with constant drag coefficient. The vertical line of fall of the body is displaced from the radar location by a distance RD. Assume that gravity is neglected, and an exponential atmosphere is assumed (see Fig. E.6.2).

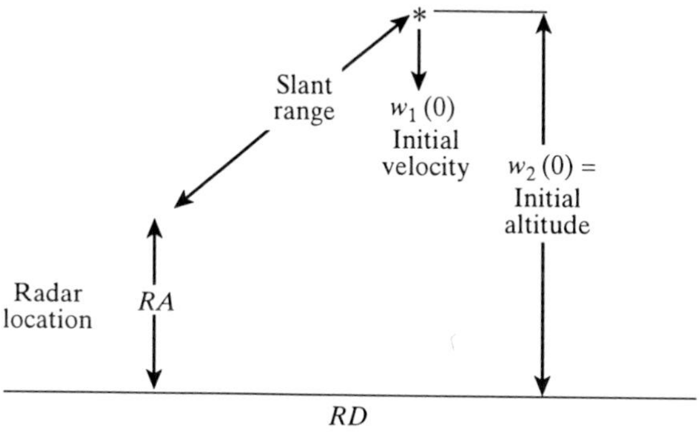

Figure E.6.2

The state equations corresponding to the altitude, velocity, and drag coefficients are modeled as

$$\dot{w}_1(t) = -w_2(t)$$
$$\dot{w}_2(t) = -k_1 \exp[-k_2 w_1(t) w_2^2(t) w_3(t)]$$
$$\dot{w}_3(t) = 0$$

The observed quantity is the slant range, so that the observations are

$$z(k) = [RD^2 + (w_1(k) - RA)^2]^{1/2} + v(k)$$

where $v(k)$ is an additive white noise sequence. Obtain the algorithm to estimate the states of the system. Simulate the system for

(a) $k_1 = 0$ (no drag), $RA = 250{,}000$ ft
$RD = 100{,}000$ ft (linear dynamics, nonlinear observations)

(b) $k_1 = 0.5$, $k_2 = 1$, but $RA = RD = 0$ (nonlinear dynamics, linear observation).

Assume the initial conditions:

$$w_1(0) = 300{,}000 \text{ ft}$$

$$w_2(0) = 20{,}000 \text{ ft/sec}$$

$$w_3(0) = 2 \times 10^{-2} \text{ ft/lb}$$

$$V_{\tilde{x}}(0) = \begin{bmatrix} 500 & 0 & 0 \\ 0 & 2 \times 10^4 & 0 \\ 0 & 0 & 6 \times 10^{-4} \end{bmatrix}$$

The noise variance is constant and equals 100 ft^2. In your simulations, compute the sampled mean, and sampled mean-square error using Monte Carlo methods. Compare your results with the actual trajectories obtained by numerically solving the system equations (refer to [33] for a detailed exposition).

6.30. Derive the extended discrete-time Kalman filter algorithms of Table 6.6.2. [*Hint:* Expand the state equations in a Taylor series about $\hat{x}(k \mid k)$, and the observation about $\hat{x}(k \mid k-1)$.]

6.31. *Adaptive filter:* Consider the discrete-time model

$$\mathbf{x}(k+1) = \boldsymbol{\phi}\mathbf{x}(k) + \boldsymbol{\Gamma}\mathbf{w}(k)$$

with the observations

$$\mathbf{z}(k) = \mathbf{C}\mathbf{x}(k) + \mathbf{v}(k)$$

with usual assumptions on $\mathbf{w}(k)$ and $\mathbf{v}(k)$. However, we do not have exact knowledge of the matrices $\boldsymbol{\phi}$, $\boldsymbol{\Gamma}$, and $\mathbf{C}$. We assume, instead, that these matrices may be denoted by a vector parameter $\boldsymbol{\theta}$ taking one of N values, $\boldsymbol{\theta}_1, \boldsymbol{\theta}_2, \ldots, \boldsymbol{\theta}_N$ with equal probabilities. In other words, we quantize the values the unknown parameters can take. Show that the conditional mean estimate of the state can be expressed as

$$\hat{\mathbf{x}}(k \mid k) = \sum_{i=1}^{N} \hat{\mathbf{x}}_i(k \mid k) P(\boldsymbol{\theta}_i \mid \mathbf{Z}(k))$$

where $\hat{\mathbf{x}}_i(k \mid k)$ is the conditional mean estimate of the system assuming that the model used corresponds to $\boldsymbol{\theta}_i$. $P(\boldsymbol{\theta}_i \mid \mathbf{Z}(k)$ is the posterior probability that $\boldsymbol{\theta}_i$ is the true vector parameter after observing $[\mathbf{z}(1), \mathbf{z}(2), \ldots, \mathbf{z}(k)] \underline{\Delta} Z(k)$. Obtain the complete set of algorithms, and draw the block diagram of the estimator structure. What difficulties do you visualize in a practical implementation?

6.32. In this exercise, we consider the smoothing problem. We recall that the smoothing problem involves finding the estimate of a signal $y(t)$ given observations over the interval $[t_0, t_f]$, where $t_0 \leq t \leq t_f$. We can distinguish three types of smoothing [5]. If t_0 and t_f are fixed, we have the fixed-interval smoothing problem. If t_0 and t are fixed, but t_f is increasing, we have the fixed-point smoothing problem. If $t = t_f - \Delta$ where $\Delta > 0$ is a fixed number and if t_f is increasing, we have fixed-lag smoothing.

We consider the message and observation models (6.4.71) and (6.4.1). We recall from Eq. (6.4.65) that we can write

$$\hat{\mathbf{x}}(t \mid t_f) = \int_{t_0}^{t_f} \mathrm{E}\{x(t)v^T(\tau)\}V_{\boldsymbol{\eta}}^{-1}(\tau)\boldsymbol{\eta}(\tau)\, d\tau$$

It follows from Eq. (6.4.73) that for $t \leq t_f$, we can write the preceding as

$$\hat{\mathbf{x}}(t \mid t_f) = \hat{\mathbf{x}}(t \mid t) + \int_{t}^{t_f} \mathrm{E}\{x(t)\boldsymbol{\eta}^T(\tau)\}V_{\boldsymbol{\eta}}^{-1}(\tau)\boldsymbol{\eta}(\tau)\, d\tau$$

Let $\psi_c(t, \tau)$ denote the state transition matrix corresponding to the matrix $[A(t) - K(t)C(t)]$ where $K(t)$ is the Kalman gain for the filtering problem for this model. Now define

$$\lambda(t, t_f) = \int_{t}^{t_f} \psi_c^T(\tau, t)C^T(\tau)V_{\boldsymbol{\eta}}^{-1}(\tau)\boldsymbol{\eta}(\tau)\, d\tau$$

(a) Show that

$$\hat{\mathbf{x}}(t \mid t_f) = \hat{\mathbf{x}}(t \mid t) + \mathbf{V}_{\tilde{\mathbf{x}}}(t \mid t)\lambda(t, t_f)$$

(b) For the three cases of smoothing mentioned earlier, use the equation above to derive explicit algorithms for obtaining $\hat{\mathbf{x}}(t \mid t_f)$. (You may find [34] useful in this regard.)

REFERENCES

1. H. Cramer and M. R. Ledbetter, *Stationary and Related Stochastic Processes*, Wiley, New York, 1967.
2. H. L. Van Trees, *Detection, Estimation and Modulation Theory*, Part I, Wiley, New York, 1968.
3. A. P. Sage and J. L. Melsa, *Estimation Theory with Applications to Communications and Control*, McGraw-Hill, New York, 1971.
4. A. H. Jazwinski, *Stochastic Processes and Filtering Theory*, Academic Press, New York, 1979.
5. T. P. McGarty, *Stochastic Systems and State Estimation*, Wiley-Interscience, New York, 1974.
6. E. Wong, *Stochastic Processes in Information and Dynamical Systems*, McGraw-Hill, New York, 1971.
7. T. Kailath, Ed., *Linear Least-Squares Estimation*, Benchmark Papers in Electrical Engineering and Computer Science, Dowden, Hutchinson & Ross, Stroudsburg, PA, 1977.
8. B. D. O. Anderson and J. B. Moore, *Optimal Filtering*, Prentice Hall, Englewood Cliffs, NJ, 1979.

9. P. S. Maybeck, *Stochastic Models, Estimation and Control*, Vol. 1, Academic Press, New York, 1979.
10. N. Wiener, *The Extrapolation, Interpolation and Smoothing of Stationary Time Series with Engineering Applications,* Wiley, New York, 1962.
11. H. W. Bode and C. E. Shannon, A simplified derivation of linear least-square smoothing and prediction theory, *Proc. IRE*, **38**, April, 417–425 (1950).
12. T. Kailath, *Lectures on Wiener and Kalman Filtering*, Springer-Verlag, New York, 1981.
13. S. S. L. Chang, *Synthesis of Optimum Control Systems*, McGraw-Hill, New York, 1961.
14. D. C. Youla, On the factorization of rational matrices, *IEEE Trans. Inf. Theory*, **IT-7**, No. 3, July, 172–189 (1961).
15. B. D. O. Anderson, An algebraic solution to the spectral factorization problem, *IEEE Trans. Autom. Control*, **AC-12**, No. 4, August, 410–414 (1967).
16. J. M. Mendel, *Lessons in Digital Estimation Theory*, Prentice Hall, Englewood Cliffs, NJ, 1987.
17. C. W. Therrien, *Discrete Random Signals and Statistical Signal Processing*, Prentice Hall, Englewood Cliffs, NJ, 1992.
18. G. J. Bierman, *Factorization Methods for Discrete Sequential Estimation*, Academic Press, New York, 1977.
19. S. Haykin, *Adaptive Filter Theory*, 2nd ed., Prentice Hall, Englewood Cliffs, NJ, 1991.
20. R. E. Kalman, A new approach to linear filtering and prediction problems, *Trans. ASME, J. Basic Eng.*, **82D**, March, 34–45 (1960).
21. R. E. Kalman and R. Bucy, New results in linear filtering and prediction theory, *Trans. ASME, J. Basic Eng.*, **83D**, March, 95–108 (1961).
22. T. Kailath, An innovations approach to least squares estimation, Part I: Linear filtering in additive white noise, *IEEE Trans. Autom. Control*, **AC-13**, No. 6, December, 646–655 (1968).
23. B. D. O. Anderson, Stability properties of Kalman–Bucy filters, *J. Franklin Inst.*, **291**, February, 137–144 (1971).
24. R. L. Stratanovich, On the theory of optimal nonlinear filtering of random functions, *Theory Prob. Appl. (USSR)*, **4**, 223–225 (1959).
25. H. J. Kushner, On the differential equations satisfied by conditional probability densities of Markov processes, with applications, *SIAM J. Control, Ser. A*, **2**, No. 1, 106–119 (1962).
26. H. J. Kushner, Nonlinear filtering: the exact dynamical equations satisfied by the conditional mode, *IEEE Trans. Autom. Control*, **AC-12**, No. 3, June, 262–267 (1967).
27. D. L. Snyder, Filtering and detection for doubly stochastic Poisson processes, *IEEE Trans. Inf. Theory*, **IT-18**, January, 91–102 (1972).
28. P. A. Frost and T. Kailath, An innovations approach to least squares estimation, Part III: Nonlinear estimation in white Gaussian noise, *IEEE Trans. Autom. Control*, **AC-16**, No. 3, June, 217–226 (1971).
29. R. S. Bucy, C. Hecht, and K. D. Senne, An engineer's guide to building nonlinear filters, *Report SRL-TR-72-004*, F.J. Seiler Research Laboratories, 19??.
30. D. J. Sakrison, *Notes on Analog Communication*, Van Nostrand Reinhold, New York, 1970.

31. J. P. Burg, Maximum entropy spectral analysis, Ph.D. thesis, Stanford University, Stanford, CA, 1975.

32. R. L. Kashyap, A new method of recursive estimation in discrete linear systems, *IEEE Trans. Autom. Control*, **AC-15**, No. 11, February, 18–24 (1970).

33. R. P. Wishner, J. A. Tabaczynski, and M. Athans, A comparison of three nonlinear filters, *Automatica*, **5**, 487–496 (1969).

34. T. Kailath and P. Frost, An innovations approach to least squares estimation, Part II: Linear smoothing in additive white noise, *IEEE Trans. Autom. Control*, **AC-13**, No. 6, December, 655–660 (1968).

7

Further Topics in Detection and Estimation

In this chapter we present some further results, beyond the introductory material of Chapters 3 and 5, on detection and estimation. We discuss nonparametric detection, locally optimal detection of weak signals in noise, and robust hypothesis testing and parameter estimation.

7.1 NONPARAMETRIC DETECTION

In Chapter 3 we discussed detection problems in which the probability distributions of the observations under either hypothesis have a known functional form. For example, in testing the mean of a random sample from a Gaussian distribution, we have stated the null hypothesis as mean equal to zero and the alternative hypothesis as mean not equal to zero. In this case the observation has a Gaussian distribution and the hypothesis concerns a single parameter of the distribution—the mean. In many cases an experimenter does not know the form of the distribution and seeks statistical techniques that are applicable regardless of the form of the distribution. These techniques are called *nonparametric* or *distribution-free methods*.

Although the terms *nonparametric* and *distribution-free* are often used interchangeably, they are not synonymous. *Nonparametric* refers to the notion that the possible distributions of the data comprise such a large class that they cannot be indexed by a finite number of real parameters. A *distribution-free* procedure is one based on a statistic computed from the observation, whose distribution is independent of the precise form of the distribution of the observations. In the sequel we will be discussing nonparametric hypotheses and the corresponding nonparametric tests. In all these cases, the test statistic under the null hypothesis will be distribution-free. That is, irrespective of the exact distribution of the samples under H_0, the test statistic has a specific distribution. The use of nonparametric detectors is recommended when either one or all of the following conditions may occur: (1) a complete statistical description of the input data is not available, (2) the statistics of the data varies with time, or (3) optimal detectors may be too complex to implement. Nonparametric or distribution-free detectors are intended to be insensitive to changes in the environment and simple to implement, at the cost of some deterioration in performance as compared to the optimal detectors.

7.1.1 Simple Nonparametric Test: The Sign Test

Consider a set of continuous observations $X_1, X_2, \ldots, X_n$, which are independent and identically distributed with a cumulative distribution function $F(\cdot)$. We wish to determine whether the median of the distribution is zero or whether it is positive. The problem can be posed as a nonparametric decision problem as follows.

Let p denote the probability that a random sample X_i exceeds zero. Clearly, the median is zero if $p = \frac{1}{2}$ and it is greater than zero if $p > \frac{1}{2}$. We thus have the two hypotheses:

$$\begin{aligned} H_0&: p = \tfrac{1}{2}, \qquad F(\cdot) \text{ otherwise arbitrary} \\ H_1&: p > \tfrac{1}{2}, \qquad F(\cdot) \text{ otherwise arbitrary} \end{aligned} \tag{7.1.1}$$

Except for the assumption that the observations are from a continuous density, no other information about the distribution is known. Thus the family of functions $F(\cdot)$ appearing in the hypotheses above is very broad.

For the problem of Eq. (7.1.1), an optimal test, in the sense of the most powerful test with type I error rate α, exists. To show this, let $f(\cdot)$ denote the density of the random sample with cdf $F(\cdot)$ and define the following [1]:

$$f^+(x_i) = f(x_i \mid X_i > 0) \tag{7.1.2}$$

$$f^-(x_i) = f(x_i \mid X_i \le 0) \tag{7.1.3}$$

$$f_0(x_i) = \tfrac{1}{2} f^+(x_i) + \tfrac{1}{2} f^-(x_i) \tag{7.1.4}$$

The likelihood ratio is given by

$$\Lambda = \prod_{i=1}^{n} \frac{f(x_i \mid H_1)}{f(x_i \mid H_0)} \tag{7.1.5}$$

It can easily be verified that

$$\frac{f(x_i \mid H_1)}{f(x_i \mid H_0)} = \begin{cases} \dfrac{p f^+(x_i)}{\frac{1}{2} f^+(x_i)} = 2p & \text{if } x_i > 0 \\[2ex] \dfrac{(1-p) f^-(x_i)}{\frac{1}{2} f^-(x_i)} = 2(1-p) & \text{if } x_i \le 0 \end{cases}$$

Hence

$$\Lambda = 2^n (p)^{\sum_{i=1}^{n} u(x_i)} (1-p)^{n - \sum_{i=1}^{n} u(x_i)} \tag{7.1.6}$$

where

$$u(x_i) = \begin{cases} 1 & \text{if } x_i > 0 \\ 0 & \text{if } x_i \le 0 \end{cases}$$

It follows that the log-likelihood ratio test based on Λ is equivalent to the following test:

$$T = \sum_{i=1}^{n} u(X_i) \underset{H_0}{\overset{H_1}{\gtrless}} c \tag{7.1.7}$$

where c is some threshold chosen to achieve type I error equal to α.

The test above is called the *sign test*, as it involves the binary function $u(x_i)$, which assigns zero value if the sample is negative and unity value if the sample is positive. The test statistic T takes on an integer value and therefore the test, Eq. (7.1.7), is equivalent to a counting rule. If the count T exceeds c, the decision is H_1. T has binomial distribution with parameters $(n, \frac{1}{2})$ under H_0 and (n, p) under H_1. The false alarm (type I error) and the probability of detection for the sign detector are given by

$$\text{Probability of false alarm:} \quad \alpha = \sum_{k=c+1}^{n} \binom{n}{k} (0.5)^n \tag{7.1.8}$$

$$\text{Probability of detection:} \quad \phi_1 = \sum_{k=c+1}^{n} \binom{n}{k} p^k (1-p)^{n-k} \tag{7.1.9}$$

Because the test statistic of Eq. (7.1.7) has a discrete distribution, an arbitrary prescribed false alarm cannot be achieved unless a randomized test is used (see Exercise 3.9). The realizable false alarms are given by Eq. (7.1.8), corresponding to different integer values of c ranging from -1 through n. Although for many nonparametric tests, the probability of detection, or equivalently, the power function, cannot be determined analytically, the sign test is an exception since the detection probability for this test can be computed from Eq. (7.1.9). Standard binomial tables can be used to compute the quantities in Eqs. (7.1.8) and (7.1.9).

The sign test is UMP for the stated hypothesis of Eq. (7.1.1). However, if the distribution of the observations is known exactly, then a likelihood ratio test designed with the known distribution would be the most powerful and would give better performance than a simple sign test. As shown in Chapter 3, for a Gaussian distribution with a positive variance, σ^2, to test whether the mean μ is zero or positive, a linear detector of the form given below is UMP (see Example 3.6.3).

$$\text{Linear detector:} \quad \sum_{i=1}^{m} X_i \underset{H_0}{\overset{H_1}{\gtrless}} d \tag{7.1.10}$$

where m is the sample size, the null hypothesis is that the mean is zero, and the threshold d is adjusted to achieve a specified false alarm probability of α. Let μ be the mean under H_1 and σ^2 be the variance of the observations. Since $\sum_{i=1}^{m} X_i$ is Gaussian under both hypotheses, the probability of detection for the linear detector is given by

$$\phi_2 = P\left(\sum_{i=1}^{m} X_i > d \mid H_1\right) = 1 - G\left(\frac{d - m\mu}{\sqrt{m}\sigma}\right) \tag{7.1.11}$$

where $d = \sqrt{m}\sigma G^{-1}(1 - \alpha)$. For a given α, ϕ_2 as determined above is greater than the probability of detection for the sign test, ϕ_1, in Eq. (7.1.9) with p for the Gaussian distribution being $G(m\mu/\sigma)$.

If the problem is to determine whether the median is zero or not, the alternative hypothesis H_1 will be two sided, that is

$$H_1: \quad p \neq \tfrac{1}{2}$$

The rejection region where the null hypothesis is not accepted will consist of values of T which are either too large or too small. Since the binomial distribution is symmetric when p is 0.5, the greatest power is achieved by choosing the two tails symmetrically. Thus the rejection region is given by $T \geq c$ or $T \leq c'$, where $c' = n - c$, and c satisfies the relation

$$\sum_{k=c}^{n} \binom{n}{k} 0.5^n = \frac{\alpha}{2} \tag{7.1.12}$$

In the above, the null hypothesis states that the median is at the origin. If it is desired to test whether the median is a constant M_0, the sign test can be applied to a new set of data $\{Z_i\}$ obtained as $Z_i = X_i - M_0, i = 1, 2, \ldots, n$.

7.1.2 The Wilcoxon Test

Consider testing whether the median of a distribution is a value M_0. The sign test assigns identical weights of 1 or zero, respectively, to all samples that exceed or fall below the median. If the distribution is symmetric, then under H_0 roughly 50% of the samples $\{Z_i = X_i - M_0\}$ will be positive. Consider the rank ordering of $|Z_i|$. That is, arrange the absolute values of Z_i in ascending order and assign the rank of 1 to the lowest value through a rank of n to the largest value. Let the rank of $|Z_i|$ be denoted as $r(|Z_i|)$. It can be seen that $\sum_{i=1}^{n} r(|Z_i|) = n(n + 1)/2$. The *Wilcoxon test*, also called the *signed rank test*, is based on either of the following statistics:

$$T^+ = \sum_{i=1}^{n} u(Z_i) r(|Z_i|) \tag{7.1.13}$$

or

$$T^- = \sum_{i=1}^{n} [1 - u(Z_i)] r(|Z_i|) \tag{7.1.14}$$

It is clear that $T^+ + T^- = n(n + 1)/2$. Therefore, a test can be based on either of these two statistics. T^+ can also be expressed equivalently as

$$T^+ = \sum_{i=1}^{n} i T_i \tag{7.1.15}$$

where $T_i = 1$ if the sample Z corresponding to the ith smallest value in the set $\{|Z_k|, k = 1, \ldots, n\}$ is positive and is zero otherwise. It can be shown that under H_0, $T_1, \ldots, T_n$ are mutually independent and that (see Exercise 7.1)

$$P(T_i = 0) = P(T_i = 1) = \tfrac{1}{2} \tag{7.1.16}$$

Hence, under H_0, the expected value and variance of T^+ are given by

$$E(T^+) = \sum_{i=1}^{n} iE(T_i) = \frac{1}{2}\sum_{i=1}^{n} i = \frac{n(n+1)}{4} \tag{7.1.17}$$

$$\text{var}(T^+) = \sum_{i=1}^{n} i^2 \text{ var}(T_i) = \frac{n(n+1)(2n+1)}{24} \tag{7.1.18}$$

For testing the one-sided hypothesis that the median exceeds M_0, the following test can be employed:

$$\text{Reject } H_0 \text{ if } T^+ - \frac{n(n+1)}{4} \geq k \tag{7.1.19}$$

where k is chosen to achieve a specified false alarm probability. This is reasonable since under the alternative hypothesis, more samples with larger ranks would be positive, making T^+ larger. For large values of n, the test statistic in Eq. (7.1.19) is approximately Gaussian. Therefore, using the mean and the variance given in Eqs. (7.1.17) and (7.1.18), the value of k required to achieve a specified false alarm can be obtained using the standard normal tables.

Example 7.1.1

A Wilcoxon test is applied to a situation with sample size 20. The null hypothesis is that the samples are from a continuous symmetric distribution with median zero and the alternative is that the median is positive. It is required to find the approximate value of k in Eq. (7.1.19) so that the false alarm probability is 0.001.

For this problem it follows that

$$P\left(T^+ - \frac{n(n+1)}{4} \geq k \mid H_0\right) \approx 1 - G\left(\frac{k}{\sqrt{\text{var}(T^+ \mid H_0)}}\right) = 1 - G\left(\frac{k}{\sqrt{717.5}}\right) = 0.001$$

Use of a table of the normal cumulative distribution function yields a value of $k = 83$.

If n is small, the exact value of k required in Eq. (7.1.19) can be obtained by enumerating all possible values of T^+ and by computing the corresponding probabilities [2]. If the symmetry assumption is satisfied, the Wilcoxon test may provide more power than a sign test for certain distributions. One measure of the performance of a test is the asymptotic relative efficiency (ARE). It is customary in signal detection applications to consider the hypothesis H_1 as the signal present condition. Therefore, the ARE compares the relative performances of two tests under the conditions of large sample sizes and weak signal (alternative tending toward the hypothesis). We shall provide ARE comparisons of the sign and Wilcoxon tests after we define ARE and discuss the means of deriving it.

7.1.3 Asymptotic Relative Efficiency

Consider testing a simple null hypothesis

$$H_0: \theta = \theta_0 \tag{7.1.20a}$$

against a one-sided alternative hypothesis of the form

$$H_1\colon \theta \in \Theta_1 = \{\theta > \theta_0\} \tag{7.1.20b}$$

For testing these hypotheses, if there exist two tests T_1 and T_2, each achieving the same false alarm probability α and detection probability β but requiring sample sizes n_1 and n_2, respectively, the relative efficiency of T_1 with respect to T_2 is defined to be

$$\text{RE}_{12} = \frac{n_2}{n_1} \tag{7.1.21}$$

When RE_{12} exceeds 1, this implies that the detector T_2 requires more samples than the detector T_1 to achieve the same performance. Hence T_1 is relatively more efficient than T_2. The difficulty with RE_{12} is that n_2 and n_1 are usually functions of α and β. It is possible that for a certain range of values of α and β, RE_{12} will be larger than 1, whereas for certain other ranges it may be smaller than 1. In other words, one test is not necessarily superior to another over all ranges of (α, β), as determined by the relative efficiency. However, in many situations, sample sizes n_1 and n_2 assume large values. Under certain restrictions, with sample sizes approaching infinity, the evaluation of RE_{12} produces a single number, independent of α and β. Such a procedure leads to the definition of *asymptotic relative efficiency* (ARE).

For a consistent test, at any given $\alpha > 0$, the probability of detection β approaches 1 as the sample size approaches infinity. Any reasonable test is consistent. Any specific $\beta < 1$ can also be obtained by a test if the alternative hypothesis H_1 is allowed to approach H_0 as the sample size n approaches infinity. That is, consider a sequence of $\theta_n \in \Theta_1$ such that $\theta_n \to \theta_0$ from above, as n increases. This situation in which the alternative hypothesis becomes closer to H_0 is of importance, because in a number of signal detection problems, the signal happens to be weak. We can formally define the asymptotic relative efficiency as

$$\text{ARE}_{12} = \lim_{\substack{n_1 \to \infty \\ n_2 \to \infty \\ H_1 \to H_0}} \text{RE}_{12} \tag{7.1.22}$$

For many standard test procedures, ARE happens to be a number independent of α and β, but this number is obtained only when θ_n approaches θ_0 in a specific manner. Nevertheless, under conditions of large sample size and weak signal strength, ARE provides a useful measure of performance.

Formal derivation of ARE. Let the hypotheses be

$$\begin{aligned} H_0&\colon \theta = \theta_0 \\ H_1&\colon \Theta_1 = \{\theta > \theta_0\} \end{aligned} \tag{7.1.23}$$

and let $\theta_n \in \Theta_1$ be such that

$$\theta_n = \theta_0 + \frac{k}{\sqrt{n}} \tag{7.1.24}$$

Consider a test T_n for the problem of Eq. (7.1.23) with the rejection region

$$\Psi_n(\mathbf{x}) = \begin{cases} 1 & \text{if } t_n(\mathbf{x}) \geq t_{n,\alpha} \\ 0 & \text{if } t_n(\mathbf{x}) < t_{n,\alpha} \end{cases} \tag{7.1.25}$$

where $\mathbf{x}$ is the observation vector of sample size n and $t_{n,\alpha}$ is the threshold to achieve a type I error of α. Let $\sigma_\theta^2(T_n)$ denote the variance of T_n. Further, let us assume that the following regularity conditions are satisfied.

1. $\dfrac{d}{d\theta}\mathrm{E}_\theta(T_n)$ exists and is positive at θ_0 and continuous at θ_0.

2. $$\lim_{n\to\infty} \left. \frac{\dfrac{d}{d\theta}\mathrm{E}_\theta(T_n)}{\sqrt{n}\,\sigma_\theta(T_n)} \right|_{\theta_0} = c.$$

3. $$\lim_{n\to\infty} \frac{\left.\dfrac{d}{d\theta}\mathrm{E}_\theta(T_n)\right|_{\theta_n}}{\left.\dfrac{d}{d\theta}\mathrm{E}_\theta(T_n)\right|_{\theta_0}} = 1.$$

$$\lim_{n\to\infty} \frac{\sigma_{\theta_n}(T_n)}{\sigma_{\theta_0}(T_n)} = 1.$$

4. T_n is asymptotically normal with mean $\mathrm{E}_{\theta_n}(T_n)$ and variance $\sigma_{\theta_n}^2(T_n)$.

Because T_n is asymptotically normal, it follows that the condition

$$\lim_{n\to\infty} P(T_n > t_{n,\alpha} \mid \theta_0) = \alpha \tag{7.1.26}$$

implies

$$\lim_{n\to\infty} G\left(\frac{t_{n,\alpha} - \mathrm{E}_{\theta_0}(T_n)}{\sigma_{\theta_0}(T_n)}\right) = 1 - \alpha \tag{7.1.27}$$

or

$$\lim_{n\to\infty} \frac{t_{n,\alpha} - \mathrm{E}_{\theta_0}(T_n)}{\sigma_{\theta_0}(T_n)} = z_\alpha \tag{7.1.28}$$

where

$$1 - G(z_\alpha) = \alpha \tag{7.1.29}$$

The limiting power of the test is $1 - G(\hat{z})$, where

$$\hat{z} = \lim_{n\to\infty} \frac{t_{n,\alpha} - \mathrm{E}_{\theta_n}(T_n)}{\sigma_{\theta_n}(T_n)} \tag{7.1.30}$$

Using regularity condition 3, we have

$$\hat{z} = \lim_{n\to\infty} \frac{t_{n,\alpha} - \mathrm{E}_{\theta_n}(T_n)}{\sigma_{\theta_0}(T_n)} \tag{7.1.31}$$

Expand $E_{\theta_n}(T_n)$ in a Taylor series around θ_0, to get

$$E_{\theta_n}(T_n) = E_{\theta_0}(T_n) + (\theta_n - \theta_0)\frac{dE_\theta(T_n)}{d\theta}\bigg|_{\theta_0} + \epsilon \tag{7.1.32}$$

where $\epsilon \to 0$ as $n \to \infty$. Hence

$$\hat{z} = \lim_{n\to\infty}\left\{\frac{t_{n,\alpha} - E_{\theta_0}(T_n)}{\sigma_{\theta_0}(T_n)} - \frac{\dfrac{(\theta_n - \theta_0)\,dE_\theta(T_n)}{d\theta}\bigg|_{\theta_0} + \epsilon}{\sigma_{\theta_0}(T_n)}\right\} \tag{7.1.33}$$

$$\begin{aligned} &= \lim_{n\to\infty}\left\{\frac{t_{n,\alpha} - E_{\theta_0}(T_n)}{\sigma_{\theta_0}(T_n)} - kc\right\} \\ &= z_\alpha - kc \end{aligned} \tag{7.1.34}$$

where the second step follows from the regularity condition 2 and Eq. (7.1.24). Thus the limiting power of the test T_n is given by $1 - G(z_\alpha - kc)$, where α satisfies

$$\alpha = 1 - G(z_\alpha) \tag{7.1.35}$$

Let T_n and T_n^* be two tests satisfying Eq. (7.1.35), with constants $(k, c), (k^*, c^*)$, respectively. The two tests will have the same sequence of alternatives, $\theta_n = \theta_{n^*}$ if

$$\frac{k}{\sqrt{n}} = \frac{k^*}{\sqrt{n^*}} \tag{7.1.36}$$

The two tests will have the same power if

$$kc = k^*c^* \tag{7.1.37}$$

Combining Eqs. (7.1.36) and (7.1.37), we have

$$\text{ARE} = \frac{n^*}{n} = \left(\frac{c}{c^*}\right)^2 \tag{7.1.38}$$

Use of regularity condition 2 then gives the expression for the ARE as

$$\text{ARE} = \lim_{i\to\infty}\frac{\dfrac{\left[\dfrac{dE_\theta(T_i)}{d\theta}\bigg|_{\theta_0}\right]^2}{\sigma_{\theta_0}^2(T_i)}}{\dfrac{\left[\dfrac{dE_\theta(T_i^*)}{d\theta}\bigg|_{\theta_0}\right]^2}{\sigma_{\theta_0}^2(T_i^*)}} \tag{7.1.39}$$

The numerator on the right-hand side of Eq. (7.1.39) divided by i is called the *efficacy* of the test T_i. Hence ARE can also be defined as the ratio of the efficacies of two tests.

Computation of ARE of some specific tests. First, let us find the ARE of the sign test with respect to the linear test. The sign test, Eq. (7.1.7), has the test statistic $T_n = \sum_{i=1}^{n} u(X_i)$, which is asymptotically normal, and it satisfies the regularity conditions given earlier. In this section we consider an equivalent form of Eq. (7.1.1):

$$\begin{aligned} H_0&: X_i \sim F_0(x), \text{ with } F_0(0) = \tfrac{1}{2} \\ H_1&: X_i \sim F_0(x - \theta), \text{ with } \theta > 0 \end{aligned} \tag{7.1.40}$$

Hence $\theta_0 = 0, \theta_n \in \Theta_1 = \{\theta > 0\}$.

$$\mathrm{E}_\theta(T_n) = \mathrm{E}_\theta\left[\sum_{i=1}^{n} u(X_i)\right] = nF_0(-\theta) \tag{7.1.41}$$

$$\sigma_\theta^2(T_n) = nF_0(-\theta)[1 - F_0(-\theta)] \tag{7.1.42}$$

The efficacy of the sign test is equal to

$$\frac{\left\{\left.\dfrac{d[nF_0(-\theta)]}{d\theta}\right|_{\theta=0}\right\}^2}{n[nF(-\theta)(1 - F(-\theta))|_{\theta=0}]} = 4f_0^2(0) \tag{7.1.43}$$

where $f_0(x) = dF_0(x)/dx$.

Whereas the sign detector is designed for testing the median, one can consider a linear detector of the type of Eq. (7.1.10) for testing the mean. With the additional assumption that the density $dF_0(x)/dx$ in Eq. (7.1.40) is symmetric about the origin, the hypotheses in Eq. (7.1.40) can be written as

$$\begin{aligned} H_0&: \text{mean} = 0 \\ H_1&: \text{mean} = \theta > 0 \end{aligned} \tag{7.1.44}$$

The efficacy of the linear detector is given by

$$\frac{\left\{\left.\dfrac{d\mathrm{E}(\sum X_i \mid \theta)}{d\theta}\right|_{\theta=0}\right\}^2}{n\ \mathrm{var}_\theta(\sum X_i)|_{\theta=0}} = \frac{\left(\left.\dfrac{d[n\theta]}{d\theta}\right|_{\theta=0}\right)^2}{n \cdot n\sigma_x^2} = \frac{1}{\sigma_x^2} \tag{7.1.45}$$

where σ_x^2 is the variance of a sample x_i. Hence the ARE of the sign detector with respect to the linear detector is given by

$$\mathrm{ARE}_{\text{sign, linear}} = 4f_0^2(0)\sigma_x^2 \tag{7.1.46}$$

If the X_i are Gaussian distributed, then $f_0(0) = (1/\sqrt{2\pi})\sigma_x$, and ARE $= 2/\pi \cong 0.63$. Hence the sign detector is less efficient than the linear detector. This is to be expected because the linear detector is uniformly most powerful in this situation. On the other hand, if X_i is Laplacian with $f_0(x) = (\lambda/2)e^{-\lambda|x|}$, then $f_0(0) = \lambda/2, \sigma_x^2 = 2/\lambda^2$, so that the ARE is 2. In this case, the sign detector is twice as efficient as the linear detector in the sense of ARE.

It should be noted that the linear detector of Eq. (7.1.10) cannot be constructed if the variance σ_x^2 is unknown. For testing the mean of a normal distribution when the variance is unknown, Student's t-test can be considered. The decision rule then becomes

$$\text{Reject the hypothesis that the mean is zero if } S_n^* = \frac{\sqrt{n}\bar{x}_n}{S_n} \text{ is large} \tag{7.1.47}$$

where $\bar{x}_n$ and S_n^2 are the sample mean and variance of the $\{x_i\}$, defined as

$$\bar{x}_n = \frac{1}{n}\sum_{i=1}^{n} x_i$$

$$S_n^2 = \frac{1}{n-1}\sum_{i=1}^{n}(x_i - \bar{x})^2$$

Asymptotically, as $n \to \infty$, the t-test is equivalent to the linear detector, which is the test for Gaussian samples with known variance. Therefore, the efficacy of the t-test is the same as that of the linear detector (see Exercise 7.2).

To compute the efficacy of the Wilcoxon test, the test statistic T^+ of Eq. (7.1.13) is written in a third equivalent form:

$$T^+ = \sum_{i=1}^{n}\sum_{j=1}^{i} u(X_i + X_j) \tag{7.1.48}$$

To show this equivalence, let us arrange the $\{X_i\}$ in increasing order so that $X_{i_1} < X_{i_2} < \cdots < X_{i_{k-1}} < 0 < X_{i_k} < \cdots < X_{i_n}$. From Eq. (7.1.15) it can be seen that T^+ equals the sum of ranks of the positive samples, where the ranking is done on the magnitude of the samples. This is equivalent to computing the number, N_j, of terms that are positive in the set $\{X_{i_j}, X_{i_j} + X_{i_{j-1}}, \ldots, X_{i_j} + X_{i_1}\}$ and writing T^+ as $\sum_{j=i_k}^{i_n} N_j$. Since N_j is zero for $j < i_k$, we can write $T^+ = \sum_{j=i_1}^{i_n} N_j$. Since the index j covers all integers 1 through n, the sum is equivalent to the right-hand side of Eq. (7.1.48). Now

$$E(T^+) = \sum_{i=1}^{n} E(u(X_i)) + \sum_{i=2}^{n}\sum_{j=1}^{i-1} E(u(X_i + X_j)) \tag{7.1.49}$$

where

$$E(u(X_i)) = \int_0^\infty f_0(x - \theta)\,dx = 1 - F_0(-\theta) = F_0(\theta) \tag{7.1.50}$$

and

$$\begin{aligned} E(u(X_i + X_j)) &= P(X_i + X_j > 0) \\ &= 1 - \int_{-\infty}^{\infty} f_0(\sigma - \theta)F_0(-\sigma - \theta)\,d\sigma \end{aligned} \tag{7.1.51}$$

Hence, assuming $f_0(x)$ is symmetric about origin,

$$\left.\frac{\partial E(T^+)}{\partial \theta}\right|_{\theta=0} = nf_0(0) + n(n-1)I \tag{7.1.52}$$

where

$$I = \int_{-\infty}^{\infty} f_0^2(\sigma)\, d\sigma \tag{7.1.53}$$

The quantity I appears frequently in the ARE of nonparametric tests. Using Eqs. (7.1.39) and (7.1.18), the efficacy of the Wilcoxon test is given by $12I^2$. The ARE of the Wilcoxon test relative to the t-test (or equivalently the linear detector) is given by $12\sigma_x^2 I^2$. The ARE of a few tests for specific distributions are shown in Table 7.1.1. Notice that the entries in the last column are obtained as the ratio of the corresponding entries in the second and first columns. It can be seen that the Wilcoxon test has a higher relative efficiency than the sign test for uniform, logistic, and normal distributions. The reverse is true for the double exponential distribution. This is because the Wilcoxon test uses the magnitude information along with the sign. In fact, the ARE of the Wilcoxon test relative to the t-test is never less than 0.864 for any continuous symmetric distribution [3]. For the Laplace density which has a heavy tail, both the sign and the Wilcoxon tests are more efficient than the t-test.

TABLE 7.1.1 ASYMPTOTIC RELATIVE EFFICIENCIES OF SOME TESTS

Distribution	ARE Wilcoxon, t	ARE Sign, t	ARE Sign, Wilcoxon
Uniform	1	$\frac{1}{3}$	$\frac{1}{3}$
Normal	$\frac{3}{\pi} = 0.955$	$\frac{2}{\pi} = 0.64$	$\frac{2}{3}$
Double exponential	$\frac{3}{2}$	2	$\frac{4}{3}$
Logistic	$\frac{\pi^2}{9}$	$\frac{\pi^2}{12}$	$\frac{3}{4}$

7.1.4 Two-Input Systems

Let us consider the example discussed in Thomas [1]. Let $\mathbf{W} = ((X_1, Y_1), (X_2, Y_2), \ldots, (X_n, Y_n))$ be the vector of paired observations. Typically, the vector $\mathbf{X} = (X_1, X_2, \ldots, X_n)$ is the output of one channel, whereas $\mathbf{Y} = (Y_1, Y_2, \ldots, Y_n)$ is the output of another channel. In many applications, such as radio astronomy, underwater sound detection, and geophysics (see references in [1]), a common signal may be transmitted over two channels perturbed by independent noises. The hypotheses to be considered are as follows.

H_0: **X** and **Y** are independent sequences with zero mean and fixed variances $E(X_i^2) = \sigma_1^2$, $E(Y_i^2) = \sigma_2^2$, which are independent of each other.

H_1: A random signal $\mathbf{S} = (S_1, S_2, \ldots, S_n)$ with zero mean, $\text{var}(S_i) = \sigma_3^2$, is present in both channels. The noise components of $\mathbf{X}$ and $\mathbf{Y}$ have distribution as in H_0 and are independent of $\mathbf{S}$.

For this problem, let us consider both a parametric and a nonparametric formulation.

Case (i). All processes are Gaussian; $\sigma_1 = \sigma_2 = \sigma$. The hypotheses become

H_0: (X_i, Y_i) are jointly Gaussian, with $\rho = 0, \sigma_x = \sigma_y = \sigma$.

H_1: (X_i, Y_i) are jointly Gaussian, with $\rho > 0, \rho = \dfrac{\sigma_3^2}{\sigma^2 + \sigma_3^2}, \sigma_x^2 = \sigma_y^2 = \dfrac{\sigma^2}{1 - \rho}$.

The Neyman–Pearson test for testing the foregoing hypothesis is

$$\sum_{i=1}^{n} (x_i + y_i)^2 \underset{H_0}{\overset{H_1}{\gtrless}} t_1 \tag{7.1.54}$$

Case (ii). The densities of X_i, Y_i, and S_i are known to be symmetric and continuous. Consider the following distributions:

$$F_1(x) = P(X_i \le x \mid H_0) \tag{7.1.55}$$

$$F_2(y) = P(Y_i \le y \mid H_0) \tag{7.1.56}$$

$$F_3(s) = P(S_i \le s \mid H_0) \tag{7.1.57}$$

Let $F(x, y)$ denote the joint CDF of (X_i, Y_i). The hypotheses become

$$H_0: F(x, y) = F_1(x)F_2(y)$$

$$H_1: F(x, y) = \int_{-\infty}^{\infty} F_1(x - s)F_2(y - s)f_3(s)\, ds$$

where $f_3(s) = dF_3(s)/ds$. Proceeding as in the case of the sign detector, the Neyman–Pearson test for the hypothesis above becomes

$$\sum_{i=1}^{n} u(x_i, y_i) \underset{H_0}{\overset{H_1}{\gtrless}} t_2 \tag{7.1.58}$$

where $u(\cdot)$ is the unit-step function.

The test statistic in Eq. (7.1.58) counts the number of polarity coincidences between x_i and y_i and hence is termed the polarity coincidence correlator (PCC). The ARE of the PCC with respect to the test in Eq. (7.1.54) can be derived (see Exercise 7.3). It turns out that this ARE is 0.202 for a Gaussian distribution, whereas it is 3.5 for a double exponential distribution. Therefore, for distributions with a heavy tail, PCC would be preferable over the test in Eq. (7.1.54) which assumes a Gaussian distribution.

Equality of distribution in two samples. A general two-sample hypothesis-testing problem is as follows. Let $\{X_i, i = 1, 2, \ldots, n_1\}, \{Y_i, i = 1, 2, \ldots, n_2\}$ be random samples from continuous distributions with CDFs $F_1(\cdot)$ and $F_2(\cdot)$, respectively. Consider the two hypotheses

$$H_0\colon F_1(\cdot) = F_2(\cdot)$$
$$H_1\colon F_1(\cdot) \neq F_2(\cdot) \tag{7.1.59}$$

Some of the well-known tests for the hypotheses above are discussed next.

Wald–Wolfowitz Run Test. In a sequence consisting of the integers 0 and 1, when several 0's occur together, we have a run of 0's. Specifically, in the sequence 00111100010100, there is a run of 0's of length 2, followed by a run of 1's of length 4, a run of 0's of length 3, a run of 1's of length 1 and so on. Let us tag 1 with the samples x_i and 0 with the samples y_i. Then, in the combined ordering (ascending) of x and y samples, when the x and y samples are replaced by their tags, we get a sequence of 1's and 0's. Let r_{1j} and r_{2j} denote the total number of runs of length j for the x and y samples, respectively. Let

$$R_1 = \sum_{j=1}^{n_1} r_{1j} \tag{7.1.60}$$

$$R_2 = \sum_{j=1}^{n_2} r_{2j} \tag{7.1.61}$$

The total number of runs is given by $R = R_1 + R_2$. If H_0 is true, then R tends to have large values. Therefore, the Wald–Wolfowitz run test is of the form:

$$\text{Reject } H_0 \text{ for } R < t_\alpha \tag{7.1.62}$$

For small n_1 and n_2, the exact distribution for R under the null hypothesis can be found [2,4]. The threshold t_α can be set for a given $P_F = \alpha$. As $n_1, n_2 \rightarrow \infty$ such that $n_2/n_1 = \lambda$, it can be shown that under H_0, R is asymptotically normal with mean $2n_1\lambda/(1 + \lambda)$ and variance $4\lambda^2 n_1/(1 + \lambda)^3$ [2,4]. Moreover, the test is consistent. That is, for any $0 < \alpha \leq 1$, the power of the test approaches 1, under H_1, as $n_1, n_2 \rightarrow \infty$. The run test is sensitive to both differences in shape and differences in location between the two distributions.

Kolmogorov–Smirnov (K-S) Test. Given a set of samples $(x_1, x_2, \ldots, x_{n_1})$, the sample CDF or the empirical CDF is defined as

$$\begin{aligned} F_{n_1}(x) &= \frac{1}{n_1}(\text{number of } x_i \leq x) \\ &= \frac{1}{n_1}\sum_{i=1}^{n_1} I_{(-\infty,x)}(x_i) \end{aligned} \tag{7.1.63}$$

When the $\{x_i\}$ are samples from the population with CDF $F_1(x)$, it can be shown that [5]

$$P\Big[\sup_{\substack{-\infty<x<\infty \\ n_1\rightarrow\infty}} |F_{n_1}(x) - F_1(x)| \rightarrow 0\Big] = 1 \tag{7.1.64}$$

For the two sets of samples $\{x_i\}, \{y_i\}$, the K-S test requires the computation of

$$D_{n_1,n_2} = \sup_x |F_{1n_1}(x) - F_{2n_2}(x)| \tag{7.1.65}$$

where $F_{1n_1}(\cdot)$ and $F_{2n_2}(\cdot)$ are the empirical CDFs of the samples $\{x_i\}$ and $\{y_i\}$. The K-S test is given by

$$\text{Reject } H_0 \text{ if } D_{n_1,n_2} > \delta_\alpha \tag{7.1.66}$$

For the K-S test, an asymptotic distribution for D_{n_1,n_2} under H_0 can be found as $n_1, n_2 \to \infty$ such that $n_2/n_1 = \lambda$. Hence an approximate value for δ_α can be found for a specified $P_F = \alpha$ [4]. This test is also consistent. The K-S test is easy to apply, but it is useful mainly for the general alternative hypotheses of Eq. (7.1.59) since the test is sensitive to all types of differences between the CDFs. Its primary application is for preliminary studies of data, as is the case with the run test.

Two-Sample Wilcoxon (or Mann–Whitney) Test. Consider a one-sided subset of the hypothesis H_1 in Eq. (7.1.59):

$$H_1: \quad F_1(\cdot) < F_2(\cdot)$$

Under H_1, the distribution F_1 is said to be stochastically larger than F_2. In this case a more powerful test based on ranks can be developed.

Let $Z_{\zeta\eta}, \zeta = 1, 2, \ldots, n_1, \eta = 1, 2, \ldots, n_2$ be a set of random variables defined by

$$Z_{\zeta\eta} = \begin{cases} 1 & \text{if } Y_\eta < X_\zeta \\ 0 & \text{if } Y_\eta \geq X_\zeta \end{cases} \tag{7.1.67}$$

Let

$$U = \sum_{\eta=1}^{n_2} \sum_{\zeta=1}^{n_1} Z_{\zeta\eta} \tag{7.1.68}$$

Thus if $(X_1, X_2, \ldots, X_{n_1})$ and $(Y_1, Y_2, \ldots, Y_{n_2})$ are arranged in a sequence of increasing order of magnitude, U denotes the total number of times an X exceeds a Y. If all the X's and the Y's in this sequence are assigned ranks from 1 to $n_1 + n_2$, the sum T of the ranks of the X's is called the *Wilcoxon statistic*. It can be shown that

$$T - U = \frac{n_1(n_1 + 1)}{2} \tag{7.1.69}$$

Therefore, the tests based on U and T are equivalent. The test based on U, called the *Mann–Whitney U test*, is of the following type:

$$\text{Reject } H_0 \text{ if } U \leq u_\alpha \tag{7.1.70}$$

While the mean and the variance of U can be found for any distribution (Exercise 7.4), for large values of n_1 and n_2, U is asymptotically normal under H_0 with mean $n_1 n_2/2$ and variance $n_1 n_2(n_1 + n_2 + 1)/12$. This test is also consistent. The Wilcoxon test is preferable to other rank tests if the underlying distribution is logistic.

In this section we have discussed some specific nonparametric tests for one- and two-sample situations. There exist a number of other tests, such as the median test, van der Waerden test, goodness-of-fit tests, tests concerning quantiles, Mood test, and so on. For a discussion of these and other tests, the reader is referred to [26]. An extensive bibliography of nonparametric tests in signal-processing applications up to the year 1980, can be found in Kassam [6]. Some application examples in the literature include the following: radar targets [7–11], spectral estimation [12], and mobile radio [13,14].

7.2 LOCALLY OPTIMAL DETECTION

In one-sided testing concerning a parameter θ, if the distribution of the observation either belongs to an exponential family or possesses a monotone likelihood ratio, a UMP test can be obtained (Chapter 3). When neither condition is satisfied, we may try to restrict the class of tests and then find an optimum within the class, such as is the case with UMP unbiased tests. In some cases such a test may not exist. Also, in several situations the problem is one of detecting weak signals in noise [15]. In this section we develop an optimum test for the detection of weak signals, assuming that we have a set of random observations. While we develop the test for the case of an absolutely continuous distribution, the result can be extended in a straightforward manner to the discrete case by replacing integrals with summations and the probability density function with the probability mass function. Let $X_1, X_2, \ldots, X_n$ be i.i.d. observations from an absolutely continuous CDF, $F(\cdot;\theta)$. The hypotheses of interest are as follows:

$$\begin{aligned} H_0&: \theta = \theta_0 \\ H_1&: \theta > \theta_0, \text{ with } \theta \text{ close to } \theta_0 \end{aligned} \tag{7.2.1}$$

Consider any test T_n for the hypothesis of Eq. (7.2.1), with type I error α_n and type II error $\beta_n(\theta)$, where the subscript n is used to explicitly indicate the dependence on sample size n. If T_n^* is any other test with the same α_n and type II error $\beta_n^*(\theta)$ such that

$$\left.\frac{\partial \beta_n^*(\theta)}{\partial \theta}\right|_{\theta=\theta_0} \le \left.\frac{\partial \beta_n(\theta)}{\partial \theta}\right|_{\theta=\theta_0} \tag{7.2.2}$$

then T_n^* is locally optimal. Notice that the type II error is related to the power function by

$$\beta_n(\theta) = 1 - \phi_n(\theta) \tag{7.2.3}$$

Hence Eq. (7.2.2) is equivalent to

$$\left.\frac{\partial \phi_n^*(\theta)}{\partial \theta}\right|_{\theta=\theta_0} \ge \left.\frac{\partial \phi_n(\theta)}{\partial \theta}\right|_{\theta=\theta_0} \tag{7.2.4}$$

The optimality implied by Eq. (7.2.4) can be seen by observing that under H_1, θ, although greater than θ_0, is very close to θ_0. Hence, for a test T_n^*, if the slope of the power function at θ_0 is greater than or equal to the slope of the power function at the same point for any other test T_n, the power (or the probability of detection) of T_n^* at a point $\theta = \theta_0 + \epsilon, \epsilon > 0, \epsilon \to 0$, will be the largest among all tests achieving the same type I error α_n.

Let the critical region of a test be denoted by I and its complement by I*. Then

$$\alpha_n = \int_{\mathrm{I}} \cdots \int \prod_{i=1}^{n} f(x_i; \theta_0)\, dx \tag{7.2.5}$$

$$\beta_n(\theta) = \int_{\mathrm{I}^*} \cdots \int \prod_{i=1}^{n} f(x_i; \theta)\, dx \tag{7.2.6}$$

$$\phi_n(\theta) = \int_{\mathrm{I}} \cdots \int \prod_{i=1}^{n} f(x_i; \theta)\, dx \tag{7.2.7}$$

When the distribution of the observations satisfies certain regularity conditions, it follows that

$$\frac{\partial \phi_n(\theta)}{\partial \theta} = \int_{\mathrm{I}} \cdots \int \frac{\partial}{\partial \theta} \prod_{i=1}^{n} f(x_i; \theta)\, dx \tag{7.2.8}$$

so that the problem is equivalent to finding a test that maximizes Eq. (7.2.8) subject to the constraint of Eq. (7.2.5). This problem is an extension of the Neyman–Pearson problem and hence the solution is readily available (see Exercise 7.5). The locally optimal test is given by

$$\left. \frac{\dfrac{\partial \left[\prod_{i=1}^{n} f(x_i; \theta) \right]}{\partial \theta}}{\prod_{i=1}^{n} f(x_i; \theta)} \right|_{\theta = \theta_0} \begin{array}{ll} > c & \text{decide } H_1 \\ = c & \text{decide } H_1 \text{ with probability } \gamma \\ < c & \text{decide } H_0 \end{array} \tag{7.2.9}$$

where c and $0 \le \gamma \le 1$ are appropriately chosen to yield the type I error α_n. The test statistic on the left-hand side of (7.2.9) can be written equivalently as

$$\sum_{i=1}^{n} \left. \frac{\partial \ln f(x_i; \theta)}{\partial \theta} \right|_{\theta = \theta_0} \triangleq \sum_{i=1}^{n} b(x_i) \tag{7.2.10}$$

Figure 7.2.1 shows an implementation of the decision rule in Eqs. (7.2.9) and (7.2.10).

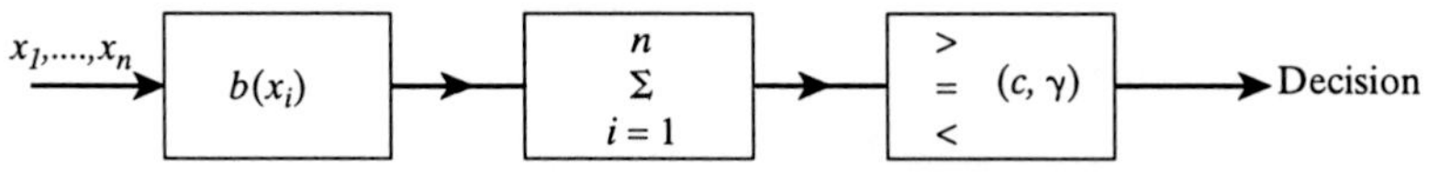

Figure 7.2.1 Locally optimal detector.

It can be noted that $b(\cdot)$ is a memoryless nonlinearity that operates on each input sample x_i independent of other samples. Thus the test statistic is obtained as a memoryless nonlinear transformation of the data sample x_i, followed by the summation operation.

It can be seen that for vanishing signal, $\theta \to \theta_0$, the locally optimal test is as efficient as the most powerful N-P test. Hence the ARE of one with respect to the other is 1 [16].

7.2.1 Examples

Example 7.2.1

Let $x_1, x_2, \ldots, x_n$ be a random sample of size n from a Cauchy distribution with location parameter θ, so that

$$f(x_1, x_2, \ldots, x_n; \theta) = \frac{1}{\pi^n} \prod_{i=1}^{n} \frac{1}{1 + (x_i - \theta)^2} \tag{7.2.11}$$

Consider testing H_0: $\theta = 0$ against H_1: $\theta > 0$. It can be verified that no UMP test with type I error α exists. However, a locally optimal test of size α can be found from Eq. (7.2.9). Since

$$\frac{\partial \ln f(x_1, x_2, \ldots, x_n; \theta)}{\partial \theta} = \sum_{i=1}^{n} \frac{2(x_i - \theta)}{1 + (x_i - \theta)^2} \tag{7.2.12}$$

the locally optimal test is

$$\sum_{i=1}^{n} \frac{2x_i}{1 + x_i^2} \underset{H_0}{\overset{H_i}{\gtrless}} k \tag{7.2.13}$$

where k is chosen to achieve the size α. In Eq. (7.2.13) no randomization is needed [that is, γ in Eq. (7.2.9) is zero] since the test statistic on the left-hand side of Eq. (7.2.13) has a continuous distribution.

For large n, we can use the central limit theorem to show that the distribution for $\sum_{i=1}^{n} 2x_i/(1 + x_i^2)$ approaches a Gaussian distribution and hence find an approximate value of the threshold k that yields a specified α [17]. It is possible that a locally optimal test can be disastrous for strong signals. The test in Eq. (7.2.13) is an example. As $x_i \to \infty$, $x_i/(1 + x_i^2) \to 0$ and since $k > 0$ for $\alpha < \frac{1}{2}$, the probability that the test statistic in Eq. (7.2.13) exceeds k goes to zero, when θ approaches ∞. Hence the power of the test of Eq. (7.2.13) goes to zero as θ approaches infinity!

In Exercise 7.6 the reader is asked to show that the locally optimal detector for the detection of a constant signal in Laplace noise is a hard limiter.

Example 7.2.2 Locally Optimal Detection of Nonconstant Signal in i.i.d. Noise

We have developed earlier a locally optimal detector for i.i.d. observations. A similar result exists for detecting a nonconstant signal, in i.i.d. noise. Consider the following hypotheses:

$$\begin{aligned} &H_0\text{: } X_i = N_i \\ &H_1\text{: } X_i = N_i + \theta s_i \qquad \theta > 0, \theta \to 0 \quad \text{for } i = 1, 2, \ldots, N \end{aligned} \tag{7.2.14}$$

Let N_i be i.i.d. with density $f_n(\cdot)$. A nonconstant signal implies that the signal samples s_i need not be identical. Now, since

$$f(x_i \mid H_1) = f_n(x_i - \theta s_i) \tag{7.2.15}$$

it follows that

$$\begin{aligned} \frac{\partial}{\partial\theta} \ln[f(x_i \mid H_1)]\big|_{\theta=0} &= -\frac{s_i}{f_n(x_i)} \frac{df_n(x_i)}{dx_i} \\ &= s_i g(x_i) \end{aligned} \tag{7.2.16}$$

where

$$g(x_i) = -\frac{1}{f_n(x_i)} \frac{df_n(x_i)}{dx_i}$$

By proceeding as in the case of i.i.d. observations, it can be seen that a locally optimal detector for Eq. (7.2.14) is similar to Fig. 7.2.1 with $b(x_i)$ replaced by $s_i g(x_i)$. This detector structure is shown in Fig. 7.2.2.

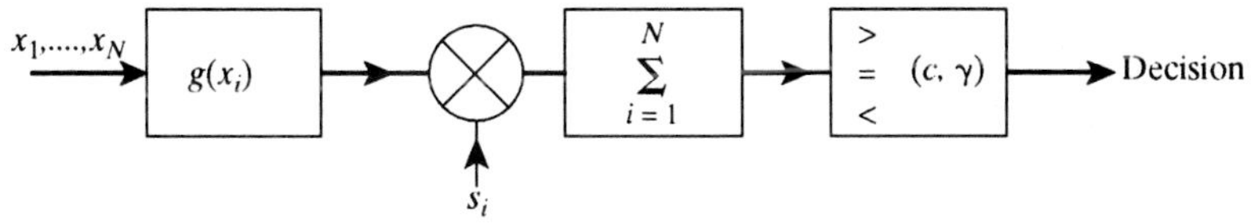

Figure 7.2.2 Locally optimal detector for nonconstant signal.

7.2.2 Generalized Gaussian Noise and Locally Optimal Detection

A random variable N is said to possess a density belonging to a generalized Gaussian family if

$$f_N(x) = \frac{c\eta(\sigma_n, c)}{2\Gamma(1/c)} e^{-[\eta(\sigma_n, c)|x|]^c} \qquad -\infty < x < \infty \tag{7.2.17}$$

where

$$\eta(\sigma_n, c) = \sigma_n^{-1} \left[\frac{\Gamma(3/c)}{\Gamma(1/c)}\right]^{1/2} \qquad c > 0 \tag{7.2.18}$$

The utility of Eq. (7.2.17) is that it has as special cases two standard distributions, and that with a proper choice of the parameter c, different distributions with varying degrees of tail decay can be obtained. For $c = 1$, $f_N(x)$ is the double exponential and for $c = 2$, it is Gaussian. For $c < 1$, one can get a heavy tail distribution, heavier than the double exponential. Similarly for $c > 2$, $f_N(x)$ has tails decaying faster than Gaussian. Several densities corresponding to different values of c are shown in Fig. 7.2.3.

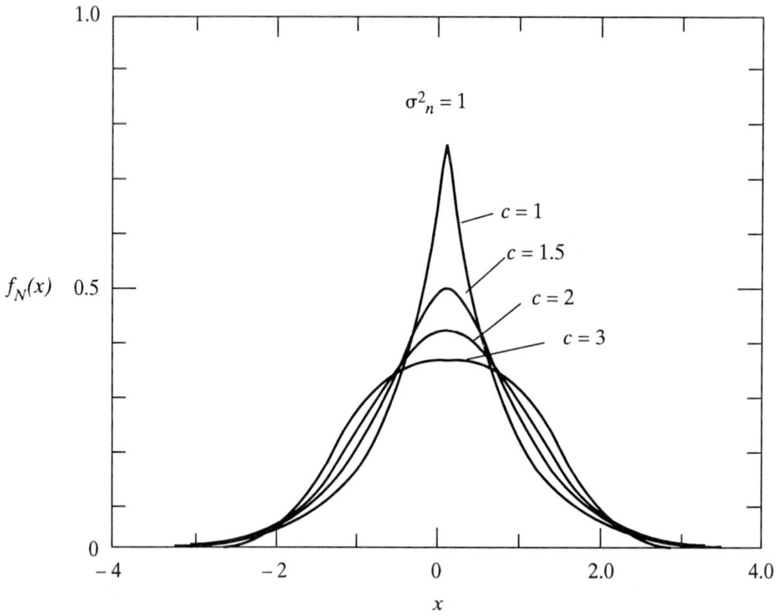

Figure 7.2.3 Generalized Gaussian density functions.

For detecting a constant positive signal in noise with sample size $n > 1$, of the generalized Gaussian family of Eq. (7.2.17), only the Gaussian density admits a UMP test. However, a locally optimal test can be found by using Eq. (7.2.9). In [18], the nonlinearity $b(x_i)$ for locally optimal tests for different values of c are given. This paper also derives ARE of locally optimal tests with respect to the linear detector.

For testing the two-sided hypotheses for weak signals, that is, H_0: $\theta = \theta_0$ versus H_1: $\theta \neq \theta_0$ but close to θ_0, a locally optimal test can be formulated. In this situation, the test will be locally best unbiased [i.e., $\phi(\theta_0) = \alpha$, $d\phi(\theta)/d\theta \,|_{\theta_0} = 0$, and the test maximizes $d^2\phi(\theta)/d\theta^2 \,|_{\theta_0}$]. For details, the reader is referred to Ferguson [17].

7.3 ROBUST DETECTION AND ESTIMATION

In Section 7.1 we considered nonparametric tests for the case where the density of the observation is not completely known, but some general property of the density, for example, that it is symmetric about the origin, could be assumed. In some cases the density function might be known to belong to a family F in the neighborhood of a nominal density f_0. As shown by Huber [19,20], in a number of such cases, techniques for detection and estimation that minimize the maximum risk for all densities in F can be obtained. Such techniques are said to be *robust*. For a detailed review of the application of Huber's result to signal detection and estimation problems, the reader is referred to the article by Kassam and Poor [21]. Here we present a brief discussion of Huber's results and their application to some problems.

7.3.1 Robust Hypothesis Testing

Let $(X_1, X_2, \ldots, X_n)$ be a set of i.i.d. observations having a common density p. The problem is to test whether $p = p_1$ or $p = p_0$ where

$$p_i = \{q \,|\, q = (1 - \epsilon_i)f_i + \epsilon_i H_i, H_i \in H\}, \qquad i = 0, 1 \tag{7.3.1}$$

$0 \le \epsilon_i < 1$ are fixed numbers and H denotes the class of all probability measures in the observation space. Usually, in practice, ϵ_i are small fractions, so that the p_1 or p_0 are close to their nominal densities f_1 and f_0, respectively. It is necessary that p_1 and p_0 do not overlap. That is, the families of densities p_1 and p_0 given by Eq. (7.3.1) have no common members. Let $R(q_i', \phi)$ be the risk or expected loss of the test $\phi(x)$ (average cost as defined in Chapter 3) when q_i' is the true underlying distribution. Consider the minimax problem

$$\text{Find } \min_{\phi} \sup_{q_1'} R(q_1', \phi), \text{ subject to } \sup_{q_0'} R(q_0', \phi) \le \alpha \tag{7.3.2}$$

Let q_0 and q_1 be two least favorable distributions, in that for any test $\phi(\cdot)$ the risk associated with q_i' is less than the risk associated with q_i for $i = 0, 1$. That is,

$$R(q_i', \phi) \le R(q_i, \phi) \qquad i = 0, 1 \tag{7.3.3}$$

It can be shown that the solution to the minimax problem of Eq. (7.3.2) can be obtained as a probability ratio test between the two least favorable distributions [19]. Therefore, once the least favorable densities q_1 and q_0 have been found, the solution to the minimax problem of Eq. (7.3.2) can be determined as the solution to the following problem:

$$\text{Find } \min_{\phi} R(q_1, \phi) \text{ subject to } R(q_0, \phi) \le \alpha \tag{7.3.4}$$

For example, when the risk equals a probability of error, the solution to the problem of Eq. (7.3.2) becomes the standard Neyman–Pearson test for deciding between q_1 and q_0.

A systematic procedure for finding the least favorable densities given in [19] results in the following expressions:

$$q_0(x) = \begin{cases} (1 - \epsilon_0) f_0(x) & \dfrac{f_1(x)}{f_0(x)} < c_0 \\[2ex] \dfrac{1}{c_0}(1 - \epsilon_0) f_1(x) & \dfrac{f_1(x)}{f_0(x)} \ge c_0 \end{cases} \tag{7.3.5a}$$

and

$$q_1(x) = \begin{cases} (1 - \epsilon_1) f_1(x) & \dfrac{f_1(x)}{f_0(x)} > c_1 \\[2ex] c_1(1 - \epsilon_1) f_0(x) & \dfrac{f_1(x)}{f_0(x)} \le c_1 \end{cases} \tag{7.3.5b}$$

The numbers $0 \le c_1 < c_0 < \infty$ have to be determined such that q_0 and q_1 are probability densities. That is,

$$(1 - \epsilon_0)\left\{P\left(\frac{f_1}{f_0} < c_0 \,|\, f_0\right) + \frac{1}{c_0} P\left(\frac{f_1}{f_0} \ge c_0 \,|\, f_1\right)\right\} = 1$$
$$(1 - \epsilon_1)\left\{P\left(\frac{f_1}{f_0} > c_1 \,|\, f_1\right) + c_1\, P\left(\frac{f_1}{f_0} \le c_1 \,|\, f_0\right)\right\} = 1 \tag{7.3.6}$$

The least favorable densities as given above are intuitively appealing. For example, wherever possible, $q_0(x)$ deviates from the nominal $f_0(x)$ and mimics $f_1(x)$!

The ϵ_i's may have to be sufficiently small so as to ensure that p_1 and p_0 do not overlap. In signal detection applications, this places a restriction that the signal level be sufficiently large [22]. The solution to Eq. (7.3.2) is given by the following probability ratio test:

$$\phi^*(\mathbf{x}) = \begin{cases} 1 & \text{if } T_n \equiv \sum_{i=1}^{n} \ln \frac{q_1(x_i)}{q_0(x_i)} > k \\ \tau & T_n = k \\ 0 & T_n < k \end{cases} \tag{7.3.7}$$

where

$$\frac{q_1(x_i)}{q_0(x_i)} = \begin{cases} bc_1 & \frac{f_1}{f_0} \le c_1 \\ b\frac{f_1(x)}{f_0(x)} & c_1 < \frac{f_1}{f_0} < c_0 \\ bc_0 & \frac{f_1}{f_0} \ge c_0 \end{cases} \tag{7.3.8}$$

and

$$b = \frac{1 - \epsilon_1}{1 - \epsilon_0}$$

Example 7.3.1

Consider the special case of Huber's model: $\epsilon_0 = \epsilon_1 = \epsilon$, $f_0(x)$ is the standard Gaussian distribution, and $f_1(x) = f_0(x - \theta)$, $\theta > 0$. This corresponds to the detection of a constant positive signal θ in nominally Gaussian noise. Here $b = 1$. From Eq. (7.3.6) it follows that c_1 and c_0 are the solutions to

$$G\left(\frac{\ln c_0}{\theta} + \frac{\theta}{2}\right) + \frac{1}{c_0}\left[1 - G\left(\frac{\ln c_0}{\theta} - \frac{\theta}{2}\right)\right] = \frac{1}{1 - \epsilon}$$
$$1 - G\left(\frac{\ln c_1}{\theta} - \frac{\theta}{2}\right) + c_1 G\left(\frac{\ln c_1}{\theta} + \frac{\theta}{2}\right) = \frac{1}{1 - \epsilon} \tag{7.3.9}$$

Since the condition $f_1/f_0 \leq c$ is equivalent to $x \leq \ln(c)/\theta + \theta/2$, Eq. (7.3.8) reduces to

$$\frac{q_1(x)}{q_0(x)} = \begin{cases} c_1 & x \leq \dfrac{\ln c_1}{\theta} + \dfrac{\theta}{2} \\ e^{\theta x - \theta^2/2} & \dfrac{\ln c_1}{\theta} + \dfrac{\theta}{2} < x < \dfrac{\ln c_0}{\theta} + \dfrac{\theta}{2} \\ c_0 & x \geq \dfrac{\ln c_0}{\theta} + \dfrac{\theta}{2} \end{cases} \tag{7.3.10}$$

The structure of the resulting detector based on the test statistic T_n in Eq. (7.3.7) is shown in Fig. 7.3.1. In order that p_1 and p_0 do not overlap, we require that $q_1 \neq q_0$. This is achieved when $\theta > \theta_\epsilon$, where θ_ϵ is the solution to

$$2G\left(\frac{\theta_\epsilon}{2}\right) = \frac{1}{1-\epsilon} \tag{7.3.11}$$

(see Exercise 7.7).

The robustness achieved by the nonlinearity of Fig. 7.3.1 can be quantitatively explained as follows. From the nominal $f_0(x)$ one does not expect large values of x, since the standard Gaussian density has a rapidly decaying tail. However with the ϵ-contaminant model, with the contaminant being arbitrary, a heavy tail density in p_0 can produce a large x. Therefore, to discriminate between p_0 and p_1, the influence of such large samples must be bounded. It is clear that in a linear detector which is optimal for testing the mean of a Gaussian distribution, a single large sample can lead to an erroneous decision of p_1 when the actual distribution is p_0. The linear detector is very sensitive and is extremely nonrobust with respect to the ϵ-contaminant model.

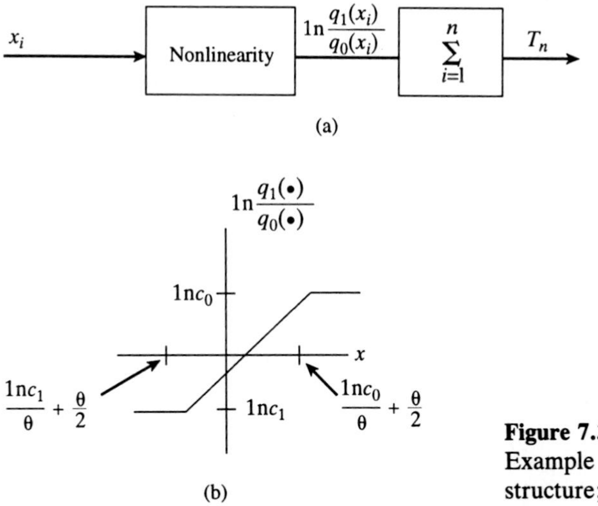

Figure 7.3.1 Robust detector for Example 7.3.1: (a) detector structure; (b) nonlinear function.

The extension of Huber's result to the detection of a nonconstant signal in nominally Gaussian noise is considered in [22]. In this case the observation model is of the form

$$X_i = \theta s_i + N_i \qquad i = 1, 2, \ldots, n \tag{7.3.12}$$

The signal amplitudes s_i are known, N_i are i.i.d. noise components with nominal standard Gaussian density. The robust test statistic for $\theta = 0$ versus $\theta > 0$ is similar to Fig. 7.3.1 except that each ith sample is passed through a specific soft limiter of the type in Fig. 7.3.1, but with the clipping levels dependent on s_i. Also, in order that the robust solution exists, it is necessary to have a strong signal $\theta > \theta_\epsilon/(s_i)_{\min}$, where θ_ϵ is given by Eq. (7.3.11).

7.3.2 Robust Estimation

As in the case of hypothesis testing, variation in the density of an observation from its nominal density can seriously affect the quality of an estimator of a parameter of the density. Consider, for example, the estimation of a location parameter θ. That is, the observations $X_1, X_2, \ldots, X_n$ are i.i.d. with density $f(x - \theta)$, where θ is the location parameter to be estimated. The sample mean $(1/n)\sum x_i$ and the sample median $\text{med}\{x_1, x_2, \ldots, x_n\}$ are common estimators of location. A maximum likelihood estimate (MLE) of θ can be found if the density of $f(x)$ is completely known. When $f(x)$ is Gaussian $\phi(x)$†, the MLE becomes the sample mean and when $f(x)$ is double exponential, the MLE becomes sample median. However, in the presence of even one or two outliers in the observations, that is, if there are even one or two samples with very large or small values that are inconsistent with the assumed $f(x)$, the quality of the sample mean degrades. This was first demonstrated by Tukey [23]. To illustrate this further, let us consider the following class of density functions [21]:

$$F = \{f \mid f = (1 - \epsilon)\phi + \epsilon h, h \in \mathrm{H}\} \tag{7.3.13}$$

where $\phi(x)$ is the standard normal, H is the class of all symmetric bounded densities and $\epsilon \in (0, 1)$. For this problem, Huber proposed a class of estimators, known as M-estimators, given by

$$\hat{\theta}(\mathbf{x}) = \arg\left\{\min_{\theta} \sum_{i=1}^{n} L(x_i - \theta)\right\} \tag{7.3.14}$$

where L is a function determining the estimator. The sample mean corresponds to choice $L(x) = x^2$ and the MLE corresponds to $L(x) = -\ln f(x)$. Assuming that the L is convex, symmetric about the origin, and sufficiently regular, the M-estimate based on L is consistent and has the property that $\sqrt{n}(\hat{\theta}(\mathbf{x}) - \theta)$ is asymptotically Gaussian with zero mean and variance $V(l, f)$, where [20]

$$V(l, f) = \frac{\displaystyle\int l^2(x) f(x)\, dx}{\left[\displaystyle\int l'(x) f(x)\, dx\right]^2} \tag{7.3.15}$$

with $l = dL(x)/dx$. Application of the Schwarz inequality to Eq. (7.3.15) shows that for a fixed f_0,

$$V(l, f_0) \geq V(l_0, f_0) = \frac{1}{I(f_0)} \tag{7.3.16}$$

† In this section $\phi(x)$ denotes standard Gaussian density.

where

$$l_0(x) = -\frac{1}{f_0(x)}\frac{df_0(x)}{dx} \quad \text{and} \quad I(f) = \int \frac{(df/dx)^2}{f(x)}\,dx$$

is the Fisher information for location (see Exercise 7.8).

For the density function of Eq. (7.3.13), the optimum estimator based on the nominal model $f = \phi$ is the sample mean, which corresponds to $l(x) = x$, so that

$$\begin{aligned} V(l,f) &= \int_{-\infty}^{\infty} x^2 f(x)\,dx \\ &= (1-\epsilon) + \epsilon\int_{-\infty}^{\infty} x^2 h(x)\,dx \end{aligned} \tag{7.3.17}$$

For $\epsilon > 0$, the asymptotic variance of the sample mean can be arbitrarily large since h is arbitrary. Hence the sample mean is a very nonrobust estimator with respect to the ϵ-contamination model. In the case of the sample median, $l(x) = \text{sgn}(x)$ [20] and

$$V(l,f) = \frac{1}{4f^2(0)} \leq \frac{1}{4(1-\epsilon)^2\phi^2(0)} = \frac{\pi}{2(1-\epsilon)^2} \tag{7.3.18}$$

Certainly, the sample median is more robust than the sample mean. However, when the density f is equal to the nominal density ϕ, the ratio of the variance of the sample median to the variance of the sample mean is as high as 1/0.64. Ideally, we would like to get an estimator that is nearly as good as it can be for the nominal density and that is also robust with respect to a contamination. Huber proposed the design of location estimators using a minimax formulation. That is, find the estimator that satisfies

$$\min_{l}\max_{f\in F} V(l,f) \tag{7.3.19}$$

With some assumptions on F, the solution to (7.3.19) is given by [20]

$$l_R = -\frac{1}{f_R(x)}\frac{df_R(x)}{dx} \tag{7.3.20}$$

where $f_R(x)$ is a least favorable density defined by

$$f_R = \arg\{\max_{f\in F} I(f)\} \tag{7.3.21}$$

The pair (l_R, f_R) is a saddle-point solution for Eq. (7.3.19) provided that the negative of $\ln f_R$ is convex.

For the case of nominal Gaussian, $l_R(x)$ is given by [24,25]

$$l_R(x) = \begin{cases} x & |x| < a \\ a\ \text{sgn}(x) & |x| \geq a \end{cases} \tag{7.3.22}$$

where

$$\int_{-a}^{a} \phi(x)\,dx + \frac{2\phi(a)}{b} = \frac{1}{1-\epsilon}$$

and

$$\frac{-\phi'(a)}{\phi(a)} = b$$

That is $l_R(x)$ is a soft limiter. A robust estimator $\hat{\theta}_R$ in the sense of Eq. (7.3.19) is given as the solution to

$$\sum_{i=1}^{n} l_R(x_i - \hat{\theta}_R) = 0 \tag{7.3.23}$$

SUMMARY

In this chapter we have presented some further topics in detection and estimation. In earlier chapters, the detection and estimation methods presented were based on the assumption that the probability distribution of the observation belongs to a parametric family. There are situations where (1) either sufficient information is not available to determine the parametric model that best fits the data or the nonstationary nature of the disturbance causes the probability distribution of the observation to belong to a wider class that cannot be indexed by a finite set of parameters, or (2) the probability distribution varies over a prescribed class in the neighborhood of a specified nominal distribution. For the first problem we have presented several nonparametric tests and their asymptotic (large sample, weak signal) relative efficiencies. Specifically, the sign test and Wilcoxon test are presented for the one-sample problem and the PCC, K-S test, run test, and Wilcoxon signed rank test are discussed as examples of two-sample testing problems. The nonparametric tests can be constructed to achieve a prescribed false alarm (type I error) without the knowledge of the exact distribution under the null hypothesis. The ARE of a nonparametric test with respect to a parametric test of a specific distribution is smaller than 1, but the ARE is greater than 1 for several distributions that deviate from the specific parametric test distribution.

For the second problem we have presented minimax robust solutions. Both hypothesis testing and parametric estimation problems are considered. There exists a voluminous amount of material on nonparametric and robust methods. Our discussion has been limited to introducing the basic ideas of these methods to the reader.

There are situations, such as underwater signal detection, where weak signals become important. Assuming known parametric model and i.i.d. samples, the detector that achieves the largest power for a given type I error for a weak signal is derived. Some examples of locally optimal detector are presented.

EXERCISES

7.1. For the Wilcoxon test, show that under H_0, $T_1, T_2, \ldots, T_n$ are mutually independent and

$$P(T_i = 0) = P(T_i = 1) = \tfrac{1}{2}$$

[*Hint:* The proof is built up in stages [26]. Let R_i^+ denote the rank of $|Z_i|$, $R^+ = (R_1^+, \ldots, R_n^+)$. The inverse permutation $D^+ = (R^+)^{-1}$, called the *vector of antiranks*, $D^+ = (D_1^+, \ldots, D_n^+)$, is such that $|Z_i| = |Z|_{(R_i^+)}, |Z|_{(j)} = |Z_{D_j^+}|$. Since R_i^+ is the rank of $|Z_i|$, in the ordering of $\{|Z_1|, |Z_2|, \ldots, |Z_n|\}$, $|Z_i| = |Z|_{(R_i^+)}$ simply states the fact that the (R_i^+)th-order statistic of $\{|Z_l| l = 1, \ldots, n\}$ is nothing but $|Z_i|$. Similarly, the sample that assumes the jth-order statistic, $|Z|_{(j)}$, is given by $Z_{D_j^+}$. Notice that R^+ and D^+ form a one-to-one transformation. (i) Show that the random variables $u(Z_1), \ldots, u(Z_n)$ and the random vector R^+ (and hence D^+) are mutually independent; R^+ is a permutation of $(1, 2, \ldots, n)$ with probability 1 and $P(u(Z_i) = 0) = P(u(Z_i) = 1) = \frac{1}{2}$. To prove this part, consider the following. Since Z_i are mutually independent by pairs, $(u(Z_i), |Z_i|), 1 \le i \le n$ are also mutually independent. If we prove that $u(Z_i)$ is independent of $|Z_i|$, it will follow that the random variables $u(Z_1), \ldots, u(Z_n), |Z_1|, \ldots, |Z_n|$ are mutually independent. Hence show that $P(u(Z_1) = k, |Z_1| \le z) = P(u(Z_i) = k)P(Z_i \le z), k = 0, 1$. (ii) Observe that $T_i = u(Z_{D_i^+})$. Show that $P(T_i = t_i, 1 \le i \le n \mid D^+ = d) = (\frac{1}{2})^n$ for any d. According to (a), $(T_1, \ldots, T_n)$, a function of $(u(Z_1), \ldots, u(Z_n))$, is independent of $(|Z_1|, \ldots, |Z_n|)$ and hence independent of R^+ and D^+.]

7.2. Show that the efficacy of the t-test equals $1/\sigma_x^2$. [*Hint:* Show that for large n, $\mathrm{E}(S_n^*) = \sqrt{n}\theta/\sigma_x$, $\mathrm{var}(S_n^*) = 1/\sigma_x^2$. Hence find the efficacy.]

7.3. Show that the ARE of PCC with respect to the normal test Eq. (7.1.54) is

$$(\gamma_1^4 + \gamma_2^4 + 4\sigma_1^2\sigma_2^2 - \sigma_1^4 - \sigma_2^4) f_1^2(0) f_2^2(0)$$

where

$$f_j(x) = \frac{dF_j(x)}{dx} \qquad j = 1, 2$$

$$\gamma_1^4 = \mathrm{E}((X_i - \mathrm{E}(X_i))^4)$$

$$\gamma_2^4 = \mathrm{E}((Y_i - \mathrm{E}(Y_i))^4)$$

[*Hint:* To find the efficacy of Eq. (7.1.54), observe that for large n, $\sum_{i=1}^n (X_i + Y_i)^2$ is normal according to the central limit theorem. Find the mean and the variance of the test statistic and hence find its efficacy. Notice that the alternative H_1 tends to H_0, when $\sigma_3^2 \to 0$ (small-signal case). Remember that under H_1, both X_i and Y_i have the same additive signal component. Also, since the convolution of two symmetric densities is symmetric, X_i and Y_i both have zero mean and zero median under H_1 and H_0. To find the efficacy of PCC, observe that $\sum_{i=1}^n u(X_i, Y_i)$ is distributed as a binomial (n, p) with $p = \frac{1}{2} + 2\int_{-\infty}^{\infty} (F_1(x) - \frac{1}{2})(F_2(x) - \frac{1}{2})\, dF_3(x)$. For small signals, $p \approx \frac{1}{2} + 2f_1(0)f_2(0)\sigma_3^2$.]

7.4. Find the mean and the variance of U. *Hint:* $\mathrm{E}(U) = \sum_i \sum_j P(X_i \ge Y_i) = n_1 n_2 p$, $p = P(X_i \ge Y_j)$. If H_0 is true, $p = \frac{1}{2}$.

$$E(U^2) = E\left[\sum_i \sum_j I_{[Y_j, \infty]}(X_i)\right]^2.$$

7.5. The fundamental Neyman–Pearson lemma (Chapter 3) is valid even if the two functions

f_1 and f_0 are not densities. That is, f_1 and f_0 can assume negative values. To derive a locally optimal test, consider this problem. Given a set of values $\mathbf{x} = (x_1, x_1, \ldots, x_n)$, a rule $\phi(\mathbf{x})$ divides the space $\mathbf{x} \in \mathcal{R}_n$ into two mutually exclusive regions, I and I*. Let $h_1(\mathbf{x})$ and $h_0(\mathbf{x})$ be two functions such that the integrals defined below exist. Among all the rules that satisfies $\int_{\mathrm{I}} \cdots \int h_0(\mathbf{x})\,d\mathbf{x} = \alpha$, the following rule

$$\mathbf{x} \in \begin{cases} \mathrm{I} & \text{if } T(\mathbf{x}) = \dfrac{h_1(\mathbf{x})}{h_0(\mathbf{x})} > c \\ \mathrm{I}^* & \text{if } T < c \\ \text{randomly assign I for a fraction } \gamma \text{ of the times when } T = c & \end{cases}$$

achieves the largest $\int_{\mathrm{I}} \cdots \int h_1(\mathbf{x})\,dx$. Taking $h_1(\mathbf{x})$ as $\dfrac{\partial}{\partial \theta} \prod_{i=1}^{n} f(X_i; \theta)\Big|_{\theta=\theta_0}$ and $h_0(\mathbf{x})$ as $\prod_{i=1}^{n} f(X_i, \theta_0)$, we obtain the desired locally optimal test (7.2.9).

7.6. Show that the locally optimal detector for testing a constant signal in Laplace noise is a hard limiter. That is, in Fig. 7.2.1, $b(X_i) = \operatorname{sgn}(X_i)$.

7.7. In Example 7.3.1 we require that $q_0 \neq q_1$. Show that this is achieved whenever $\theta > \theta_\epsilon$, where θ_ϵ satisfies

$$2G\left(\frac{\theta_\epsilon}{2}\right) = \frac{1}{1-\epsilon}$$

7.8. Establish the inequality in Eq. (7.3.16).

REFERENCES

1. J. B. Thomas, Nonparametric detection, *Proc. IEEE*, May, 623–631 (1970).
2. J. D. Gibbons and S. Chakraborti, *Nonparametric Statistical Inference*, 3rd ed., Marcel Dekker, New York, 1992.
3. J. L. Hodges and E. L. Lehmann, The efficiency of some nonparametric competitors to the *t*-test, *Ann. Math. Stat.*, **27**, 324–335 (1956).
4. A. M. Mood, F. A. Greybill, and D. C. Boes, *Introduction to the Theory of Statistics*, 3rd ed., McGraw Hill, New York, 1974.
5. M. Fisz, *Theory of Probability and Mathematical Statistics*, Wiley, New York, 1963.
6. S. A. Kassam, Bibliography on nonparametric tests, *IEEE Trans. Inf. Theory*, September, 596–602 (1980).
7. G. W. Zeoli and T. S. Fong, Performance of a two sample Mann–Whitney nonparametric detector in a radar application, *IEEE Trans. Aerosp. Electron. Syst.*, September, 951–959 (1971).
8. V. G. Hansen and B. A. Olsen, Nonparametric radar extraction using a generalized sign test, *IEEE Trans. Aerosp. Electron. Syst.*, September, 942–950 (1971).
9. E. K. Al-Hussaini, Trimmed generalized sign and modified median detectors for multiple target situations, *IEEE Trans.*, July, 573–575 (1979).
10. E. K. Al-Hussaini and L. F. Turner, The asymptotic performance of two sample non-parametric detectors when detecting fluctuating signals in non-Gaussian noise, *IEEE Trans. Inf. Theory*, January, 124–127 (1979).

11. G. M. Dillard and C. E. Antoniak, A practical distribution-free detection procedure for multiple-range-bin radar, *IEEE Trans. Aerosp. Electron. Syst.*, September, 629–635 (1970).

12. M. N. Woinsky, Nonparametric detection using spectral data, *IEEE Trans. Inf. Theory*, January, 110–118 (1972).

13. R. Viswanathan and S. C. Gupta, Nonparametric receiver for FH-MFSK mobile radio, *IEEE Trans. Commun.*, February, 178–184 (1985).

14. D. P. Grybos, Nonparametric detectors with applications to spread spectrum, Ph.D. dissertation, Purdue University, West Lafayette, IN, 1980.

15. S. A. Kassam, *Signal Detection in Non-Gaussian Noise*, Springer-Verlag, New York, 1988.

16. J. Capon, On the asymptotic efficiency of locally optimum detectors, *IRE Trans. Inf. Theory*, April, 67–71 (1961).

17. T. S. Ferguson, *Mathematical Statistics*, Academic Press, New York, 1967.

18. J. H. Miller and J. B. Thomas, Detectors for discrete time signals in non-Gaussian noise, *IEEE Trans. Inf. Theory*, March, 241–250 (1972).

19. P. J. Huber, A robust version of the probability ratio test, *Ann. Math. Stat.*, **36**, 1753–1758 (1965).

20. P. J. Huber, Robust estimation of a location parameter, *Ann. Math. Stat.*, **35**, 73–104 (1964).

21. S. A. Kassam and H. V. Poor, Robust techniques for signal processing: a survey, *Proc. IEEE*, **73**, March, 433–481 (1985).

22. R. D. Martin and S. C. Schwartz, Robust detection of a known signal in nearly Gaussian noise, *IEEE Trans. Inf. Theory*, **IT-17**, January, 50–56 (1971).

23. J. W. Tukey, A survey of sampling from contaminated distributions, in *Contributions to Probability and Statistics*, I. Olkin et al., Eds., Stanford University Press, Stanford, CA, 1960, pp. 448–485.

24. P. J. Huber, *Robust Statistics*, Wiley, New York, 1981.

25. H. V. Poor, *An Introduction to Signal Detection and Estimation*, Springer-Verlag, New York, 1988.

26. J. Hájek, *Nonparametric Statistics*, Holden-Day, San Francisco, 1969.

8

Applications to Communication and Radar Systems

8.1 INTRODUCTION TO COMMUNICATION SYSTEMS

A communication system can be loosely defined as one designed to transmit information in an efficient manner. The usual goal of a communication system is to eliminate or minimize the effects of unwanted disturbances while maximizing the amount of information being transmitted. Communication systems may be digital or analog. In analog systems, the information to be transmitted (message signal) is a continuous waveform. While the message signal can be transmitted directly, in most applications, particularly those involving transmission over long distances, the message signal modulates a carrier, a sinusoidal signal of known amplitude and frequency. The message may modulate either the amplitude or angle of the carrier waveform.

In a digital communication system, the information to be transmitted is discrete, as, for example, with the output of a digital computer. In many instances, analog data are sampled and quantized before transmission, so that the information is in a discrete form. In both cases, the data are usually encoded into a sequence of symbols from a finite alphabet. With each symbol of the alphabet, we associate a pulse $y_i(t), i = 1, \ldots, M$ of duration T, where $\{y_i(t)\}$ are a set of known waveforms. The transmission of a symbol is then accomplished by the transmission of the associated waveform.

The transmitted signals in most cases undergo distortion over the transmitting medium. This distortion can usually be modeled as additive noise at the receiving end. In the case of an analog communication system, the receiver therefore must be designed to recover (estimate) the message signal. In a digital communication system, the receiver must decide which waveform was transmitted over each interval of duration T and thus decide which symbol was actually sent. The sequence of symbols at the receiver is then decoded to recover the message.

In this chapter we consider the application of the techniques of detection and estimation to the design of optimal communication systems. We present a brief discussion of spread spectrum communication systems. The problem of encoding and decoding discrete information is a significant step in digital communication systems. We will not, however, discuss coding in this chapter, as it is outside the scope of the present book. The interested reader is referred to [1, 2] for a treatment of this subject.

We start our discussions in this chapter with a consideration of various digital transmission schemes. We derive optimal receivers for these systems and evaluate their performance. We also discuss the problem of intersymbol interference which arises when the duration of a pulse waveform at the receiver exceeds the spacing between transmitted pulses so that two or more pulses interact to produce distortion at the receiver.

8.2 DIGITAL COMMUNICATION

Figure 8.2.1 illustrates the basic components of a digital communication system. In this section we discuss some basic digital schemes and derive the associated receiver structures. We restrict our discussions to binary channels in which the alphabet consists of the symbols 0 and 1. Thus the output of the encoder will be a sequence of 1's and 0's appearing at a rate of one every T seconds. Associated with a 0 is the signal $y_0(t)$ and with a 1, the signal $y_1(t)$.

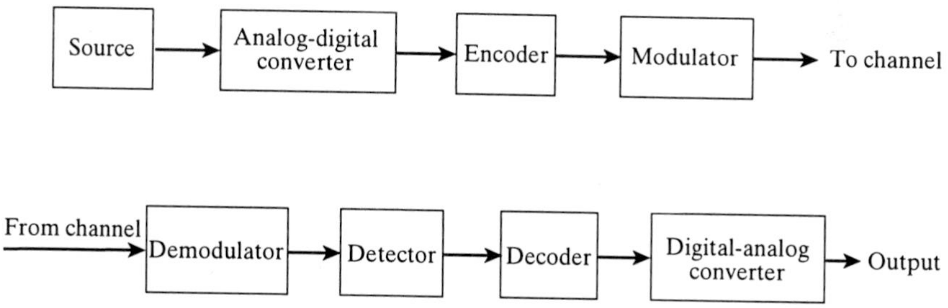

Figure 8.2.1 Block diagram of digital communication system.

We first consider baseband transmission schemes in which the waveforms $y_0(t)$ and $y_1(t)$ are transmitted directly. For communication over long distances, as indicated earlier, the waveforms modulate a high-frequency carrier. In the latter case, if a local reference is available at the receiver for demodulation which is in phase with the carrier component of the received signal, the system is said to be *coherent*. Otherwise, it is said to be *noncoherent*. Similarly, if a periodic signal (referred to as a *clock*) is available at the receiver, which is in synchronism with the transmitted sequence of pulses, the system is said to be *synchronous*. If such a clock is not available or necessary, the system is said to be *asynchronous*. Material related to this section can be found in [3–5].

8.2.1 Baseband Digital Communication

Consider the binary digital communication system in which the signal pulse $y_1(t)$ represents a 1 and $y_0(t)$ represents a 0. At the receiver, the transmitted signal is assumed to be observed in additive white Gaussian noise $v(t)$ of two-sided spectral density $N_0/2$. Over each signaling interval of T seconds (bit interval), the receiver must make a decision as to whether $y_1(t)$ or $y_0(t)$ was transmitted.

If we cast the problem into the decision framework of Chapters 3 and 4, the observations can be modeled as

$$\begin{aligned} H_1&: z(t) = y_1(t) + v(t) \qquad 0 \leq t \leq T \\ H_0&: z(t) = y_0(t) + v(t) \qquad 0 \leq t \leq T \end{aligned} \tag{8.2.1}$$

In most communication problems an error in detecting either 1 or 0 is equally important. If we assume that the prior probabilities for the two are also equal, a natural choice is the minimum probability of error criterion discussed in Section 3.3. We recall that the threshold for the test under this criterion is unity. In terms of the log-likelihood ratio, the test is [see Eq. (4.2.28)]

$$\int_0^T z(t)[y_1(t) - y_0(t)]\,dt \underset{H_0}{\overset{H_1}{\gtrless}} \frac{1}{2}\int_0^T [y_1^2(t) - y_0^2(t)]\,dt \tag{8.2.2}$$

The test can thus be implemented by correlating the observation $z(t)$ with the difference signal $[y_1(t) - y_0(t)]$ and comparing the correlator output at time T with a threshold. If the signals are of equal energy, the threshold is zero. In this case we only need to sense the sign of the correlator output at time T. The receiver structure is shown in Fig. 8.2.2.

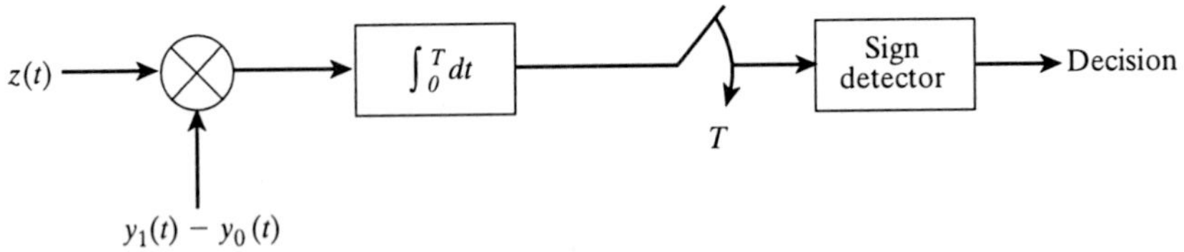

Figure 8.2.2 Receiver structure for baseband binary digital communication system with equal energy signals.

Let us now consider the polar binary system in which the signals are

$$\begin{aligned} y_1(t) &= \sqrt{\frac{E}{T}} \qquad 0 \leq t \leq T \\ y_0(t) &= -\sqrt{\frac{E}{T}} \qquad 0 \leq t \leq T \end{aligned} \tag{8.2.3}$$

The waveforms of Eq. (8.2.3) are referred to as *non-return-to-zero* (NRZ) waveforms. Other formats include the *return to zero* (RZ), in which

$$y_1(t) = \begin{cases} \sqrt{\dfrac{E}{T}} & 0 \leq t \leq \tau \\ 0 & \tau < t \leq T \end{cases}$$

$$y_0(t) = \begin{cases} -\sqrt{\dfrac{E}{T}} & 0 \leq t \leq \tau \\ 0 & \tau < t \leq T \end{cases}$$

while in the split-phase (Manchester) format,

$$y_1(t) = \begin{cases} \sqrt{\dfrac{E}{T}} & 0 \leq t < T/2 \\ -\sqrt{\dfrac{E}{T}} & T/2 \leq t \leq T \end{cases}$$

$$y_0(t) = \begin{cases} -\sqrt{\dfrac{E}{T}} & 0 \leq t < T/2 \\ \sqrt{\dfrac{E}{T}} & T/2 \leq t \leq T \end{cases}$$

We generally assume the NRZ format in all our subsequent discussions. The receiver of Fig. 8.2.2 now reduces to the structure of Fig. 8.2.3. The average energy of the two signals is given by

$$\epsilon = \tfrac{1}{2}\int_0^T y_1^2(t)\,dt + \tfrac{1}{2}\int_0^T y_0^2(t)\,dt = E \tag{8.2.4}$$

while their correlation is

$$\bar{\rho} = \frac{1}{\epsilon}\int_0^T y_1(t)y_0(t)\,dt = -1 \tag{8.2.5}$$

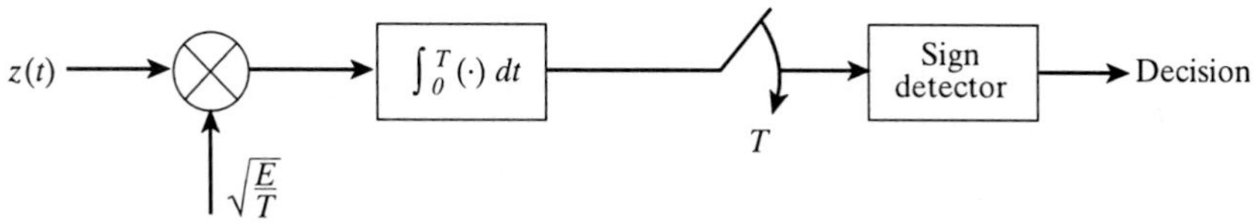

Figure 8.2.3 Optimum receiver for polar binary signals.

so that the system is an ideal binary communication system. We recall that the probability of error for this problem is just P_F or P_M and is obtained from Eq. (4.2.52) as

$$P_e = \int_{[\epsilon(1-\bar{\rho})/N_0]^{1/2}}^{\infty} \frac{1}{\sqrt{2\pi}} \exp\left(-\frac{u^2}{2}\right) du \tag{8.2.6}$$

Substitution of Eqs. (8.2.4) and (8.2.5) thus yields the expression for the probability of error as

$$P_e = \int_{[2\epsilon/N_0]^{1/2}}^{\infty} \frac{1}{\sqrt{2\pi}} \exp\left(-\frac{u^2}{2}\right) du = Q\left[\left(\frac{2\epsilon}{N_0}\right)^{1/2}\right] \tag{8.2.7}$$

Thus for binary polar signals, the probability of error decreases as $2\epsilon/N_0$, the ratio of the average energy per digit to noise power spectral density increases.

For the case of unipolar binary signals, 1 is represented by transmitting a rectangular pulse of height $\sqrt{E/T}$ and 0 by nontransmission. The decision rule is given by Eq. (8.2.2), where the threshold is now given by

$$\gamma = \frac{1}{2}\int_0^T [y_1^2(t) - y_0^2(t)]\, dt = \frac{E}{2} \tag{8.2.8}$$

The average signal energy ϵ is now equal to $E/2$, while the correlation coefficient ρ is zero. The probability of error follows from Eq. (8.2.6) as

$$P_e = \int_{(E/2N_0)^{1/2}}^{\infty} \frac{1}{\sqrt{2\pi}} e^{-u^2/2}\, du = Q\left[\left(\frac{\epsilon}{N_0}\right)^{1/2}\right] \tag{8.2.9}$$

Note that for a given ratio of average energy per digit to the noise spectral density, the polar binary signal performs better than the unipolar signal. For a given probability of error, unipolar signal transmission requires 3 dB greater average energy per bit-to-noise power density ratio than bipolar transmission. In fact, we note that the binary polar set is an optimal signal set since the correlation between the two signals is -1 and therefore yields the lowest probability of error for a given ϵ/N_0 ratio. The binary polar set belongs to a class of signals known as *simplex signals*. A set of M signals are said to be *simplex* if they are equally likely with correlation coefficient $-[1/(M-1)]$ [3,5]. Even though such a signal set is optimum, orthogonal signal sets are more widely used in practice when $M > 2$.

8.2.2 Radio-Frequency Digital Transmission

Long-haul data transmission systems rarely employ baseband transmission. Usually, the digital data modulate a sinusoidal carrier, and regenerative repeaters are used at intervals along the transmission path to maintain a certain signal-to-noise ratio. The basic digital modulation schemes are amplitude-shift keying (ASK), frequency-shift keying (FSK), and phase-shift keying (PSK). In the binary case ASK takes the form of transmission for 1 and no transmission for 0, and is called on–off keying (OOK).

In this section we derive the structure and performance of the optimum receivers for these various cases. We consider the operation of some of these receivers in fading media and indicate methods to improve their performance. The coherent detection systems to be considered first follow from our development of correlation receivers for binary communication in Section 4.2. Our discussion of noncoherent systems follows as an extension of the results of Section 4.6.

Coherent systems. We recall that in a coherent system, a local reference is available at the receiver which is in phase with the incoming carrier. Since the carrier phase is known, for purposes of analyzing coherent systems we can, without loss of generality, set the carrier phase equal to zero. We now proceed to analyze the various schemes.

The on–off keying system (OOK) employs $y_1(t) = \sqrt{2E/T} \sin \omega_c t$ for *mark* signal (digital 1) and $y_0(t) = 0$ for *space* signal (digital 0) over each one digit interval of T seconds. We again assume that the received signal $z(t)$ consists of the transmitted signal $y_i(t), i = 0, 1$ corrupted by additive white Gaussian noise of power spectral density $N_0/2$. The optimum receiver is the correlation receiver and takes the form shown in Fig. 8.2.4. For the case of equal probability of occurrence of mark and space and minimum probability of error criterion, it can be seen from Eq. (4.2.28) that the threshold is $E/2$. This case is similar to unipolar baseband transmission. Again, the average energy per digit is $E/2$ and its ratio to the noise power spectral density is E/N_0. Since the correlation coefficient $\bar{\rho}$ is zero, the probability of error is obtained from Eq. (8.2.6) as

$$\begin{aligned} P_e &= \int_{(E/2N_0)^{1/2}}^{\infty} \frac{1}{\sqrt{2\pi}} \exp\left(\frac{-u^2}{2}\right) du \\ &= Q\left[\left(\frac{\epsilon}{N_0}\right)^{1/2}\right] \end{aligned} \tag{8.2.10}$$

The receiver in Fig. 8.2.4 can be practically implemented as a multiplier (mixer) followed by a lowpass filter, or as a bandpass filter with center frequency ω_c.

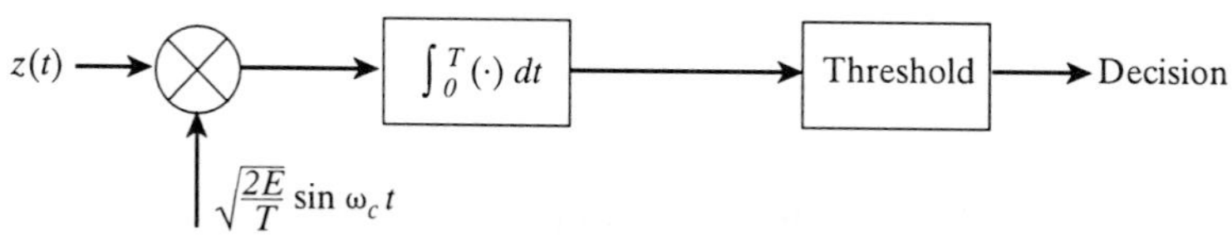

Figure 8.2.4 Optimum receiver for OOK.

The frequency-shift keying system (FSK) employs

$$y_1(t) = \sqrt{\frac{2E}{T}} \sin \omega_1 t$$

for mark, and $y_0(t) = \sqrt{2E/T}\sin\omega_0 t$ for space signals, over each interval of T seconds. Usually, the carrier frequencies are chosen such that they are much greater than $2\pi/T$. In this case the correlation coefficient $\bar{\rho}$ is given by

$$\bar{\rho} = \frac{1}{\epsilon}\int_0^T y_1(t)y_0(t)\,dt = \frac{1}{\epsilon}\frac{2E}{T}\int_0^T \sin\omega_1 t\,\sin\omega_0 t\,dt \approx 0 \tag{8.2.11}$$

so that the two signals are orthogonal. This digital transmission system is known as orthogonal FSK.

The received signal $z(t)$ is either $y_1(t)$ or $y_0(t)$ additively corrupted by white noise of spectral density $N_0/2$. The optimum receiver consists of two correlators and the difference signal is compared to 0 as shown in Fig. 8.2.5. The probability of error can be obtained from Eq. (8.2.6) by noting that the average energy is

$$\begin{aligned}\epsilon &= \frac{1}{2}\int_0^T [y_1^2(t) + y_0^2(t)]\,dt \\ &= \frac{E}{T}\int_0^T (\sin^2\omega_1 t + \sin^2\omega_0 t)\,dt \\ &= E\end{aligned} \tag{8.2.12}$$

so that

$$P_\epsilon = Q\left[\left(\frac{E}{N_0}\right)^{1/2}\right] \tag{8.2.13}$$

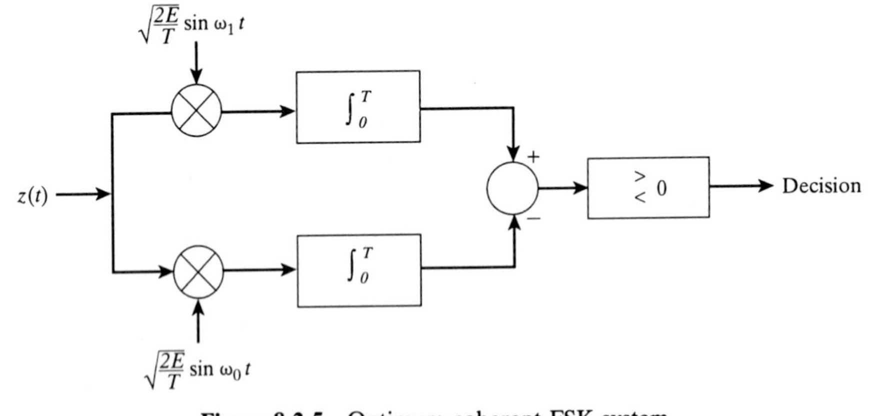

Figure 8.2.5 Optimum coherent FSK system.

The average energy per digit to noise spectral density ratio for FSK is $2E/N_0$ as compared to E/N_0 for OOK. Comparison of Eqs. (8.2.10) and (8.2.13), therefore, shows that for the same ratio of average energy per digit to noise spectral density, coherent OOK performs as well as orthogonal FSK. However, orthogonal FSK is preferred in practice because of its performance in fading channels. This is discussed in the latter part of this section.

In the case of phase-shift keying, the signals for mark and space are $y_1(t) = \sqrt{2E/T}\sin\omega_c t$ and $y_0(t) = -\sqrt{2E/T}\sin\omega_c t$, respectively. This is also referred to as *phase reversal keying* (*PRK*) *or binary phase-shift keying* (*BPSK*). Since the correlation coefficient is -1, the system is an ideal system. The average signal energy is E. The optimum receiver in the presence of additive white noise of spectral density $N_0/2$ is shown in Fig. 8.2.6.

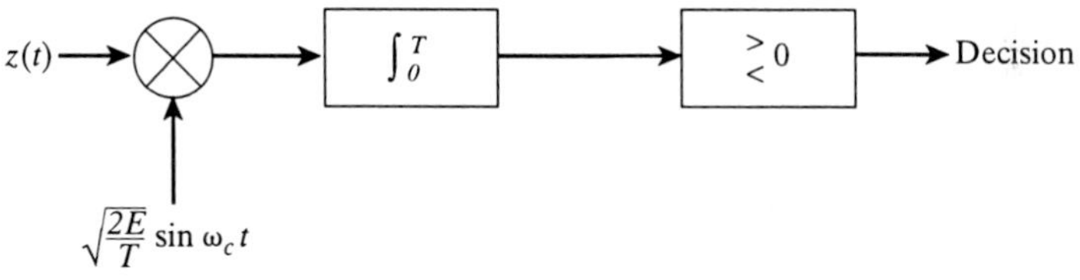

Figure 8.2.6 Optimum receiver for PSK.

We point out that a PSK system, by its very nature, must be coherent, since it would be impossible to convey information in the phase of a carrier which is completely random. The probability of error easily follows from Eq. (8.2.6) as

$$P_e = Q\left[\left(\frac{2E}{N_0}\right)^{1/2}\right] \tag{8.2.14}$$

For a given acceptable level of probability of error, the PRK system requires 3 dB less average energy per digit than orthogonal FSK as can be seen by comparing Eqs. (8.2.14) and (8.2.13).

Noncoherent systems. In noncoherent systems, the phase of the carrier is not available at the receiver, or cannot be extracted from the incoming signal. As such, we may expect the performance of a noncoherent system to be degraded in comparison to the corresponding coherent system.

In the case of noncoherent OOK, the received signals under the two hypotheses are

$$H_1\colon z(t) = \sqrt{\frac{2E}{T}}\sin(\omega_c t + \theta) + v(t)$$

and

$$H_0\colon z(t) = v(t) \tag{8.2.15}$$

over each time interval of T seconds. In the case of noncoherent systems θ is an unknown parameter. The additive noise is again assumed to be zero mean, Gaussian, white with spectral density $N_0/2$.

If θ is a random variable with known probability density function $p(\theta)$, we can first determine the conditional likelihood ratio $\Lambda(z \mid \theta)$ and then integrate over the known density $p(\theta)$ to obtain the likelihood $\Lambda(z)$ as

$$\Lambda(z(t)) = \int_{-\pi}^{\pi} \Lambda(z(t) \mid \theta) p(\theta)\, d\theta \tag{8.2.16}$$

The conditional likelihood ratio easily follows from Eq. (4.2.26) as

$$\Lambda(z(t) \mid \theta) = \exp\left\{\frac{2}{N_0}\left[\int_0^T z(t)\sqrt{\frac{2E}{T}} \sin(\omega_c t + \theta)\, dt\right] - \frac{E}{N_0}\right\} \tag{8.2.17}$$

Let us now define the following quantities:

$$L_s \triangleq \int_0^T \sqrt{\frac{2}{T}} z(t) \sin \omega_c t\, dt$$

and

$$L_c \triangleq \int_0^T \sqrt{\frac{2}{T}} z(t) \cos \omega_c t\, dt \tag{8.2.18}$$

It can easily be verified that L_s and L_c are Gaussian random variables under the two hypotheses with means and variances as follows:

$$\mathrm{E}\{L_s \mid H_0\} = \mathrm{E}\{L_c \mid H_0\} = 0 \qquad \mathrm{var}\{L_s \mid H_0\} = \mathrm{var}\{L_c \mid H_0\} = \frac{N_0}{2}$$

$$\mathrm{E}\{L_s \mid H_1, \theta\} = \sqrt{E} \cos\theta \qquad \mathrm{E}\{L_c \mid H_1, \theta\} = \sqrt{E} \sin\theta$$

$$\mathrm{var}[L_s \mid H_1, \theta] = \mathrm{var}[L_c \mid H_1, \theta] = \frac{N_0}{2}$$

Let us now assume that θ is uniformly distributed in $[-\pi, \pi]$. It then follows that in terms of L_s and L_c, we can write the likelihood ratio as

$$\begin{aligned}\Lambda(z(t)) &= \frac{1}{2\pi}\int_{-\pi}^{\pi}\left\{\exp\left[\frac{2\sqrt{E}}{N_0}(L_c \sin\theta + L_s \cos\theta) - \frac{E}{N_0}\right]\right\} d\theta \\ &= I_0\left[\frac{2\sqrt{E}}{N_0}(L_c^2 + L_s^2)^{1/2}\right] \exp\left(-\frac{E}{N_0}\right)\end{aligned} \tag{8.2.19}$$

where $I_0(\cdot)$ is the modified Bessel function of the first kind. The decision rule for equiprobable mark and space, therefore, is

$$\ln\left\{I_0\left[\frac{2\sqrt{E}}{N_0}(L_c^2 + L_s^2)^{1/2}\right]\right\} \underset{H_0}{\overset{H_1}{\gtrless}} \frac{E}{N_0} \tag{8.2.20}$$

For small values of x, $\ln I_0(x) \simeq x^2/4$, whereas for large values of x, $\ln I_0(x) \simeq x$. It therefore follows that for either low signal-to-noise ratios or high signal-to-noise ratios the decision rule is approximately given by

$$(L_c^2 + L_s^2)^{1/2} \underset{H_0}{\overset{H_1}{\gtrless}} \gamma \tag{8.2.21}$$

where γ is an appropriate threshold. The receiver can be implemented as a bandpass filter in cascade with an envelope detector [3,5]. The structure of the receiver is shown in Fig. 8.2.7. The new threshold γ is calculated according to the approximation used. We note that $L = (L_c^2 + L_s^2)^{1/2}$ is a sufficient statistic.

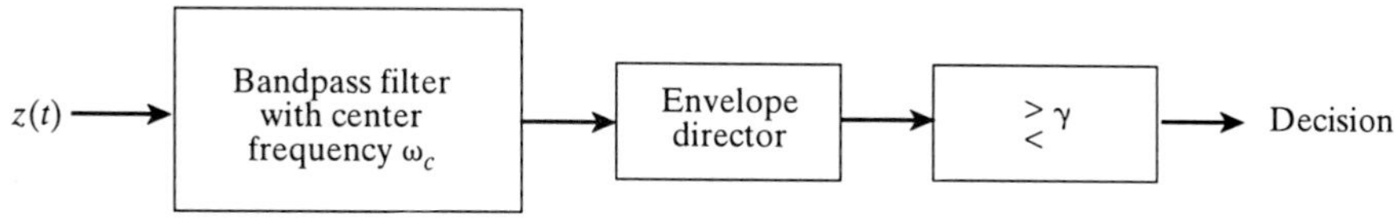

Figure 8.2.7 Noncoherent OOK receiver.

To determine the probability of error, we need to know $p(l\,|\,H_0)$ and $p(l\,|\,H_1)$. Since L_s and L_c are Gaussian random variables, it easily follows that

$$p(l\,|\,H_0) = \begin{cases} \dfrac{2l}{N_0} e^{-l^2/N_0} & l \geqslant 0 \\ 0 & l < 0 \end{cases} \tag{8.2.22a}$$

and

$$p(l\,|\,H_1) = \int_{-\pi}^{\pi} p(l\,|\,H_1, \theta) p(\theta)\, d\theta \tag{8.2.22b}$$

where

$$p(l\,|\,H_1, \theta) = \begin{cases} \dfrac{2l}{N_0} \exp\left(-\dfrac{l^2 + E}{N_0}\right) I_0\left(\dfrac{2l\sqrt{E}}{N_0}\right) & l \geqslant 0, -\pi \leqslant \theta \leqslant \pi \\ 0 & l < 0 \end{cases} \tag{8.2.23}$$

The probability of error P_{e_1} when a mark was transmitted is then given by

$$\begin{aligned} P_{e_1} &= \int_0^{\gamma} p(l\,|\,H_1)\, dl = 1 - \int_{\gamma}^{\infty} p(l\,|\,H_1)\, dl \\ &= 1 - \int_{\gamma}^{\infty} \frac{2}{N_0} l \exp\left(-\frac{l^2 + E}{N_0}\right) I_0\left(\frac{2l\sqrt{E}}{N_0}\right) dl \\ &= 1 - Q\left[\left(\frac{2E}{N_0}\right)^{1/2}, \left(\frac{2}{N_0}\right)^{1/2} \gamma\right] \end{aligned} \tag{8.2.24}$$

where $Q(a, b)$ are Marcum's Q-functions, defined as

$$Q(a, b) = \int_b^{\infty} z \exp\left(-\frac{z^2 + a^2}{2}\right) I_0(az)\, dz \tag{8.2.25}$$

and are tabulated in [6–8]. Similarly, the probability of error when space was transmitted can be determined as

$$P_{e_0} = \int_\gamma^\infty p(l \mid H_0)\, dl = \exp\left(\frac{-\gamma^2}{N_0}\right) \tag{8.2.26}$$

The probability of error can therefore be written as

$$P_e = \frac{1}{2}\left\{1 + \exp\left(\frac{-\gamma^2}{N_0}\right) - Q\left[\left(\frac{2E}{N_0}\right)^{1/2}, \left(\frac{2}{N_0}\right)^{1/2}\gamma\right]\right\} \tag{8.2.27}$$

In order for the probability of error to be minimum, the threshold γ has to be chosen depending on the factor $2E/N_0$. This optimum threshold can be obtained by setting the partial derivative of P_e in Eq. (8.2.27) with respect to γ equal to zero. We then obtain the transcendental equation

$$\exp\left(-\frac{E}{N_0}\right) I_0\left[\left(\frac{2E}{N_0}\right)^{1/2}\left(\frac{2}{N_0}\right)^{1/2}\gamma\right] = 1 \tag{8.2.28}$$

from which we can solve for $(2/N_0)^{1/2}\gamma$, the normalized threshold. An excellent analytic approximation to the solution of Eq. (8.2.28) is [4]

$$\left(\frac{2}{N_0}\right)^{1/2}\gamma = \left(2 + \frac{E}{2N_0}\right)^{1/2} \tag{8.2.29}$$

The expression for probability of error, Eq. (8.2.27) is rather complex. However, for high values of the signal-to-noise ratio E/N_0, we can obtain an approximate expression for P_e by noting that

$$Q(a, b) \simeq 1 - Q(a - b) \tag{8.2.30}$$

for large values of a and b [4,5]. Here, the $Q(\cdot)$ function on the right hand side of the above equation refers to one minus the Gaussian CDF defined in Section 3.3.1. Substitution in Eq. (8.2.27) yields

$$P_e \simeq \frac{1}{2}\left\{\exp\left[\frac{-\gamma^2}{N_0}\right] + Q\left[\left(\frac{E}{2N_0}\right)^{1/2}\right]\right\} \tag{8.2.31}$$

For large values of signal-to-noise ratio, it follows from Eq. (8.2.29) that the optimum threshold satisfies

$$\sqrt{\frac{2}{N_0}}\gamma = \sqrt{\frac{E}{2N_0}} \tag{8.2.32}$$

Furthermore, we can approximate $Q(x)$ for large values of x as

$$Q(x) \simeq \frac{1}{\sqrt{2\pi}x} \exp\left(-\frac{x^2}{2}\right) \tag{8.2.33}$$

Substitution of these results in Eq. (8.2.31) yields the probability of error for noncoherent OOK for high signal-to-noise ratios as

$$P_e \simeq \frac{1}{2} \exp\left(-\frac{E}{4N_0}\right) \tag{8.2.34}$$

Thus the error rate decreases exponentially with SNR. Note that for a given probability of error and a given noise spectral density, the optimum noncoherent OOK requires about 1 dB more average energy than coherent OOK.

For binary, noncoherent FSK, the received signals under the two hypotheses are

$$H_1\colon z(t) = \sqrt{\frac{2E}{T}} \sin(\omega_1 t + \theta_1) + v(t) \qquad 0 \leq t \leq T$$

$$H_0\colon z(t) = \sqrt{\frac{2E}{T}} \sin(\omega_0 t + \theta_0) + v(t) \qquad 0 \leq t \leq T$$

Usually the frequencies ω_1 and ω_2 are chosen so that $\sin(\omega_1 t + \theta_1)$ and $\sin(\omega_0 t + \theta_0)$ are orthogonal for any θ_1, θ_0. The optimum receiver for this case can easily be derived by introducing a "no signal" hypothesis, H_{ns}, and following the same reasoning as in the case of noncoherent OOK. This involves computing the two likelihood ratios, $\Lambda_1 = [p(z(t)\,|\,H_1)/p(z(t)\,|\,H_{ns})]$ and $\Lambda_0 = [p(z(t)\,|\,H_0)/p(z(t)\,|\,H_{ns})]$ and deciding H_1 or H_0 according as to which of Λ_1 or Λ_2 is larger.

Under the assumption that the phase is uniformly distributed in $(-\pi, \pi)$, it follows that the optimum receiver can be implemented by two bandpass filters with center frequencies corresponding to the mark and space frequencies, followed by envelope detectors. The envelopes are compared once every bit interval (T seconds), and a decision is made in favor of the larger envelope. Figure 8.2.8 shows the noncoherent FSK receiver. We derive the performance of the receiver under the assumption that the mark and space signals are orthogonal with equal energy. For purposes of analysis, let us assume that a mark signal $[y_1(t)]$ was transmitted. Then the output of the upper detector represents the envelope v_1 of the mark signal plus noise, while the output of the lower detector is the envelope v_0 of noise alone. An error is committed if v_0 exceeds v_1 when mark is transmitted. Let P_{e_1} denote this error probability. Similar to Eq. (8.2.22), we can obtain the following expressions for the probability density functions $p(v_1\,|\,H_1)$ and $p(v_0\,|\,H_1)$:

$$p(v_1\,|\,H_1) = \frac{2v_1}{N_0} \exp\left(-\frac{v_1^2 + E}{N_0}\right) I_0\left(\frac{2\sqrt{E}}{N_0} v_1\right) \qquad v_1 \geq 0 \tag{8.2.35a}$$

and

$$p(v_0\,|\,H_1) = \frac{2v_0}{N_0} \exp\left(\frac{-v_0^2}{N_0}\right) \qquad v_0 \geq 0 \tag{8.2.35b}$$

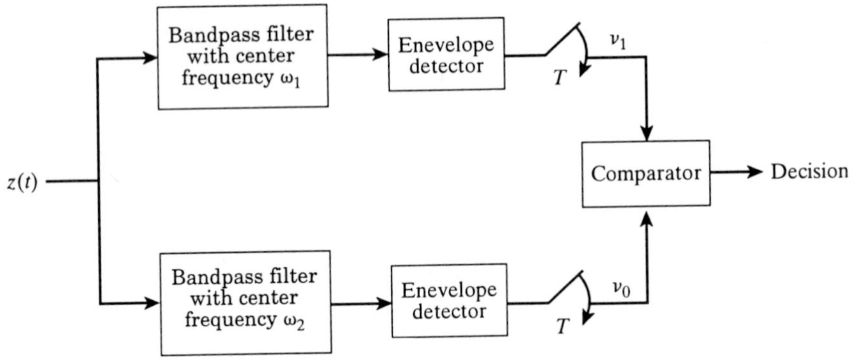

Figure 8.2.8 Noncoherent FSK receiver.

so that we have

$$\begin{aligned} P_{e_1} &= \text{Prob}[V_0 > V_1 | H_1] \\ &= \int_0^\infty \left[\int_{v_1}^\infty p(v_0 | H_1)\, dv_0 \right] p(v_1 | H_1)\, dv_1 \end{aligned} \tag{8.2.36}$$

Substituting for the density functions in Eq. (8.2.36) from Eq. (8.2.35) and using integral relations of Marcum Q functions [4], we obtain

$$P_{e_1} = \frac{1}{2} \exp\left(\frac{-E}{2N_0}\right)$$

From symmetry, it follows that

$$P_{e_0} = \frac{1}{2} \exp\left(\frac{-E}{2N_0}\right)$$

Therefore, the probability of error P_e is

$$P_e = \frac{1}{2} P_{e_1} + \frac{1}{2} P_{e_0} = \frac{1}{2} \exp\left(\frac{-E}{2N_0}\right) \tag{8.2.37}$$

We note that for a given E/N_0, the noncoherent FSK performs better than noncoherent OOK. But one has to be careful in stating this, because the FSK system uses double the energy. Therefore, for a given ratio of average energy per digit to noise power spectral density, noncoherent FSK and OOK perform almost identically, except at low values of this ratio.

We can obtain a comparison of the performance of a noncoherent FSK system with the coherent FSK system for large SNR by using the approximation of Eq. (8.2.33) in Eq. (8.2.13). This yields

$$P_e^{\text{FSK-COH}} \simeq \frac{1}{(2\pi E/N_0)^{1/2}} \exp\left(\frac{-E}{2N_0}\right) \tag{8.2.38}$$

Comparison of Eqs. (8.2.38) and (8.2.37) shows that for high SNR, coherent FSK performs better than noncoherent FSK. However, for increasingly large SNR, since the behavior of P_e is dominated by the exponential term, the difference in performance between the two becomes negligibly small.

In practice, noncoherent FSK is widely used because (1) it has a simple structure; (2) its threshold does not depend on the E/N_0 ratio; (3) its performance is comparable to coherent methods at high SNR; and as we shall see later, (4) its performance is better in fading channels than that of noncoherent OOK.

Differential phase-shift keying. We have seen earlier that PRK offers the best performance among all binary digital transmission schemes. However, PRK requires a phase reference at the receiver. If the phase of the carrier is essentially constant from interval to interval or bears a known relationship, we can obtain a reference for demodulating PRK by using the carrier phase of the previous signaling interval. A system that employs such a scheme is *differential phase-shift keying* (DPSK). Given a message sequence of binary digits, the encoded sequence is formed as follows. An arbitrary reference binary digit, say 1, is chosen as the first digit of the encoded sequence. Subsequent digits in the encoded sequence are formed by comparing a digit in the message sequence with the previous digit of the encoded sequence. For example, if the first digit of the message sequence is a 1, this is compared with the reference digit of the encoded sequence. If the two are the same, the second digit of the encoded sequence is chosen as 1. If they are different, the second digit of the encoded sequence is chosen as 0. This now becomes the reference digit for the next digit of the message sequence. The encoded message sequence then phase-shift keys a carrier with phases 0 and π. Table 8.2.1 shows a typical message sequence and the corresponding encoded sequence and carrier phase. The receiver for DPSK is shown in Fig. 8.2.9. In order to illustrate the functioning of the receiver, let us consider the encoded message sequence of Table 8.2.1, assuming noise-free transmission. Since both the reference bit and the first encoded bit are 1, the signal input to the correlator as well as the reference input will both be $\sqrt{2E/T}\sin\omega_c t$. The output of the correlator is

$$l = \int_0^T \frac{2E}{T}\sin^2\omega_c t = E$$

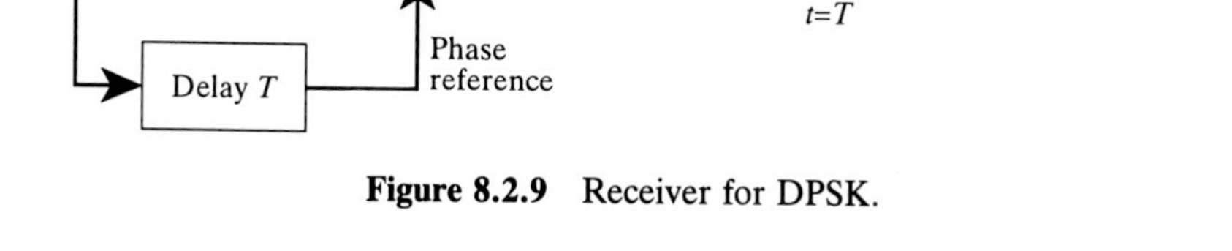

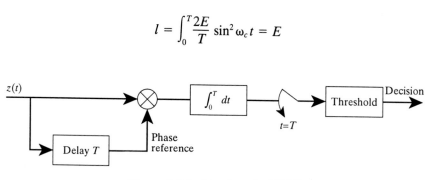

Figure 8.2.9 Receiver for DPSK.

TABLE 8.2.1 DIFFERENTIAL PHASE-SHIFT KEYED SEQUENCE

Message sequence:		1	0	1	0	1	1	1	0	0	1
Encoded sequence:	1 Reference digit	1	0	0	1	1	1	1	0	1	1
Carrier phase:	0	0	π	π	0	0	0	0	π	0	0

and the receiver decides that a 1 was transmitted. During the next bit interval, since the transmitted bit is a 0, the signal input to the correlator is $-\sqrt{2E/T}\sin\omega_c t$, whereas the reference input is $\sqrt{2E/T}\sin\omega_c t$. The correlator output now is

$$l = -\int_0^T \frac{2E}{T}\sin^2\omega_c t\,dt = -E$$

and the receiver decides that a 0 was transmitted, and so on, for the rest of the encoded sequence.

Thus, in the absence of noise, use of DPSK yields the same performance as PRK while obviating the need for a reference phase at the receiver. Noise in the received signal, however, affects both the signal input to the correlator and the reference, whereas in PRK only the signal input is corrupted by noise. We might therefore expect that the performance of DPSK in the presence of channel noise is not comparable to that of PRK. In fact, the probability of error for DPSK is given by [4]

$$P_e = \frac{1}{2}\exp\left(-\frac{E}{N_0}\right) \tag{8.2.39}$$

Comparison with Eq. (8.2.37) for noncoherent FSK shows that DPSK requires 3 dB less SNR for the same error rate.

It is also useful at this point to consider the performance of PRK when the reference phase at the receiver is not correctly known. Let us assume that the transmitted signals are $\pm\sqrt{2E/T}\sin\omega_c t$, and that the reference signal is out of phase synchronism by an amount $\Delta\phi$. The reference signal is therefore $\sqrt{2E/T}\sin(\omega_c t + \Delta\phi)$. We can then show (see Exercise 8.8) that the probability of error is

$$P_e = Q\left[\left(\frac{2E\cos^2\Delta\phi}{N_0}\right)^{1/2}\right] \tag{8.2.40}$$

This expression is useful in designing synchronization circuits.

8.2.3 Fading Channel Performance

Long-distance radio-frequency (RF) communication systems utilizing ionospheric/tropospheric reflections are often subject to fading, a phenomenon caused by multipath propagation and addition of random phasors. The received signals in mobile cellular communication systems or in indoor wireless communication systems are also subject to multipath propagation. Fortunately, it is possible to obtain a reasonable statistical model for this phenomenon and to calculate the performance of digital systems on this basis.

The fading model often utilized for ionospheric reflections is that of the nonselective, slow-fading Rayleigh model. However, depending on the bandwidths of the signals and the surrounding topography (for example, urban area), a fast-fading, frequency-selective Rayleigh model may be more appropriate in the case of mobile cellular communications. The nonselective, slow-fading model assumes that the envelope of an otherwise known sinusoid is Rayleigh distributed and the phase uniformly distributed between $-\pi$ and π. Slow fading means that the envelope, though random from bit to bit, assumes a constant value over each bit interval. For a detailed discussion of the phenomenon of fading and its modeling, the reader is referred to [4,9,10]. In the case of OOK and nonselective slow Rayleigh fading, the received waveforms can be modeled as

$$\begin{aligned} H_1&: z(t) = \sqrt{\frac{2E}{T}} \sin(\omega_c t + \theta) + v(t) \\ H_0&: z(t) = v(t) \end{aligned} \tag{8.2.41}$$

where E and θ are now random variables. The additive noise $v(t)$ is again assumed to be zero mean, Gaussian, white with spectral density $N_0/2$.

As indicated earlier, we will assume that the envelope $\eta = \sqrt{E}$ of the received signal is Rayleigh distributed and that the phase is uniformly distributed in $(-\pi, \pi)$. Let the mean-squared value of the received signal be $\overline{E}$, so that $E\{\eta^2\} = \overline{E}$. It is obvious that this corresponds to the case of noncoherent OOK except that the received envelope is now random with probability density function

$$p(\eta) = \frac{2\eta}{\overline{E}} \exp\left(\frac{-\eta^2}{\overline{E}}\right) \tag{8.2.42}$$

Therefore, we can expect the receiver structure to be the same as for noncoherent OOK, except that its performance will vary depending on η. An optimum receiver will use a *varying threshold*, depending on the received envelope (recall that the optimum threshold for noncoherent OOK depends on the ratio E/N_0) and will have to be adaptive. We can, however, use the ratio of mean received energy per bit to noise spectral density, $\overline{E}/N_0$, to calculate a *fixed optimum threshold*.

The performance of the OOK receiver in fading media can be analyzed by determining the probability of error for a given received envelope and then averaging over the known density function $p(v)$. Let E/N_0 be denoted by R for convenience. Then

$$p(r) = \begin{cases} \dfrac{1}{R_0} \exp\left(\dfrac{-r}{R_0}\right) & 0 < r < \infty \\ 0 & \text{otherwise} \end{cases} \tag{8.2.43}$$

where $R_0 = \bar{E}/N_0$. Then the probability of error expression becomes

$$P_e = \int_0^\infty P_e(r)\,p(r)\,dr \tag{8.2.44}$$

where $P_e(r)$ is the probability of error for any particular value of $R = r$. Using Eq. (8.2.27) in the preceding expression and evaluating the integral, we have

$$P_e = \frac{1}{2}\left[1 - \exp\left(-\frac{\gamma^2}{N_0}\frac{1}{1 + \bar{E}/N_0}\right)\right] + \frac{1}{2}\exp\left(\frac{-\gamma^2}{N_0}\right) \tag{8.2.45}$$

The optimum fixed threshold is obtained by differentiating Eq. (8.2.45) with respect to γ, and setting the result to zero. We then obtain

$$\gamma_{\text{opt. fixed}} = \left(\frac{N_0}{2}\right)^{1/2}\left[2\left(1 + \frac{N_0}{\bar{E}}\right)\ln\left(1 + \frac{\bar{E}}{N_0}\right)\right]^{1/2} \tag{8.2.46}$$

For large values of $\bar{E}/N_0$, this can be approximated as

$$\gamma_{\text{opt. fixed}} \simeq \left(\frac{N_0}{2}\right)^{1/2}\left[2\ln\left(\frac{\bar{E}}{N_0}\right)\right]^{1/2} \tag{8.2.47}$$

Using this value of the optimum fixed threshold in Eq. (8.2.45) yields the probability of error for high SNR as

$$P_e^{\text{opt. fixed}} = \frac{1}{2}\frac{\ln(\bar{E}/N_0)}{\bar{E}/N_0} \tag{8.2.48}$$

Note that the probability of error varies inversely with the ratio of average energy per digit to the noise spectral density for large values of this ratio. Unfortunately, this is true of all systems operating in fading media. Even a varying optimum threshold for large E/N_0 ratio does not help to improve this characteristic. The probability of error with a varying threshold can be approximated [4] as

$$P_e^{\text{varying opt.}} \simeq \frac{1.63}{\bar{E}/N_0} \tag{8.2.49}$$

Similarly, in the case of binary noncoherent orthogonal FSK, we can obtain the probability of error by averaging P_e in Eq. (8.2.37) over the density function in Eq. (8.2.43). We then obtain

$$\begin{aligned} P_e &= \int_0^\infty \frac{1}{R_0}\exp\left(\frac{-r}{R_0}\right)\frac{1}{2}\exp\left(\frac{-r}{2}\right)dr \\ &= \frac{1}{2 + \bar{E}/N_0} \\ &= \frac{1}{2 + \bar{E}/N} \end{aligned} \tag{8.2.50}$$

Note that again the error rate varies inversely with $\bar{E}/N_0$ for large values of $\bar{E}/N_0$, as compared to the nonfading case where an exponential decrease in error rate is achieved.

If it is possible to operate coherently in a slowly fading medium, that is, the phase is estimated perfectly by means such as phase locking, it is possible to get approximately 3 dB improvement over noncoherent techniques. The corresponding results for the coherent schemes can be obtained by averaging the relevant probability of error expressions over the density function of Eq. (8.2.42). Thus the probability of error in the case of coherent FSK obtained by averaging P_e in Eq. (8.2.13) is

$$P_e = \frac{1}{2}\left[1 - \left(\frac{1}{1 + 2N_0/\overline{E}}\right)^{1/2}\right] \tag{8.2.51}$$

By averaging P_e in Eq. (8.2.14) we obtain the expression for PRK as

$$P_e = \frac{1}{2}\left[1 - \left(\frac{1}{1 + N_0/\overline{E}}\right)^{1/2}\right] \tag{8.2.52}$$

We note that the performance of PRK is 3 dB better than that of FSK, reemphasizing the fact that if coherent detection is at all possible, the optimum system is PRK.

Thus, at large values of the ratio of average energy per bit to noise spectral density, all these systems have an error rate inversely proportional to the ratio $\overline{E}/N_0$. In comparison, with steady signals, the error rates decrease exponentially with this ratio. We can qualitatively argue that the degradation due to fading is due to the not so small probability of very low signal levels even when the mean energy is high, causing the probability of error to be large. The obvious solution is to modify the statistics of the received signal envelope such that the probability density function is small for small values of the envelope. Diversity combining [9–13] has proved to be a very practical tool in this context.

Comparison of digital transmission systems. We can use the error probability expressions derived in the preceding discussions to obtain a comparison of the different digital transmission schemes. To do so in a convenient fashion the probabilities of error for various systems are tabulated in Table 8.2.2 with δ, the ratio of average energy per bit to noise spectral density as a parameter.

Figure 8.2.10 shows the error rate curves for various δ for the systems shown in Table 8.2.2. It can be observed from Table 8.2.2 that OOK and FSK (orthogonal signaling) systems perform identically for almost all δ. The differences arise only in noncoherent and fading situations. We can explain this by noting that digital frequency modulation is not really a wideband noise reduction system but can be thought as two interleaved OOK signals of differing carrier frequencies. It is also seen that the noncoherent FSK performs only as well as noncoherent OOK. However, the former is preferred because its threshold is signal independent, a distinct advantage in fading conditions. The curves in Fig. 8.2.10 also emphasize that whenever possible, it is worthwhile measuring the phase in both fading and nonfading situations.

TABLE 8.2.2 PROBABILITIES OF ERROR FOR VARIOUS SYSTEMS

System	Mode	Channel	Probability of Error, P_e
(*a*) OOK	Coherent	Nonfading	$Q\left[\left(\frac{\delta}{2}\right)^{1/2}\right]$
(*b*) OOK	Noncoherent†	Nonfading	$\frac{1}{2}\exp\left(\frac{-\delta}{4}\right)$
(*c*) OOK	Noncoherent*	Fading	$\frac{1}{2}\frac{\ln\delta}{\delta}$
(*d*) FSK$^{\triangle}$	Coherent	Nonfading	$Q\left[\left(\frac{\delta}{2}\right)^{1/2}\right]$
(*e*) FSK$^{\triangle}$	Coherent$	Fading	$\frac{1}{2}\left(1-\frac{1}{\sqrt{1+4/\delta}}\right)$
(*f*) FSK$^{\triangle}$	Noncoherent	Nonfading	$\frac{1}{2}\exp\left(\frac{-\delta}{4}\right)$
(*g*) FSK$^{\triangle}$	Noncoherent	Fading	$\frac{1}{2+\delta/2}$
(*h*) PSK	Coherent	Nonfading	$Q(\delta^{1/2})$
(*i*) PSK	Coherent#	Fading	$\frac{1}{2}\left(1-\frac{1}{\sqrt{1+2/\delta}}\right)$

† Uses optimum fixed threshold, and P_e is the asymptotic value.
* Uses optimum fixed threshold, based on Rayleigh slow fading.
△ Uses orthogonal signals.
$ Assumes perfect measurement of phase in fading media.
\# Rayleigh slow fading.

8.2.4 Synchronization

A coherent communication system requires various levels of synchronization for efficient operation. In baseband transmission with known signal shapes, the beginning and terminating times of the signals must be known at the receiver. This is referred to as *symbol synchronization*. For RF systems, the carrier phase must also be known; in addition, if the bits are grouped into words, the beginning and termination of the words must be known. This is referred to as *frame synchronization*.

We first consider carrier synchronization. If the transmitted signal contains a component at carrier frequency, we can obtain carrier coherence very simply by locking onto this component. A commonly used device for achieving locking is the

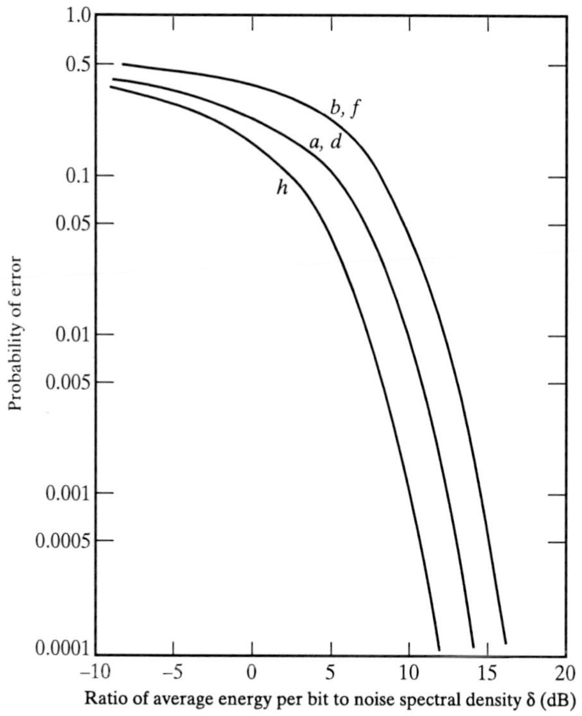

Figure 8.2.10a Error rate curves for various RF digital communication schemes in nonfading channels (see Table 8.2.2 for key to index).

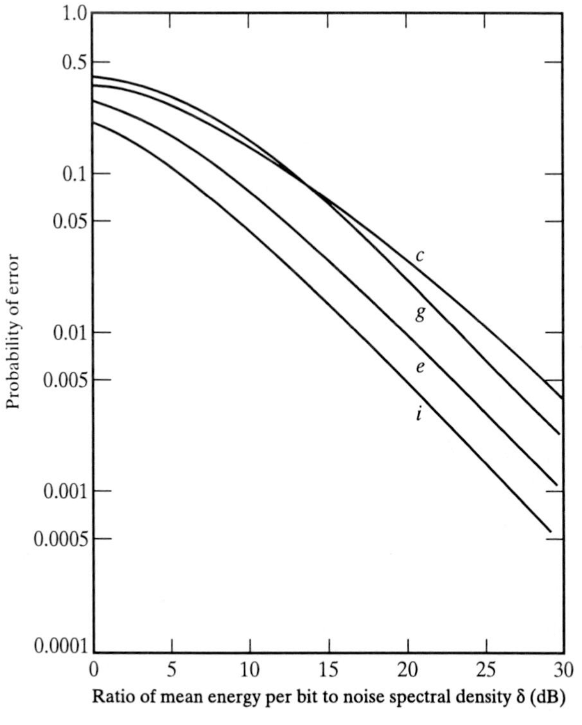

Figure 8.2.10b Error rate curves for various RF digital communication schemes in Rayleigh fading channel (see Table 8.2.2 for key to index).

phase-locked loop (PLL) shown in Fig. 8.2.11. The system generally consists of a phase detector, loop filter, loop amplifier, and a voltage-controlled oscillator (VCO). To understand the operation of the PLL, let us assume that the input signal is given by

$$y_r(t) = E_r \sin[\omega_c t + \theta(t)] \tag{8.2.53}$$

and that the VCO output is

$$e_0(t) = E_c \cos[\omega_c t + \phi(t)] \tag{8.2.54}$$

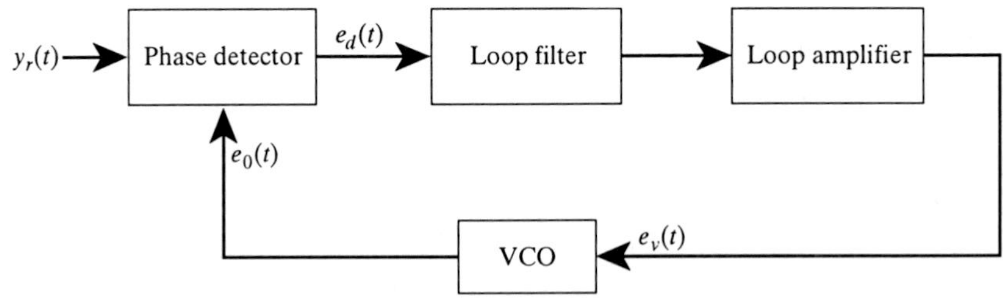

Figure 8.2.11 Phase-locked loop.

While there are many types of phase detectors, we assume for our purposes that the phase detector is a multiplier followed by a lowpass filter which removes the second harmonic in the output of the multiplier. The phase detector output then becomes

$$e_d(t) = \tfrac{1}{2} E_c E_r K_d \sin[\theta(t) - \phi(t)] \tag{8.2.55}$$

where K_d is a constant associated with the multiplier. The VCO is essentially a frequency modulator, with the frequency deviation of the output $d\phi/dt$ being proportional to the VCO input $e_v(t)$. Thus

$$\frac{d\phi(t)}{dt} = K_v e_v(t) \tag{8.2.56}$$

where K_v is the VCO constant. If we assume that the loop filter and amplifier in the PLL can be replaced by a gain, we can combine Eqs. (8.2.55) and (8.2.56) to obtain

$$\frac{d\phi(t)}{dt} = K \sin[\theta(t) - \phi(t)] \tag{8.2.57}$$

From Eq. (8.2.57) it is clear that if $\theta(t) > \phi(t)$, $d\phi(t)/dt$ will be positive, so that $\phi(t)$ will increase. Similarly when $\theta(t) < \phi(t)$, $d\phi(t)/dt$ will be negative and $\phi(t)$ decreases. In either case, $\phi(t)$ is driven in the direction of $\theta(t)$ until $\phi(t) = \theta(t)$ and lock is achieved. The VCO phase $\phi(t)$ thus provides a good estimate of the carrier phase $\theta(t)$. It can also be seen from Eq. (8.2.57) that if $\theta(t)$ and $\phi(t)$ differ by more than $\pi/2$ radians, the VCO phase will not track the carrier phase. A detailed analysis of the PLL is beyond the scope of this book and the interested reader is referred to [14,15] for discussions of its performance and design.

In the absence of a carrier component in the transmitted signal (as, for example, with PRK), we can use a Costas loop for synchronization. Figure 8.2.12 shows a block diagram of a Costas loop. For an assumed input of the type $y_r = E_r \cos \omega_c t$, the signals at the various points in the loop shown in Figure 8.2.12 can easily be derived. The output from the lowpass filter preceding the VCO is essentially the dc value of the input. This signal drives the VCO such that θ is driven to zero and locking occurs. For a detailed discussion of these and other synchronization devices, such as the decision-feedback loop, and an analysis of their performance in the presence of noise, the reader is referred to [16,17].

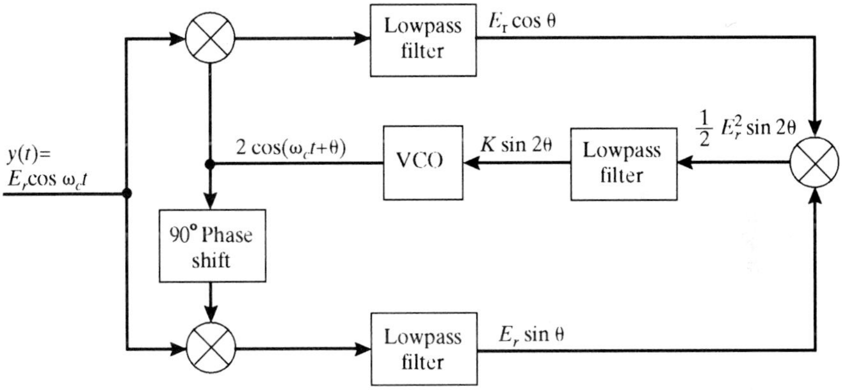

Figure 8.2.12 Costas loop.

For symbol or bit synchronization, three general methods are used. Synchronization can be derived, for example, by slaving the transmitter and receiver to the same clock, by using a separate synchronization signal (pilot clock), and by self-synchronization based on the modulation itself. Again the reader is referred to [16,17] for detailed analysis of the timing misalignment in these systems. We can also use a MAP estimation technique for estimating the timing misalignment. We illustrate this for the case of a binary PSK system following the approach in [17].

MAP estimation for symbol synchronization. Consider the BPSK system, in which over each bit interval of T seconds, the transmitted signal is a pulse $\pm p(t)$ which is zero outside the interval $(0, T)$, with $p(t)$ corresponding to mark and $-p(t)$ to space. We will assume that the time origin at the transmitter coincides with the beginning of a transmitted bit. The start time of the bit is unknown at the receiver and must be estimated from the received signal. We obtain this estimate by considering the received signal $z(t)$ over an interval of NT seconds, corresponding to a string of N transmitted bits. With reference to the time origin at the transmitter, we can write the received signal over this interval as

$$z(t) = \sum_{n=1}^{N} s(t, a_n) + v(t) \tag{8.2.58}$$

where

$$s(t, a_n) = a_n p(t - (n - 1)T) \tag{8.2.59}$$

and a_n is either $+1$ or -1, depending on whether the nth transmitted bit is a mark or space. We assume that the $\{a_n\}$ are independent, identically distributed random variables with $P(a_n = 1) = P(a_n = -1) = \frac{1}{2}$ and that the channel noise $v(t)$ is additive, Gaussian, white with power spectral density $N_0/2$.

We note that in writing Eq. (8.2.58), we have assumed the propagation delay to be zero. This entails no loss of generality since the results derived in the sequel do not depend on the actual propagation delay. Our formulation of the problem above also assumes that the successive bits (pulses) arrive at the receiver at intervals of T seconds. In general, depending on the transmission channel and the propagation conditions, the arrival times of successive bits may actually vary. If this change in arrival times occurs slowly enough, we can choose N so that our assumption of equally spaced arrival times for the bits is valid.

Let the initial timing misalignment of the synchronization clock at the receiver be equal to τ. That is, the receiver clock initially estimates the start of the transmitted bit as being $t = 0$ instead of $t = \tau$. We can then reasonably assume that τ is uniformly distributed between $-T/2$ and $+T/2$. Thus with reference to the receiver synchronization clock we can rewrite Eq. (8.2.58) as

$$z(t) = \sum_{n=0}^{N} s(t - \tau, a_n) + v(t) \tag{8.2.60}$$

Note that the sum in Eq. (8.2.60) is over values of n ranging from 0 to N. The reason that the pulse corresponding to a_0 appears in this equation can be understood from Fig. 8.2.13, which represents the timing diagram for a positive value of the timing error τ*. As can be seen from the diagram, in this case a portion of the signal corresponding to bit a_0 will appear in the received signal. If τ had been negative, a part of the signal corresponding to a_1 would appear in $z(t)$ as well as a part corresponding to a_{N+1}. Since the a_n are independent, identically distributed, and τ is uniformly distributed in $(-T/2, T/2)$, the statistical analysis of the received signal will be the same for either $\tau > 0$ or $\tau < 0$: the synchronization problem is simply to estimate τ from the observed signal $z(t)$. In general, the higher the signal-to-noise ratio and the longer the observation interval, the closer will the estimate be to the true timing misalignment. We can find the MAP estimate of τ by following a procedure similar to our discussions in earlier chapters. To do so, let us denote the nth subinterval in $(0, NT)$ as (see Fig. 8.2.13)

$$T_n(\tau) = [\max(0, (n - 1)T + \tau), \min(nT + \tau, NT)] \qquad n = 0, 1, \ldots, N \tag{8.2.61}$$

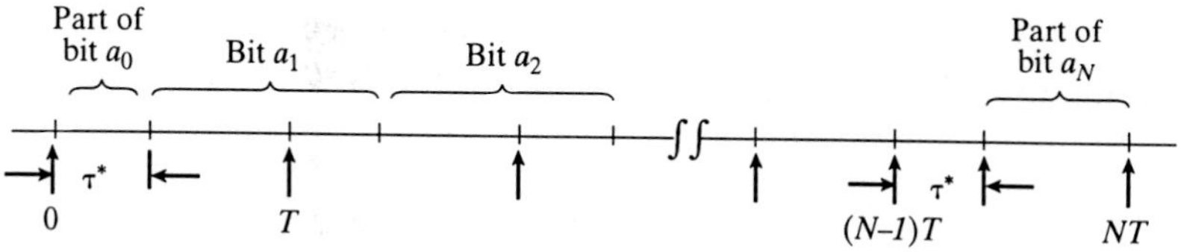

Figure 8.2.13 Timing diagram indicating misalignment of receiver clock. Arrows point to receiver clock epochs. Actual symbol (bit) arrives τ^* seconds later.

and expand the received signal $z(t)$ in an orthonormal set of functions $g_i(t)$. We choose $g_1(t) = p(t)/\sqrt{E}$, where $E = \int_0^T p^2(t)\,dt$. The other functions can be chosen to be arbitrary since $v(t)$ is a white process. We can thus write

$$z(t) = \sum_{i=1}^{\infty} z_{in} g_i(t - (n-1)T - \tau) \qquad t \in T_n(\tau) \tag{8.2.62}$$

where

$$z_{in} = \int_{T_n(\tau)} z(t) g_i(t - (n-1)T - \tau)\,dt \tag{8.2.63}$$

Clearly, the useful information regarding τ is contained in $z_{1n}, n = 0, 1, 2, \ldots, N$, since these are the only variables (coefficients) that depend on the transmitted pulse. Thus the MAP estimate of τ can be found as the value that maximizes $p(\tau \mid z_{10}, z_{11}, \ldots, z_{1N})$. Use of Eqs. (8.2.59) and (8.2.60) in Eq. (8.2.63) gives

$$\begin{aligned} z_{1n} &= \int_{T_n(\tau)} [a_n p(t - (n-1)T - \tau) + v(t)] g_1(t - (n-1)T - \tau)\,dt \\ &= \sqrt{E}\, a_n \int_{T_n(\tau)} g_1^2(t - (n-1)T - \tau)\,dt \\ &\quad + \int_{T_n(\tau)} v(t) g_1(t - (n-1)T - \tau)\,dt \\ &= \sqrt{E}\, a_n q_n(\tau) + v_n \qquad n = 0, 1, 2, \ldots, N \end{aligned} \tag{8.2.64}$$

where

$$q_n(\tau) = \int_{T_n(\tau)} g_1^2(t - (n-1)T - \tau)\,dt \tag{8.2.65a}$$

and

$$v_n = \int_{T_n(\tau)} v(t) g_1(t - (n-1)T - \tau)\,dt \tag{8.2.65b}$$

The v_n are independent zero-mean Gaussian random variables, with variance $N_0/2$ for $n = 1, 2, \ldots, N - 1$. Because of the shorter lengths of the subintervals $T_0(\tau)$ and

$T_N(\tau)$, the variances of v_{10} and v_{1N} will be somewhat smaller. However, we will assume that $\text{var}(v_n) = N_0/2$ for all n. Similarly, from Eq. (8.2.65), we note that $q_n(\tau) = 1$ for $n = 1, 2, \ldots, N-1$. We will assume that $q_0(\tau)$ and $q_N(\tau)$ are also approximately equal to 1. The errors introduced by these approximations in the estimate of τ should be small for large N.

To find the MAP estimate $\hat{\tau}_{\text{MAP}}$, we write

$$p(\tau \mid z_{10}, z_{11}, \ldots, z_{1N}) = \frac{p(z_{10}, z_{11}, \ldots, z_{1N} \mid \tau) p(\tau)}{p(z_{10}, z_{11}, \ldots, z_{1N})} \tag{8.2.66}$$

Since τ is uniformly distributed in $(-T/2, T/2)$ and $p(z_{10}, z_{11}, \ldots, z_{1N})$ does not depend on τ, the MAP estimate of τ can be obtained by maximizing $p(z_{10}, z_{11}, \ldots, z_{1N} \mid \tau)$. Now z_{1n} conditioned on τ and a_n is Gaussian with mean $\pm\sqrt{E}q_n(\tau)$ and variance $N_0/2$, so that we can write

$$\begin{aligned} p(z_{1n} \mid \tau) = \frac{1}{\sqrt{2\pi}\sqrt{N_0/2}} \biggl[\frac{1}{2} \exp\left(-\frac{(z_{1n} - \sqrt{E}q_n(\tau))^2}{N_0} \right) \\ + \frac{1}{2} \exp\left(-\frac{(z_{1n} + \sqrt{E}q_n(\tau))^2}{N_0} \right) \biggr] \end{aligned} \tag{8.2.67}$$

Therefore,

$$\begin{aligned} p(z_{10}, z_{11}, \ldots, z_{1N} \mid \tau) &= \prod_{n=0}^{N} p(z_{1n} \mid \tau) \\ &= \left(\frac{1}{\sqrt{N_0 \pi}} \right)^{N+1} \exp\left[-\sum_{n=0}^{N} \frac{z_{1n}^2 + Eq_n^2(\tau)}{N_0} \right] \prod_{n=0}^{N} \cosh\left(\frac{2\sqrt{E}}{N_0} z_{1n} q_n(\tau) \right) \end{aligned} \tag{8.2.68}$$

It can be shown that $\sum_{n=0}^{N} z_{1n}^2$ is independent of τ. Maximizing the right side of Eq. (8.2.68) with respect to τ then gives

$$\hat{\tau}_{\text{MAP}} = \arg\max_{\tau} \prod_{n=0}^{N} \cosh\left[\frac{2\sqrt{E}}{N_0} z_{1n} q_n(\tau) \right] \tag{8.2.69}$$

where, as explained earlier, $q_n(\tau) = 1$ for $n = 0, 1, 2, \ldots, N$. It therefore follows that $\hat{\tau}_{\text{MAP}}$ can be found as

$$\begin{aligned} \hat{\tau}_{\text{MAP}} &= \arg\max_{\tau} \sum_{n=0}^{N} \ln\cosh\left(\frac{2\sqrt{E}}{N_0} z_{1n} \right) \\ &= \arg\max_{\tau} \sum_{n=0}^{N} \ln\cosh\left[\frac{2}{N_0} \int_{T_n(\tau)} p(t - (n-1)T - \tau) z(t)\, dt \right] \end{aligned} \tag{8.2.70}$$

where the last step follows from Eq. (8.2.63) and the definition of $g_1(t)$.

For a practical implementation, τ is quantized to a finite number of values $\tau_1, \tau_2, \ldots, \tau_L$. Basically, the scheme consists of generating $p(t - (n-1)T - \tau)$, $n = 0, 1, 2, \ldots, N$, for each quantized value of τ in the range $(-T/2, T/2)$ and computing the right side of Eq. (8.2.70). The value of τ that gives a maximum corresponds to the MAP estimate $\hat{\tau}_{\text{MAP}}$. The implementation of this scheme as given in Figure 8.2.14

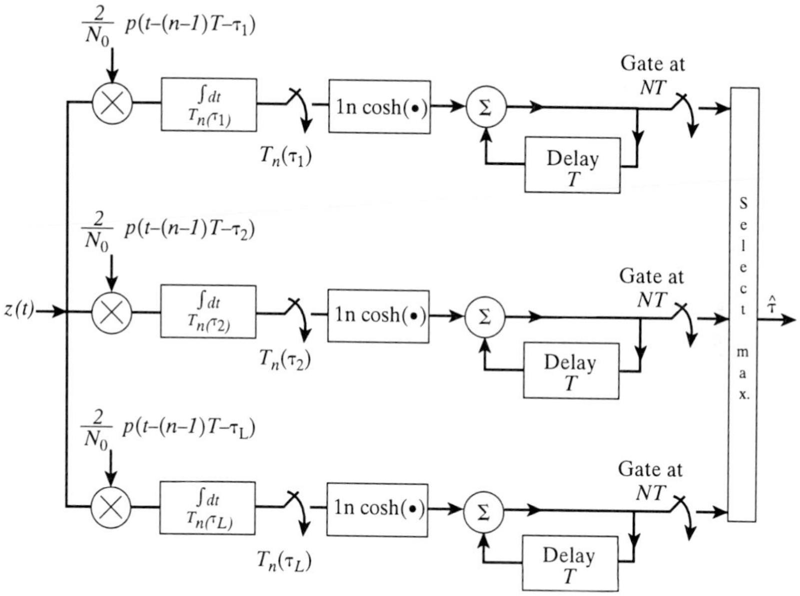

Figure 8.2.14 MAP estimator for timing misalignment.

represents an open-loop implementation. From a practical viewpoint, a closed-loop implementation that tracks changes in the epoch arrivals would be desirable. A number of schemes, such as the early–late gate, are of this type. These schemes are still motivated by the MAP estimation scheme. For details, the reader is referred to [17].

8.2.5 Intersymbol Interference in Digital Communication

In our discussions so far, we have considered two sources of error in digital communication systems: additive noise and fading over the transmission medium. A significant assumption underlying our discussions was that the channel did not possess any memory, so that the received signal over any bit interval could be considered independently of the signals in other intervals. Such an assumption is not valid in the case of high-speed transmission of digital data, as, for example, in communication between computers. The received signal over any bit interval will have contributions from transmissions over several intervals. This smearing of the transmitted signal affects the decision process rather detrimentally. This phenomenon is known as *intersymbol interference* (ISI).

In this section we formulate the model for a basic digital communication system with ISI and determine the increase in the probability of error in detecting a bit. To counter this effect, the observed signal may first be passed through a filter called the *equalizer*, whose characteristics are the inverse of the channel characteristics. If the equalizer is exactly matched to the channel, the combination of channel and equalizer is just a gain, so that there is no intersymbol interference present at the output of the equalizer. We confine our discussions to bit-by-bit detection even though many of our results can be extended to sequences of bits. An efficient algorithm due to Viterbi has been successfully used for this purpose [18]. Since this technique is outside the scope of the present book, the reader is invited to the cited reference.

To illustrate the various concepts involved, we consider only a baseband system. The extensions to modulated systems carry over in principle, and details can be found in [19]. Let $\{a_n\}, n = -\infty, \ldots, -2, -1, \ldots, \infty$ be a binary white sequence taking values 1 or -1 with equal probability. This sequence can be thought of as a representation of the input bits of a binary baseband digital communication system shown in Fig. 8.2.15. In the figure $h(t)$ represents the impulse response of the channel and $v(t)$ the additive white noise. For purposes of analysis we can consider the input to be an impulse train with weights a_n. Then the received signal can be synchronously sampled every T seconds to decide whether the bit transmitted was 1 or -1. It follows that the received signal is

$$\begin{aligned} z(t) &= y(t) + v(t) \\ &= \sum_{n=-\infty}^{\infty} a_n h(t - nT) + v(t) \end{aligned} \tag{8.2.71}$$

Consider an instant of sampling, say t_0, corresponding to the transmission of a_0. This sample is represented by

$$z_0 = a_0 h_0 + \sum_{n=-\infty}^{\infty}{}' a_n h_{-n} + v_n \tag{8.2.72}$$

where we represent $h(t_0)$ by h_0 and $h(t_0 - nT)$ by h_{-n}, and Σ' indicates the exclusion of the a_0 term. Equation (8.2.72) brings out the effects of channel memory and additive noise distinctly. The desired signal corresponds to the first term, the second term due to channel memory is the ISI effect, and the last term is the additive white noise. Even in the absence of additive noise (which is rarely the case), the second term can be significant enough to cause an erroneous decision. Thus we see that ISI is a major source of error in high-speed data transmission over memory channels, and its effect needs to be reduced to have a reliable system.

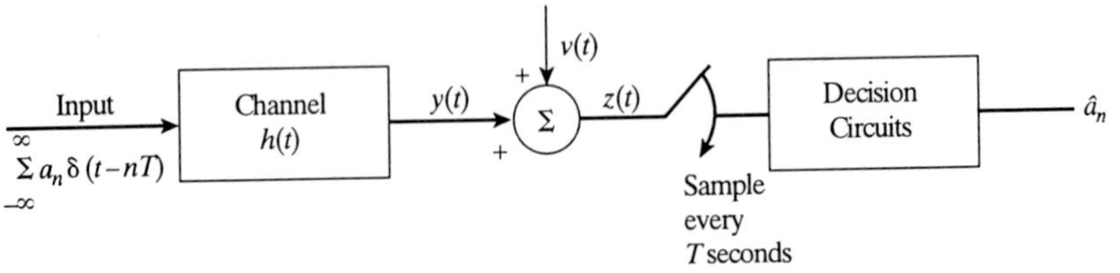

Figure 8.2.15 Baseband digital communication system with intersymbol interference.

An interesting way of viewing Eq. (8.2.72) is to note that the ISI term

$$\sum_{n=-\infty}^{\infty}{}' a_n h_{-n}$$

is a correlated random sequence and the detection problem can be considered as one of detecting a known signal in colored noise, discussed in Chapter 4. Therefore, we can expect the "optimum" receiver structure to be a whitening filter and a suitable matched filter. However, the design of such receivers in practice is subject to several constraints and certain approximations may have to be resorted to in order to obtain a design that minimizes the effect of ISI.

We recall from our discussions in Chapter 4 that the performance of a receiver in the presence of colored noise is dependent on the waveforms of the signals being transmitted. Thus by proper design of the transmitted waveforms, we can reduce the probability of error. Because we can consider ISI as colored noise, we can use the approach to advantage. If we evaluated the eigenfunction corresponding to the smallest eigenvalue of the colored noise process $\Sigma' a_n h(t - nT)$ and use it to transmit our binary data, detectability should indeed increase. In practice, this evaluation is rather difficult, and what is more, we may end up with a waveform too complex to be implemented. Usually, the waveforms are designed to have null response at sampling instants other than the one under consideration. An example is the raised cosine waveform [19].

We now turn to the problem of designing an equalizer for reducing the effect of ISI. A criterion for designing equalizers is to minimize the peak distortion D_p defined as

$$D_p = \frac{1}{h_0} \sum_{n=-\infty}^{\infty} |h_n| \tag{8.2.73}$$

Alternatively, we can also seek to minimize the mean-square distortion D_{ms}, defined as

$$D_{\text{ms}} = \frac{1}{|h_0|^2} \sum_{n=-\infty}^{\infty} |h_n|^2 \tag{8.2.74}$$

In practice, the channel memory can be assumed to be finite. Thus the infinite sums in Eqs. (8.2.73) and (8.2.74) can be replaced by finite sums, say from $n = -N$ to $n = N$. The equalizer is chosen to be a tapped delay line with, say, L taps, and the weights of the taps are adjusted such that either D_p or D_{ms} is minimum.

In the presence of additive noise at the receiver, we may choose the equalizer taps so as to minimize the mean-square error between the input sequence and the equalizer output. The main components of this scheme are shown in Fig. 8.2.16. Thus, given the observations

$$z_k = \sum_{n=-\infty}^{\infty} h_n a_{k-n} + v_k \tag{8.2.75}$$

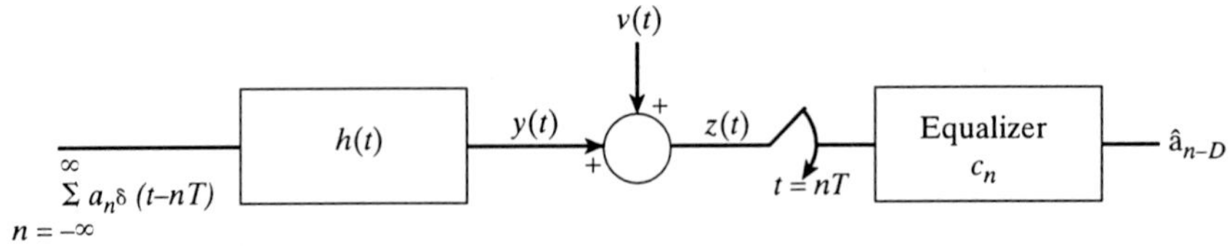

Figure 8.2.16 Equalized digital communication system.

our problem is to design the equalizer gains $c_l, l = -L, -L+1, \ldots, 0, \ldots, L$ such that the equalizer output

$$\hat{a}_{n-D} = \sum_{l=-L}^{L} c_l z_{n-l} \tag{8.2.76}$$

approximates the transmitted symbol a_{n-D} in the mean-square sense. If we denote the estimation error by $\tilde{a}_{n-D}$, we can use the orthogonal projection principle to obtain

$$\mathrm{E}[\tilde{a}_{n-D} z_{n-q}] = 0 \qquad q = -L, \ldots, 0, \ldots, L \tag{8.2.77}$$

where

$$\begin{aligned} \tilde{a}_{n-D} &= a_{n-D} - \hat{a}_{n-D} \\ &= a_{n-D} - \sum_{l=-L}^{L} c_l z_{n-l} \end{aligned} \tag{8.2.78}$$

Substitution of Eq. (8.2.76) in Eq. (8.2.75) yields the set of equations

$$\sum_{l=-L}^{L} c_l \,\mathrm{E}\{z_{n-l} z_{n-q}\} = \mathrm{E}\{a_{n-D} z_{n-q}\} \tag{8.2.79}$$

The right side of Eq. (8.2.79) is the cross-correlation between the delayed input and the unequalized channel output (observation), while the expectation on the left side corresponds to the autocorrelation function of the unequalized channel output. We can evaluate these expressions by using Eq. (8.2.75). We thus have

$$\begin{aligned} \mathrm{E}\{a_{n-D} z_{n-q}\} &= \mathrm{E}\left\{a_{n-D} \sum_{m=-N}^{N} h_m a_{n-m-q}\right\} \\ &= \sum_{m=-N}^{N} h_m \,\mathrm{E}\{a_{n-D} a_{n-m-q}\} \\ &= h_{D-q} \end{aligned} \tag{8.2.80}$$

where the last step follows from the assumed independence of the transmitted signals. Similarly, we can show that

$$\mathrm{E}\{z_{n-l} z_{n-q}\} = \left(\sum_{m=-N}^{N} h_m h_{l+q-m}\right) + \sigma_v^2 \tag{8.2.81}$$

Substitution of Eqs. (8.2.80) and (8.2.81) in Eq. (8.2.79) yields

$$\sum_{l=-L}^{L} c_l\left(\sum_{m=-N}^{N} h_m h_{l+q-m} + \sigma_v^2\right) = h_{D-q} \qquad q = -L, \ldots, L \tag{8.2.82}$$

Equation (8.2.82) constitutes a set of $(2L + 1)$ simultaneous algebraic equations which can be solved to obtain the optimal tap gains $\{c_l\}$.

It is left as an exercise to the reader to show that in the absence of additive white noise, the MMSE approach is the same as that of minimizing the mean-square distortion.

We can also obtain a sequential solution to the problem of equalizer design by using the Kalman filter discussed in Chapter 6. Let us for the moment assume that the channel model is the same as in Eq. (8.2.75) but that the interference is one-sided and finite. That is, the interference is due to a finite number of previously transmitted symbols. In this case, h_n in Eq. (8.2.75) is nonzero only for $n = 0, 1, \ldots, N$. The channel model is shown in Fig. 8.2.17. The input to the delay line is a string of independent, random, binary digits $u(k)$ where $u(k) = a_k$. The output is corrupted by stationary additive white noise $v(k)$ with variance V_v. We assume that $u(k)$ and $v(k)$ are independent for all time. The dynamic equations of the channel follow from the figure and can be represented in state-variable form as

$$\begin{aligned} \mathbf{x}(k+1) &= \mathbf{F}\mathbf{x}(k) + \mathbf{G}u(k+1) \\ z(k) &= \mathbf{h}^T\mathbf{x}(k) + v(k) \end{aligned} \tag{8.2.83}$$

where $\mathbf{x}(k)$ is an n-vector of states, $u(k)$ is the scalar input, $z(k)$ is the noise corrupted output, and the matrix $\mathbf{F}$ and vectors $\mathbf{G}$ and $\mathbf{h}$ are

$$\mathbf{F} = \begin{bmatrix} 0 & 0 & \cdots & & 0 \\ 1 & 0 & \cdots & & 0 \\ 0 & 1 & \cdots & & 0 \\ & \vdots & & & \\ 0 & 0 & \cdots & 1 & 0 \end{bmatrix} \qquad \mathbf{G} = \begin{bmatrix} 1 \\ 0 \\ 0 \\ \vdots \\ 0 \end{bmatrix} \qquad \mathbf{h} = \begin{bmatrix} h_1 \\ h_2 \\ \vdots \\ h_N \end{bmatrix}$$

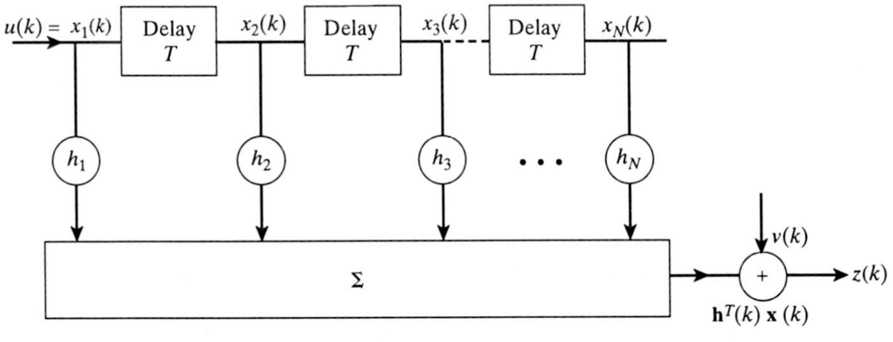

Figure 8.2.17 ISI channel model for Kalman equalizer.

The problem is to obtain an on-line equalization algorithm, given the input statistics and noise statistics, which yields an estimate of the message $u(k)$ at some delayed time $(k + D), 0 \leq D \leq N - 1$.

Since at any time k, we have the observation record $\mathbf{Z}(k) = \{z(j) : j < k\}$, the preceding problem statement can be rephrased as one of on-line estimation of the $(D + 1)$st component of $\mathbf{x}(k)$, that is, $x_{1+D}(k) = u(k - D)$, from these measurements. The Kalman filter algorithms discussed in Section 6.4 can be used to yield the minimum variance estimate of the state vector $\mathbf{x}(k)$, thereby giving a direct solution to our problem. The structure of the equalizer (Kalman filter) easily follows from our discussions in Chapter 6 and is shown in Fig. 8.2.18. The detailed derivation of the structure and the associated equations is left as an exercise for the reader. Note that an advantage of this equalizer is that the estimates $\hat{a}_{n-D}$ are obtained simultaneously for all D such that $0 \leq D \leq N - 1$.

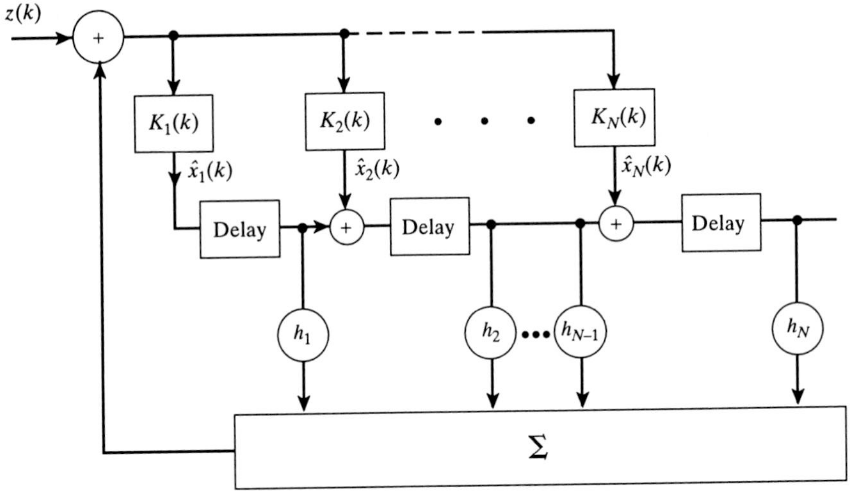

Figure 8.2.18 Kalman equalizer for ISI channels.

In our discussions so far, we have assumed that the impulse response of the channel $h(t)$ or its sample values h_n are known. In practice, this may be only approximately true because the channel characteristics may vary with time. In many cases, the samples h_n may not be known. In such cases the design of the equalizer will have to be based on some adaptive technique that identifies the samples h_n directly or indirectly, while simultaneously providing equalization. We now discuss two methods of adaptive equalization. In the first method, a maximum likelihood estimate of the channel weighting sequence is used. The technique has been widely used in several applications. The other uses an extended Kalman filter formulation.

Adaptive equalization: the Lucky equalizer. We recall from our earlier discussion that, ideally, the equalizer must be chosen such that the equalized system (channel and equalizer in cascade) represents a pure gain (or a pure delay). If we denote by $\{h_k\}$ the weighting sequence of the equalized system, we must have

$$h_k = \begin{cases} h_0 & k = 0 \\ 0 & \text{otherwise} \end{cases} \tag{8.2.84}$$

The Lucky equalizer is a tapped delay line filter with variable taps in which the tap gains are adjusted so that the weighting sequence of the equalized system approximates h_k in Eq. (8.2.84). Since the channel characteristics are usually varying, the tap gains will have to be readjusted at frequent intervals of time. To obtain an algorithm for adjusting the tap gains periodically, let us assume that the channel characteristics are slowly varying so that they may be assumed to be constant over an interval of K observations. Figure 8.2.19 shows a block diagram of the scheme. If we denote the transmitted sequence by a_n, the observations at the receiver (at the output of the equalizer) can be modeled as

$$z_k = \sum_{n=-\infty}^{\infty} h_n a_{k-n} + v_k \tag{8.2.85}$$

where the noise samples v_k are assumed to be independent, identically distributed Gaussian random variables with mean zero and variance σ^2. Let $\mathbf{h}$ denote the entire weighting sequence $\{h_i\}$ of the equalized channel. Our aim is to obtain the ML estimate of $\mathbf{h}$, given the set of observations $\mathbf{z} = [z_1, z_2, \ldots, z_K]^T$, over any K signaling intervals. From our assumptions, the conditional density $p(\mathbf{z}\,|\,\mathbf{h})$ can easily be written as

$$p(\mathbf{z}\,|\,\mathbf{h}) = \prod_{k=1}^{K} \frac{1}{\sqrt{2\pi}\,\sigma} \exp\left[-\frac{\left(z_k - \sum_{n=-\infty}^{\infty} h_n a_{k-n}\right)^2}{2\sigma^2} \right] \tag{8.2.86}$$

The log-likelihood function $\ln p(\mathbf{z}\,|\,\mathbf{h})$ is

$$\ln p(\mathbf{z}\,|\,\mathbf{h}) = -\sum_{k=1}^{K} \frac{1}{2\sigma^2}\left(z_k - \sum_{n=-\infty}^{\infty} h_n a_{k-n}\right)^2 + \ln\left(\frac{1}{2\pi\sigma^2}\right)^{K/2} \tag{8.2.87}$$

The maximum likelihood estimates $\hat{h}_j$ are obtained by setting $\{\partial \ln p(\mathbf{z}\,|\,\mathbf{h})/\partial h_j\}|_{h_j=\hat{h}_j}$ equal to 0 and are determined as the solution to the set of equations

$$\sum_{k=1}^{K} a_{k-j}\left(z_k - \sum_{n=-\infty}^{\infty} \hat{h}_n a_{k-n}\right) = 0 \qquad \text{for } j = -N, \ldots, N \tag{8.2.88}$$

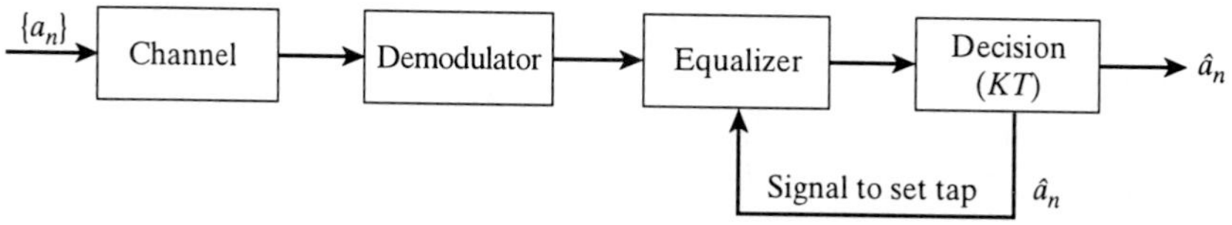

Figure 8.2.19 Adaptive equalizer scheme.

or, equivalently,

$$\sum_{n=-\infty}^{\infty} \hat{h}_n \sum_{k=1}^{K} [a_{k-j} a_{k-n}] = \sum_{k=1}^{K} a_{k-j} z_k \tag{8.2.89}$$

Equation (8.2.89) is a set of simultaneous linear algebraic equations that must be solved to obtain the estimates $\hat{h}_n$.

We can simplify the equation if we assume that the transmitted symbols are independent. In this case, for sufficiently large K, we can write

$$\frac{1}{K} \sum_{k=1}^{K} a_{k-j} a_{k-n} \simeq S\delta_{nj} \tag{8.2.90}$$

where S denotes the average power in the transmitted sequence. The estimates $\hat{h}_n$ are then obtained as

$$\hat{h}_n = \frac{1}{KS} \sum_{k=1}^{K} a_{k-n} z_k \qquad n = -N, \ldots, N \tag{8.2.91}$$

The algorithm for adjusting the tap gains c_n in the equalizer is now straightforward. For perfect equalization, we require that $h_0 = 1$ and $h_n = 0$ for $n \neq 0$. Therefore, for $n = 0$, if h_n is positive we decrease the corresponding tap gain by a fixed amount Δ whereas if h_n is negative, we increase the tap gain by Δ. Figure 8.2.20 shows the schematic of the adaptive equalization arrangement. The determination of $\hat{h}_n$ requires that we have the transmitted sequence available at the receiver. If the probability of error is small, the output of the detector $\hat{a}_n$ can be used in place of a_n in Eq. (8.2.91).

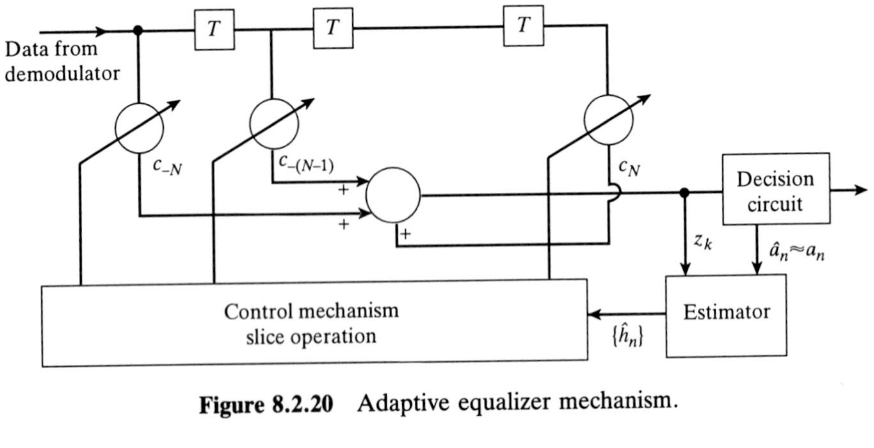

Figure 8.2.20 Adaptive equalizer mechanism.

Ideally $h_0 = 1$ and $h_n = 0, n \neq 0$, so that the estimates $\hat{h}_n$ for $n \neq 0$ will be fairly small. Thus in order to obtain reasonably converging estimates, the number of observations K must be fairly large. We can overcome this problem by estimating $h_0' = h_0 - 1$ rather than estimating h_0 itself. If we now define $h_n' = h_n$ for $n \neq 0$, the estimates $\hat{h}'_n$ will be of the same order of magnitude for all n. The likelihood function of Eq. (8.2.87) can easily be written in terms of h'_n as

$$\ln p(\mathbf{z} \mid \mathbf{h}') = \sum_{k=1}^{K} \frac{1}{2\sigma^2}\left(z_k - a_k - \sum_{n=-\infty}^{\infty} h'_n a_{k-n}\right)^2 + \ln\left(\frac{1}{2\pi\sigma^2}\right)^{K/2} \tag{8.2.92}$$

so that the ML estimates are given by

$$\hat{h}'_n = \frac{1}{KS}\sum_{k=1}^{K} a_{k-n} e_k \qquad n = -N, \ldots, N \tag{8.2.93}$$

where

$$e_k = z_k - a_k \tag{8.2.94}$$

The moments of the estimates can be easily determined. We have, from Eqs. (8.2.93) and (8.2.85),

$$\begin{aligned} \mathrm{E}\{\hat{h}'_n\} &= \mathrm{E}\left\{\frac{1}{KS}\sum_{k=1}^{K} a_{k-n} e_k\right\} \\ &= \frac{1}{KS}\sum_{k=1}^{K} \mathrm{E}\left\{a_{k-n}\left[\sum_{m=-\infty}^{\infty} h_m a_{k-m} + v_k - a_k\right]\right\} \\ &= \begin{cases} h_n & n \neq 0 \\ h_n - 1 & n = 0 \end{cases} \end{aligned} \tag{8.2.95}$$

where the last step follows from the assumed independence of the sequence $\{a_k\}$. It follows, therefore, that the estimates are unbiased.

The variance of the estimates is given by

$$\mathrm{var}[\hat{h}'_n] = \mathrm{E}\{(\hat{h}'_n)^2\} - \mathrm{E}^2\{h'_n\} \tag{8.2.96}$$

and can be reduced, after considerable algebraic manipulation, to

$$\mathrm{var}[\hat{h}'_n] = \frac{1}{K}\sum_{\substack{m=-\infty \\ m \neq n}}^{\infty} h_m^2 + \frac{2}{K^2}\sum_{j=1}^{K-1} (K - j) h_{n+j} h_{n-j} + \frac{\sigma^2}{KS} \tag{8.2.97}$$

As the interval over which the tap gains remain constant increases, that is, as K becomes increasingly large, the gains $\hat{h}'_n$ become Gaussian with mean and variance given by Eqs. (8.2.95) and (8.2.97).

The quantity fundamental to the accuracy and setting time of this equalizer is the probability that the sign of $\hat{h}_n$ is the same as that of h_n, and is denoted by P_c. For large K, since $\hat{h}_n$ is Gaussian with mean h_n, it is easy to see that $P_c \simeq 1$ when $|h_n|$ is large and $P_c \simeq 0.5$ when $|h_n|$ is small. In the latter case, the tap gain "wanders" because of the oscillatory corrective nature of the gain setting algorithm. The adaptive equalizer is usually designed to keep this "wander" of the tap gain within bounds imposed by accuracy requirements. Equation (8.2.97) can be used as a design relation in determining the value of K required for any value of N. Generally, each tap gain will have an error of about 0.5Δ, where Δ is the magnitude of the adjustment every K samples. Thus each of the gains h_n will have a magnitude of about 0.5Δ. If $|h_n| = 0.5\Delta$ and the adjustment is made in the wrong direction, the corresponding error will be approximately 1.5Δ. We should design the equalizer so that the probability of each h_n being 1.5Δ is fairly small while the probability of h_n being close to 0.5Δ is large. It can be shown [20] from Eq. (8.2.97) that for $h_n = \pm 0.5\Delta$ with equal probability, for $P_c = 0.99$, K must be chosen to be approximately equal to $10.8N$ under no-noise conditions (see also Exercises 8.18–8.20 at the end of the chapter). A detailed discussion of different equalization schemes can be found in [10].

Adaptive equalization using extended Kalman filter. We earlier formulated the design of an equalizer using a Kalman filter, where we assumed that the channel sample response $\{h_n\}$ was known. If this knowledge is not available and if $\{h_n\}$ are varying with time, the equalizer algorithms must be adaptive.

By treating the coefficients (tap gains) of the channel as additional states, we can formulate a nonlinear estimation problem. Since the tap gains are assumed constant, they can be modeled by the difference equation

$$\mathbf{h}(k+1) = \mathbf{h}(k) \tag{8.2.98}$$

If we now augment the message model of Eq. (8.2.83) by this equation, we obtain the set of equations

$$\begin{bmatrix} \mathbf{x}(k+1) \\ \hline \mathbf{h}(k+1) \end{bmatrix} = \left[\begin{array}{c|c} \mathbf{F} & \mathbf{0} \\ \hline \mathbf{0} & \mathbf{I} \end{array}\right] \begin{bmatrix} \mathbf{x}(k) \\ \hline \mathbf{h}(k) \end{bmatrix} + \begin{bmatrix} \mathbf{G} \\ \hline \mathbf{0} \end{bmatrix} u(k)$$

and

$$z(k) = \mathbf{h}^T(k)\mathbf{x}(k) + v(k)$$

We note that our message model is still linear in the state equations but is nonlinear in the observations. We can use the extended Kalman filter algorithms of Section 6.6 to estimate the states $\mathbf{x}(k)$ and the coefficients $\mathbf{h}$. The resulting algorithms are as follows:

$$\begin{bmatrix} \hat{\mathbf{x}}(k|k) \\ \hat{\mathbf{h}}(k|k) \end{bmatrix} = \begin{bmatrix} \hat{\mathbf{x}}(k|k-1) \\ \hat{\mathbf{h}}(k|k-1) \end{bmatrix} + \mathbf{K}(k)[z(k) - \hat{\mathbf{h}}^T(k|k-1)\hat{\mathbf{x}}(k|k-1)] \tag{8.2.99}$$

$$\begin{aligned} \mathbf{K}(k) = {} & \mathbf{V}_{\tilde{\mathbf{x}}_a}(k|k-1)[\hat{\mathbf{h}}(k|k-1) \,\vdots\, \hat{\mathbf{x}}(k|k-1)]^T \\ & \times \{[\hat{\mathbf{h}}(k|k-1) \,\vdots\, \hat{\mathbf{x}}(k|k-1)]\mathbf{V}_{\tilde{\mathbf{x}}_a}(k|k-1) \\ & [\hat{\mathbf{h}}(k|k-1) \,\vdots\, \hat{\mathbf{x}}(k|k-1)]^T + V_v(k)\}^{-1} \end{aligned} \tag{8.2.100}$$

$$\begin{aligned} [\mathbf{V}_{\tilde{\mathbf{x}}_a}(k|k-1) = {} & \left[\begin{array}{c|c} \mathbf{F} & \mathbf{0} \\ \hline \mathbf{0} & \mathbf{I} \end{array}\right] \\ & \times \mathbf{V}_{\tilde{\mathbf{x}}_a}(k-1|k-1) \left[\begin{array}{c|c} \mathbf{F}^T & \mathbf{0} \\ \hline \mathbf{0} & \mathbf{I} \end{array}\right] + \begin{bmatrix} \mathbf{G} \\ \hline \mathbf{0} \end{bmatrix} [\mathbf{G}^T \,\vdots\, \ 0] V_u(k) \end{aligned} \tag{8.2.101}$$

$$\mathbf{V}_{\tilde{\mathbf{x}}_a}(k|k) = \{\mathbf{I} - \mathbf{K}(k)[\hat{\mathbf{h}}(k|k-1) \,\vdots\, \hat{\mathbf{x}}(k|k-1)]\}\mathbf{V}_{\tilde{\mathbf{x}}_a}(k|k-1) \tag{8.2.102}$$

where $\mathbf{V}_{\tilde{\mathbf{x}}_a}(k\,|\,k)$ is the filtered error covariance of the estimate of the augmented state vector.

Equations (8.2.99)–(8.2.102) are the sequential adaptive equalizer algorithms. Reference [21] presents an example of equalization using an extended Kalman filter demonstrating an error percentage almost equaling that of equalization with known channel. The algorithms are necessarily complex, but the advantage again is that a sequence of input symbols can be estimated simultaneously, but with different amount of delays. We point out, however, that the adaptive Kalman equalizer can become unstable.

8.3 SPREAD SPECTRUM COMMUNICATIONS

Spread spectrum techniques have been widely used in military and civilian applications in recent years [22]. The term *spread spectrum communications* refers to communication systems that meet the following two conditions: (1) the transmitted signal bandwidth is much greater than the bandwidth of the information bearing signal, and (2) the spectral spreading is achieved not by the information-bearing signal but by a signal that is independent of it. The basic philosophy behind spectral spreading involves two principles. First, since resolution accuracy in the time domain depends directly on the bandwidth of the measured signal, spread signals can be used for highly accurate range/position measurements. Second, since the dimensionality of a signal depends on its bandwidth, spectral spreading can be effectively used to hide a low-bandwidth information signal in a high-dimensional space. This concept leads to communication systems designed for antijamming capability/low probability of intercept/multiple access capability. A spread spectrum system in a multiuser environment is usually referred to as a code division multiple access (CDMA) system—the name refers to the fact that each user has a unique spectral spreading signal (code) and that each user fully occupies the available bandwidth. CDMA systems are employed in satellite communications. They have also been proposed for indoor wireless personal communication systems and mobile cellular systems.

Spread spectrum systems may be broadly divided into two classes, *frequency-hopping* (FH) systems and *direct-sequence* (DS) systems. In frequency-hopping systems, the data signal $d(t)$, which is assumed to be a binary signal at a bit rate of R_b, is modulated on to a carrier whose frequency changes (hops) at periodic intervals among a set of frequencies over a chosen band. The bandwidth of this band of frequencies defines the spread bandwidth of the FH system. In direct-sequence systems, a pseudonoise (PN) generator is used to generate a pseudorandom binary sequence $c(t)$ at a rate R_c, which is much larger than R_b. The *code signal* or *code sequence* $c(t)$ is multiplied with the signal $d(t)$ and modulated on to a carrier. The rate R_c is referred to as the *chip rate* and $T_c = 1/R_c$ is referred to as the *chip duration*. Since $c(t)$ changes at a much faster rate than $d(t)$, the bandwidth of the spread signal $d(t)c(t)$ is nearly equal to that of $c(t)$.

The generation and reception of spread spectrum signals have received considerable attention in recent years and the literature in this area is quite extensive. Thus we do not attempt to discuss this subject here but refer the interested reader to some recent texts in this area [22,23]. In the next section we illustrate application of the statistical techniques we have discussed in previous chapters to determine the performance of spread spectrum systems. Specifically, we will consider a direct-sequence modulated binary phase-shift keying system (DS-BPSK) in the presence of interference.

8.3.1 Performance of DS-BPSK Receiver in Interference

Consider the DS-BPSK system, in which the transmitted signal is given by

$$s(t) = d(t)c(t) \tag{8.3.1}$$

where

$$d(t) = \sqrt{\frac{2E_b}{T_b}} \sum_{i=-\infty}^{\infty} d_i p_{T_b}(t - iT_b) \cos(\omega_0 t + \theta) \tag{8.3.2}$$

and

$$c(t) = \sum_{j=-\infty}^{\infty} c_j p_{T_c}(t - jT_c) \tag{8.3.3}$$

where $p_T(t)$ is the rectangular pulse of unit height over the interval $[0, T]$ and zero elsewhere, E_b is the energy per bit of the data $d_i, d_i \in \{\pm 1\}$. $c_j, c_j \in \{\pm 1\}$ is the code sequence used by the transmitter, and T_b and T_c refer to the duration of the data bit and the chip interval, respectively. Figure 8.3.1 shows an example of a data sequence $d(t)$, code sequence $c(t)$, and the spread signal $d(t)c(t)$.

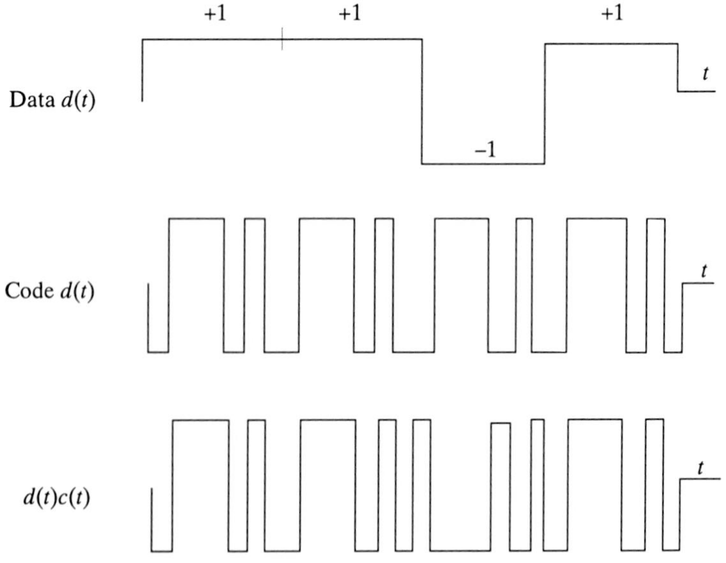

Figure 8.3.1 Typical signals in a DS spread system.

The PN sequence $c(t)$ is usually chosen to be a *maximal-length* sequence, because its properties resemble those of a random binary sequence, especially for large sequence lengths [22]. These properties make maximal-length sequences attractive for a variety of applications. Maximal-length sequences are generated by a m-stage shift register in which the binary output from some of the stages are fed to a modulo-2 adder. The modulo-2 adder output is then fed back to the first-stage input of the shift register. The binary outputs of the register at any time define the *state* of the register. The sequence generated by the register is called a maximal-length sequence since the register goes through all possible states, except the all-zero state, corresponding to a total of $2^m - 1$ states. Given a value for m, only certain combinations of the register stages being fed back will generate a maximal-length sequence. These combinations have been found for several values of m and are readily available [23].

For reasons that will become apparent later, the ratio $N = T_b/T_c$ is called the processing gain (PG) of the system. The ratio N is usually an integer and thus represents the number of chips contained in a bit interval. Although it is possible to have a scenario in which the transitions in $d(t)$ and $c(t)$ are not aligned but occur with a time shift between the two, we will assume that the transitions are aligned to happen at the same time. Let the signal at the receiver be

$$r(t) = s(t - \tau) + I(t) + v(t) = c(t - \tau)d(t - \tau) + I(t) + v(t) \tag{8.3.4}$$

where $v(t)$ is additive white Gaussian noise with power spectral density $N_0/2$, $I(t)$ is some form of interference, and τ is the time taken for the transmitted signal to reach the receiver. Let us for the moment neglect $v(t)$ and $I(t)$. If $c(t - \tau)$ is known at the receiver, then $d(t - \tau)$ can be recovered simply by multiplying $r(t)$ with $c(t - \tau)$, since $c^2(t - \tau)$ is equal to 1 for all t. The recovery of $c(t - \tau)$ from $r(t)$ is generally referred to as *code acquisition and tracking*. Typically, standard statistical techniques such as noncoherent detection schemes and the sequential probability ratio test are employed by code acquisition [22,23]. In the sequel we assume that an estimate (replica) of $c(t - \tau)$ is available at the receiver and concentrate only on the detection of the data bits d_i from $r(t)$.

When the interference signal $I(t)$ is absent, $r(t) = s(t - \tau) + v(t)$, so that

$$r(t)c(t - \tau) = d(t - \tau) + v(t)c(t - \tau) \tag{8.3.5}$$

The process of multiplying the received signal by the code sequence $c(t)$ [or $c(t - \tau)$] is referred to as *despreading*. Since $c(t)$ has a wide spectrum and takes on only ± 1 values, we can assume that $v(t)c(t - \tau)$ is nearly Gaussian with spectral density $N_0/2$. Thus the BPSK signal $d(t - \tau)$ can be assumed to be observed in white Gaussian noise. As we have seen earlier, the minimum probability of error receiver is therefore the correlator (or matched filter). The block diagram of the receiver is shown in Fig. 8.3.2. The probability of bit error is the same as in the case of BPSK and is given by $P_b = Q(\sqrt{2E_b/N_0})$.

In the presence of interference, the problem becomes more complicated. We first consider the case where $I(t)$ is a narrowband Gaussian signal with two-sided power spectral density $N_I/2$. For simplicity, let us assume that $v(t)$ can be neglected

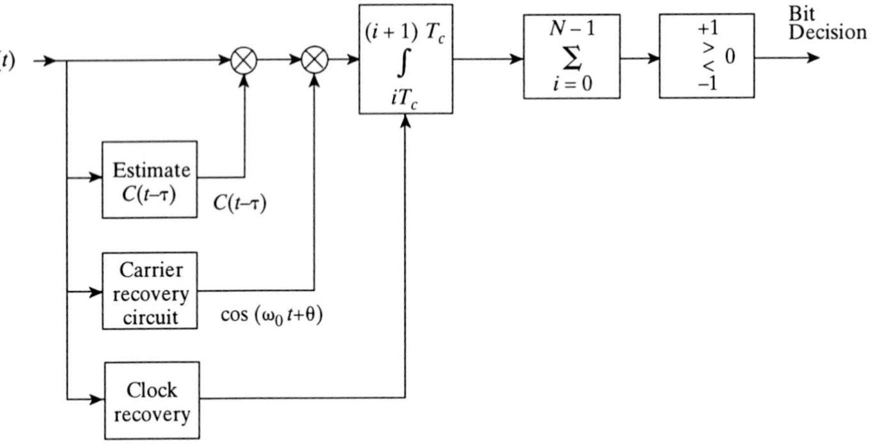

Figure 8.3.2 DS-BPSK receiver in the absence of interference.

in comparison with the interfering signal $I(t)$. If the signal $I(t)$ has a bandwidth equal to the data bandwidth W_d, the average power of $I(t)$ is $N_I W_d$. Then $I(t)c(t - \tau)$ is nearly wideband, Gaussian with power spectral density $(N_I/2)(W_{ss}/W_d)$, where W_{ss} is the bandwidth of the DS-BPSK signal. Therefore, at the receiver, after despreading, we have a BPSK signal in additive white Gaussian noise. Figure 8.3.3 shows the power spectra of the various signals in the transmitter and receiver.

The receiver again is a correlator or matched filter with the probability of bit error being given by $P = Q(\sqrt{(2E_b/N_I)(W_{ss}/W_d)})$. Since $W_{ss}/W_d = (1/T_c)/(1/T_b) = N$, the effective signal-to-noise ratio is increased by a factor of N, the processing gain (PG). The increase in SNR is due to a reduction, at the input to the correlator, of the power spectral density of the interferer, caused by the despreading code sequence $c(t - \tau)$. That is, despreading serves not only to recover $d(t)$ from the signal component $s(t)$ in the received signal $r(t)$, it also spreads the interference over a larger bandwidth to effectively produce a reduced spectral density of N_I/PG. If the interference is a tone at the carrier frequency, the effective SNR is still increased approximately by PG [22,23].

When the interference is a peaked narrowband signal, we can decrease the probability of error further by using an interference rejection filter prior to the despreading operation. The rejection filter essentially serves to predict the interference $I(t)$ and subtract it from the received signal. Even though the rejection filter introduces some distortion in the signal component of the received signal, since the DS-BPSK signal is wideband, the resulting distortion will be small. In any case, the reduction in interference greatly outweighs whatever distortion is introduced. A review of interference rejection techniques in DS systems can be found in [24].

In multiple-access systems with K users, $I(t)$ is due to signal-like interference in the spread bandwidth, from the other users. Let $c^k(t)$ and $d^k(t)$ denote the code

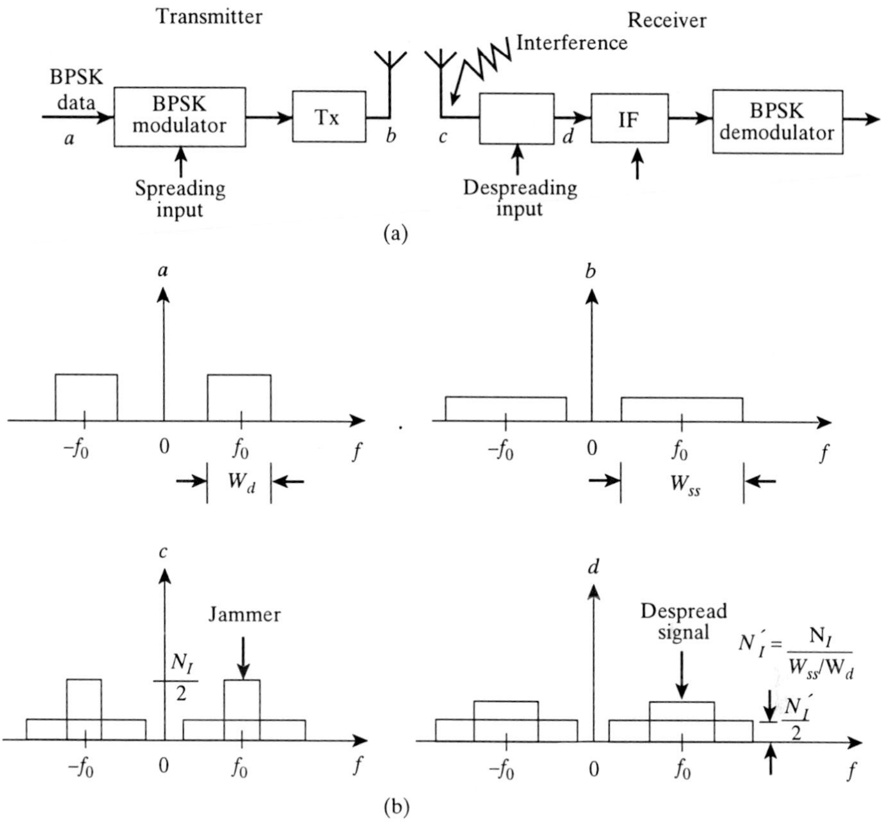

Figure 8.3.3 (a) Block diagram of typical BPSK spread spectrum system; (b) spectra of various signals.

sequence and data signal, respectively, of the kth user, and let $k = 1$ correspond to the signal of interest. We can write

$$r(t) = d^1(t - \tau_1)c^1(t - \tau_1) + v(t) + I(t) \tag{8.3.6}$$

where

$$I(t) = \sum_{k=2}^{K} d^k(t - \tau_k)c^k(t - \tau_k) \tag{8.3.7}$$

If $I(t)$ were Gaussian, we could again use the *single-user* receiver structure of Fig. 8.3.2. However, as has been shown in [25], $I(t)$ is not Gaussian. Any multiuser receiver must take into account the nature of $I(t)$. Clearly, therefore, there may be other receiver structures which may yield a lower probability of error [26,27]. For example, we could estimate $d^k(t), k = 2, 3, \ldots, K$ and hence estimate and subtract $I(t)$ from the received signal $r(t)$. This requires that all the code sequences

$c^k(t), k = 1, 2, \ldots, K$ be known at each of the receivers and that all the receivers decode all the data sequences $d^k(t), k = 1, 2, \ldots, K$. The complexity of the receiver structures therefore grows exponentially as the number of users increases. The scheme also implies a loss of privacy. Other (suboptimal) detectors can be designed in which the complexity increases only linearly with the number of users [28].

8.4 INTRODUCTION TO RADAR SYSTEMS

Radar is an acronym for *radio detection and ranging*. It is an active device that operates by radiating electromagnetic energy and detecting the presence and character of the echo returned from reflecting objects. The primary objectives of earlier radars were to detect the presence of a target at large distances and to locate the position of the target. These capabilities have been extended to obtain information regarding target velocity, shape, and size, and to include mechanisms of automatic detection.

Radars can generally be classified broadly into two functional categories: search radars and tracking radars. As the name implies, a *search radar* is planned and designed with emphasis on detecting target presence. It may also provide some coarse information regarding the target. A *tracking radar* emphasizes information extraction, that is, parameter estimation, from the target echo. It tries to estimate with sufficient accuracy the significant target parameters, such as range, range rate, angle, and angular velocity once the target presence is established. However, in present-day radars with electronic scanning arrays performing both functions adequately, such a binary classification of radars is not always possible.

In this chapter we apply the results of our development of detection and estimation theory to the search and track problems of radar. We first discuss the radar detection problem and subsequently, consider the problem of estimating certain parameters of interest. Related material may be found in [29–33].

To detect the presence or absence of a target, the radar transmitter sends a pulse of electromagnetic energy to illuminate the target. The volume of space covered by the pulse is determined by the azimuth and elevation beamwidths of the transmitting antenna and is referred to as *angular beamwidth*. By scanning any volume in space in steps of angular beamwidths, the radar tries to search for the target. The total volume scanned is called *search volume* or *surveillance region*. Usually, the search radars transmit a sequence of narrow RF pulses every T_p seconds (referred to as *interpulse interval*), and receive the delayed echoes from the target. The target range is determined in terms of the delay τ as $R = c\tau/2$, where c is the velocity of electromagnetic radiation in free space. For a pulse width of T seconds and a delay of T seconds, the range $cT/2$ defines a cell called the range resolution cell.† A search radar effectively examines or interrogates every resolution cell in the

†The range resolution cell is defined more precisely in terms of the effective bandwidth T_{eff} of the basic waveforms (see [34] for details).

surveillance region. In other words, every range element of $cT/2$ in each angular beamwidth is tested for target presence.

If a target is detected in a resolution cell, we know its location within the accuracies of a range element (or range cell $cT/2$) and the beamwidths in elevation and azimuth. This information is rather gross for many tactical applications. By further processing the echo return of a resolution cell, the various parameters can be estimated. The usual parameters of interest are range, range rate, angular position, and angular velocities and can be estimated using maximum likelihood or maximum posterior probability estimation techniques. In recent times, sequential estimation algorithms have been used for dynamic target tracking. Such tracking can also take into account target maneuvers.

We start our discussions in Section 8.5 with radar target models. We use the complex envelope representations of bandpass signals (see Chapter 2) in our model, as they are convenient in representing the various physical phenomena. We show that depending on the target model, the return signal can be characterized as completely known, completely known except for some random parameters, or as a realization of a stochastic process.

8.5 RADAR TARGET MODELS

We assume that a radar transmits an RF signal which is modulated in both amplitude and phase. We can represent this signal as

$$y_T(t) = \sqrt{2E_T}\, f(t) \cos[\omega_c t + \theta_1(t)] \qquad 0 \leqslant t \leqslant T \tag{8.5.1}$$

Here $f(t)$ is the amplitude modulation, which is usually a shaped pulse of finite duration T seconds. $\theta_1(t)$ is a phase modulation which is often used for considerations of increased resolution, ω_c is the carrier frequency, and E_T is the transmitted energy. We can represent this bandpass signal in terms of the complex envelope $\tilde{y}(t)$ as

$$y_T(t) = \mathrm{Re}\{\sqrt{2E_T}\tilde{y}(t) \exp(j\omega_c t)\} \tag{8.5.2}$$

where

$$\tilde{y}_T(t) = \sqrt{2E_T}\tilde{y}(t) = \sqrt{2E_T} f(t) \exp(j\theta_1(t_1)$$

For convenience, we assume that the complex envelope is normalized such that

$$\int_{-\infty}^{\infty} |\tilde{y}(t)|^2\, dt = 1$$

We recall from our discussions in Chapter 2 that the complex envelope $\tilde{y}(t)$ is a baseband signal and corresponds to the in-phase and quadrature components of the modulation (information-bearing) portion of the transmitted signal. We can therefore use the complex envelope $\tilde{y}(t)$ in our discussions without actually using $y_T(t)$.

In what follows we discuss basically three types of models for the radar target.

In the first type, the return signal is essentially completely known, in the second it is known except for some random parameters, and in the third it is a realization of a stochastic process.

8.5.1 Steady Point Target

We assume that the reflecting characteristic of the radar target is linear. The reflected signal energy from a target depends on its electrical echoing area, called *radar cross section*† [30]. By a point target we mean that all this echoing area is concentrated at a single point in space with an appropriate reflection gain. Suppose that the target is located at a range R and is stationary. In this case the returned signal $y_r(t)$ will be a delayed version of the transmitted signal with a deterministic gain $\tilde{b}$, so that we can write

$$y_r(t) = \mathrm{Re}\{\sqrt{2E_T}\tilde{b}\tilde{y}(t-\tau)\exp(j\omega_c t)\} \tag{8.5.3}$$

where τ is the time delay and corresponds to a target range $R = c\tau/2$. The reflection gain is a complex quantity, so that we can write

$$\tilde{b} = |\tilde{b}|\exp(j\theta_2) \tag{8.5.4}$$

Here the phase θ_2 includes the term due to carrier delay, $\omega_c\tau$. It is obvious from Eq. (8.5.3) that the received signal energy is

$$E_r = E_T|\tilde{b}|^2 \tag{8.5.5}$$

For nonstationary targets, the effect of target velocity can be taken into account by introducing a Doppler shift of ω_d rad/sec in the carrier frequency ω_c. While this represents only a first-order approximation, it is adequate in many situations. The radar return signal can then be represented as

$$y_r(t) = \mathrm{Re}\{\sqrt{2E_T}\tilde{b}\tilde{y}(t-\tau)\exp[j(\omega_c+\omega_d)t]\} \tag{8.5.6}$$

Again, we have absorbed the term $\exp(j\omega_d\tau)$ in the phase θ_2 of the reflection gain $\tilde{b}$. It follows that the complex envelope of the return signal from a steady point target at a range $c\tau/2$ and moving with a velocity v corresponding to a Doppler shift of $\omega_d = (2v/c)\omega_c$ can be written as [31]

$$\tilde{y}_r(t) = \sqrt{2E_T}\tilde{b}\tilde{y}(t-\tau)\exp(j\omega_d t) \tag{8.5.7}$$

Equation (8.5.7) can be used to model the return from a steady point target for the case where $\tilde{b}$ is known as well as the case where $|\tilde{b}|$ is known but θ_2 is unknown.

8.5.2 Slowly Fluctuating Point Target

For the case in which the target has many different physical reflecting surfaces with arbitrary orientation, we can model it as a point target whose reflecting characteristic is random. That is, the reflection gain $\tilde{b}$ is a complex random variable. We further

†The radar cross section (RCS) is defined as the absolute value of the ratio of the reflected electric field to the incident electric field, and is generally different from the physical area.

assume that $\tilde{b}$ takes a particular value for the time duration of interest. That is, if we are observing one pulse return, $\tilde{b}$ remains constant between τ and $\tau + T$ seconds. This is usually valid when the target orientation does not change significantly during the time duration under consideration. Such a target is called a *slowly fluctuating point target*. The return signal for such a target can therefore be written as in Eq. (8.5.6) but with the modification that $\tilde{b}$ is random. The complex envelope of the return signal is again given by Eq. (8.5.7) with $\tilde{b}$ random. The average return signal energy is now given by

$$E_r = E_T \mathrm{E}[|\tilde{b}|^2] \tag{8.5.8}$$

where $\mathrm{E}[|\tilde{b}|^2]$ is the mean-square value of $\tilde{b}$.

A number of statistical descriptions have been used to characterize $\tilde{b}$ in the radar literature to account for various physical situations. Almost all of them assume that θ_2, the angle of $\tilde{b}$, is uniformly distributed in $[0, 2\pi]$, and postulate different probability density functions to describe the amplitude $|\tilde{b}|$ [31].

One of the most widely used models assumes that the target is composed of a collection of independent identically distributed random scatterers. In this case, the reflection gain $\tilde{b}$ will be Gaussian distributed, or equivalently, the amplitude $|\tilde{b}|$ will have a Rayleigh distribution and the phase θ_2 will be uniformly distributed in $[0, 2\pi]$. If we denote the mean-square value of $\tilde{b}$ as $2\sigma_b^2$ (or equivalently, the real and imaginary parts of $\tilde{b}$ as having zero mean with variance σ_b^2 each), the probability density of the amplitude of the reflection gain can be written explicitly as

$$p(|\tilde{b}|) = \begin{cases} \dfrac{|\tilde{b}|}{\sigma_b^2} \exp\left(-\dfrac{|\tilde{b}|^2}{2\sigma_b^2}\right) & |\tilde{b}| \geq 0 \\ 0 & \text{otherwise} \end{cases} \tag{8.5.9}$$

This model is due to Swerling [35], who has also presented various other models for radar returns. For example, if we assume that the target is composed of one large specular reflector and a number of random scatterers, we obtain the *one dominant plus Rayleigh density function* (see Exercise 8.28).

As we will see later, for a given average received energy, the detection performance for the case of a fluctuating target is degraded compared to that of a steady target. The performance of the detector can be improved by transmitting a number of pulses, say N, and processing the resulting N returns. In this case, the return signal can be modeled in one of three different ways, depending on the physical situation and the assumptions. First let us assume that the reflection gain remains constant, albeit random, during the entire pulse burst length of NT_p seconds, where T_p is the interpulse interval. This means that the radar has a pulse-to-pulse phase coherence. We write the complex envelope of the return signal corresponding to the ith pulse as

$$\tilde{y}_r(t) = \sum_{i=1}^{N} \tilde{b}\sqrt{2E_T}\,\tilde{y}(t - iT_p - \tau) \exp[j\omega_d(t - iT_p)] \tag{8.5.10}$$

with the average signal energy

$$E_r = NE_T \mathrm{E}[|\tilde{b}|^2]$$

It can thus be seen that by transmitting a coherent pulse train of length N, we can increase the received signal energy N-fold to achieve better performance.

There are situations in which pulse-to-pulse coherence does not exist. This leads to the model

$$\tilde{y}_r(t) = \sum_{i=1}^{N} |\tilde{b}|\sqrt{2E_T}\tilde{y}(t - iT_p - \tau) \exp[j\omega_d(t - iT_p) + j\theta_i] \qquad (8.5.11)$$

where $\{\theta_i\}$ are a set of independent random variables, uniformly distributed in $[0, 2\pi]$. Note, however, that the amplitude $|\tilde{b}|$ is constant through the N-pulse period. This corresponds to a slowly (scan to scan) fluctuating target [32], where one scan corresponds to transmitting N pulses.

The third situation arises when the target reflecting characteristics change with each pulse, but remain constant within a pulse period. In this case we can write the complex envelope of the returned signal as

$$\tilde{y}_r(t) = \sum_{i=1}^{N} \tilde{b}_i\sqrt{2E_T}\tilde{y}(t - iT_p - \tau) \exp[j\omega_d(t - iT_p)] \qquad (8.5.12)$$

where the reflection gains, $\tilde{b}_i$, are independent but identically distributed. That is the phase θ_{2i} are independent and uniformly distributed in $[0, 2\pi]$, and the amplitudes $|\tilde{b}_i|$ are independent and distributed according to Eq. (8.5.9) or some other density function [35]. This is the pulse-to-pulse or rapidly fluctuating model of Swerling's classical work. We later find that with proper modifications, Eq. (8.5.12) is a good approximation to range spread targets to be discussed in the next section.

8.5.3 Spread Targets

In this section we concern ourselves with targets that cannot be modeled as slowly fluctuating. Typically, these targets exhibit time and/or frequency-selective reflecting properties. Consequently, the complex envelope of the returned signal is a stochastic process. Since we have not developed the techniques for the detection of a stochastic process in noise, we discuss these models only briefly and refer the reader to selected references available in the literature [31].

Let us first consider the case of a target with *time-selective reflecting characteristic*. This can happen when a target is composed of many different physical shapes and its orientation changes rapidly with time. At different times the target exhibits different random reflecting gains, with the result that the returned signal is no longer an attenuated and delayed replica of the transmitted signal. In this case the reflection gain will be time varying and can be modeled as a realization $\tilde{b}(t)$ of a stochastic process so that the return signal can be written as

$$y_r(t) = \text{Re}\left\{\sqrt{2E_T}\tilde{b}\left(t - \frac{\tau}{2}\right)\tilde{y}(t - \tau) \exp(j\omega_c t)\right\} \qquad (8.5.13)$$

where τ is the delay corresponding to the target range. The delay of $\tau/2$ is introduced in the reflection gain to show explicitly that the reflected signal traversed the target range only once [31]. Equation (8.5.13) tacitly assumes that the phase of the complex stochastic process $\tilde{b}(t)$ is uniform and includes the delay in the carrier term, $\omega_c \tau$. We can now represent the complex envelope of the return signal as

$$\tilde{y}_r(t) = \sqrt{2E_T}\tilde{b}\left(t - \frac{\tau}{2}\right)\tilde{y}(t - \tau) \qquad \tau \leq t \leq \tau + T \tag{8.5.14}$$

As can be seen from Eq. (8.5.13) the effect of the time-varying reflection gain can be thought of as an amplitude modulation, which spreads or widens the energy spectrum of the reflected signal. To make this point clearer, consider the energy spectrum of the return from a slowly fluctuating point target. It is centered at a frequency offset from the carrier by the target Doppler frequency ω_d and has the same shape as that of the transmitted energy spectrum. In the case of a return from a target with time-selective reflecting characteristic, the effect of the multiplicative term $\tilde{b}(t)$ is to attenuate and spread the transmitted energy spectrum. The center frequency is now given by $\omega_c + \overline{\omega}_d$, where $\overline{\omega}_d$ is the average Doppler frequency to be defined shortly. Because of this spreading effect of the energy spectrum, which is due primarily to rapid changes in target orientation and hence velocity, we call such a target a *Doppler spread target*. That is, its Doppler frequency is spread about an average frequency $\overline{\omega}_d$.

If we assume that $\tilde{b}(t)$ is a complex stationary process with zero mean and covariance function $\phi_{\tilde{b}}(\tau)$; then the power spectrum $\Phi_{\tilde{b}}(f)$ is known as the *Doppler scattering function*. The name derives from the fact that the energy spectrum of the transmitted signal is convolved with $\Phi_{\tilde{b}}(f)$ to produce the spectral spreading described earlier. The average Doppler frequency is then given by $\overline{\omega}_d = 2\pi\bar{f}_d$ with

$$\bar{f}_d \triangleq \frac{1}{2\sigma_b^2}\int_{-\infty}^{\infty} f\Phi_{\tilde{b}}(f)\,df \tag{8.5.15}$$

The average received energy is given by

$$\overline{E}_r = \phi_{\tilde{b}}(0)E_T = 2\sigma_b^2 E_T \tag{8.5.16}$$

As noted earlier, the return signal from a Doppler spread target is a sample function of a stochastic process. Detection of such signals requires techniques beyond the scope of the present text. However, if we approximate the reflecting gain $\tilde{b}(t)$ by a piecewise constant function, we can use the techniques discussed in earlier chapters to obtain a solution. This is necessarily suboptimum. We proceed as follows. Assume that $\tilde{b}(t)$ is band-limited to w Hz. We can then approximate $\tilde{b}(t)$ as

$$\tilde{b}(t) = \sum_{i=1}^{N} \tilde{b}_i p\left(t - \frac{i}{w}\right) \tag{8.5.17}$$

where $p(t)$ is the pulse function which is equal to 1 over the range $0 < t < 1/w$ and is zero elsewhere, and $\tilde{b}_i$ are a set of identically distributed independent random variables. With this approximation, the return signal from a Doppler spread target can be seen to be similar to the signal in Eq. (8.5.7), with obvious correspondences and modifications.

We now consider a target that is extended in range but fluctuates slowly. That is, we assume that the target has a physical length L, and each point on the length of the target acts as a slowly fluctuating, independent point target. In other words, the reflection gain depends on which point of the total target is reflecting, and the return signal will be a superposition of all such reflections. If we parametrically represent the reflection gain as $\tilde{b}(\tau)$ to denote dependence on τ, the delay corresponding to range, we can write the return signal as

$$y_r(t) = \text{Re}\left\{\sqrt{2E_T}\int_{\tau_1}^{\tau_2} \tilde{b}(\tau)\tilde{y}(t-\tau)\exp(j\omega_c t)\,dt\right\} \qquad \tau_1 \leq t \leq T + \tau_2 \qquad (8.5.18)$$

where τ_1 and τ_2 are, respectively, the delays in the return signals from the target ends nearest and farthest from the transmitter. Note that the return signal in Eq. (8.5.18) is spread out in time. This is due to the convolution of the range-dependent reflection gain with the transmitted signal. Such targets are called *range spread targets*. It is left to the reader to show that range spread targets indeed cause frequency-selective reflecting characteristics. More detailed discussions can be found in [31].

The most general target model is both range and Doppler spread. From our previous discussions, we know that the reflecting gain is time varying, in addition to being dependent on the range. Therefore, we can write the complex envelope of the return signal as

$$\tilde{y}_r(t) = \sqrt{2E_T}\int_{\tau_1}^{\tau_2} \tilde{b}\left(t - \frac{\tau}{2}, \tau\right)\tilde{y}(t-\tau)\,d\tau \qquad (8.5.19)$$

which is a zero-mean complex stochastic process. We note that the singly spread target (Doppler or range) is only a special case of the doubly spread target model given by Eq. (8.5.19).

We are now in a position to use these models in developing the receiver structures for the detection of targets. We present only certain basic results and develop related special results in problems.

8.6 TARGET DETECTION

Target detection is a major function of a radar. The reflected signal from a target is a delayed attenuated version of the transmitted signal, which may also be offset in frequency by a Doppler shift. The radar observation of the reflected signal is corrupted by additive noise. The noise presents uncertainty in establishing the presence or absence of the target. Thus we are led to a hypothesis-testing problem. We use the results developed in Chapters 3 and 4 to solve this problem. We formulate the hypothesis-testing problem in terms of complex envelopes of the bandpass signals involved. Further, we also assume that the target to be detected is at zero range with zero Doppler offset (i.e., $\tau = 0$ and $\omega_d = 0$) without any loss of generality. The extension to nonzero range and Doppler frequency is fairly straightforward and is not discussed further in this section.

Let us denote the observed signal by $z(t)$. Then, under hypothesis H_1 with target present, we have

$$z(t) = y_r(t) + v(t) \qquad 0 \leq t \leq T$$

where $y_r(t)$ is the reflected pulse modeled as in Section 8.5 and $v(t)$ is the observation noise. We assume that $v(t)$ is bandpass, Gaussian white noise with mean zero and spectral density $N_0/2$ and centered at the carrier frequency ω_c. For simplicity $v(t)$ can be considered as having a flat spectrum everywhere, without affecting the final results.

We can equivalently represent the received signal in terms of its complex envelope as

$$\tilde{z}(t) = \tilde{y}_r(t) + \tilde{v}(t) \qquad 0 \leq t \leq T \tag{8.6.1}$$

where $\tilde{y}_r(t)$ is the complex envelope of the reflected signal,

$$z(t) = \text{Re}\{\tilde{z}(t) \exp(j\omega_c t)\}$$

and

$$v(t) = \text{Re}\{\tilde{v}(t) \exp(j\omega_c t)\}$$

In the alternative hypothesis H_0, the observation consists of pure noise only and hence

$$\tilde{z}(t) = \tilde{v}(t) \qquad 0 \leq t \leq T \tag{8.6.2}$$

In the sequel we are concerned with the hypothesis-testing problem when $\tilde{y}_r(t)$ is the complex envelope corresponding to the different target models discussed earlier. Specifically, we consider targets that can be modeled as (1) steady point targets (known and unknown random phase), and (2) slowly fluctuating point targets. We also consider multiple pulse detection of such targets for achieving improved performance. We conclude the section with a brief discussion on the detection of spread targets. Related material can be found in [35–38].

In deriving the decision rules for the radar problem we use exclusively the Neyman–Pearson criterion. This is mainly because of the difficulty or lack of significance in assigning prior probabilities to target presence at a point in space. We are now ready to discuss the simplest radar detection problem, the completely known steady point target.

8.6.1 Steady Point Target: Known Phase

In this case we model the complex envelope $\tilde{y}_r(t)$ of Eq. (8.6.1) as in Eq. (8.5.7) where the reflecting gain $\tilde{b}$ is completely known including its phase θ_2. The problem then corresponds to the problem of detecting a completely known signal discussed in Section 4.2. As discussed therein, the solution can be obtained by choosing a set of orthonormal functions $\{\tilde{g}_i(t)\}$ as basis functions and expanding $\tilde{z}(t)$ in terms of these functions. If we choose our first basis function to be

$$\tilde{g}_1(t) = \tilde{y}(t)$$

it follows from our discussions in Section 4.2 that the decision can be based entirely on the coefficient corresponding to $\tilde{g}_1(t)$. That is,

$$\tilde{y}_1 = \int_0^T \tilde{z}(t)\tilde{y}^*(t)\,dt \tag{8.6.3}$$

is a sufficient statistic for this problem.

The correlation operation of Eq. (8.6.3) can also be implemented by using a matched filter with impulse response $\tilde{y}^*(T - t)$.

We can now write the likelihood ratio as

$$\Lambda(\tilde{y}_1) = \frac{p(\tilde{y}_1 | H_1)}{p(\tilde{y}_1 | H_0)} \tag{8.6.4}$$

and the test as

$$\Lambda(\tilde{y}_1) \underset{H_0}{\overset{H_1}{\gtrless}} \gamma \tag{8.6.5}$$

where γ is a threshold adjusted according to the level of P_F required.

The sufficient statistic is a linear functional of a complex Gaussian process $\tilde{z}(t)$ and is, therefore, a complex Gaussian random variable. The reader can easily establish that $\tilde{y}_1$ under both hypotheses is distributed normally with [see Eq. (2.5.28)]

$$p(\tilde{y}_1 | H_1) = \frac{1}{2\pi N_0} \exp\left\{ -\frac{|\tilde{y}_1 - \sqrt{2E_r}\exp(j\theta_2)|^2}{2N_0} \right\} \tag{8.6.6}$$

and

$$p(\tilde{y}_1 | H_0) = \frac{1}{2\pi N_0} \exp\left(-\frac{|\tilde{y}_1|^2}{2N_0} \right) \tag{8.6.7}$$

where E_r was defined in Eq. (8.5.5). Substitution of these probability expressions in Eq. (8.6.4), enables us to write the likelihood ratio test as

$$\frac{\sqrt{2E_r}\,\mathrm{Re}[\tilde{y}_1 \exp(-j\theta_2)]}{N_0} \underset{H_0}{\overset{H_1}{\gtrless}} \ln\gamma + \frac{E_r}{N_0} \tag{8.6.8}$$

Denoting the left side of Eq. (8.6.8) by y_s, it follows from Eqs. (8.6.6) and (8.6.7) that

$$\begin{aligned} \mathrm{E}[y_s | H_1] &= \frac{2E_r}{N_0} \quad \text{and} \quad \mathrm{var}[y_s | H_1] = \frac{2E_r}{N_0} \\ \mathrm{E}[y_s | H_0] &= 0 \qquad \text{and} \quad \mathrm{var}[y_s | H_0] = \frac{2E_r}{N_0} \end{aligned} \tag{8.6.9}$$

The probability of false alarm is the probability that with the target not present, the test statistic y_s will exceed the threshold

$$\gamma_1 = \ln\gamma + \frac{E_r}{N_0}$$

Thus we have

$$P_F = \int_{\gamma_1}^{\infty} p(y_s)\,dy_s = \int_{\gamma_1}^{\infty} \frac{1}{\sqrt{2\pi}\,\eta} \exp\left(\frac{-y_s^2}{2\eta^2}\right) dy_s$$

where

$$\eta^2 \triangleq \frac{2E_r}{N_0}$$

This can be evaluated by a simple transformation as in Example 3.3.2 to yield

$$P_F = Q\left(\frac{\gamma_1}{\eta}\right) = Q\left(\frac{\eta}{2} + \frac{\ln\gamma}{\eta}\right) \tag{8.6.10}$$

The probability of detection is the probability that y_s exceeds the threshold γ_1 when the target is present and is given by

$$\begin{aligned} P_D &= \int_{\gamma_1}^{\infty} \frac{1}{\sqrt{2\pi}\,\eta} \exp\left[-\frac{(y_s - \eta^2)^2}{2\eta^2}\right] dy_s \\ &= Q\left(\frac{\ln\gamma}{\eta} - \frac{\eta}{2}\right) \end{aligned} \tag{8.6.11}$$

The radar system design usually utilizes the expression for P_D and P_F to calculate the threshold γ and the required predetection peak signal-to-noise ratio η^2. For example, let us assume that we require that $P_F \leq 10^{-6}$ and $P_D = 0.6$ (or 60% as usually expressed in radar terminology). We can then solve Eqs. (8.6.10) and (8.6.11) simultaneously to arrive at values of γ and η. In this case the predetection peak signal-to-noise ratio required is 14.5 dB. In general, it is convenient to plot P_D in percentage against η^2 in dB with P_F as a parameter on the curve. Such plots are presented in Fig. 8.6.1. The predetection peak signal-to-noise ratio required for a

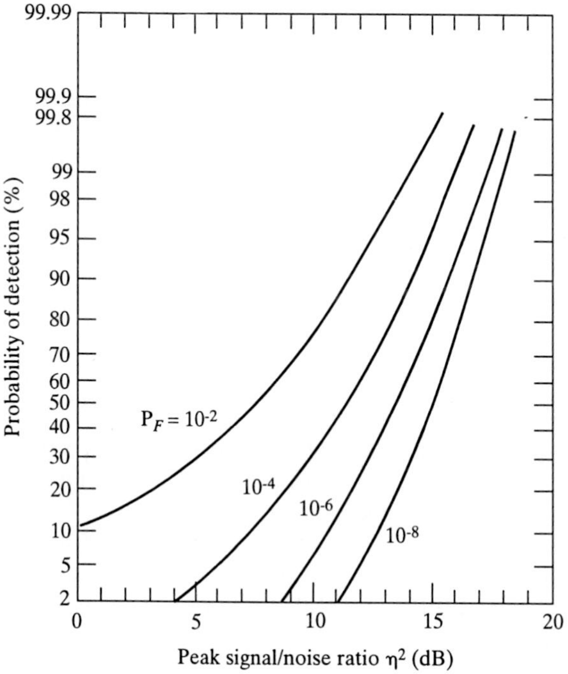

Figure 8.6.1 Performance curves for single-pulse coherent detection.

given P_D and P_F specification in turn determines the transmitted energy requirements through a fundamental equation called the *radar equation* [30] (see exercise 8.26).

8.6.2 Steady Point Target: Unknown Random Phase

In this case we again use the complex envelope of the received signal, $\tilde{y}_r(t)$ as in Eq. (8.5.7) but with the assumption that θ_2, the phase of the reflection gain is random and is uniformly distributed in $[0, 2\pi]$. The detection problem is now one of detecting a signal with unknown random parameter discussed in Section 4.6. Since $p(\tilde{y}_1 | H_0)$ does not involve θ_2, we can obtain the likelihood ratio as

$$\Lambda(\tilde{y}_1) = \int_{\chi_{\theta_2}} \Lambda(\tilde{y}_1 | \theta_2) p(\theta_2)\, d\theta_2 \tag{8.6.12}$$

where χ_{θ_2} is the space of the parameter θ_2.

From our discussions previously, we can easily obtain from Eqs. (8.6.6) and (8.6.7) that

$$\Lambda(\tilde{y}_1 | \theta_2) = \exp\left(-\frac{E_r}{N_0}\right) \exp\left[\frac{\sqrt{2E_r}}{N_0}(\text{Re } \tilde{y}_1 \cos \theta_2 + \text{Im } \tilde{y}_1 \sin \theta_2)\right] \tag{8.6.13}$$

Denoting $\text{Re}\, \tilde{y}_1$ as L_c and $\text{Im}\, \tilde{y}_1$ as L_s in the preceding expression, and using Eq. (8.6.12), we can write the likelihood ratio as

$$\begin{aligned}\Lambda(\tilde{y}_1) &= \exp\left(-\frac{E_r}{N_0}\right) \int_0^{2\pi} \frac{1}{2\pi} \exp\left[\frac{\sqrt{2E_r}}{N_0}(L_c \cos \theta_2 + L_s \sin \theta_2)\right] d\theta \\ &= \exp\left(-\frac{E_r}{N_0}\right) I_0\left[\frac{\sqrt{2E_r}}{N_0}(L_c^2 + L_s^2)^{1/2}\right]\end{aligned} \tag{8.6.14}$$

In Eq. (8.6.14), $I_0(\cdot)$ is a modified Bessel function of the first kind and order zero. The decision rule is to compare the likelihood ratio with a threshold. Alternatively, $\ln \Lambda(\tilde{y}_1)$ can be compared with a threshold γ, so that the decision rule becomes

$$\ln I_0\left[\frac{\sqrt{2E_r}}{N_0}(L_c^2 + L_s^2)^{1/2}\right] \underset{H_0}{\overset{H_1}{\gtrless}} \frac{E_r}{N_0} + \gamma \underline{\Delta} \gamma_1$$

or

$$\ln I_0\left[\frac{\sqrt{2E_r}}{N_0}|\tilde{y}_1|\right] \underset{H_0}{\overset{H_1}{\gtrless}} \gamma_1 \tag{8.6.15}$$

The optimum receiver structure using complex signals is shown in Fig. 8.6.2. Actually, sine $I_0(\cdot)$ is monotonic, an equivalent test is to compare $|\tilde{y}|$ to an appropriate threshold. A block diagram using real signals is shown in Fig. 8.6.3.

As in the case of noncoherent digital communication, the decision rule of Eq. (8.6.15) can be approximated. The logarithm of the modified Bessel function $\ln I_0(x)$ can be approximated by $x^2/4$ for very small values of x. For large values of x, $\ln I_0(x)$ can be approximated by x.

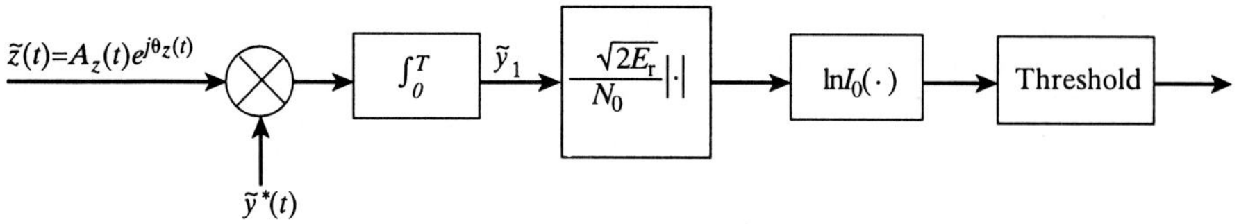

Figure 8.6.2 Optimum receiver for detection of target return with unknown phase.

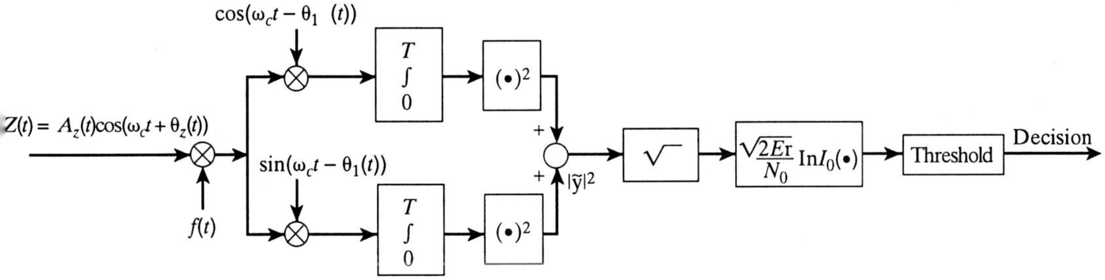

Figure 8.6.3 Optimal receiver of Fig. 8.6.2 with real signals.

We now derive the performance of our radar receiver. To do this, we note that the decision law is

$$|\tilde{y}_1| \underset{H_0}{\overset{H_1}{\gtrless}} \gamma_2 \tag{8.6.16}$$

where γ_2 is a modified threshold. Let $l = |\tilde{y}_1|$. It is easy to show that the statistic l under the hypothesis H_0 is Rayleigh distributed as

$$p(l \mid H_0) = \begin{cases} \dfrac{l}{N_0}\exp\left(-\dfrac{l^2}{2N_0}\right) & l \geq 0 \\ 0 & \text{otherwise} \end{cases} \tag{8.6.17}$$

The probability of false alarm can then be obtained from Eqs. (8.6.16) and (8.6.17) as

$$\begin{aligned} P_F &= \Pr\{l > \gamma_2 \mid H_0\} \\ &= \int_{\gamma_2}^{\infty} p(l \mid H_0)\, dl = \int_{\gamma_2}^{\infty} \frac{l}{N_0}\exp\left(-\frac{l^2}{2N_0}\right) dl \\ &= \exp\left(-\frac{\gamma_2^2}{2N_0}\right) \end{aligned} \tag{8.6.18}$$

To compute the probability of detection, we need to determine $p(l \mid H_1)$. To do this, we note that L_c and L_s conditioned on θ_2, are Gaussian under hypothesis H_1 with the following moments:

$$E[L_c \mid \theta_2, H_1] = \sqrt{2E_r}\cos\theta_2, E[L_s \mid \theta_2, H_1] = \sqrt{2E_r}\sin\theta_2 \tag{8.6.19}$$

and

$$\text{var}[L_c \mid \theta_2, H_1] = \text{var}[L_s \mid \theta_2, H_1] = N_0 \tag{8.6.20}$$

Therefore, $p(l \mid H_1, \theta_2)$ is a Rician density given by

$$p(l \mid \theta_2, H_1) = \begin{cases} \dfrac{l}{N_0}\exp\left(-\dfrac{l^2 + 2E_r}{2N_0}\right) I_0\left(\dfrac{l\sqrt{2E_r}}{N_0}\right) & l \geq 0 \\ 0 & \text{otherwise} \end{cases} \tag{8.6.21}$$

The density function $p(l \mid H_1)$ is obtained by averaging out $p(l \mid \theta_2, H_1)$ of Eq. (8.6.21) over the density function of θ_2. Because the expression in Eq. (8.6.21) is independent of θ_2, $p(l \mid H_1)$ is given by the same expression. The probability of detection is given by

$$\begin{aligned} P_D &= \Pr\{l \geq \gamma_2 \mid H_1\} \\ &= \int_{\gamma_2}^{\infty} p(l \mid H_1)\, dl = \int_{\gamma_2}^{\infty} \frac{l}{N_0}\exp\left(-\frac{l^2 + 2E_r}{2N_0}\right) I_0\left(\frac{l\sqrt{2E_r}}{N_0}\right) dl \\ &= Q\left[\left(\frac{2E_r}{N_0}\right)^{1/2}, \left(\frac{1}{N_0}\right)^{1/2} \gamma_2\right] \end{aligned} \tag{8.6.22}$$

where $Q[a, b]$ is Marcum's Q function defined in Eq. (8.2.25). The values of the function are tabulated in [6] or can be computed using a recursive formula [39].

A relation between P_D and P_F can be obtained by expressing the threshold γ_2 in terms of P_F. We then have, from Eq. (8.6.18),

$$\gamma_2 = \sqrt{-2N_0 \ln P_F}$$

Substituting the preceding in the expression for P_D, we obtain

$$P_D = Q[\eta, (-2\ln P_F)^{1/2}] \tag{8.6.23}$$

where η^2 is the peak signal-to-noise ratio $2E_r/N_0$. The curves of P_D versus η^2 with P_F as parameter is presented in Fig. 8.6.4. This curve is useful in determining the predetection peak signal-to-noise ratio required to attain a given performance level of P_D and P_F.

8.6.3 Slowly Fluctuating Point Target

For slowly fluctuating point targets, we model the reflection from the target according to Eq. (8.5.6), where $\tilde{b}$ is a complex random variable. Again the complex envelope of the target return $\tilde{y}_r(t)$ is given by Eq. (8.5.7).

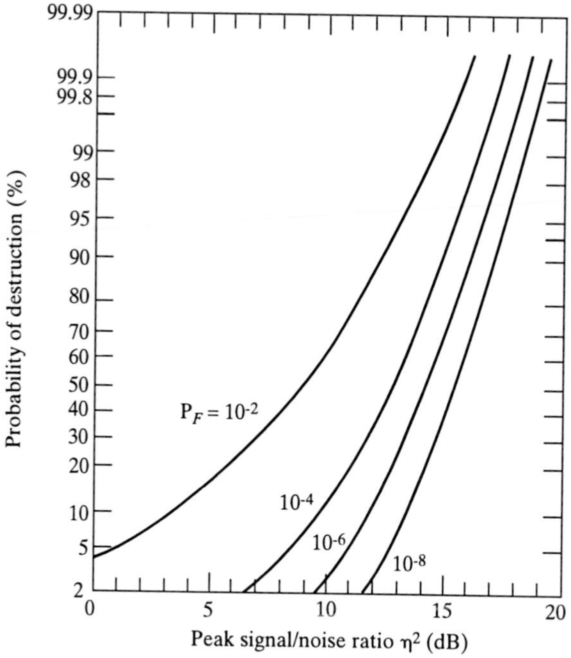

Figure 8.6.4 Performance curves for single-pulse noncoherent detection.

We assume that the random variable $\tilde{b}$ is complex Gaussian with zero mean. In other words, we consider the target to be an assembly of random scatterers, and hence the absolute value of the gain $|\tilde{b}|$ is Rayleigh distributed as in Eq. (8.5.9), and the phase θ_2 is uniformly distributed in $[0, 2\pi]$.

For convenience, we let

$$e = \sqrt{E_T}|\tilde{b}| = \sqrt{E_r}$$

Then e is also Rayleigh distributed as follows:

$$p(e) = \begin{cases} \dfrac{2e}{\overline{E}_r} \exp\left(-\dfrac{e^2}{\overline{E}_r}\right) & e \geqslant 0 \\ 0 & \text{otherwise} \end{cases} \tag{8.6.24}$$

where $\overline{E}_r$ is the average received energy and equals $2\sigma_b^2 E_T$.

The problem of detecting a target is now one of obtaining the likelihood ratio and comparing it with a proper threshold.

Thus the optimum decision test can be implemented by first obtaining a likelihood ratio conditioned on the unknown random quantities e and θ_2, and then averaging it over e and θ_2 using appropriate density functions. In order to do this, we recognize that the likelihood ratio expression in Eq. (8.6.14) with $\sqrt{E_r}$ replaced by e gives us the likelihood ratio conditioned on e and averaged over θ_2. Therefore, we write

$$\Lambda(\tilde{y}_1 \,|\, e) = \exp\left(-\frac{e^2}{N_0}\right) I_0\left(\frac{\sqrt{2}e}{N_0}|\tilde{y}_1|\right) \tag{8.6.25}$$

The likelihood ratio is obtained by averaging $\Lambda(\tilde{y}_1 | e)$ over e as

$$\Lambda(\tilde{y}_1) = \int_0^\infty \exp\left(-\frac{e^2}{N_0}\right) I_0\left(\frac{\sqrt{2}e}{N_0} l\right) p(e)\, de \tag{8.6.26}$$

where $l = |\tilde{y}_1|$. Let us denote the integral on the right side of Eq. (8.6.26) by $f(l)$. We note that $f(l)$ is monotonic in l. This is so irrespective of what the distribution of e is, because $I_0(\cdot)$ is monotonic in its argument, and $p(e)$ is positive as it is a density function. This simply shows that l, the envelope of the received waveform is indeed a sufficient statistic. Therefore, we can implement the optimum test by comparing $l = |\tilde{y}_1|$ with a threshold set according to P_F requirements. In other words, the test of Eq. (8.6.16) is indeed optimum even if the amplitude of the echo is a random variable.

We can obtain the probability of detection by averaging the expression for P_D in Eq. (8.6.22) over the density function for the random variable e, so that

$$P_D = \int_0^\infty P_D(e) p(e)\, de \tag{8.6.27}$$

where $P_D(e)$ is obtained from Eq. (8.6.22) as

$$P_D(e) = \int_{\gamma_2}^\infty \frac{l}{N_0} \exp\left(-\frac{l^2 + 2e^2}{2N_0}\right) I_0\left(\frac{\sqrt{2}el}{N_0}\right) dl \tag{8.6.28}$$

Here γ_2 is the threshold of the test. The evaluation of P_D is most conveniently carried out by defining the normalized variables

$$a = \left(\frac{1}{N_0}\right)^{1/2} l, \quad \eta^2 = \frac{2e^2}{N_0} \quad \text{and} \quad \gamma_3 = \left(\frac{1}{N_0}\right)^{1/2} \gamma_2$$

We then have, for the probability of detection

$$P_D = \int_0^\infty P_D(\eta) p(\eta)\, d\eta \tag{8.6.29}$$

where

$$P_D(\eta) = \int_{\gamma_3}^\infty a \exp\left(-\frac{a^2 + \eta^2}{2}\right) I_0[\eta a]\, da \tag{8.6.30}$$

For Rayleigh fluctuating targets [Eq. (8.6.24)] it easily follows that the density function of the variable η is

$$p(\eta) = \begin{cases} \dfrac{2\eta}{\overline{\eta}^2} \exp\left(-\dfrac{\eta^2}{\overline{\eta}^2}\right) & \eta \geq 0 \\ 0 & \text{otherwise} \end{cases} \tag{8.6.31}$$

where $\overline{\eta}^2 = 2\overline{E}_r/N_0$, the average peak signal-to-noise ratio. On substituting Eqs. (8.6.30) and (8.6.31) in Eq. (8.6.29) and interchanging the orders of integration, we obtain the expression for P_D as

$$P_D = \int_{\gamma_3}^\infty \frac{2a}{\overline{\eta}^2} \exp\left(-\frac{a^2}{2}\right) \int_0^\infty \eta \exp\left(-\frac{2 + \overline{\eta}^2}{2\overline{\eta}^2} \eta^2\right) I_0[\eta a]\, d\eta\, da \tag{8.6.32}$$

The inner integral can be evaluated using standard formulas† to yield the expression for P_D as

$$P_D = \int_{\gamma_3}^{\infty} \frac{2a}{2+\overline{\eta}^2} \exp\left(-\frac{a^2}{2+\overline{\eta}^2}\right) da$$
$$= \exp\left(-\frac{\gamma_3^2}{2+\overline{\eta}^2}\right)$$
$$= \exp\left[-\frac{\gamma_2^2}{N_0(2+\overline{\eta}^2)}\right] \tag{8.6.33}$$

The probability of false alarm is the same as for noncoherent detection and is given by Eq. (8.6.18). Repeating the expression for convenience, we have

$$P_F = \exp\left(-\frac{\gamma_2^2}{2N_0}\right) \tag{8.6.34}$$

so that P_D can be written as

$$P_D = \exp\left[\frac{2 \ln P_F}{2+\overline{\eta}^2}\right] = P_F^{\frac{1}{1+\overline{\eta}^2/2}} \tag{8.6.35}$$

A plot of the probability of detection, P_D versus $\overline{\eta}^2$, the average peak signal-to-noise ratio with the probability of false alarm for the Rayleigh fluctuating target, P_F, as parameter, is shown in Fig. 8.6.5. From Eq. (8.6.34), it is clear that the threshold

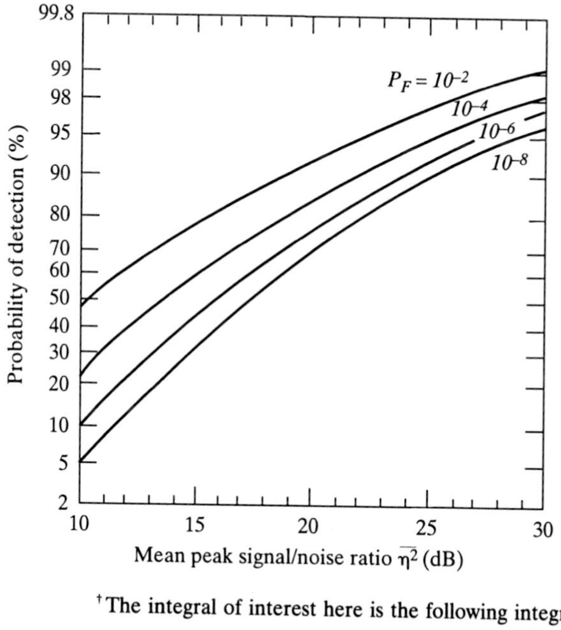

Figure 8.6.5 Performance curves for single-pulse detection of a Rayleigh fluctuating (pulse-to-pulse) target.

†The integral of interest here is the following integral [38]:

$$\int_0^{\infty} t^{\mu-1} I_\nu(\alpha t) \exp(-p^2 t^2)\, dt = \frac{\Gamma\left(\frac{\mu+\nu}{2}\right)\left(\frac{\alpha}{2p}\right)^{\nu}}{2p^{\mu}\Gamma(\nu+1)} \exp\left(\frac{\alpha^2}{4p^2}\right) {}_1F_1\left(\frac{\nu-\mu}{2}+1, \nu+1; \frac{-\alpha^2}{4p^2}\right)$$

where $I_\nu(\cdot)$ is the modified Bessel function of the first kind and order ν, $\Gamma(\cdot)$ is the gamma function, and ${}_1F_1(x_1, x_2, x_3)$ is the confluent hypergeometric function.

γ_2 can be fixed to achieve a specified P_F provided that N_0 is known. Any fixed threshold detection scheme will yield changing P_F values if N_0 changes. A scheme that guarantees a fixed value for P_F when the noise level changes is called a *constant false alarm rate detection scheme* and was briefly discussed in Section 3.8. The reader is referred to [33,40] for further discussion of these schemes.

8.6.4 Multiple-Pulse Detection

We now consider radar systems that use several reflected echoes to make a decision regarding the presence of a target. We refer to this as multiple-pulse detection. Because we are using a number of pulses, the average signal-to-noise ratio is significantly greater than that for a single pulse,† resulting in better detectability. Further, in the case of fluctuating targets, multiple-pulse detection provides better performance in the useful ranges of P_D and P_F. This can be explained by observing that multiple-pulse detection corresponds to time diversity in the case of digital communication.

In the echo pulse train, each of the pulses may be coherently related. That is, the initial phases at successive pulses are fixed and known at the receiver. In this case the optimum filter is the same as for coherent detection based on a single pulse, except that the matched filter is matched to the pulse train. Such a processing technique is known as *coherent processing* or *predetection integration*. If the average peak signal-to-noise ratio per pulse is $\overline{\eta}^2$, the average peak signal-to-noise ratio for a train of N coherent pulses is $N\overline{\eta}^2$. Therefore, for performance calculations, we can use the results of coherent single-pulse detection with $\overline{\eta}^2$ replaced by $N\overline{\eta}^2$. We can calculate the peak SNR required for a given P_D and P_F from Fig. 8.6.1 and divide by N to obtain the peak SNR required per pulse, for the same performance.

In many cases we will have a pulse train in which the pulses are coherently related, but the initial phase of the pulse train is unknown and random. However, the different pulses of the pulse train maintain a known relationship. An optimum receiver for this case will then consist of a bandpass filter (or equivalently, two quadrature correlators) matched to the pulse train and followed by an envelope detector and threshold comparator.

An important case of multiple-pulse detection arises when the target reflecting characteristics are varying according to Eq. (8.5.12). Following arguments similar to the case of a single pulse and using the independence of the reflection gains $\tilde{b}_i$, we can write the likelihood ratio for this case as

$$\begin{aligned} \Lambda(\tilde{y}_1) &= \prod_{i=1}^{N} \int_0^\infty \Lambda_i(y_{1i} \mid e_i) p(e_i)\, de_i \\ &= \prod_{i=1}^{N} \Lambda_i(\tilde{y}_{1i}) \end{aligned} \tag{8.6.36}$$

†The signal-to-noise ratio will be increased only if the pulses are on the target. Usually, most targets stay within the radar beamwidth for the entire scan.

where e_i is the magnitude of $\tilde{b}_i$, $\tilde{y}_{1i}$ is the sufficient statistic (correlator output) for the ith pulse, and $\Lambda_i(\tilde{y}_{1i} | e_i)$ is the likelihood ratio for the ith pulse conditioned on e_i but averaged over the phase θ_{2i}. Assuming that e_i values are independent and identically Rayleigh distributed† according to Eq. (8.6.24), we can write with $l_i = |\tilde{y}_{1i}|$

$$\Lambda_i(\tilde{y}_{1i}) = \int_0^\infty \frac{2e_i}{\overline{E}_r} \exp\left(-\frac{e_i^2}{\overline{E}_r}\right) I_0\left(\frac{\sqrt{2}e_i}{N_0} l_i\right) \exp\left(-\frac{e_i^2}{N_0}\right) de_i \tag{8.6.37}$$

The use of normalized variables as shown below Eq. (8.6.28) yields a simplified form for the per pulse likelihood ratio $\Lambda_i(\tilde{y}_{1i})$. Thus, in terms of the variable $a_i = \sqrt{(1/N_0)}l_i$, we obtain

$$\Lambda_i(\tilde{y}_{1i}) = \frac{1}{2p^2\overline{\eta}^2} \exp\left(\frac{a_i^2}{4p^2}\right) \tag{8.6.38}$$

where

$$p^2 = \left(\frac{1}{2} + \frac{1}{\overline{\eta}^2}\right) \quad \text{and} \quad \overline{\eta}^2 = \frac{2\overline{E}_r}{N_0}$$

Thus $\overline{\eta}^2$ represents the mean peak SNR.

Substitution in Eq. (8.6.36) yields the likelihood ratio for pulse to pulse Rayleigh fluctuating targets as

$$\Lambda(\tilde{y}_1) = \left(\frac{1}{2p^2\overline{\eta}^2}\right)^N \exp\left(\frac{1}{4p^2}\sum_{i=1}^N a_i^2\right) \tag{8.6.39}$$

It follows that the decision rule can be implemented as

$$\sum_{i=1}^N a_i^2 \underset{H_0}{\overset{H_1}{\gtrless}} \gamma_1^2$$

or, equivalently,

$$\sum_{i=1}^N |\tilde{y}_{1i}|^2 \underset{H_0}{\overset{H_1}{\gtrless}} \gamma_2^2 \tag{8.6.40}$$

where γ_2^2 is the threshold. We conclude that the square law detector, followed by a pulse train sampler and integrator, is indeed the optimum receiver structure for all SNR in the case of pulse-to-pulse fluctuating targets.

We can determine the performance of this detector by applying the results of Example 3.4.2. Clearly, $\tilde{y}_{1i}$ is complex Gaussian under H_0. Now $p(\tilde{y}_{1i} | H_1) = \int\int p(\tilde{y}_{1i} | H_1, e_i, \theta_{2i}) p(e_i) p(\theta_{2i})\, de_i\, d\theta_{2i}$. By explicitly evaluating this integral, it can be shown that $\tilde{y}_{1i}$ is complex Gaussian under H_1 also. Its moments under the two hypotheses follow as

$$\mathrm{E}\{\tilde{y}_{1i} | H_1\} = 0 \quad \text{and} \quad \mathrm{var}\{\tilde{y}_{1i} | H_1\} = 2\overline{E}_r + 2N_0$$

† This corresponds to the Swerling II model in radar literature [38].

Under hypothesis H_0, $\tilde{y}_{1i}$ is purely noise, so that

$$E\{\tilde{y}_{1i} \mid H_0\} = 0 \quad \text{and} \quad \text{var}\{\tilde{y}_{1i} \mid H_0\} = 2N_0$$

The left side of Eq. (8.6.40) represents the sum of the magnitude squares of N independent complex Gaussian random variables (or equivalently, $2N$ real random variables) with the same mean but different variances under the two hypotheses. We can, therefore, use the results of Example 3.4.2 to write the probabilities of detection and of false alarm as

$$P_D = 1 - I\left[\frac{\gamma_2^2/(2N_0)}{\sqrt{N}(1 + \overline{\eta}^2/2)}, N - 1\right] \tag{8.6.41}$$

and

$$P_F = 1 - I\left[\frac{\gamma_2^2/(2N_0)}{\sqrt{N}}, N - 1\right] \tag{8.6.42}$$

where $I(\cdot,\cdot)$ is the incomplete gamma function defined in Chapter 3 as

$$I(m,n) = \int_0^{m\sqrt{1+n}} \frac{\exp(-u)u^n}{n!}\,du \tag{8.6.43}$$

and is tabulated [53].

Curves of P_D versus $\overline{\eta}^2$ for $N = 20$ and various values of P_F are shown in Fig. 8.6.6. The threshold levels required to plot these curves were obtained from the tables of Pachares [41].

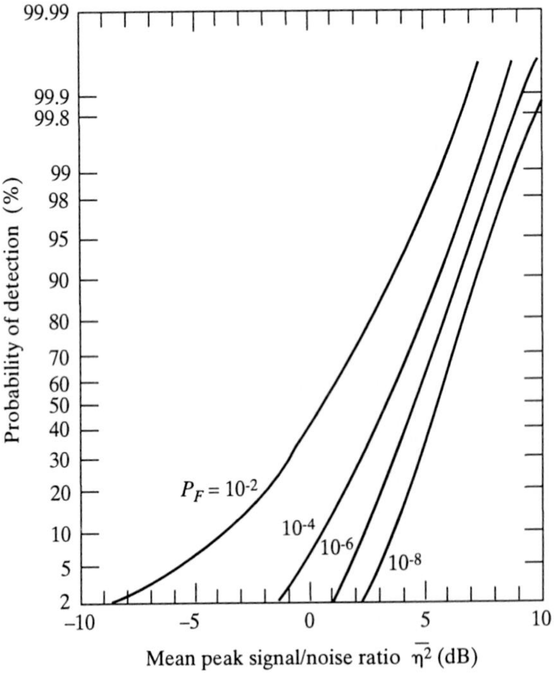

Figure 8.6.6 Performance curves for multiple-pulse detection of a Rayleigh fluctuating (pulse-to-pulse) target.

8.6.5 Detection of Spread Targets: A Brief Discussion

From our earlier discussion on target models spread in range and/or Doppler, we find that the return signal is a stochastic process. This return signal is corrupted by additive Gaussian noise. Thus we are led to the problem of detecting stochastic signals in noise, which we have not considered in our development so far. The solution usually calls for an estimate of the stochastic process to be correlated with the received signal to generate a sufficient statistic [31,42,43].

However, if we make a piecewise constant (albeit random) approximation to the reflecting characteristic $\tilde{b}(t)$ or the stochastic process $\tilde{y}_r(t)$, we can use the techniques developed to obtain a solution that is suboptimum. For example, assume that the target is Doppler spread and the reflecting gain is approximated by Eq. (8.5.17). Then we can consider the complex envelope of the return signal as made up of N components:

$$\begin{aligned}\tilde{y}_r(t) &= \sqrt{E_T}\tilde{y}(t-\tau)\sum_{i=1}^{N}\tilde{b}_i p\left(t-\frac{i}{W}\right)\\ &= \sum_{i=1}^{N}\frac{\sqrt{E_T}}{N}\tilde{b}_i\tilde{y}_i(t-\tau)\end{aligned} \tag{8.6.44}$$

where

$$\tilde{y}_i(t) = \tilde{y}(t) \qquad \frac{i-1}{W} \leqslant t \leqslant \frac{i}{W} \qquad i = 1,2,\ldots,N$$

If we now consider $\tilde{y}_i(t)$ to be the ith pulse in a sequence of N subpulses forming the pulse $\tilde{y}(t)$, we can use techniques similar to that used for multiple-pulse detection of slowly fluctuating Rayleigh point targets discussed in Section 8.6.4. An important difference, however, is that the test statistic will now be a weighted sum of the statistics generated by correlating the received signal with the subpulse $\tilde{y}_i(t)$. The calculation of these weights and obtaining the receiver structure for this case as well as for range spread targets is left as an important exercise to the reader.

This concludes our discussion of the detection of radar targets. We have been selective in our material so as to illustrate the application of techniques to practical cases. A number of problems at the end deal with the application of detection theory to develop additional results on radar detection.

8.7 PARAMETER ESTIMATION IN RADAR SYSTEMS

In Section 8.6, where we considered detection of targets, we assumed that the target range and velocity (or Doppler frequency shift) were known. Any such knowledge of these parameters is usually coarse, and proper techniques have to be used to obtain fine estimates. In addition to these parameters, one may be interested in obtaining estimates of the angular coordinates of the target position.

In this section we consider maximum likelihood estimation of the delay and Doppler frequency (equivalently, range and velocity) of a steady point target with unknown random phase. We also obtain the Cramér–Rao bounds on the estimates. Exercises at the end of the chapter extend and develop additional results for the cases of slowly fluctuating point target and multiple pulse reception.

In the process of our development, we obtain an important quantity, called the *radar ambiguity function*. We briefly discuss this function and indicate the significant role it plays in waveform design.

8.7.1 Likelihood Function

The return signal from a steady point target is usually corrupted by additive Gaussian bandpass noise of spectral height $N_0/2$. We can represent the complex envelope of the noisy observation as

$$\tilde{z}(t) = \tilde{y}_r(t) + \tilde{v}(t) \qquad -\infty < t < \infty \tag{8.7.1}$$

where $\tilde{y}_r(t)$ was defined in Eq. (8.5.7). Further, we assume that θ_2, the phase of the reflecting gain is a random variable uniformly distributed over $[0, 2\pi]$.

We can expand $\tilde{z}(t)$ in an orthonormal series expansion as in Section 8.6. In this case, however, we choose the first orthonormal function as

$$\tilde{g}_1(t) = \tilde{y}(t - \tau)\exp(j\omega_d t) \tag{8.7.2}$$

To obtain the likelihood function, we determine the probability density function of the coefficient of $\tilde{g}_1(t)$ assuming that τ and ω_d are known. Once we obtain this function, we vary τ and ω_d to seek a maximum of the function. The values of τ and ω_d for which the maximum is obtained are the maximum likelihood estimates $\hat{\tau}$ and $\hat{\omega}_d$. As in Chapters 4 and 5, we can use the log-likelihood function in determining the estimates. We can readily write the log-likelihood function from Eq. (8.6.15) as

$$L_1(\tau, \omega_d) = K + \ln I_0\left(\frac{\sqrt{2E_r}}{N_0}|\tilde{y}_1|\right) \tag{8.7.3}$$

where $I_0(\cdot)$ is the modified Bessel function of the first kind and order zero, K is a constant not depending on either τ or ω_d, $E_r = E_T|\tilde{b}|^2$, and

$$\tilde{y}_1 = \int_{-\infty}^{\infty} \tilde{z}(t)\tilde{y}^*(t - \tau)\exp(-j\omega_d t)\,dt \tag{8.7.4}$$

The maximum likelihood estimates of τ and ω_d are then obtained by maximizing $L_1(\tau, \omega_d)$. Equivalently, we can maximize the function

$$L(\tau, \omega_d) = \ln I_0\left(\frac{\sqrt{2E_r}}{N_0}|\tilde{y}_1|\right) \tag{8.7.5}$$

to obtain our estimates.

Let us now consider a practical implementation of the maximum likelihood estimator. Because $\ln I_0(\cdot)$ is a monotonic function of its argument, we can seek the vector parameter $\boldsymbol{\alpha} = [\tau \,\vdots\, \omega_d]^T$ such that $|\tilde{y}_1|$ is maximum. That is, we implement the correlator of Eq. (8.7.4), and obtain the envelope of the output for each value of

$\boldsymbol{\alpha}$ in its range. The estimate $\hat{\boldsymbol{\alpha}}$ will then correspond to the value of $\boldsymbol{\alpha}$ for which this is a maximum. This is clearly impractical except in special cases. To obtain a practical structure, let us first assume that ω_d is known. Then we can implement the correlator of Eq. (8.7.4) and monitor the envelope of its output, which is a function of τ. The time at which it peaks gives $\hat{\tau}$. However, if τ is known, but ω_d is unknown, we have to build a continuum of correlators to correspond to the values of ω_d in its expected range. We can overcome this problem by quantizing the range of ω_d into a set of M intervals centered around the frequencies $\omega_{d_1}, \omega_{d_2}, \ldots, \omega_{d_M}$. We perform M simultaneous correlation operations on the observation and observe the envelopes of the outputs at time τ. The estimate $\hat{\omega}_d$ will correspond to the maximum of these values. The practical implementation of the receiver when both τ and ω_d have to be estimated follows easily. We quantize the range of ω_d into M intervals to implement M correlators followed by envelope detectors. The outputs of the envelope detectors are monitored and a two-dimensional search for the maximum is conducted. The values of Doppler frequency and time for which this maximum occurs yield maximum likelihood estimates.

An important question that arises is: How fine should the quantization of the range of ω_d be? We can obtain an answer to this as well as the theoretical maximum accuracy possible in terms of the Cramér–Rao bounds. We will find from our subsequent discussions that the quantization interval should not exceed one standard deviation of the estimate $\hat{\omega}_d$.

8.7.2 Cramér–Rao Bounds for Estimates of Delay and Doppler Frequency

We now derive the Cramér–Rao bounds on the estimates of τ and ω_d. We assume a high signal-to-noise ratio to simplify the derivation. This is not unreasonable, since as we have seen earlier, we require a fairly high signal-to-noise ratio to obtain reliable detection.

To determine the bounds, we proceed as in Section 8.7.1. That is, we first obtain the bound assuming that either τ or ω_d is known but not both and then extend our results to the case where both parameters are unknown.

Suppose that the target has an actual delay of τ_a and Doppler frequency ω_a, and we correlate the received signal envelope $z(t)$ as in Eq. (8.7.4). Then we can write

$$|\tilde{y}_1| = \int_{-\infty}^{\infty} [\sqrt{E_r}\tilde{y}(t-\tau_a)\exp(j\omega_a t)\exp(j\theta_2) + \tilde{v}(t)] \cdot [\tilde{y}^*(t-\tau)\exp(-j\omega_d t)]\,dt \tag{8.7.6}$$

Under the assumption of high signal-to-noise ratio we can write the function of Eq. (8.7.5) as

$$\begin{aligned} L(\tau, \omega_d) &= \ln I_0\left(\frac{\sqrt{2E_r}}{N_0}|\tilde{y}_1|\right) \\ &\simeq \frac{\sqrt{2E_r}}{N_0}|\tilde{y}_1| \end{aligned} \tag{8.7.7}$$

In general, we can also assume that the signal and noise are uncorrelated so that

$$\frac{\sqrt{2E_r}}{N_0}|\tilde{y}_1| \simeq \frac{2E_r}{N_0}\left|\int_{-\infty}^{\infty} \tilde{y}(t-\tau_a)\tilde{y}^*(t-\tau)\exp[j(\omega_d-\omega_a)t]\,dt\right| \tag{8.7.8}$$

Let $\tau' = \tau - \tau_a$ denote the error in the estimate of delay and $\omega_d' = \omega_d - \omega_a$ denote the error in the estimate of the Doppler shift.

Let $r = t - \tau + \tau'/2$. We can then rewrite Eq. (8.7.7) in the form

$$L(\tau, \omega_d) = \frac{\sqrt{2E_r}}{N_0}|\tilde{y}_1| \simeq \frac{2E_r}{N_0}|\chi(\tau', \omega_d')| \tag{8.7.9a}$$

where

$$\chi(\tau', \omega_d') = \int_{-\infty}^{\infty} \tilde{y}\left(r + \frac{\tau'}{2}\right)\tilde{y}^*\left(r - \frac{\tau'}{2}\right)\exp(j\omega_d' r)\,dr \tag{8.7.9b}$$

The function $\chi(\tau', \omega_d')$ is sometimes called the *time frequency autocorrelation function* of $y(t)$ and sometimes called the *ambiguity function*† [29,31,44]. The function $|\chi(\tau', \omega_d')|^2$ is the response due to the signal mismatched in delay by τ' and in Doppler by ω_d' of a matched filter [matched to the signal $\tilde{y}(t-\tau)\exp(j\omega_d t)$]. As such, the ambiguity function $\chi(\tau', \omega_d')$ can be expected to play an important role in distinguishing between a number of close objects. To illustrate this, consider two targets, one at the matched delay and Doppler with $|\tilde{b}| = 1$, and another mismatched by τ' and ω_d' with $|\tilde{b}| = p$. Then the "mismatched" target contributes an interfering energy of $p^2|\chi(\tau', \omega_d')|^2$, which may be larger or comparable to that of the true signal. This makes the problem of resolution difficult. In many modern radar systems, the transmitted signal $\tilde{y}(t)$ is carefully designed such that the ambiguity function exhibits desired nulls or minima at the locations corresponding to values of delay and Doppler frequency where interfering targets are to be expected. Thus the ambiguity function is a practical tool to study the design of waveforms for target resolution. As we shall see later, waveform design can also improve the accuracy bounds of our estimates. In fact, we relate to these design parameters of waveforms through the ambiguity function $\chi(\tau', \omega_d')$.

We now return to the problem of deriving the accuracy bounds for delay and Doppler estimates. If our interest is to estimate the delay when the Doppler frequency shift ω_d is known, we can set $\omega_d = \omega_a$ or $\omega_d' = 0$ in Eqs. (8.7.9a) and (8.7.9b) without loss of generality. Similarly, if ω_d is to be estimated, we can set $\tau = \tau_a$ or $\tau' = 0$ in the foregoing equations.

We will first derive the bound on the estimate of the delay. Note that this is the same as the variance of τ'. In Chapter 5 we derived the Cramér–Rao bound for an unbiased estimate. Applying this bound for obtaining the variance of τ, we have

$$\operatorname{var}\tau = \operatorname{var}\tau' \geqslant \frac{-1}{E\left\{\dfrac{\partial^2 L(\tau', 0)}{\partial \tau'^2}\right\}} \tag{8.7.10}$$

† We purposely digress here from our main course to briefly discuss the significance of $\chi(\tau', \omega_d')$. A number of its properties are developed in exercises at the end of the chapter.

Assuming high SNR, we can approximate the bound above by using the function of Eq. (8.7.9). This yields

$$\operatorname{var}\tau' \geqslant \frac{-1}{\left(\dfrac{2E_r}{N_0}\right)\dfrac{\partial^2|\chi(\tau',0)|}{\partial\tau'^2}} \tag{8.7.11}$$

In Eq. (8.7.11) the derivative must be evaluated at the true value of τ', that is, $\tau' = 0$.

To evaluate the derivative, we note that

$$|\chi(\tau',0)| = [\chi(\tau',0)\chi^*(\tau',0)]^{1/2}$$

so that

$$\begin{aligned}\frac{\partial|\chi(\tau',0)|}{\partial\tau'} = \frac{1}{2|\chi(\tau',0)|}\bigg[&\frac{\partial}{\partial\tau'}\chi(\tau',0)\cdot\chi^*(\tau',0)\\ &+\frac{\partial}{\partial\tau'}\chi^*(\tau',0)\chi(\tau',0)\bigg]\end{aligned} \tag{8.7.12}$$

and

$$\begin{aligned}\frac{\partial^2|\chi(\tau',0)|}{\partial\tau'^2} = \frac{1}{|\chi(\tau',0)|}\operatorname{Re}\bigg[&\chi(\tau',0)\frac{\partial^2}{\partial\tau'^2}\chi^*(\tau',0)\\ &+\frac{\partial}{\partial\tau'}\chi(\tau',0)\frac{\partial}{\partial\tau'}\chi^*(\tau',0)\bigg]\\ &-\frac{1}{|\chi(\tau',0)|^3}\left[\operatorname{Re}\left(\chi^*(\tau',0)\frac{\partial}{\partial\tau'}\chi(\tau',0)\right)\right]^2\end{aligned} \tag{8.7.13}$$

Evaluating the derivative in Eq. (8.7.13) at $\tau' = 0$, we obtain

$$\begin{aligned}\frac{\partial^2|\chi(\tau',0)|}{\partial\tau'^2}\bigg|_{\tau'=0} = \bigg[&\operatorname{Re}\left\{\frac{\partial^2}{\partial\tau'^2}\chi^*(\tau',0)\right\} + \left|\frac{\partial}{\partial\tau'}\chi(\tau',0)\right|^2\\ &-\operatorname{Re}\left[\frac{\partial}{\partial\tau'}\chi(\tau',0)\right]^2\bigg]_{\tau'=0}\end{aligned} \tag{8.7.14}$$

To compute the right-hand side of Eq. (8.7.14), we note that

$$\begin{aligned}\frac{\partial}{\partial\tau'}\chi(\tau',0)\bigg|_{\tau'=0} &= \frac{\partial}{\partial\tau'}\left\{\int_{-\infty}^{\infty}\tilde{y}\left(r-\frac{\tau'}{2}\right)\tilde{y}^*\left(r+\frac{\tau'}{2}\right)dr\right\}\bigg|_{\tau'=0}\\ &= \frac{\partial}{\partial\tau'}\left\{\int_{-\infty}^{\infty}\tilde{y}(t-\tau+\tau')\tilde{y}^*(t-\tau)\,dt\right\}\bigg|_{\tau'=0}\\ &= \int_{-\infty}^{\infty}\frac{\partial}{\partial\tau'}\{[\tilde{y}(t-\tau+\tau')]\tilde{y}^*(t-\tau)\,dt\}\bigg|_{\tau'=0}\\ &= \int_{-\infty}^{\infty}\left[\frac{d}{dt}\tilde{y}(t)\right]\tilde{y}^*(t)\,dt\end{aligned} \tag{8.7.15}$$

Similarly, we can show that

$$\frac{\partial^2}{\partial\tau'^2}\{\chi(\tau',0)\}\bigg|_{\tau'=0} = \int_{-\infty}^{\infty}\left[\frac{d^2}{dt^2}\tilde{y}(t)\right]\tilde{y}^*(t)\,dt \tag{8.7.16}$$

Let $G(j\omega)$ be the Fourier transform of $\tilde{y}(t)$. It then follows that

$$\frac{\partial}{\partial\tau'}\chi(\tau',0)\bigg|_{\tau'=0} = j\int_{-\infty}^{\infty}\omega|G(j\omega)|^2\frac{d\omega}{2\pi} \tag{8.7.17}$$

and

$$\frac{\partial^2}{\partial\tau'^2}\chi(\tau',0)\bigg|_{\tau'=0} = -\int_{-\infty}^{\infty}\omega^2|G(j\omega)|^2\frac{d\omega}{2\pi} \tag{8.7.18}$$

Using Eqs. (8.7.17) and (8.7.18) in Eq. (8.7.14), we get

$$\frac{\partial^2}{\partial\tau'^2}|\chi(\tau',0)|\bigg|_{\tau'=0} = -\left\{\int_{-\infty}^{\infty}\omega^2|G(j\omega)|^2\frac{d\omega}{2\pi} - \left|\int_{-\infty}^{\infty}\omega|G(j\omega)|^2\frac{d\omega}{2\pi}\right|^2\right\} \tag{8.7.19}$$

By assumption,

$$\int_{-\infty}^{\infty}|\tilde{y}(t)|^2\,dt = \int_{-\infty}^{\infty}|G(j\omega)|^2\frac{d\omega}{2\pi} = 1 \tag{8.7.20}$$

so that we can write Eq. (8.7.19) as

$$\frac{\partial^2}{\partial\tau'^2}|\chi(\tau',0)|\bigg|_{\tau'=0} = -\left\{\frac{\displaystyle\int_{-\infty}^{\infty}\omega^2|G(j\omega)|^2\frac{d\omega}{2\pi} - \left[\int_{-\infty}^{\infty}\omega|G(j\omega)|^2\frac{d\omega}{2\pi}\right]^2}{\displaystyle\int_{-\infty}^{\infty}|G(j\omega)|^2\frac{d\omega}{2\pi}}\right\} \tag{8.7.21}$$

The quantity inside the brackets on the right-hand side of Eq. (8.7.21) is a measure of the bandwidth of the signal $\tilde{y}(t)$ and is denoted by β^2. Using Eq. (8.7.21) in Eq. (8.7.11), we finally obtain

$$\text{var}(\tau) = \text{var}(\tau') > \left(\frac{2E_r}{N_0}\beta^2\right)^{-1} \tag{8.7.22}$$

We see from Eq. (8.7.22) that we can increase the accuracy of our estimate by increasing the energy E_r or making β^2 as large as possible. That is, signals with larger bandwidth can provide more accurate estimates of time delay.

For estimating the Doppler frequency, we assume $\tau' = 0$, that is, the delay is known exactly, and use the Cramér–Rao bound to obtain

$$\text{var}\,\omega_d = \text{var}\,\omega_d' \geqslant \frac{-1}{\left(\dfrac{2E_r}{N_0}\right)\dfrac{\partial^2}{\partial\omega_d'^2}|\chi(0,\omega_d')|\bigg|_{\omega_d'=0}} \tag{8.7.23}$$

From Eq. (8.7.14) we can write

$$\frac{\partial^2}{\partial \omega_d'^2}|\chi(0,\omega_d')|\bigg|_{\omega_d'=0} = \left[\mathrm{Re}\left\{\frac{\partial^2}{\partial \omega_d'^2}\chi(0,\omega_d')\right\} + \left|\frac{\partial}{\partial \omega_d'}\chi(0,\omega_d')\right|^2 - \mathrm{Re}\left[\frac{\partial}{\partial \omega_d'}\chi(0,\omega_d')\right]^2\right]_{\omega_d'=0} \tag{8.7.24}$$

Since

$$\begin{aligned}\frac{\partial}{\partial \omega_d'}\chi(0,\omega_d') &= \frac{\partial}{\partial \omega_d'}\left\{\int_{-\infty}^{\infty}|\tilde{y}(t-\tau)|^2\exp[j\omega_d'(t-\tau)]d(t-\tau)\right\}\\ &= j\int_{-\infty}^{\infty}(t-\tau)|\tilde{y}(t-\tau)|^2\exp[j\omega_d'(t-\tau)]d(t-\tau)\end{aligned} \tag{8.7.25}$$

and

$$\frac{\partial^2}{\partial \omega_d'^2}\chi(0,\omega_d') = -\int_{-\infty}^{\infty}(t-\tau)^2|\tilde{y}(t-\tau)|^2\exp[j\omega_d'(t-\tau)]d(t-\tau) \tag{8.7.26}$$

it follows that

$$\frac{\partial}{\partial \omega_d'}\chi(0,\omega_d')\bigg|_{\omega_d'=0} = j\int_{-\infty}^{\infty}t|\tilde{y}(t)|^2\,dt \tag{8.7.27}$$

and

$$\frac{\partial^2}{\partial \omega_d'^2}\chi(0,\omega_d')\bigg|_{\omega_d'=0} = -\int_{-\infty}^{\infty}t^2|\tilde{y}(t)|^2\,dt \tag{8.7.28}$$

Similar to our definition of β^2, we can define the parameter t_d^2 as a measure of the duration of the transmitted waveform $\tilde{y}(t)$ by

$$t_d^2 = \frac{\int_{-\infty}^{\infty}t^2|\tilde{y}(t)|^2\,dt - \left[\int_{-\infty}^{\infty}t|\tilde{y}(t)|^2\,dt\right]^2}{\int_{-\infty}^{\infty}|\tilde{y}(t)|^2\,dt} \tag{8.7.29}$$

Use of Eqs. (8.7.27) to (8.7.29) in Eq. (8.7.24) yields

$$\frac{\partial^2}{\partial \omega_d'^2}|\chi(0,\omega_d')|\bigg|_{\omega_d'=0} = -t_d^2 \tag{8.7.30}$$

The bound on the estimate of Doppler frequency obtained by substituting Eq. (8.7.30) in Eq. (8.7.23) is

$$\mathrm{var}\,\omega_d = \mathrm{var}\,\omega_d' \geqslant \left(\frac{2E_r}{N_0}t_d^2\right)^{-1} \tag{8.7.31}$$

From Eq. (8.7.31) we see that to increase the accuracy of our estimate we must increase either the energy E_r or the parameter t_d^2 or both. The transmitter power available places a limit on the maximum value of E_r. However, the parameter t_d^2 can be increased by proper design of the waveform $\tilde{y}(t)$.

In the case of joint estimation of τ and ω_d, we need to obtain the Fisher information matrix discussed in Section 5.5 to obtain the variance bounds on the estimates. It is obvious from the previous development that

$$J_{11} = -\mathrm{E}\left\{\frac{\partial^2 L(\tau, \omega_d)}{\partial \tau'^2}\right\} = \frac{2E_r}{N_0}\beta^2 \tag{8.7.32}$$

and

$$J_{22} = -\mathrm{E}\left\{\frac{\partial^2}{\partial \omega_d'^2} L(\tau, \omega_d)\right\} = \frac{2E_r}{N_0} t_d^2 \tag{8.7.33}$$

It only remains to determine

$$J_{12} = \mathrm{E}\left\{\frac{\partial^2}{\partial \tau' \, \partial \omega_d'} L(\tau, \omega_d)\right\} \tag{8.7.34}$$

to be able to calculate the variance bounds on the estimates using Eq. (5.5.53). Using the approximation in Eq. (8.7.8) J_{12} can be expressed as

$$J_{12} \simeq -\frac{2E_r}{N_0} \frac{\partial^2}{\partial \tau' \, \partial \omega_d'} |\chi(\tau', \omega_d')| \tag{8.7.35}$$

Using a derivation similar to the ones developed previously, we can show after some algebraic manipulations that J_{12} can be evaluated as

$$J_{12} = \frac{2E_r}{N_0}(\overline{\omega t} - \overline{\omega}\bar{t}) \tag{8.7.36}$$

where the quantities $\overline{\omega t}$, $\overline{\omega}$, and $\bar{t}$ are defined as

$$\overline{\omega t} = \mathrm{Re}\left\{\frac{\partial^2}{\partial \tau' \, \partial \omega_d'} \chi(\tau', \omega_d') \Big|_{\substack{\tau'=0 \\ \omega_d'=0}}\right\} = \int_{-\infty}^{\infty} t \frac{d\theta_1(t)}{dt} |\tilde{y}(t)|^2 \, dt \tag{8.7.37}$$

$$\overline{\omega} = -j\frac{\partial}{\partial \tau'} \chi(\tau', \omega_d') \Big|_{\substack{\tau'=0 \\ \omega_d'=0}} = \int_{-\infty}^{\infty} \omega |G(j\omega)|^2 \frac{d\omega}{2\pi} \tag{8.7.38}$$

and

$$\bar{t} = -j\frac{\partial}{\partial \omega_d'} \chi(\tau', \omega_d') \Big|_{\substack{\tau'=0 \\ \omega_d'=0}} = \int_{-\infty}^{\infty} t |\tilde{y}(t)| \, dt \tag{8.7.39}$$

The inverse of the information matrix, $J^{-1} = \psi$, is then obtained as

$$\psi = \left(\frac{2E_r}{N_0}\right)^{-1} \begin{bmatrix} t_d^2 & \overline{\omega}\bar{t} - \overline{\omega t} \\ \overline{\omega}\bar{t} - \overline{\omega t} & \beta^2 \end{bmatrix} [\beta^2 t_d^2 - (\overline{\omega t} - \overline{\omega}\bar{t})^2]^{-1} \tag{8.7.40}$$

so that the bounds for the variance of the estimates follow as

$$\mathrm{var}(\tau) \geq \left(\frac{2E_r}{N_0}\right)^{-1} \frac{t_d^2}{\beta^2 t_d^2 - (\overline{\omega t} - \overline{\omega}\bar{t})^2} \tag{8.7.41}$$

and

$$\mathrm{var}(\omega_d) \geq \left(\frac{2E_r}{N_0}\right)^{-1} \frac{\beta^2}{\beta^2 t_d^2 - (\overline{\omega t} - \overline{\omega}\overline{t})^2} \tag{8.7.42}$$

From Eqs. (8.7.41) and (8.7.42), we see that for a given level of transmitted signal energy E_r, better estimation accuracy is obtained if the "time–bandwidth" product βt_d is large and the difference $(\overline{\omega t} - \overline{\omega}\overline{t})$ is small.

8.8 DYNAMIC TARGET TRACKING

The application of discrete-time recursive filtering and prediction techniques for tracking purposes in a radar system has been of significant interest in recent times [44–46]. Such techniques are well suited for tracking both maneuvering and constant-velocity targets. The radar provides noisy measurements of the target coordinates once every T seconds, where the scan rate is $1/T$ revolutions per second. These measurements are processed by the tracking filter to yield estimates of the target coordinates and their derivatives. Typical applications include airport surveillance radars (ASR) and air route surveillance radars (ARSR) and reentry vehicle tracking. These techniques can also be applied in a multitarget environment and with multisensor returns, but decision techniques will have to be used to correlate the correct return to the correct target. This is a rather involved situation and is not discussed here.

In the first part of this section we model the target by a linear system obtained by linearizing the actual target dynamics about the nominal range and angle coordinates. We then use the discrete Kalman filter algorithms of Chapter 6 to track the target. We consider both constant-velocity and linear acceleration models.

Maneuver initiation in a target that has been flying a constant-velocity trajectory causes loss of track since the track filter will no longer be matched to the actual target dynamics. In the second part of this section we formulate the problem of maneuver detection as a composite hypothesis-testing problem. Once maneuvering has been established, the parameters of the filter can be suitably updated to maintain track.

8.8.1 Tracking of Maneuvering and Constant-Velocity Targets

We recall that a radar system measures the range and angle every T seconds. Let us assume that at time kT, the target is at a range $R_0 + r(k)$ and angle $\theta_0 + \theta(k)$. One scan time later at time $(k + 1)T$, the range and angle will be $R_0 + r(k + 1)$ and $\theta_0 + \theta(k + 1)$. Therefore, $r(k)$ and $\theta(k)$ are the deviations from a nominal value R_0 and θ_0. From the range and angle measurements, which are generally noisy, we would like to estimate the present range, range rate, bearing, and bearing rate. In many applications we would also like to predict these quantities for tactical control purposes. Let us denote by $\mathbf{x}$ the six-component vector whose elements are the target range, velocity, acceleration, bearing, bearing rate, and bearing acceleration. To derive a set of state equations for this problem, we approximate the velocity and acceleration at time kT as [45]

$$\dot{x}_1(k) = \frac{x_1(k+1) - x_1(k)}{T} = x_2(k)$$

and

$$\dot{x}_2(k) = \frac{x_2(k+1) - x_2(k)}{T} = x_3(k)$$

so that

$$x_1(k+1) = x_1(k) + Tx_2(k)$$

and

$$x_2(k+1) = x_2(k) + Tx_3(k)$$

We similarly model the acceleration at time $(k+1)T$ as

$$x_3(k+1) = \rho x_3(k) + w_1(k)$$

where $w_1(k)$ is a zero-mean, white sequence with variance $\sigma_{w_1}^2$. This represents a maneuvering target with linear acceleration.

With similar approximations for bearing, bearing rate, and bearing acceleration, the model for maneuvering targets can be written as

$$\mathbf{x}(k+1) = \mathbf{A}\mathbf{x}(k) + \mathbf{w}(k) \tag{8.8.1}$$

where

$$\mathbf{A} = \begin{bmatrix} 1 & T & 0 & 0 & 0 & 0 \\ 0 & 1 & T & 0 & 0 & 0 \\ 0 & 0 & \rho & 0 & 0 & 0 \\ 0 & 0 & 0 & 1 & T & 0 \\ 0 & 0 & 0 & 0 & 1 & T \\ 0 & 0 & 0 & 0 & 0 & \rho \end{bmatrix} \tag{8.8.2}$$

and

$$\mathbf{w}(k) = \begin{bmatrix} 0 \\ 0 \\ w_1(k) \\ 0 \\ 0 \\ w_2(k) \end{bmatrix} \tag{8.8.3}$$

where $w_2(k)$ is also a white sequence with variance $\sigma_{w_2}^2$ and uncorrelated with $w_1(k)$. These inputs lead to correlated acceleration terms $x_3(k)$ and $x_6(k)$ with variances $\sigma_{w_1}^2/(1-\rho^2)$ and $\sigma_{w_2}^2/(1-\rho^2)$.

The model of Eq. (8.8.1) is quite general in the sense that we can derive constant-velocity and uncorrelated acceleration models from it. If the target is on a constant-velocity path, the acceleration is zero, so that $\rho = 0$ and $\sigma_{w_1}^2 = \sigma_{w_2}^2 = 0$. Wind gusts, and so on, usually affect the constant-velocity path of targets and can be modeled as white noise acceleration inputs. In this case, $\rho = 0$, making $x_3(k) = w_1(k)$ and $x_6(k) = w_2(k)$. In both the constant-velocity and white noise acceleration cases, the system order reduces to four. Further, note that the equations for range variables and bearing variables are decoupled. We can use this decoupling to advantage to reduce the numerical computations involved.

The radar measures noisy versions of the range and bearing. The measurement model is thus given by

$$\mathbf{z}(k) = \mathbf{H}\mathbf{x}(k) + \mathbf{v}(k) \tag{8.8.4}$$

where

$$\mathbf{H} = \begin{bmatrix} 1 & 0 & 0 & 0 & 0 & 0 \\ 0 & 0 & 0 & 1 & 0 & 0 \end{bmatrix}$$

and

$$\mathbf{v}(k) = \begin{bmatrix} v_1(k) \\ v_2(k) \end{bmatrix}$$

In Eq. (8.8.4), $v_1(k)$ and $v_2(k)$ are assumed to be independent noise sequences uncorrelated with each other. Further, they are uncorrelated with $w_1(k)$ and $w_2(k)$.

The tracking algorithms are obtained in a straightforward manner by applying the discrete-time Kalman filter algorithms of Chapter 6, with message and observation models given by Eqs. (8.8.1) and (8.8.4). To do this we need to initialize the Kalman filter algorithms. In other words, we need the values of $\mathbf{V}_{\tilde{\mathbf{x}}}(k \mid k)$ and $\hat{\mathbf{x}}(k \mid k)$ for some k to start with. These are generally not available. We can, however, generate a suboptimal estimate and compute the associated error variance. These values can then be used to initialize the optimal algorithms.

To illustrate the initiation procedure, consider that the target is essentially on a constant-velocity path, except for disturbance due to wind gust acceleration, that is, $\rho = 0$. In this case, the message model has a dimension of only 4. We wait for the first two measurements $\mathbf{z}(1)$ and $\mathbf{z}(2)$ to obtain the "estimate" $\hat{\mathbf{x}}(2 \mid 2)$ (which is ad hoc or suboptimal) as

$$\begin{aligned}
\hat{x}_1(2 \mid 2) &= z_1(2) \\
\hat{x}_2(2 \mid 2) &= \frac{1}{T}[z_1(2) - z_1(1)] \\
\hat{x}_4(2 \mid 2) &= z_2(2) \\
\hat{x}_5(2 \mid 2) &= \frac{1}{T}[z_2(2) - z_2(1)]
\end{aligned} \tag{8.8.5}$$

That is, the estimates are obtained by considering the measurements to be noise-free, and the velocities to be constant. The error is then given by

$$\tilde{\mathbf{x}}(2|2) = \mathbf{x}(2) - \hat{\mathbf{x}}(2|2)$$

so that the error variance $\mathbf{V}_{\tilde{\mathbf{x}}}(2|2)$ as obtained from Eqs. (8.8.1), (8.8.4), and (8.8.5) is

$$\mathbf{V}_{\tilde{\mathbf{x}}}(2|2) = \begin{bmatrix} \sigma_{v_1}^2 & \sigma_{v_1}^2/T & 0 & 0 \\ \sigma_{v_1}^2/T & \sigma_{w_1}^2 + (2\sigma_{v_1}^2/T^2) & 0 & 0 \\ 0 & 0 & \sigma_{v_2}^2 & \sigma_{v_2}^2/T \\ 0 & 0 & \sigma_{v_2}^2/T & \sigma_{w_2}^2 + (2\sigma_{v_2}^2/T^2) \end{bmatrix} \tag{8.8.6}$$

Equations (8.8.5) and (8.8.6) can be used to initialize the discrete Kalman filter algorithms to process the measurements from $k = 3$ onward. A similar initialization procedure can be adopted for maneuvering or constant-velocity targets.

Because of the difficulty in implementing the time-varying gain needed for the Kalman filter, often in practice a constant gain is used. Furthermore, a constant-velocity model rather than a linear acceleration model is usually used in simple tracking systems. Such filters are called $\alpha - \beta$ filters, where α is the gain for range update and β the gain for velocity update. Optimum values of α and β have also been determined [44]. However, for maneuvering targets the $\alpha - \beta$ tracking filters may perform poorly.

8.8.2 Detection of Target Maneuver

A target can be tracked accurately if we know the correct state representation of the target. If we use a constant-velocity model to track a maneuvering target, we are likely to obtain poor tracking, particularly for tactical control purposes. It therefore becomes necessary to detect the time at which the target started maneuvering. Once this is known, we can switch the tracking filter to match the maneuvering state model to obtain proper track.

For simplicity of illustration, we consider only the range coordinate of the model given by Eq. (8.8.1). We assume a constant-velocity path initially, and therefore let $\rho = 0$ and $\sigma_{w_1}^2 = 0$. A Kalman filter can be matched to this model to yield us the one-stage prediction estimates $\hat{x}_1(k|k-1)$ and $\hat{x}_2(k|k-1)$. From our discussions in Chapter 6 we know that the innovation sequence

$$\nu(k) = z(k) - \hat{x}_1(k|k-1) \tag{8.8.7}$$

is white with zero mean and covariance

$$V_\nu(k) = \sigma_{v_1}^2 + V_{\tilde{x}_1}(k|k-1) \tag{8.8.8}$$

where $V_{\tilde{x}_1}(k|k-1)$ is the error covariance of one-stage prediction estimate of range. In steady state, that is, after the tracking transients have died out, the tracking can be assumed exact, and consequently, $\mathbf{V}_{\tilde{\mathbf{x}}}(k|k-1) \simeq 0$. Then we can write

$$V_\nu(k) \simeq \sigma_{v_1}^2 \tag{8.8.9}$$

Now, suppose that the target undergoes a maneuver starting at time $k_0 T$. This shows up as an addition of the acceleration term to the dynamical equations, and consequently manifests itself as an addition to range measurement. That is, for $k \geqslant k_0$, we can write the measurement equation as

$$z(k) = z_0(k) + \xi(k - k_0) \tag{8.8.10}$$

where $z_0(k)$ is the measurement that would have been obtained if there had been no maneuver and $\xi(k - k_0)$ is an unknown time function due to maneuver which is zero for $k \leqslant k_0$. Because the Kalman filter is linear, its range prediction estimate can be written as

$$\hat{x}_1(k \mid k-1) = \hat{x}_{10}(k \mid k-1) + \hat{x}_{1\xi}(k \mid k-1) \tag{8.8.11}$$

where $\hat{x}_{10}(k \mid k-1)$ is the estimate based on the measurement $z_0(k-1)$ and $\hat{x}_{1\xi}(k \mid k-1)$ is the estimate based on $\xi[(k-1) - k_0]$. The residual or innovation sequence, in turn, can be written as

$$\nu(k) = \nu_0(k) + \zeta(k - k_0) \tag{8.8.12}$$

where $\nu_0(k)$ is the innovation sequence that would have been obtained had there been no maneuver and

$$\zeta(k - k_0) = \xi(k - k_0) - \hat{x}_{1\zeta}(k \mid k-1) \tag{8.8.13}$$

Equation (8.8.12) indicates that whenever there is maneuver, it adds a bias term to the zero-mean innovation sequence $\nu_0(k)$. Thus, to detect the occurrence of the maneuver, it is only necessary to monitor the bias. As long as no bias develops, we continue to use the constant-velocity tracker; when a bias is detected, we would switch to the tracker matched to the appropriate dynamics [47,48]. The decision scheme in [47] uses the sequence $\nu(k)$ not merely to decide whether a maneuver has occurred but also to determine the time $(k - j)T, j = 1, 2, \ldots, m$ at which it occurred. In [48] the decision is based on a weighted average of the innovations as given by

$$\mu(k) = \alpha\mu(k-1) + \delta(k) \tag{8.8.14}$$

with

$$\delta(k) = V_\nu^{-1}(k)\nu^2(k) \tag{8.8.15}$$

where $V_\nu(k)$ is the covariance of $\nu(k)$ and the fading parameter α is chosen empirically between 0 and 1. If $\nu(k)$ is assumed to be Gaussian, $\delta(k)$ is chi-square distributed with n_2 degrees of freedom, so that

$$\lim_{k\to\infty} \mathrm{E}\{\mu(k)\} = \frac{n_2}{1-\alpha} \tag{8.8.16}$$

Thus we can look at $1/(1 - \alpha)$ as the effective window length over which the presence of a maneuver is detected. We decide that a maneuver is taking place if $\mu(k)$ exceeds a threshold.

8.9 SUMMARY

In this chapter we first considered the application of detection and estimation techniques to the study of digital communication systems. We derived optimal receiver structures and obtained expressions for the probability of error for both baseband and carrier systems. For coherent binary communication systems, the optimum system is PSK for a given average bit energy-to-noise ratio. The receiver consists of correlators followed by a threshold detector.

In the case of noncoherent systems, we essentially have a problem of detecting a signal with unknown parameters in the presence of noise. The unknown parameter in this case is the phase parameter θ. The optimum receiver is obtained by determining the likelihood ratio. The receiver can generally be implemented by bandpass filters followed by envelope detectors.

As can be expected, the performance of noncoherent systems is not as good as that of the corresponding coherent systems. However, for very large SNR, the difference in performance between coherent FSK and noncoherent FSK becomes vanishingly small.

A system that does not require a separate phase reference is DPSK. While the performance of DPSK is not as good as PSK, it is better than that of noncoherent FSK. It hence represents a useful compromise between a fully coherent system and a noncoherent system.

In many communication channels, fading occurs due to random fluctuations of the transmission medium. This case is similar to noncoherent communication except that in addition to the phase the amplitude of the transmitted signals is also a random variable. A particularly useful channel model is the Rayleigh model. Expressions for P_e were derived for this model. The problem of synchronization was discussed briefly. The use of the phase-lock loop for carrier synchronization was indicated and a MAP technique for estimating timing misalignment was discussed.

An important problem in digital communication is that of intersymbol interference, which is caused by the fact that the channel acts as a system with memory. Thus the received pulse over any interval is distorted by contributions from previous pulses. An effective way of overcoming this problem is to pass the observations through an equalizer that is matched to the channel characteristics. The Lucky equalizer uses a tapped delay model of the channel, while the Kalman filter equalizer uses a state-variable description of the signal and channel characteristics. The design of adaptive equalizers was also considered.

We concluded our discussion of digital communication systems with a brief introduction to spread spectrum communications in Section 8.3 and a discussion of the performance of a DS-BPSK system in the presence of three different types of interference.

In the second part of this chapter, we presented certain applications of detection theory, maximum likelihood estimation, and Kalman filtering to the radar detection and tracking problems. To do so, we developed some target models in Section 8.5 which used the complex representation of signals. The specific models considered were those for a steady point target, a slowly fluctuating point target, and spread targets.

In Section 8.6 we discussed the detection of radar targets in the presence of additive white Gaussian noise. The target models of Section 8.5 were used to obtain receiver structures and determine their performance. Because detection of stochastic signals in noise is outside the scope of the present book, we discussed the detection of spread targets only briefly. The results were again derived in terms of complex envelopes of bandpass signals.

In Section 8.7 we introduced the problem of parameter estimation in radar. Typical parameters of interest are target range and velocity, or equivalently, the time delay and Doppler frequency shift. We used the maximum likelihood approach and obtained the likelihood function in terms of complex envelopes. We derived accuracy bounds on our estimates and related them to certain design parameters of the transmitted waveform. The concept of ambiguity function was introduced and its significance was discussed briefly.

Section 8.8 was concerned with continuous tracking of target range velocity and angular position. We illustrated the application of the Kalman filter to this problem. When a target begins maneuvering, track maintenance is difficult mainly because the tracker dynamics are no longer matched to that of the target. In such a case maneuver detection becomes important so that the dynamics may be modified properly to keep the target in track.

For reasons of space, we have not covered in this chapter many interesting topics to which the theory developed in the text can be applied. Some of these have been left as exercises at the end of the chapter for the interested reader. The first is the detection of targets in the presence of colored noise, which arises naturally in the form of clutter in radar (Exercise 8.33). The techniques of Section 8.5 can be used with certain approximations to the case of spread targets (Exercise 8.32). Parameter estimation in the case of slowly fluctuating point targets and spread targets was not discussed. An important topic that has also not been covered is that of array processing, in which several spatially distributed observations are available.

EXERCISES

8.1. In a baseband binary polar system, E is 20 mV. Let $N_0 = 10^{-3}$ W/Hz.

(a) What is P_e if the binary digits (bits) are transmitted at a rate of 5000 per second?

(b) If the bit rate is increased to 10,000 bps, what should E be to give the same probability of error as in part (a)? Assume NRZ waveforms.

8.2. In a baseband binary polar system, the amplitude of the transmitted signal over each T-second interval is either $+1$ or -1 with equal probability.

(a) If $N_0 = 10^{-3}$, for the minimum probability of error receiver, what is the minimum value of T such that $P_e = 10^{-5}$?

(b) If we now want to halve the value of T so that we double transmission rate, what should the amplitude of the signal be for the same P_e?

8.3. We want to build a digital data transmission system with a desired P_e of 10^{-3}. Compare the necessary signal-to-noise ratios required for PRK, OOK, FSK, noncoherent FSK, noncoherent OOK, and DPSK.

8.4. Derive the optimum receiver structure of a DPSK system under the usual assumptions of additive white noise. Prove that the bit error probability is given by $P_e = \frac{1}{2}\exp(-E/N_0)$. Note that 3-dB advantage of DPSK over noncoherent FSK arises because the total energy in the message bearing waveforms is twice that of FSK.

8.5. In Chapter 4 we derived the minimum probability of error receiver for M-ary signal detection in white noise under the assumption that the signals were uncorrelated. In this problem we seek to extend our results to the case of correlated signals. The observations are

$$H_1:\quad z(t) = y_i(t) + v(t) \qquad 0 \leq t \leq T \quad i = 0, 1, 2, \ldots, M$$

where $v(t)$ is zero-mean, white, Gaussian with spectral density $N_0/2$. The correlation coefficients are

$$\rho_{ij} = \frac{1}{\epsilon}\int_0^T y_i(t)y_j(t)$$

where ϵ is the average energy of the signals. Find the optimum receiver and determine the probability of error assuming that the hypotheses are equally likely and the criterion is minimum probability of error.

8.6. In Exercise 8.5 assume that the signals $y_i(t)$ are

$$y_i(t) = \sqrt{\frac{2E_i}{T}}\sin\omega_c t \qquad 0 \leq t \leq T \quad i = 1, 2, 3, 4$$

Find the minimum probability of error if there are an integral number of cycles of the sinusoidal waveform over the observation interval.

8.7. Specialize the results of Exercise 8.5 to M-ary ASK and M-ary PSK. In M-ary ASK, the transmitted signals are of the form

$$y_i(t) = \sqrt{E_i}\,s(t) \qquad i = 0, 1, 2, \ldots, M$$

where

$$\sqrt{E_i} = (i - 1)A, \qquad \int_0^T s^2(t)\,dt = 1$$

In M-ary PSK, the signals are

$$y_i(t) = \sqrt{\frac{2E}{T}}\cos\left(\omega_c t + \frac{2\pi i}{M}\right) \qquad 0 \leq t \leq T, \quad i = 0, 1, 2, \ldots, M - 1$$

Assume that there are an integral number of cycles of the carrier in each signaling interval.

8.8. In a PRK system, the receiver has a phase error $\Delta\phi$ radians. Prove that the bit error probability assuming equal prior probabilities and minimum probability of error criterion is

$$P_e = Q\left[\left(\frac{2E\cos^2\Delta\phi}{N_0}\right)^{1/2}\right]$$

8.9. A binary communication system using orthonormal signals undergoing Rician fading can be modeled as follows:

$$\text{Mark: } z(t) = \sqrt{2}\,\alpha f_1(t)\cos(\omega_c t + \delta) + \sqrt{2}\,\eta f_1(t)\cos(\omega_c t + \theta) + v(t)$$

$$\text{Space: } z(t) = \sqrt{2}\,\alpha f_0(t)\cos(\omega_c t + \delta) + \sqrt{2}\,\eta f_0(t)\cos(\omega_c t + \theta) + v(t) \qquad 0 \leqslant t \leqslant T$$

α and δ are the amplitude and phase of the specular component of the signal and are assumed known. η is Rayleigh distributed with $E[\eta^2] = 2\sigma^2$, and θ is uniformly distributed. Obtain the optimum receiver structure, and prove that the bit error probability is

$$P_e = Q\left[\frac{\gamma}{(\beta + 1/2)^{1/2}\beta^{1/2}}, \frac{\gamma(\beta + 1)}{(\beta + 2)^{1/2}\beta^{1/2}}\right]$$

$$- \left(\frac{\beta + 1}{\beta + 2}\right)\exp\left[-\frac{\gamma^2}{2}\left(\frac{\beta^2 + 2\beta + 2}{\beta^2 + 2\beta}\right)\right] I_0\left[\gamma^2 \frac{\beta + 1}{\beta(\beta + 2)}\right]$$

where $\beta \underline{\Delta}\, 2\sigma^2/N_0$ and $\gamma^2 = \alpha^2/\sigma^2$ (ratio of the energy in the specular component to the average energy in the random component). Note that when $\gamma = 0$, we have Rayleigh channel, and when $\gamma = \infty$, we have completely known signal.

8.10. In transmitting phase information to the receiver for a coherent communication system, the information is usually corrupted by noise and must be estimated at the receiver. This uncertainty in phase may be modeled as a random variable. Consider partially coherent OOK in which the observations are

$$H_1: \quad z(t) = \sqrt{2E/T}\sin(\omega_c t + \theta) + v(t) \qquad 0 \leqslant t \leqslant T$$

$$H_0: \quad z(t) = v(t) \qquad 0 \leqslant t \leqslant T$$

where $v(t)$ is zero mean, white, Gaussian with spectral density $N_0/2$. Here θ is a random variable. It is clear that the decision rule can be obtained by finding the conditional likelihood ratio $\Lambda(z(t)\,|\,\theta)$ and averaging over $p(\theta)$. A useful model for θ is to assume that

$$p(\theta\,|\,\Lambda_m) = \frac{\exp(\Lambda_m \cos\theta)}{2\pi I_0(\Lambda_m)} \qquad -\pi \leqslant \theta \leqslant \pi$$

where $I_0(\Lambda_m)$ is a modified Bessel function of the first kind. For $\Lambda_m = 0$, we see that $p(\theta) = 1/2\pi$, so that θ is uniformly distributed. For $\Lambda_m \to \infty$, $p(\theta)$ tends to an impulse so that we approach the known signal case. Find the optimum receiver for this problem assuming that the two hypotheses are equally likely. What is the minimum probability of error?

8.11. Consider the dual-diversity FSK system in Rayleigh medium. If mark was transmitted, the received signals are

$$z_1(t) = \sqrt{\frac{2}{T}E_{11}}\sin(\omega_1 t + \theta_1) + v_1(t)$$

$$z_2(t) = \sqrt{\frac{2}{T}E_{21}}\sin(\omega_1 t + \theta_2) + v_2(t)$$

If space was transmitted, the received signals are

$$z_1(t) = \sqrt{\frac{2}{T}E_{10}}\sin(\omega_0 t + \phi_1) + v_1(t)$$

$$z_2(t) = \sqrt{\frac{2}{T}E_{20}}\sin(\omega_0 t + \phi_2) + v_2(t)$$

Over any observation interval $(0, T)$ the amplitude and phase are constant, but may vary from pulse to pulse (slow fading). The additive noises are Gaussian, independent, and zero mean with spectral density $N_0/2$. The amplitudes $\sqrt{E_{ji}}$ are identically and independently Rayleigh distributed with mean-square value $\overline{E}$. The phase terms, θ's and ϕ's, are independent and identically uniformly distributed over $[0, 2\pi]$. Obtain the likelihood ratio test and the optimum receiver structure. Show that the probability of error under the usual assumptions of equal prior probability and the criterion of minimum probability of error is

$$P_e = \frac{4 + 3\delta}{(2 + \delta)^3} \qquad \text{where} \quad \delta = \frac{\overline{E}}{N_0}$$

Compare this result to the no-diversity reception of FSK signaling in Rayleigh medium.

8.12. Extend the results of Exercise 8.11 for M-diversity operation.

8.13. Consider a baseband equiprobable binary (± 1) channel, whose impulse response samples are given by

$$h_{-3} = -0.077 \qquad h_{-2} = -0.355 \qquad h_{-1} = 0.059$$

$$h_0 = 1 \qquad h_1 = 0.059 \qquad h_2 = -0.273$$

and all other samples are zero. The additive noise samples are Gaussian, white, with zero mean and variance σ_v^2. For each case, consider different signal-to-noise ratios. (The signal-to-noise ratio may be defined as $1/\sigma_v^2$, where unity in the numerator represents the variance of the binary input sequence and therefore is the signal power.) Design a tap delay equalizer

(a) With three taps and delay $D = 2$
(b) With five taps and delay $D = 1$
(c) With five taps and delay $D = 2$
(d) With five taps and delay $D = 3$

Compute the mean-square distortion and mean-square error of the equalized system. Present your results in a graphical form involving the various parameters. Discuss your results.

8.14. Apply the discrete Kalman filtering algorithms to design the equalizer for the channel described by Eq. (8.2.83), and derive the structure in Fig. 8.2.17.

8.15. Design a Kalman equalizer for the channel defined in Exercise 8.13. Obtain the steady-state mean-square error using numerical computation. How do the results compare with the results of Exercise 8.13? (*Hint:* Redefine the channel sample response index for convenience. For example, change h_{-3} to h_1, h_{-2} to h_2, etc.)

8.16. Show that the minimum mean-square error design of a tapped delay line equalizer discussed in Section 8.2.5 reduces to the design for minimum mean-square distortion under no-noise conditions.

8.17. Assume that in an equiprobable binary (± 1) baseband digital communication system, the channel has a memory of $(2N + 1)$ symbols. The additive noise sequence is white, Gaussian, with zero mean and variance σ_v^2. Obtain a design procedure for a MMSE equalizer (no constraint on structure, such as nonrecursive or recursive) based on the discrete Weiner filtering technique presented in Chapter 6. Show that the equalizer has the structure shown in Fig. E.8.1 when D represents the delay. Identify the coefficients of the equalizer in the spectral factorization involved. (Refer to [49] if you have difficulty.)

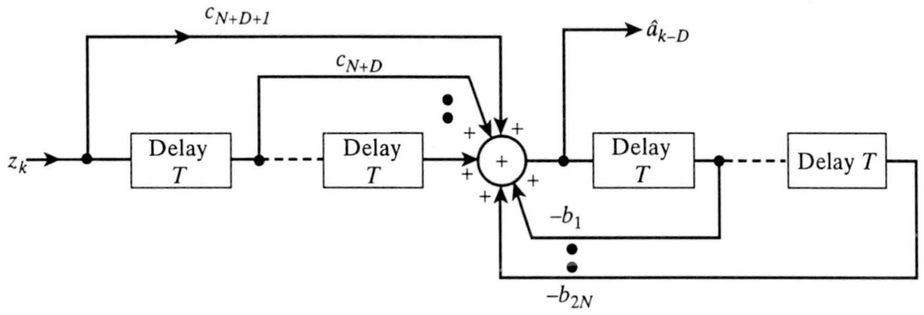

Figure E.8.1

8.18. In the development of the maximum likelihood estimation approach to the adaptive equalizer, prove (Eq. 8.2.97)

8.19. In the maximum likelihood estimation approach to adaptive equalization discussed in Section 8.2.5, assume that we require $P_e = 0.95$ when $h_j = \pm 0.5\Delta$ and $\Delta = 0.025$. Plot the number of samples required for averaging as a function of SNR (σ^2/S) in dB for different numbers of taps.

8.20. In Exercise 8.19 we require that $P_e = 0.98$ when $h_j = \pm 0.5\Delta$. Assume a SNR of 10 dB. Obtain a plot of K, the number of samples, versus Δ, step size for different N.

8.21. Consider the effect of a tone interferer on a DS-BPSK system. That is, assume that $I(t) = I\cos(\omega_0 t + \theta_1)$ in Eq. (8.3.4) and neglect the thermal noise $v(t)$. If the PG is large, by invoking the central limit theorem, the test statistic at the output of the summer in Fig. 8.3.2 can be taken to be approximately Gaussian. Find the approximate probability of bit error as a function of signal power I, θ_1, and N.

8.22. Consider a single-shot multiuser synchronous system in which decisions on the bits of all the K users are to be made simultaneously, based on the received signal $r(t)$ over the interval $(0, T)$ [see Eq. (8.3.6)]. Here, *single-shot* means that the bit decision is made using only the received signal over the corresponding bit interval and synchronous system refers to the fact that the signals corresponding to all the users arrive at the same time at the receiver. That is, $\tau_i = 0, i = 1, 2, \ldots, K$ in Eqs. (8.3.6) and (8.3.7). Derive the maximum likelihood rule for deciding the bits of the K users. The computational complexity of evaluating the bit error probability can be somewhat reduced by exploiting certain properties. See [50] for details.

8.23. Show that the range spread target modeled in Eq. (8.5.18) exhibits frequency-selective reflecting characteristics. You may assume that $\tilde{b}(\tau)$ is a complex stochastic process in τ with zero mean and is an uncorrelated process; that is, $E[\tilde{b}(\tau_1)\tilde{b}^*(\tau_2)] = \phi_{\tilde{b}}(\tau_1, \tau_2)\delta(\tau_1 - \tau_2)$. The corresponding power spectrum is known as the *range scattering function*.

8.24. In Section 8.5.3 we obtained piecewise constant approximation to the Doppler spread target. Following an analogous development, obtain a piecewise constant approximation model of the range spread target. You may assume that $\tilde{b}(\tau)$ is a complex, zero mean, uncorrelated process, as in Exercise 8.23.

8.25. In a radar receiver employing single-pulse detection, we require that $P_F = 10^{-3}$ and $P_D = 0.8$. Calculate the average peak signal-to-noise ratio $\overline{\eta}^2$ to satisfy the requirements, and also the threshold for (a) a completely known signal, (b) a signal with unknown random phase, and (c) a signal from a Rayleigh fluctuating target. What is the increase in $\overline{\eta}^2$ required if a $P_D = 0.95$ is specified?

8.26. The radar equation [30,38] relates the range of the target, transmitted energy (power), system parameters like antenna gains, target cross section, and system losses to the received peak signal-to-noise ratio, $\overline{\eta}^2$, and is given by

$$\overline{\eta}^2 = 2\frac{E_T G_T G_R \lambda^2 \overline{\sigma}_c}{(4\pi)^3 R^4 N_0 L}$$

where

$$\begin{aligned}
\overline{\eta}^2 &= \text{average peak signal-to-noise ratio} \\
E_T &= \text{transmitted pulse energy} \\
G_T &= \text{transmitter antenna gain} \\
G_R &= \text{receiving antenna gain} \\
\lambda &= \text{wavelength of transmission, corresponds to } f_c \\
\overline{\sigma}_c &= \text{average target cross section} \\
R &= \text{target range} \\
N_0 &= k_B T_{eq} \\
k_B &= \text{Boltzmann constant, } 1.38 \times 10^{-23} \text{ joule/K} \\
T_{eq} &= \text{equivalent input noise temperature} \\
L &= \text{system loss factor}
\end{aligned}$$

Design a radar system, assuming single-pulse detection and exactly known signal, to meet $P_F = 10^{-3}$ and $P_D = 0.8$ at a range of 200 nautical miles. The data available are

$$\begin{aligned}
G_T = G_R &= 30 \text{ dB} \\
f_c &= 1000 \text{ MHz} \\
T_{eq} &= 1000°\text{F} \\
L &= 3 \text{ dB} \\
\overline{\sigma}_c &= 1 \text{ m}^2
\end{aligned}$$

Repeat the problem if the signal is of unknown random phase, and for the case of Rayleigh target fluctuation.

8.27. Design a radar system to meet $P_F = 10^{-2}$ and $P_D = 0.9$ at a range of 300 nautical miles. The system losses are 3 dB and the effective input noise temperature is 2000°F. Assume an average cross section of 1 m^2 for the target and the transmitted rectangular pulse has at most 40% duty cycle. Consider the cases of fluctuating and nonfluctuating targets.

8.28. Consider a slowly fluctuating point target in which the reflecting gain $\tilde{b}$ is modeled as due to one specular or dominant scatterer plus a number of independent small scatterers. This gives rise to the one dominant plus Rayleigh model [35] for $|\tilde{b}|$ and a uniform density function for the phase θ_2. Let $c \triangleq |\tilde{b}|^2$. The density function of c is approximated for the dominant ray case by

$$p(c) = \frac{2c}{\sigma_c^2} \exp\left(-\frac{\sqrt{2}c}{\sigma_c}\right) \qquad c \geq 0$$

Obtain the receiver structure for the detection of such a target from bandlimited white-noise-corrupted observations. Set up the performance equations and try to solve them. Consult Swerling [35] or DiFranco and Rubin [38] if you have difficulty. Obtain the graphs relating P_D and $\overline{\eta}^2$ with P_F as parameter.

8.29. Consider a slowly fluctuating point target in which we model $\tilde{b}$ as a complex Gaussian random variable but with nonzero mean and variance $\sigma_{\tilde{b}}^2$. This leads to the Rician density model for $|\tilde{b}|$ and a uniform density function for θ_2. Obtain the optimum Neyman–Pearson receiver under the usual assumptions. Set up and solve the performance equations. Draw plots of P_D versus $\overline{\eta}^2$ with P_F as parameter. Also study the effect of varying the mean value of $\tilde{b}$ and compare with the results obtained for Rayleigh fluctuating target (mean is zero for this case), exactly known steady point target, and steady point target with unknown random phase.

8.30. In Section 8.6.2 we studied detection of point targets that had unknown random phase and targets with random amplitude and random phase (Rayleigh fluctuating target, for example). Very often it is desirable to measure the phase as accurately as possible and use it to improve performance. Consider the point target models in which $|\tilde{b}|$ is modeled as a Rayleigh or Rician density function and the phase θ_2 is exactly known. Obtain the optimum Neyman–Pearson receiver for detection under usual assumptions. Evaluate the expressions for P_D and P_F. Compare them with the results for Rayleigh fluctuating target and target with unknown random phase. What do you conclude?

8.31. We considered essentially the optimum schemes for radar detection. A suboptimum radar detection technique is shown in Fig. E.8.2. It is called the *binary integrator* [51] or *double-threshold detection* [52] and consists of a single pulse-matched filter, followed by an envelope detector, an amplitude quantizer (first threshold), and a range gate. (This corresponds to sampling the output of the first threshold, so that the output is due to the return from a particular range resolution cell.) Typically, the first threshold output is 1 or 0. The binary counter counts the number of 1's in a total of N pulses of the train. If the counter equals or exceeds a fixed number (say, $M, M \leq N$), a target is declared present in the particular range cell. Show that this system has the following performance:

$$P_D = \sum_{k=M}^{N} \binom{N}{k} P_S^k (1 - (P_S)^{N-k})$$

and

$$P_F = \sum_{k=M}^{N} \binom{N}{k} P_N^k (1 - (P_N)^{N-k})$$

where P_S is the probability that the output of the envelope detector exceeds the first threshold when a target is present (single-pulse basis); and P_N is the probability that the output of the envelope detector exceeds the first threshold when noise alone is present. Also obtain expressions for P_S and P_N. Consider the cases of nonfluctuating and Rayleigh fluctuating targets. Plot P_D versus $\overline{\eta}^2$ with $P_N = 10^{-2}$, and $N = 5, 10, 100$, and for different values of M.

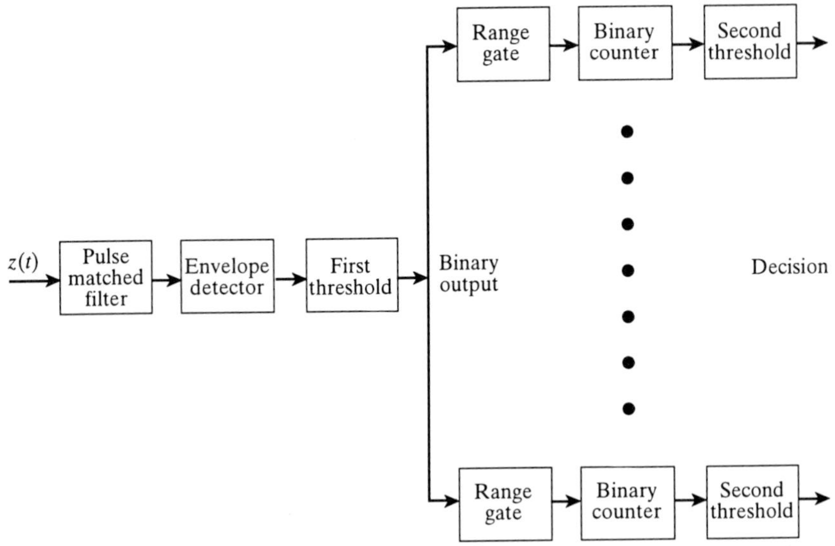

Figure E.8.2

8.32. Equation (8.5.17) represents a piecewise constant approximation to a Doppler spread target. We had indicated that an approach similar to multiple-pulse detection of Rayleigh fluctuating targets discussed in Section 8.6.4 can be used to obtain an optimum receiver. Obtain this receiver under usual assumptions. Bring out the difference between this receiver and multiple-pulse detection of Rayleigh fluctuating point target. Assume that $\tilde{b}_i$'s in Eq. (8.5.17) are identically distributed zero-mean complex Gaussian random variables with variance σ_b^2.

8.33. In many radar detection problems, the target return signal is masked by colored noise $\tilde{v}_c(t)$ in addition to the white noise $\tilde{v}(t)$. The colored noise is usually due to the returns from unwanted or interfering targets and is called clutter. Develop, using results from Chapter 4, the optimum Neyman–Pearson receivers for the detection of the different target models in the presence of clutter. Assume that the clutter $\tilde{v}_c(t)$ is Gaussian and has a covariance $V_{\tilde{v}_c}(t_1, t_2)$.

8.34. Develop the maximum likelihood estimates for τ and ω_d for a slowly fluctuating Rayleigh target. Obtain the accuracy bounds for individual as well as joint estimation of τ and ω_d.

8.35. In this problem we develop some interesting properties of the ambiguity function discussed in Section 8.7. Prove the following properties:

(a) $|\chi(\tau', \omega_d')| \leq \chi(0,0) = 1$. This shows that the peak of the ambiguity function is at its origin.

(b) $\dfrac{1}{2\pi}\int_{-\infty}^{\infty}\int_{-\infty}^{\infty}|\chi(\tau', \omega_d')|^2\, d\tau'\, d\omega_d' = 1$. This shows that the volume under the ambiguity surface $|\chi(\tau', \omega_d')|^2$ is a constant for all waveforms. We can shape $\chi(\tau', \omega_d')$ but we will always satisfy this invariance of volume. In other words, we can suppress the ambiguity at a particular point, but it redistributes to other points.

(c) If $\tilde{y}(t)$ has an ambiguity function $\chi(\tau', \omega_d')$, then $\tilde{y}(kt)$ has an ambiguity function $(1/\sqrt{k})\chi(k\tau', \omega_d'/k)$.

(d) Let $\tilde{y}(t)$ and $\tilde{y}_1(t)$ be two functions such that $\tilde{y}_1(f)$ is the Fourier transform of $\tilde{y}(t)$. If $\chi(\tau', \omega_d')$ and $\chi_1(\tau', \omega_d')$ represent the ambiguity functions of $\tilde{y}(t)$ and $\tilde{y}_1(t)$, respectively, then

$$\chi_1(\omega_d', -\tau) = \chi(\tau, \omega_d)$$

(e) If $\chi_1(\tau', \omega_d')$ and $\chi_2(\tau', \omega_d')$ represent the ambiguity functions of $\tilde{y}_1(t)$ and $\tilde{y}_2(t)$, respectively, the ambiguity function $\chi(\tau', \omega_d')$ of the function $\tilde{y}(t) = \tilde{y}_1(t)\tilde{y}_2(t)$ is given by

$$\chi(\tau', \omega_d') = \int_{-\infty}^{\infty} \chi_1(\tau', z)\chi_2(\tau', \omega_d - z)\, dz$$

8.36. In radar parameter estimation and resolution problems, the ambiguity function of Section 8.7 plays a significant role. Derive and plot the ambiguity functions of the following signals (you may use the properties derived in Exercise 8.35):

(a) Rectangular pulse:

$$\tilde{y}(t) = \begin{cases} \dfrac{1}{T} & -\dfrac{T}{2} < t < \dfrac{T}{2} \\ 0 & \text{otherwise} \end{cases}$$

(b) Gaussian pulse:

$$\tilde{y}(t) = \left(\frac{1}{\pi T^2}\right)^{1/4} \exp\left(-\frac{t^2}{2T^2}\right) \qquad -\infty < t < \infty$$

(c) Gaussian pulse with linear frequency modulation:

$$\tilde{y}(t) = \left(\frac{1}{\pi T^2}\right)^{1/4} \exp\left(-\frac{t^2}{2T^2}\right) \exp(-jkt^2)$$

where k is a constant.

(d) Rectangular pulse with linear frequency modulation:

$$\tilde{y}(t) = \begin{cases} \dfrac{1}{T}\exp(jkt^2) & -\dfrac{T}{2} < t < \dfrac{T}{2} \\ 0 & \text{otherwise} \end{cases}$$

where k is a constant.

REFERENCES

1. R. G. Gallager, *Information Theory and Reliable Communication*, Wiley, New York, 1968.
2. G. C. Clark, Jr., and J. B. Cain, *Error-Correction Coding for Digital Communications*, Plenum Press, New York, 1981.
3. J. M. Wozencraft and I. M. Jacobs, *Principles of Communication Engineering*, Wiley, New York, 1965.
4. M. Schwartz, W. R. Bennett, and S. Stein, *Communication Systems and Techniques*, McGraw-Hill, New York, 1966.
5. H. L. Van Trees, *Detection, Estimation and Modulation Theory*, Part I, Wiley, New York, 1968.
6. J. Marcum, Table of Q-functions, *Report RM-339*, Rand Corporation, January 1950.
7. A. R. Di Donato and M. I. Jarnagin, A method for computing the circular coverage function, *Math. Comp.*, **16**, July, 347–355 (1955).
8. D. E. Johansen, New techniques for machine computation of the Q-functions, truncated normal derivatives and matrix eigenvalues, *Scientific Report 2*, Sylvania ARL, Waltham, MA, July 1961.
9. W. C. Jakes, *Microwave Mobile Communications*, IEEE Press, New York, 1994.
10. J. C. Proakis, *Digital Communications*, McGraw-Hill, New York, 1989.
11. D. G. Brennan, Linear diversity combining techniques, *Proc. IRE*, **47**, No. 6, June, 1075–1102 (1959).
12. G. L. Turin, On optimal diversity reception I, *IRE Trans. Inf. Theory*, **IT-7**, No. 3, July, 311–329 (1960).
13. G. L. Turin, on optimal diversity reception II, *IRE Trans. Commun. Syst.*, **CS-10**, No. 1, March, 22–31 (1962).
14. A. J. Viterbi, *Principles of Coherent Communication*, McGraw-Hill, New York, 1966.
15. S. C. Gupta, Phase-locked loops, *Proc. IEEE*, **63**, No. 2, February, 291–306 (1975).
16. J. J. Stiffler, *Theory of Synchronous Communications*, Prentice Hall, Englewood Cliffs, NJ, 1971.
17. W. C. Lindsey and M. K. Simon, *Telecommunication Systems Engineering*, Prentice Hall, Englewood Cliffs, NJ, 1973.
18. G. D. Forney, Maximum likelihood sequence estimation of digital sequences in the presence of intersymbol interference, *IEEE Trans. Inf. Theory*, **IT-19**, No. 3, May, 363–377 (1972).
19. R. W. Lucky, J. Salz, and E. J. Weldon, *Principles of Data Communication*, McGraw-Hill, New York, 1968.
20. R. W. Lucky, Techniques for adaptive equalization of digital communication systems, *Bell Syst. Tech. J.*, February, 255–286 (1966).
21. R. E. Lawrence and H. Kaufmann, The Kalman filter for the equalization of a digital communications channel, *IEEE Trans. Commun. Technol.*, **COM-19**, No. 6, December, 1137–1141 (1971).
22. M. K. Simon, J. K. Omura, R. A. Scholtz, and B. K. Levitt, *Spread Spectrum Communications*, Vols. I, II, and III, Computer Science Press, Rockville, MD, 1985.

23. R. E. Ziemer and R. L. Peterson, *Digital Communications and Spread Spectrum Systems*, Macmillan, New York, 1985.
24. L. B. Milstein, Interference rejection techniques in spread spectrum communications, *Proc. IEEE*, **76**, No. 6, June, 657–671 (1988).
25. M. B. Pursley, D. V. Sarwate, and W. E. Stark, Error probability for direct sequence spread spectrum multiple-access communications, *IEEE Trans. Commun.*, Part I, May, 975–984 (1982).
26. S. Verdu, Minimum probability of error for asynchronous Gaussian multiple access channels, *IEEE Trans. Inf. Theory*, January, 85–96 (1986).
27. H. V. Poor and S. Verdu, Single-user detectors for multi-user channels, *IEEE Trans. Commun.*, **36**, January, 50–60 (1988).
28. M. K. Varanasi and B. Aazhang, Multistage detection in asynchronous code division multiple access systems, *IEEE Trans. Commun.*, **38**, April, 509–519 (1990).
29. P. M. Woodward, *Probability and Information Theory, with Applications to Radar*, Pergamon Press, Long Island City, NY, 1953.
30. M. I. Skolnik, Introduction to radar, in *Radar Handbook*, McGraw-Hill, New York, 1990.
31. H. L. Van Trees, *Detection, Estimation and Modulation Theory*, Part III, Wiley, New York, 1971.
32. M. I. Skolnik, *Introduction to Radar Systems*, McGraw-Hill, New York, 1980.
33. F. E. Nathanson, *Radar Design Principles*, McGraw-Hill, New York, 1991.
34. A. W. Rihaczek, *Principles of High Resolution Radar*, McGraw-Hill, New York, 1969.
35. P. Swerling, Probability of detection for fluctuating targets, *Report RM-1217*, Rand Corporation, March 1954. Also appeared in a special issue, *IRE Trans. Inf. Theory*, **IT-6**, April, 269–308 (1960).
36. J. I. Marcum, A statistical theory of target detection by pulsed radar, *Report RM-753*, Rand Corporation, December 1947. Also appeared in a special issue, *IRE Trans. Inf. Theory*, **IT-6**, April, 59–144 (1960),.
37. J. I. Marcum, A statistical theory of target detection by pulsed radar, *Mathematical Appendix, Rand Research Report RM-753*, July 1948. Also appeared in a special issue, *IRE Trans. Inf. Theory*, **IT-6**, April, 145–267 (1960).
38. J. V. DiFranco and W. L. Rubin, *Radar Detection*, Prentice Hall, Englewood Cliffs, NJ, 1968.
39. L. E. Brennan and I. S. Reed, A recursive method of computing the Q function, *IEEE Trans. Inf. Theory*, **IT-1**, No. 2, April, 312–313 (1965).
40. R. Nitzberg, *Adaptive Signal Processing for Radar*, Artech House, Dedham, MA, 1991.
41. J. Pachares, A table of bias levels useful in radar detection problems, *IRE Trans. Inf. Theory*, **IT-14**, No. 1, March, 38–45 (1958).
42. F. C. Schweppe, Evaluation of likelihood functions for Gaussian signals, *IEEE Trans. Inf. Theory*, **IT-11**, No. 1, January, 61–70 (1965).
43. T. Kailath, A general likelihood ratio formula for random signals in Gaussian noise, *IEEE Trans. Inf. Theory*, **IT-15**, No. 3, May, 350–361 (1969).
44. S. S. Blackman, *Multiple Target Tracking with Radar Applications*, Artech House, Dedham, MA, 1986.

45. R. A. Singer and K. W. Behnke, Real time tracking filter evaluation and selection for tactical applications, *IEEE Trans. Aerosp. Electron. Syst.*, **AES-7**, January, 100–110 (1971).

46. Y. Bar-Shalom and T. E. Fortmann, *Tracking and Data Association*, Academic Press, 1988.

47. R. J. McAulay and E. Denlinger, A decision-directed adaptive tracker, *IEEE Trans. Aerosp. Electron. Syst.*, **AES-9**, March, 229–236 (1973).

48. Y. Bar-Shalom and K. Birmiwal, Variable dimension filter for maneuvering target tracking, *IEEE Trans. Aerosp. Electron. Syst.*, **AES-18**, September, 621–629 (1982).

49. H. L. Van Trees, *Detection, Estimation and Modulation Theory*, Part II, *Nonlinear Modulation Theory*, Wiley, New York, 1971.

50. M. K. Varanasi and B. Aazhang, Near-optimum detection in synchronous code-division multiple access systems, *IEEE Trans. Commun.*, **39**, May, 725–736 (1991).

51. J. V. Harrington, An analysis of the detection of repeated signals in noise by binary integration, *IRE Trans. Inf. Theory*, **IT-1**, No. 1, March, 1–9 (1955).

52. P. Swerling, The double threshold method of detection, *Memo RM-1008*, Rand Corporation, December 1952.

53. K. Pearson, *Tables of the Incomplete Gamma Function*, Cambridge University Press, New York, 1946.

9

Miscellaneous Applications

9.1 INTRODUCTION

The basic problems of detection and estimation arise in several areas besides the communication and radar applications discussed in the previous chapter. Solutions to the detection and estimation problems discussed earlier can be applied in areas such as control, pattern recognition, system identification, and speech and picture processing. In this chapter we consider two specific applications, statistical pattern classification and system identification.

9.2 APPLICATION TO STATISTICAL PATTERN CLASSIFICATION

Automatic pattern recognition covers a wide variety of problems, such as speech recognition, character recognition, fingerprint classification, aerial and microphotograph processing, seismic signal analysis for geophysical exploration, and analysis of biomedical signals such as the electrocardiogram (EKG) or the electroencephelogram (EEG) for use in medical and physiological applications. We can roughly describe the pattern recognition problem as follows. We are given several groups of objects such that the objects within each group have certain common attributes. We refer to the representation of the object in terms of these attributes as a *pattern* and the description of the attributes as the *features* of the pattern. The several groups are called *pattern classes*. The basic pattern recognition problem is to classify or label an object as belonging to one of the pattern classes by processing the features associated with the pattern.

Features may be symbolic or numerical or both. Examples of symbolic descriptions are color (black, green, etc.) or feel (rough or smooth). Attributes such as voltage and intensity which are measurable (quantifiable) lead to numerical features. Since our interest is in statistical methods, we restrict ourselves to numerical features which can be described either completely or partially in terms of probability density functions.

The design of an automatic pattern recognition system involves several steps. The first is to represent the input data in a form suitable for machine classification of the object. We thus measure certain distinctive attributes of the object being classified and arrange the measurements in the form of a *pattern vector*. For example, in character recognition the area surrounding the character may be divided into a grid pattern and a value of 0 or 1 may be assigned to each cell, depending on whether the cell contains a portion of the character. If the measurements are continuous waveforms as in the case of ECG analysis or speech recognition, the sample values of the waveform at a set of discrete instants of time can serve as the pattern vector. Alternatively, the waveforms can be expanded in a complete set of orthonormal functions and the coefficients of the expansion can be used as the pattern vector.

In most cases, not all the elements of a pattern vector have the same significance in the classification of the pattern. The measurements usually contain many irrelevant data as far as the classification problem is concerned. We may therefore seek to order the elements of the pattern vector in decreasing order of importance and retain only those elements that are most useful. This is usually referred to as *feature selection*.

The third step in pattern recognition systems is the determination of optimum procedures for classification of the patterns. If there are M pattern classes, we can consider the *pattern space* to be the sum of M regions each of which contains the pattern vectors of that class. The recognition problem thus consists of generating the decision boundaries that separate the M pattern classes. The functions defining the decision boundaries are referred to as *discriminant functions*.

We can determine the discriminant functions in several ways. If we have complete prior information abut the pattern classes, the determination of the discriminant functions is relatively straightforward. If there is little or no information regarding the classes, we can use a training or learning procedure to determine the characteristics of the patterns. There are several approaches to the classification of patterns, and both deterministic and statistical algorithms are available in the literature. Here we restrict ourselves to statistical methods. The reader is referred to the literature for other approaches [1–10]. In particular, we show how the techniques of detection and estimation discussed in earlier chapters can be used. Even though we are concerned mainly with the classification problem, we briefly discuss the feature extraction problem in Section 9.2.5.

9.2.1 Basic Problem of Statistical Pattern Classification

In this section we discuss the basic statistical pattern classification problem. As stated in Section 9.1, the problem is to classify patterns into one of several classes on the basis of certain measurements. Usually, the number M of the pattern classes is known, but it may not be known in some cases. The pattern is characterized by the feature vector, which forms the basis for classification. In practice, within a particular class the features of a pattern show statistical variability, so that it is appropriate to treat the patterns (pattern vectors) as random vectors described by multivariate (vector) probability distribution or density functions. The simplest problem arises when these underlying distributions are exactly known and our machine is to classify a test pattern into one of the M pattern classes.

Often, the form of these distributions is known or can be assumed, but certain parameters associated with the distributions may be unknown. In this case, if certain patterns with known classifications are available, we can use these sample patterns to estimate or learn the unknown parameters of the distributions involved. These patterns with known classification which are used for learning purposes are referred to as *training samples*. As can be expected, a machine that needs to learn the associated distribution functions will, in general, not perform as well as a machine classifying patterns with completely known distributions. The performance of the machine is usually evaluated by determining the probability of misclassifying the patterns. This mode of learning using training samples of known classification is termed *learning with a teacher* or *supervised learning*.

If training samples with known classification are not available, the samples with unknown classification can be used to learn or estimate the parameters of interest to design the classifier. The classifier is then said to work in a *nonsupervised mode* or *learn without a teacher*.

It is helpful to formulate mathematically the various aspects of the problem described previously. Let $\mathbf{x}$ denote the pattern vector

$$\mathbf{x} = [x_1, x_2, \ldots, x_n]^T$$

If $\mathbf{x}$ belongs to a pattern from the jth class H_j of M possible classes, we denote it as $\mathbf{x}^j$, and its components as x_i^j. The probability density function of $\mathbf{x}$ when it belongs to class H_j is represented as $p(\mathbf{x} \mid H_j)$. Patterns presented in sequence or as a batch are denoted $\mathbf{x}_1, \mathbf{x}_2, \mathbf{x}_3, \ldots$. If the probability density function is known except for certain parameters $\boldsymbol{\theta}_j$, it is represented as $p(\mathbf{x} \mid H_j, \boldsymbol{\theta}_j)$.

The simplest problem arises when $p(\mathbf{x} \mid H_j), j = 1, 2, \ldots, M$ are known, and a test pattern $\mathbf{x}$ has to be classified into one of the M categories. The decision rule (discriminant function) may be linear, quadratic, or otherwise, depending on $p(\mathbf{x} \mid H_j)$.

The problem of supervised learning pattern classification may be stated as follows. For each class we know either the density function $p(\mathbf{x} \mid H_j)$ or the conditional density function $p(\mathbf{x} \mid H_j, \boldsymbol{\theta}_j)$, where $\boldsymbol{\theta}_j$ represents a vector of unknown parameters. For each class in which the parameter $\boldsymbol{\theta}_j$ has to be learned, we are given a set of training patterns $\mathbf{Z}_{N_j} = [\mathbf{x}_1^j, \mathbf{x}_2^j, \ldots, \mathbf{x}_{N_j}^j]$ belonging to that class. We can assume the parameters $\boldsymbol{\theta}_j$ to be either deterministic or random variables. In the latter case, we assume a prior density function for the parameter. We then arrive at a decision rule to classify the test pattern $\mathbf{x}$. This is discussed in Section 9.2.3. The machine essentially operates in two modes. In the learning mode the unknown parameters are estimated using the samples with known classification. In the testing mode new samples are classified into the various classes on the basis of decision rules formulated at the end of the learning mode.

In nonsupervised learning, we are given a set of patterns with unknown classification, $\mathbf{Z}_N = [\mathbf{x}_1, \mathbf{x}_2, \ldots, \mathbf{x}_N]$. Learning the vector $\boldsymbol{\theta}_j$ for those classes with unknown parameters must proceed using these unclassified samples. Any new patterns that are presented to the machine may also be used to refine the estimates of the unknown parameters. In this case there is no clear distinction between the learning and testing modes. Unfortunately, as we shall see later, the computational and storage requirements for unsupervised learning become exceedingly large as the number of sample patterns increases. We discuss this in Section 9.2.4, where we also present some techniques to overcome this difficulty.

We start our discussions with the problem of classification with known probability density functions. For clarity of presentation, we restrict most of our discussions to the case where we have only two pattern classes. The extension to M classes is fairly straightforward.

9.2.2 Pattern Classification with Known Densities

When the underlying density functions for the pattern classes are completely known, we can obtain the discriminant functions from our discussion of hypothesis-testing problems in Section 3.3. In practice the discriminant functions are chosen so as to minimize the average probability of misclassification. This corresponds to the minimum probability of error criterion discussed in Section 3.3.1. If the prior probabilities of the pattern classes are not known, we can use either the Neyman–Pearson criterion or a minimax approach in designing our classifiers. In any case, the results can be obtained by a straightforward application of our discussions in Chapter 3.

Example 9.2.1

Let us consider the classification of a test pattern $\mathbf{x}$ into one of two classes. The densities $p(\mathbf{x}|H_j), j = 1, 2$ are assumed to be Gaussian with means $\mathbf{m}_1$ and $\mathbf{m}_2$, and covariance matrices $\mathbf{V}_1$ and $\mathbf{V}_2$. We seek the decision rule so as to minimize the probability of misclassification. The rule follows easily from Eq. (3.3.26) as

$$\frac{p(\mathbf{x}|H_1)}{p(\mathbf{x}|H_2)} \underset{H_2}{\overset{H_1}{\gtrless}} \frac{P_2}{P_1} \tag{9.2.1}$$

where P_1 and P_2 are the prior probabilities of pattern classes 1 and 2, respectively. If we assume that $P_1 = P_2 = \frac{1}{2}$, then taking the logarithm of Eq. (9.2.1) and rearranging yields the decision rule as

$$\tfrac{1}{2}(\mathbf{x} - \mathbf{m}_2)^T \mathbf{V}_2^{-1}(\mathbf{x} - \mathbf{m}_2) - \tfrac{1}{2}(\mathbf{x} - \mathbf{m}_1)^T \mathbf{V}_1^{-1}(\mathbf{x} - \mathbf{m}_1) + \tfrac{1}{2}(\ln|\mathbf{V}_2| - \ln|\mathbf{V}_1|) \underset{H_2}{\overset{H_1}{\gtrless}} 0 \tag{9.2.2}$$

We now consider some special cases. When the covariances are equal, that is $\mathbf{V}_1 = \mathbf{V}_2 = \mathbf{V}$, Eq. (9.2.2) reduces to

$$(\mathbf{m}_1 - \mathbf{m}_2)^T \mathbf{V}^{-1}\mathbf{x} - \tfrac{1}{2}(\mathbf{m}_1^T \mathbf{V}^{-1}\mathbf{m}_1 - \mathbf{m}_2^T \mathbf{V}^{-1}\mathbf{m}_2) \underset{H_2}{\overset{H_1}{\gtrless}} 0 \tag{9.2.3}$$

We note that the left-hand side of Eq. (9.2.3) is linear in the features **x**. Let us define the vector **w** and the constant w_0 as

$$\mathbf{w} = \mathbf{V}^{-1}(\mathbf{m}_1 - \mathbf{m}_2) \quad \text{and} \quad w_0 = \tfrac{1}{2}(\mathbf{m}_1^T\mathbf{V}^{-1}\mathbf{m}_1 - \mathbf{m}_2^T\mathbf{V}^{-1}\mathbf{m}_2) \tag{9.2.4}$$

The decision rule of Eq. (9.2.3) therefore consists in computing the discriminant function $f(\mathbf{x}) = \mathbf{w}^T\mathbf{x}$ and comparing it with the threshold w_0 and is given as

$$f(\mathbf{x}) = \mathbf{w}^T\mathbf{x} \underset{H_2}{\overset{H_1}{\gtrless}} w_0 \tag{9.2.5}$$

Because of their simplicity, linear discriminant functions are widely used in many classifiers, even though such operation may be nonoptimal.

The probability of misclassification easily follows. The sufficient statistic l equals $f(\mathbf{x})$ and is hence one-dimensional. It is Gaussian distributed under both hypotheses with

$$\mu_j = \mathrm{E}\{l \mid H_j\} = \mathbf{w}^T\mathbf{m}_j$$

and

$$\sigma_j^2 = \mathrm{var}\{l \mid H_j\} = \mathbf{w}^T\mathbf{V}\mathbf{w} \tag{9.2.6}$$

We can therefore write the probability of error as

$$\begin{aligned} P_e &= \frac{1}{2}\int_{-\infty}^{w_0} p(l \mid H_1)\,dl + \frac{1}{2}\int_{w_0}^{\infty} p(l \mid H_2)\,dl \\ &= \frac{1}{2}Q\left(\frac{\mu_1 - w_0}{\sigma_l}\right) + \frac{1}{2}Q\left(\frac{w_0 - \mu_2}{\sigma_l}\right) \end{aligned} \tag{9.2.7}$$

where

$$Q(x) = \frac{1}{\sqrt{2\pi}}\int_x^{\infty} \exp\left(-\frac{x^2}{2}\right) dx \tag{9.2.8}$$

Let us now assume that $\mathbf{V} = \sigma^2\mathbf{I}$ where σ^2 is a scalar and **I** is the $n \times n$ identity matrix. The assumption is valid when the physical features are statistically independent with equal variances σ^2. In such a case, the classification rule of Eq. (9.2.3) reduces to

$$(\mathbf{m}_1^T\mathbf{x} - \tfrac{1}{2}\mathbf{m}_1^T\mathbf{m}_1) - (\mathbf{m}_2^T\mathbf{x} - \tfrac{1}{2}\mathbf{m}_2^T\mathbf{m}_2) \underset{H_2}{\overset{H_1}{\gtrless}} 0 \tag{9.2.9}$$

By adding and subtracting $\frac{1}{2}\mathbf{x}^T\mathbf{x}$ to the left side of Eq. (9.2.9) and rearranging terms, we can write the classification rule as

$$(\mathbf{x} - \mathbf{m}_2)^T(\mathbf{x} - \mathbf{m}_2) - (\mathbf{x} - \mathbf{m}_1)^T(\mathbf{x} - \mathbf{m}_1) \underset{H_2}{\overset{H_1}{\gtrless}} 0 \tag{9.2.10}$$

Thus the classifier is a minimum-distance classifier. That is, given a pattern vector **x**, we measure the Euclidean distance $\|\mathbf{x} - \mathbf{m}_i\|$ from **x** to the mean vector of each class and assign **x** to that class which corresponds to the least distance.

The probability of error follows from Eq. (9.2.7) by noting that for this case

$$2(\mu_1 - w_0) = 2(w_0 - \mu_2) = \sigma_l^2 = \frac{(\mathbf{m}_1 - \mathbf{m}_2)^T(\mathbf{m}_1 - \mathbf{m}_2)}{\sigma^2} \tag{9.2.11}$$

so that

$$P_e = Q\left(\frac{\mu_1 - w_0}{\sigma_l}\right)$$
$$= Q\left(\frac{1}{2\sigma}\sqrt{\sum_{i=1}^{n}(m_{1i} - m_{2i})^2}\right) \tag{9.2.12}$$

The significance of each feature can be determined from an examination of the preceding expression. Those features for which the difference between the two means is largest are the most useful, while those features for which the mean is the same do not play a role in reducing P_e. It follows from Eq. (9.2.12) that we can decrease the probability of error by increasing n, that is, by introducing new independent features. In theory at least, P_e can be reduced to an arbitrarily small quantity by increasing n without limit. In practice it has been found that P_e does decrease with increasing n up to a certain point. However, the inclusion of additional features beyond this point tends to degrade performance. This seems to be related to the fact that the number of samples in the design of any classifier is always finite. Since a detailed discussion of this is beyond the scope of this book, the interested reader is referred to the literature for details [2].

While the linear discriminant function is optimum for the case of Gaussian class conditional densities with equal covariance matrices, as noted earlier, linear discriminant functions are often used because of their simplicity. We note that the discriminant function of Eq. (9.2.5) represents a projection of the pattern vector onto a line in the direction of the weight vector. Thus the main advantage of using a linear discriminant function is in reducing a n-dimensional problem to a one-dimensional problem, which is presumably easier to solve. Even though the resulting classifiers may be suboptimal, if the weight vector is chosen such that the projections of patterns belonging to different classes are well separated, the resulting classifier will have good performance. The weight vector $\mathbf{w}$ is determined by minimizing a suitably chosen criterion function $J(\mathbf{w})$ [11]. One such is the *Fisher linear discriminant* for which $J(\mathbf{w})$ is chosen based on the observation that we can use the difference between the means of the pattern vectors in the two classes as a measure of the separation between them. Let $\hat{\mathbf{m}}_i, i = 1, 2$ be the sample means of the pattern vectors under the two classes. Then the criterion function for the Fisher discriminant is

$$J(\mathbf{w}) = \frac{\mathbf{w}^T \mathbf{S}_b \mathbf{w}}{\mathbf{w}^T \mathbf{S}_w \mathbf{w}} \tag{9.2.13}$$

where the matrices $\mathbf{S}_b$ and $\mathbf{S}_w$ are defined as

$$\mathbf{S}_b = \sum_{i=1}^{2}(\hat{\mathbf{m}}_i - \hat{\mathbf{m}}_0)(\hat{\mathbf{m}}_i - \hat{\mathbf{m}}_0)^T$$

where $\mathbf{m}_0$ is the sample mean of the vectors of the two classes put together.

$$\mathbf{S}_w = \sum_{\mathbf{x}\in H_1}(\mathbf{x} - \hat{\mathbf{m}}_1)(\mathbf{x} - \hat{\mathbf{m}}_1)^T + \sum_{\mathbf{x}\in H_2}(\mathbf{x} - \hat{\mathbf{m}}_2)(\mathbf{x} - \hat{\mathbf{m}}_2)^T \tag{9.2.14}$$

The matrix $\mathbf{S}_w$, called the *within-class scatter matrix*, is a measure of the scatter or variance of the pattern vectors around the respective class means and is expressed as the sum of the sample covariance matrices of the vectors under the two classes. The matrix $\mathbf{S}_b$, called the between class scatter matrix is the scatter of the sample mean vectors of the two classes around the sample mean of the mixture distribution. $J(\mathbf{w})$ therefore maximizes the ratio of between-class scatter to within-class scatter.

It can be shown that the weight vector which maximizes $J(\mathbf{w})$ is given by

$$\mathbf{w} = \mathbf{S}_w^{-1}(\hat{\mathbf{m}}_1 - \hat{\mathbf{m}}_2) \tag{9.2.15}$$

Comparison with Eq. (9.2.4) shows that the optimal discriminant function for the case when the pattern vectors under the two classes are Gaussian distributed with equal covariance matrices is the same as the Fisher discriminant function if the means and covariances are replaced by sample means and sample covariances.

For the case when the means of the two classes are equal but the variances are different, with $\mathbf{m}_1 = \mathbf{m}_2 = \mathbf{m}$, the decision rule becomes

$$\tfrac{1}{2}(\mathbf{x}-\mathbf{m})^T\mathbf{V}_2^{-1}(\mathbf{x}-\mathbf{m}) - \tfrac{1}{2}(\mathbf{x}-\mathbf{m})^T\mathbf{V}_1^{-1}(\mathbf{x}-\mathbf{m}) + \tfrac{1}{2}(\ln|\mathbf{V}_2| - \ln|\mathbf{V}_1|) \underset{H_1}{\overset{H_2}{\gtrless}} 0 \tag{9.2.16}$$

The discriminant function $f(\mathbf{x})$ is quadratic in $\mathbf{x}$ and can be written as

$$f(\mathbf{x}) = \mathbf{x}^T\mathbf{A}\mathbf{x} + \mathbf{x}^T\mathbf{b} + c \tag{9.2.17}$$

where $\mathbf{A}$, $\mathbf{b}$, c, are appropriate quantities. The probability of misclassification can be computed as in Example 3.4.2.

9.2.3 Supervised Learning Pattern Recognition

In supervised learning pattern recognition, the class conditional densities are assumed to be known except for certain parameters. To facilitate learning of the unknown parameters, we have access to a certain number of patterns corresponding to each class for which the parameters have to be estimated. For simplicity, let us assume that we have two classes. Then the conditional densities of interest are $p(\mathbf{x}\,|\,H_j, \boldsymbol{\theta}_j), j = 1, 2$. We also assume that we are given N_1 patterns from class 1 and N_2 patterns from class 2. The classification rule is obtained by computing the conditional likelihood ratio

$$\Lambda(\mathbf{x}\,|\,\mathbf{Z}_{N_1}, \mathbf{Z}_{N_2}) = \frac{p(\mathbf{x}\,|\,\mathbf{Z}_{N_1}, H_1)}{p(\mathbf{x}\,|\,\mathbf{Z}_{N_2}, H_2)} = \frac{p(\mathbf{x}\,|\,\mathbf{x}_1^1, \mathbf{x}_2^1, \ldots, \mathbf{x}_{N_1}^1, H_1)}{p(\mathbf{x}\,|\,\mathbf{x}_1^2, \mathbf{x}_2^2, \ldots, \mathbf{x}_{N_2}^2, H_2)} \tag{9.2.18}$$

We consider the case where $\boldsymbol{\theta}_j$ are random vectors with known a priori density $p(\boldsymbol{\theta}_j)$.† To determine $p(\mathbf{x}\,|\,\mathbf{Z}_{N_i}, H_j), j = 1, 2$, we use the conditional densities in each pattern class to write

$$p(\mathbf{x}\,|\,\mathbf{Z}_{N_j}, H_j) = \int_{\chi_{\boldsymbol{\theta}_j}} p(\mathbf{x}\,|\,\boldsymbol{\theta}_j, H_j)p(\boldsymbol{\theta}_j\,|\,\mathbf{Z}_{N_j})\,d\boldsymbol{\theta}_j \qquad j = 1, 2 \tag{9.2.19}$$

where $\chi_{\boldsymbol{\theta}_j}$ is the space of the parameter vector.

There remains the problem of finding $p(\boldsymbol{\theta}_j\,|\,\mathbf{Z}_{N_j})$. We can determine $p(\boldsymbol{\theta}_j\,|\,\mathbf{Z}_{N_j})$ in a recursive manner from $p(\boldsymbol{\theta}_j\,|\,\mathbf{Z}_{N_j-1})$ by applying Bayes' rule:

$$\begin{aligned} p(\boldsymbol{\theta}_j\,|\,\mathbf{Z}_{N_j}) &= p(\boldsymbol{\theta}_j\,|\,\mathbf{Z}_{N_j-1}, \mathbf{x}_{N_j}^j) \\ &= \frac{p(\mathbf{x}_{N_j}^j\,|\,\mathbf{Z}_{N_j-1}, \boldsymbol{\theta}_j)p(\boldsymbol{\theta}_j\,|\,\mathbf{Z}_{N_j-1})}{\int_{\chi_{\boldsymbol{\theta}_j}} p(\mathbf{x}_{N_j}^j\,|\,\mathbf{Z}_{N_j-1}, \boldsymbol{\theta}_j)p(\boldsymbol{\theta}_j\,|\,\mathbf{Z}_{N_j-1})\,d\boldsymbol{\theta}_j} \end{aligned} \tag{9.2.20}$$

† For convenience we write $p(\boldsymbol{\theta}_j\,|\,H_j)$ as $p(\boldsymbol{\theta}_j)$ and $p(\boldsymbol{\theta}_j\,|\,\mathbf{z}_{N_j}, H_j)$ as $p(\boldsymbol{\theta}_j\,|\,\mathbf{z}_{N_j})$.

If we assume that the patterns are conditionally independent, that is,

$$p(\mathbf{x}^j_{N_j} \mid \mathbf{Z}_{N_j-1}, \boldsymbol{\theta}_j) = p(\mathbf{x}^j_{N_j} \mid \boldsymbol{\theta}_j) \tag{9.2.21}$$

we can simplify the algorithm of Eq. (9.2.20) as

$$p(\boldsymbol{\theta}_j \mid \mathbf{Z}_{N_j}) = \frac{p(\mathbf{x}^j_{N_j} \mid \boldsymbol{\theta}_j)}{\displaystyle\int_{\chi_{\boldsymbol{\theta}_j}} p(\mathbf{x}^j_{N_j} \mid \boldsymbol{\theta}_j) p(\boldsymbol{\theta}_j \mid \mathbf{Z}_{N_j-1})\, d\boldsymbol{\theta}_j} p(\boldsymbol{\theta}_j \mid \mathbf{Z}_{N_j-1}) \tag{9.2.22}$$

Equation (9.2.22) represents an algorithm for updating the density function for $\boldsymbol{\theta}_j$ as more patterns from class j are presented to the classifier in a recursive manner, starting from the known density $p(\boldsymbol{\theta}_j \mid H_j)$. As such it is usually referred to as a learning algorithm for $\boldsymbol{\theta}_j$. The implementation of the classification algorithm is now straightforward, at least conceptually. We determine the conditional density $p(\boldsymbol{\theta}_j \mid \mathbf{Z}_{N_j})$ recursively from Eq. (9.2.22) and use it in Eq. (9.2.19) to determine $p(\mathbf{x} \mid \mathbf{Z}_{N_j}, H_j), j = 1, 2$ for each pattern to be classified. We then compute the likelihood ratio of Eq. (9.2.18) and compare it with the threshold to determine the classification.

The computational feasibility of Eq. (9.2.22) depends critically on the existence of a finite dimensional sufficient statistic for the relevant prior and posterior density functions. Otherwise, one needs to quantize the parameter vector $\boldsymbol{\theta}_j$ to implement the algorithm practically. Prior and posterior density functions that satisfy the requirement of a finite dimensional sufficient statistic are called conjugate or reproducing pairs, as stated in Chapter 5. In this case, assuming a reproducing form for $p(\boldsymbol{\theta})$ leads to considerable simplification of the computations. If an infinite number of training samples are available, the reproducing pair converges to the correct parameter value [12]. Again, in the evaluation of Eq. (9.2.19) we may need to quantize $\boldsymbol{\theta}_j$ over its range.

Example 9.2.2

We consider the case where the density function $p(\mathbf{x} \mid H_j, \boldsymbol{\theta}_j)$ is n-variate Gaussian with mean $\mathbf{m}$ and variance $\mathbf{V}$. For simplicity, we suppress the dependence of the various quantities on j, the category. We assume that this has to be carried out for each class, when relevant. The unknown parameter $\boldsymbol{\theta}$ can be either the mean $\mathbf{m}$ or the variance $\mathbf{V}$ or both and must be learned from the training pattern set $\mathbf{Z}_N = \{\mathbf{x}_1, \mathbf{x}_2, \ldots, \mathbf{x}_N\}$. In every case, we will assume an appropriate reproducing form for the prior density function for $\boldsymbol{\theta}$.

When $\boldsymbol{\theta}$ is the mean vector, we have

$$p(\mathbf{x} \mid \mathbf{m}) = \frac{1}{(2\pi)^{n/2} |\mathbf{V}|^{1/2}} \exp\left[-\frac{1}{2} (\mathbf{x} - \mathbf{m})^T \mathbf{V}^{-1} (\mathbf{x} - \mathbf{m}) \right]$$

We choose for $\mathbf{m}$, a Gaussian prior probability function $p(\mathbf{m})$ of the form (see Section 5.7).

$$p(\mathbf{m}) = \frac{1}{(2\pi)^{n/2} |\boldsymbol{\phi}_0|^{1/2}} \exp\left[-\frac{1}{2} (\mathbf{m} - \mathbf{m}_0)^T \boldsymbol{\phi}_0^{-1} (\mathbf{m} - \mathbf{m}_0) \right]$$

where $\mathbf{m}_0$ and $\boldsymbol{\phi}_0$ are the expected value and covariance matrix (usually, an educated guess of the designer) of the parameter vector $\mathbf{m}$. The density function $p(\mathbf{m}|\mathbf{Z}_k)$ can then be determined by successive application of Bayes' rule as in Eq. (9.2.22). Because of the reproducing property of the Gaussian density, the density function $p(\mathbf{m}|\mathbf{Z}_k)$ is also Gaussian. The mean and variance of the density function $p(\mathbf{m}|\mathbf{Z}_k)$ can be computed using either the MAP approach or by application of the orthogonal projection principle and a modest amount of manipulations of matrix equations. These parameters can be sequentially updated using the following algorithms [13] (see also Example 5.5.7):

$$\mathbf{m}_k = \mathbf{V}(\boldsymbol{\phi}_{k-1} + \mathbf{V})^{-1}\mathbf{m}_{k-1} + \boldsymbol{\phi}_{k-1}(\boldsymbol{\phi}_{k-1} + \mathbf{V})^{-1}\mathbf{x}_k \tag{9.2.23a}$$

$$\boldsymbol{\phi}_k = \mathbf{V}(\boldsymbol{\phi}_{k-1} + \mathbf{V})^{-1}\boldsymbol{\phi}_{k-1} \tag{9.2.23b}$$

Use of Eq. (9.2.19) then yields the density function of interest $p(\mathbf{x}|\mathbf{Z}_N, H_j)$ as a Gaussian density with mean $\mathbf{m}_N$ and covariance $(\mathbf{V} + \boldsymbol{\phi}_N)$.

When $\boldsymbol{\theta}$ is the unknown covariance matrix $\mathbf{V}$, the appropriate prior density function to use is the Wishart density function with parameters $\mathbf{V}_0$ and $\mathbf{c}_0$. Let us without loss of generality, assume that $\mathbf{m} = 0$. With $\mathbf{Q} = \mathbf{V}^{-1}$, we can write for the prior density

$$p(\mathbf{Q}) = \begin{cases} k_{n,c_0}\left|\dfrac{c_0}{2}\mathbf{V}_0\right|^{(c_0-1)/2} |\mathbf{Q}|^{(c_0-n-2)/2} \exp\left\{\dfrac{1}{2}\,\mathrm{tr}[c_0\mathbf{V}_0\mathbf{Q}]\right\} & \text{on } \chi_{\mathbf{Q}} \\ 0 & \text{otherwise} \end{cases}$$

where $\chi_{\mathbf{Q}}$ denotes the subset of the Euclidean space of dimension $\frac{1}{2}n(n+1)$. The matrix $\mathbf{Q}$ is positive definite, and k_{n,c_0} is the normalizing constant

$$k_{n,c_0} = \left[(\pi)^{n(n-1)/4}\prod_{\alpha=1}^{n}\Gamma\left(\frac{c_0-\alpha}{2}\right)\right]^{-1}$$

Here $\mathbf{V}_0$ is a positive definite matrix reflecting the initial knowledge of $\mathbf{V}^{-1}$, c_0 is a scalar reflecting the confidence of the initial estimate $\mathbf{V}_0$, and $\Gamma(\cdot)$ is the gamma function. By successive applications of Bayes' formula, we can show that the posterior density function $p(\mathbf{Q}|\mathbf{Z}_N)$ is also Wishart with the parameters $\mathbf{V}_N$ and c_N replacing $\mathbf{V}_0$ and c_0, where

$$\mathbf{V}_N = \frac{c_0\mathbf{V}_0 + \sum_{i=1}^{N}\mathbf{x}_i\mathbf{x}_i^T}{c_0 + N} \tag{9.2.24a}$$

$$c_N = c_0 + N \tag{9.2.24b}$$

Again the density function $p(\mathbf{x}|\mathbf{Z}_{N_j}, H_j)$ is calculated using Eq. (9.2.19).

When both the mean $\mathbf{m}$ and the covariance matrix $\mathbf{V}$ (or more appropriately, its inverse $\mathbf{V}^{-1}$) are unknown, we choose the prior density as Gauss–Wishart, that is, the mean $\mathbf{m}$ is Gaussian with prior mean $\mathbf{m}_0$ and covariance matrix $\boldsymbol{\phi}_0 = d_0^{-1}\mathbf{V}$, and $\mathbf{V}^{-1} = \mathbf{Q}$ is distributed according to Wishart distribution with parameters c_0 and $\mathbf{V}_0$. Successive application of Bayes' formula yields the posterior density function to be Gauss–Wishart with parameters c_N, d_N, $\mathbf{m}_N$, and $\mathbf{V}_N$ replacing c_0, d_0, $\mathbf{m}_0$, and $\mathbf{V}_0$. The respective expressions are

$$c_N = c_0 + N \tag{9.2.25a}$$

$$d_N = d_0 + N \tag{9.2.25b}$$

$$\mathbf{m}_N = \frac{d_0\,\mathbf{m}_0 + \sum_{i=1}^{N} \mathbf{x}_i}{d_0 + N} \tag{9.2.25c}$$

$$\begin{aligned}\mathbf{V}_N = \frac{1}{c_0 + N}\Bigg\{&(c_0\,\mathbf{V}_0 + d_0\,\mathbf{m}_0\,\mathbf{m}_0^T) + (N-1)\mathbf{S} \\ &+ \frac{1}{N}\left[\sum_{i=1}^{N} \mathbf{x}_i\right]\left[\sum_{j=1}^{N} \mathbf{x}_j\right]^T - d_N\,\mathbf{m}_N\,\mathbf{m}_N^T\Bigg\}\end{aligned} \tag{9.2.25d}$$

where

$$S = \frac{1}{N-1}\sum_{i=1}^{N}\left(\mathbf{x}_i - \frac{1}{N}\sum_{j=1}^{N}\mathbf{x}_j\right)\left(\mathbf{x}_i - \frac{1}{N}\sum_{j=1}^{N}\mathbf{x}_j\right)^T \tag{9.2.25e}$$

Again the density function $p(\mathbf{x}\,|\,H_j, \mathbf{Z}_{N_j})$ can be calculated by using Eq. (9.2.19).

If the vectors $\boldsymbol{\theta}_j$ are nonrandom parameters, we can find a maximum-likelihood estimate of $\boldsymbol{\theta}_j$ for each j and use these as the true values of the parameters in computing the likelihood ratio. The decision rule then becomes

$$\Lambda(\mathbf{x}) = \frac{p(\mathbf{x}|H_1, \boldsymbol{\theta}_1)|_{\boldsymbol{\theta}_1=\hat{\boldsymbol{\theta}}_{1\text{ml}}}}{p(\mathbf{x}|H_2, \boldsymbol{\theta}_2)|_{\boldsymbol{\theta}_2=\hat{\boldsymbol{\theta}}_{2\text{ml}}}} \tag{9.2.26}$$

The ML estimates based on the training samples are obtained in the usual manner (see Chapter 5) by maximizing the conditional density $p(\mathbf{Z}_{N_j}\,|\,\boldsymbol{\theta}_j), j = 1, 2$. Since the samples are independent, we can write

$$p(\mathbf{Z}_{N_j}\,|\,\boldsymbol{\theta}_j) = \prod_{k=1}^{N_j} p(\mathbf{x}_k^j\,|\,\boldsymbol{\theta}_j) \qquad j = 1, 2 \tag{9.2.27}$$

Taking logarithms and setting the gradient with respect to $\boldsymbol{\theta}_j$ equal to zero yields

$$\sum_{k=1}^{N_j} \frac{1}{p(\mathbf{x}_k^j\,|\,\boldsymbol{\theta}_j)} \nabla_{\boldsymbol{\theta}_j} p(\mathbf{x}_k^j\,|\,\boldsymbol{\theta}_j)\Bigg|_{\boldsymbol{\theta}_j=\boldsymbol{\theta}_{j\text{ml}}} = 0 \qquad j = 1, 2 \tag{9.2.28}$$

These usually constitute a set of nonlinear equations in the components of $\boldsymbol{\theta}_j$, which can be solved to obtain the ML estimates.

9.2.4 Nonsupervised Learning Pattern Recognition

In Section 9.2.3 we assumed that the classification of the training samples used to learn the parameters were known. In this section we consider the problem of unsupervised learning, in which the classifications of the training samples $\mathbf{Z}_N = \{\mathbf{x}_1, \mathbf{x}_2, \ldots, \mathbf{x}_N\}$ are not known. Again we assume that the form of the class conditional densities are known and only certain parameters associated with these densities are unknown, that is, that $p(\mathbf{x}\,|\,H_j, \boldsymbol{\theta}_j), j = 1, \ldots, M$ are known. If we denote by $\boldsymbol{\theta}$ the vector of all unknown parameters,

$$\boldsymbol{\theta} = [\boldsymbol{\theta}_1, \boldsymbol{\theta}_2, \ldots, \boldsymbol{\theta}_M]^T$$

we can write the density function of a pattern $\mathbf{x}$ conditioned on the (unknown) parameter vector $\boldsymbol{\theta}$, by expanding in a *mixture density* as

$$p(\mathbf{x}\,|\,\boldsymbol{\theta}) = \sum_{j=1}^{M} p(\mathbf{x}\,|\,H_j, \boldsymbol{\theta}_j)P_j \tag{9.2.29}$$

where P_j denotes the prior probabilities of the various classes. The conditional densities $p(\mathbf{x}\,|\,H_j, \boldsymbol{\theta}_j)$ are referred to as the *component densities* and the prior probabilities P_j as the *mixing parameters*.

The problem of nonsupervised learning can now be stated as follows. Given samples corresponding to the mixture density of Eq. (9.2.29), we are required to estimate the unknown parameter vector, using these samples. Once $\boldsymbol{\theta}$ is known, the mixture can be decomposed into its components so that the problem reduces to pattern classification using known densities considered in Section 9.2.2.

A question that arises is whether we can estimate the vector $\boldsymbol{\theta}$ uniquely from the mixture. We say that the density $p(\mathbf{x}\,|\,\boldsymbol{\theta})$ is *identifiable* if for $\boldsymbol{\theta} \neq \boldsymbol{\theta}'$, there exists a pattern $\mathbf{x}$ for which $p(\mathbf{x}\,|\,\boldsymbol{\theta}) \neq p(\mathbf{x}\,|\,\boldsymbol{\theta}')$ [14–16]. In practice, most mixtures of commonly encountered distributions are identifiable. However, with discrete distributions such is not always the case. In the sequel we assume only identifiable mixtures.

We now turn to the problem of designing the classifier. Again, we restrict ourselves to two pattern classes. We then write the decision rule for classifying the samples as

$$\Lambda(\mathbf{x}\,|\,\mathbf{Z}_N) \underset{H_2}{\overset{H_1}{\gtrless}} \gamma \tag{9.2.30}$$

where

$$\Lambda(\mathbf{x}\,|\,\mathbf{Z}_N) = \frac{p(\mathbf{x}\,|\,H_1, \mathbf{Z}_N)}{p(\mathbf{x}\,|\,H_2, \mathbf{Z}_N)} \tag{9.2.31}$$

and γ is the threshold.

In the case where $\boldsymbol{\theta}$ is a random vector with known prior density $p(\boldsymbol{\theta})$, the density functions $p(\mathbf{x}\,|\,H_j, \mathbf{Z}_N)$ can be written as

$$p(\mathbf{x}\,|\,H_j, \mathbf{Z}_N) = \int_{\chi_\theta} p(\mathbf{x}\,|\,\boldsymbol{\theta}, H_j, \mathbf{Z}_N)p(\boldsymbol{\theta}\,|\,H_j, \mathbf{Z}_N)\,d\boldsymbol{\theta} \tag{9.2.32}$$

Since the selection of the test pattern $\mathbf{x}$ is independent of the set of training samples, $p(\mathbf{x}\,|\,\boldsymbol{\theta}, H_j, \mathbf{Z}_N) = p(\mathbf{x}\,|\,\boldsymbol{\theta}_j, H_j)$. We can also assume that $p(\boldsymbol{\theta}\,|\,H_j, \mathbf{Z}_N) = p(\boldsymbol{\theta}\,|\,\mathbf{Z}_N)$. We thus obtain

$$p(\mathbf{x}\,|\,H_j, \mathbf{Z}_N) = \int_{\chi_\theta} p(\mathbf{x}\,|\,\boldsymbol{\theta}_j, H_j)p(\boldsymbol{\theta}\,|\,\mathbf{Z}_N)\,d\boldsymbol{\theta} \tag{9.2.33}$$

It only remains to determine $p(\boldsymbol{\theta}\,|\,\mathbf{Z}_N)$ to complete the design of the classifier.

We can use Bayes' rule to write

$$p(\boldsymbol{\theta} \mid \mathbf{Z}_N) = \frac{p(\mathbf{Z}_N \mid \boldsymbol{\theta}) p(\boldsymbol{\theta})}{\int_{\chi_\theta} p(\mathbf{Z}_N \mid \boldsymbol{\theta}) p(\boldsymbol{\theta}) \, d\boldsymbol{\theta}} \tag{9.2.34}$$

where

$$p(\mathbf{Z}_N \mid \boldsymbol{\theta}) = \prod_{i=1}^{N} p(\mathbf{x}_i \mid \boldsymbol{\theta}) \tag{9.2.35}$$

This can also be written in a recursive form analogous to Eq. (9.2.22) in the case of supervised learning as

$$p(\boldsymbol{\theta} \mid \mathbf{Z}_N) = \frac{p(\mathbf{x}_N \mid \boldsymbol{\theta}) p(\boldsymbol{\theta} \mid \mathbf{Z}_{N-1})}{\int_{\chi_\theta} p(\mathbf{x}_N \mid \boldsymbol{\theta}) p(\boldsymbol{\theta} \mid \mathbf{Z}_{N-1}) \, d\boldsymbol{\theta}} \tag{9.2.36}$$

The computation of $p(\boldsymbol{\theta} \mid \mathbf{Z}_N)$ is again formidable. In the supervised learning problem, at least in the cases where a finite-dimensional sufficient statistic exists, we can find solutions which are computationally feasible. With unsupervised learning, since the samples are obtained from a mixture distribution, we cannot expect that a simple sufficient statistic exists. Hence, even if we assume a reproducing form for the prior density $p(\boldsymbol{\theta})$, because the multiplying factor $p(\mathbf{x}_N \mid \boldsymbol{\theta})$ is a mixture density, the reproducing form of $p(\boldsymbol{\theta} \mid \mathbf{Z}_N)$ is not usually preserved. This means that, in practice, we have to store the density functions $p(\boldsymbol{\theta} \mid \mathbf{Z}_k)$ for all values of $\boldsymbol{\theta}$, almost an impossible task [17]. We can overcome this by storing the density only at a discrete set of values for $\boldsymbol{\theta}$ and use the expression in Eq. (9.2.36) to obtain the updated probabilities for the various levels of $\boldsymbol{\theta}$. The computations can be quite cumbersome for even moderate-sized problems.

We can also obtain an idea of the problems involved in the computation of $p(\boldsymbol{\theta} \mid \mathbf{Z}_N)$ by writing this density function in a slightly different manner [18]. We note that since there are two hypotheses and N patterns, there are 2^N possible ways of obtaining a sequence of N training samples. Let us order these sequences in any manner and denote the jth sequence by α_j. We can then write $p(\boldsymbol{\theta} \mid \mathbf{Z}_N)$ in a mixture distribution as

$$p(\boldsymbol{\theta} \mid \mathbf{Z}_N) = \sum_{j=1}^{2^N} p(\boldsymbol{\theta} \mid \mathbf{Z}_N, \alpha_j) P(\alpha_j \mid \mathbf{Z}_N) \tag{9.2.37}$$

As an example, let us assume that $N = 2$. Then the training set $\{\mathbf{x}_1, \mathbf{x}_2\}$ could be obtained in any of the following ways:

α_1: (H_1, H_1) with probability $P(\alpha_1 \mid \mathbf{Z}_2)$
α_2: (H_1, H_2) with probability $P(\alpha_2 \mid \mathbf{Z}_2)$
α_3: (H_2, H_2) with probability $P(\alpha_3 \mid \mathbf{Z}_2)$
α_4: (H_2, H_1) with probability $P(\alpha_4 \mid \mathbf{Z}_2)$

For any particular sequence α_j, since the classifications of the patterns are known, learning takes place in a supervised manner. Once the density functions $p(\boldsymbol{\theta} \mid \mathbf{Z}_N, \alpha_j)$ are determined for each j, Eq. (9.2.37) can be used to average them over the probability $P(\alpha_j \mid \mathbf{Z}_N)$, which is the probability that sequence α_j occurred, given the data set $\mathbf{Z}_N$, to obtain the density $p(\boldsymbol{\theta} \mid \mathbf{Z}_N)$. This average density is used in Eq. (9.2.33) to obtain $p(\mathbf{x} \mid \mathbf{Z}_N, H_j), j = 1, 2$ for computing the likelihood ratio.

Even though this solution appears elegant, we encounter difficulties in implementing it in practice. As the number of samples in the training set increases, that is, N increases, the total number of sequences increases exponentially. Therefore, the computational and storage requirements "explode" for large N. In the sequel we discuss two suboptimal schemes which overcome this problem.

If $\boldsymbol{\theta}$ is a nonrandom parameter, as in the case of supervised learning, we can obtain a maximum likelihood estimate and use this as the true value in designing the classifier. The classification rule now becomes

$$\Lambda(\mathbf{x}) = \frac{p(\mathbf{x} \mid H_1, \boldsymbol{\theta}_1) \mid_{\boldsymbol{\theta}_1 = \hat{\boldsymbol{\theta}}_{1\text{ml}}}}{p(\mathbf{x} \mid H_2, \boldsymbol{\theta}_2) \mid_{\boldsymbol{\theta}_2 = \hat{\boldsymbol{\theta}}_{2\text{ml}}}} \tag{9.2.38}$$

To determine the ML estimates, we note from Eq. (9.2.29) that

$$p(\mathbf{x} \mid \boldsymbol{\theta}) = \sum_{j=1}^{2} p(\mathbf{x} \mid H_j, \boldsymbol{\theta}_j) P_j \tag{9.2.39}$$

Assuming independence of samples, we have

$$p(\mathbf{Z}_N \mid \boldsymbol{\theta}_j) = \prod_{k=1}^{N} p(\mathbf{x}_k \mid \boldsymbol{\theta})$$

Taking logarithms and setting the gradient with respect to $\boldsymbol{\theta}_j$ equal to zero yields

$$\sum_{k=1}^{N} \frac{1}{p(\mathbf{x}_k \mid \boldsymbol{\theta})} \nabla_{\boldsymbol{\theta}_j} \left[\sum_{j=1}^{2} p(\mathbf{x}_k \mid H_j, \boldsymbol{\theta}_j) P_j \right] \Bigg|_{\hat{\boldsymbol{\theta}}_{j\text{ml}}} = 0 \tag{9.2.40}$$

where we have used Eq. (9.2.39).

By assuming that $\boldsymbol{\theta}_1$ and $\boldsymbol{\theta}_2$ are independent, and noting that $P(H_j) = P_j$ can be written as

$$P(H_j) = \frac{P(H_j \mid \mathbf{x}_k, \boldsymbol{\theta}_j) p(\mathbf{x}_k \mid \boldsymbol{\theta})}{p(\mathbf{x}_k \mid H_j, \boldsymbol{\theta}_j)} \qquad j = 1, 2, \tag{9.2.41}$$

we can write Eq. (9.2.40) as

$$\sum_{k=1}^{N} P(H_j \mid \mathbf{x}_k, \hat{\boldsymbol{\theta}}_{j\text{ml}}) \nabla_{\boldsymbol{\theta}_j} \log p(\mathbf{x}_k \mid H_j, \hat{\boldsymbol{\theta}}_{j\text{ml}}) = 0 \tag{9.2.42}$$

These again constitute a set of nonlinear equations for obtaining the ML estimates of $\boldsymbol{\theta}_1$ and $\boldsymbol{\theta}_2$.

The exploding memory problem encountered in the determination of $p(\theta \mid \mathbf{Z}_N)$ arises because the number of sequences in Eq. (9.2.37) over which one needs to average increases exponentially as the number of samples in the training set increases. One way of reducing the computations is to limit the summation in Eq. (9.2.42) to a fixed number of sequences, preferably to a single sequence. As can be expected, such a technique is necessarily suboptimal in most cases.

Such an approach is the *decision-directed* technique [19], in which when a sample (pattern) is presented, the classifier makes a decision as to which class the sample belongs. The decision can be made using any of the criteria discussed in Chapter 3. For example, we could use the maximum a posteriori criterion. This decision by the classifier is then accepted as the true class for the sample for purposes of learning. The learning sequence is thereby partitioned into two sets, $\mathbf{Z}_{N_1}$ and $\mathbf{Z}_{N_2}$, with known classification. Learning can then proceed in the normal supervised mode and in a sequential manner if desired. Although the method is not too complicated to implement, it leads to asymptotically biased results.

An alternative approach to unsupervised classification is to determine a self-consistent procedure to *cluster*, or group together, a set of unlabeled training vectors, into different classes. Starting with an initial labeling of the vectors into the two classes, a supervised classification procedure is used to determine the corresponding discriminant functions and decision regions. If the density functions contain unknown parameters, these are estimated from the training vectors by assuming that the initial classifications represent the true labels. The training vectors are reclassified using the discriminant functions and decision regions so determined. The procedure is repeated until there is no change in the class of the training vectors. Since these approaches are outside the scope of this book, we will not consider them in detail. We will, however, illustrate their use by the following example.

Suppose that we are given a set of unlabeled patterns drawn from two Gaussian densities with unknown means but known covariances. We do an initial assignment of class labels (clustering) of the training vectors and use as an estimate of the mean under each class, the sample means, μ_i, of the vectors assigned to that class. The decision rule follows from Eq. (9.2.2) by replacing $\mathbf{m}_i$ by μ_i. Reclassify each training vector $\mathbf{x}$ by computing $(\mathbf{x} - \mu_i)^T \mathbf{V}_i^{-1}(\mathbf{x} - \mu_i)$, $i = 1, 2$ and assigning $\mathbf{x}$ to the class for which this quantity is larger. That is, we are forming new clusters by assigning $\mathbf{x}$ to the class whose mean vector μ_i is closest to $\mathbf{x}$. The sample means are referred to as *cluster centers*. Update the cluster centers by recomputing the sample means. Stop when the cluster centers do not change.

9.2.5 Nonparametric Approaches

The approaches to pattern classification discussed so far assume that the class conditional densities, or at least their functional forms, except for a few parameters, are known. Often we cannot assume a parametric form for the density functions and they must be directly estimated locally based on a small number of samples. As such, the estimates are far less reliable than the corresponding parametric forms [11], particularly for multivariate random variables. However, our interest is in the use of these estimates for pattern classification and not in the determination of the density functions themselves. We discuss two nonparametric techniques, the *Parzen estimate* and the *k-nearest neighbor* estimate.

Parzen density estimate. The basic idea in obtaining nonparametric estimates of density functions is that the probability mass in a small local region $L(\mathbf{x})$ around a point $\mathbf{x}$ can be approximated as $p(\mathbf{x})V$, where V is the volume of $L(\mathbf{x})$. By drawing a large number, N, of n-dimensional samples $\mathbf{X}_i, i = 1, 2, \ldots, N$ from $p(\mathbf{X})$ and counting the number of samples $k(\mathbf{x})$ that fall in the region $L(\mathbf{x})$, we can obtain an estimate for the density function as

$$\hat{p}(\mathbf{x})V = \frac{k(\mathbf{x})}{N} \quad \text{or} \quad \hat{p}(\mathbf{x}) = \frac{k(\mathbf{x})}{NV} \tag{9.2.43}$$

We can interpret this result differently as follows. Let us define a *kernel function* $\phi(\mathbf{x} - \mathbf{X}_i)$ around each sample $\mathbf{X}_i$ as

$$\phi(\mathbf{x} - \mathbf{X}_i) = \begin{cases} \dfrac{1}{V} & \mathbf{x} \in L(\mathbf{X}_i) \\ 0 & \text{otherwise} \end{cases} \tag{9.2.44}$$

Then, clearly, we have

$$\sum_{i=1}^{N} \phi(\mathbf{x} - \mathbf{X}_i) = \frac{k(\mathbf{x})}{V} \tag{9.2.45}$$

so that we can write

$$\hat{p}(\mathbf{x}) = \frac{1}{N}\sum_{i=1}^{N} \phi(\mathbf{x} - \mathbf{X}_i) \tag{9.2.46}$$

The Parzen density estimate represents a generalization of this concept wherein the kernel function is chosen to be of a more general form as long as it satisfies $\int \phi(\mathbf{x})\, d\mathbf{x} = 1$.

Because of practical reasons, particularly in the case of multivariate densities, the kernel function is usually restricted to be either uniform or Gaussian. The Gaussian kernel is given by

$$\phi(\mathbf{x}) = \frac{1}{(2\pi|\Sigma|)^{n/2}} \exp(-\mathbf{x}^T \Sigma^{-1} \mathbf{x}) \tag{9.2.47}$$

For the uniform kernel, let us define a neighborhood $L(\mathbf{x})$ of $\mathbf{x}$ as the set of vectors $\mathbf{X}$ satisfying

$$L(\mathbf{x}) = \{\mathbf{X}: d(\mathbf{x}, \mathbf{X}) \leq \rho\sqrt{n + 2}\} \tag{9.2.48}$$

where

$$d^2(\mathbf{x}, \mathbf{X}) = (\mathbf{x} - \mathbf{X})^T \mathbf{A}^{-1} (\mathbf{x} - \mathbf{X}) \tag{9.2.49}$$

and ρ is a scalar. Thus $L(\mathbf{x})$ defines a hyperellipsoidal neighborhood around the vector $\mathbf{x}$ whose shape and orientation is determined by the matrix $\mathbf{A}$ and whose size (volume) is determined by ρ.

Often, for convenience, $L(\mathbf{x})$ is defined as $L(\mathbf{x}) = \{\mathbf{X}, |x_i - X_i| \leq d_i\}$ where d_i are constants. The uniform kernel is given by

$$\phi(\mathbf{x}) = \begin{cases} 1 & \mathbf{X} \in L(\mathbf{x}) \\ V & \text{otherwise} \end{cases} \tag{9.2.50}$$

where $V = \int_{L(\mathbf{x})} d\mathbf{x}$. Typically, V is fixed and the number of vectors k falling inside the neighborhood $L(\mathbf{X})$ is treated as a random variable.

Approximate expressions for the mean and variance of the density estimates using the two kernels can be derived [11]. For both the Gaussian and uniform kernels, we have

$$\mathrm{E}\{\hat{p}(\mathbf{x})\} = p(\mathbf{x}) \tag{9.2.51}$$

so that the estimates are approximately unbiased. The variance of the estimate for the Gaussian kernel is

$$\operatorname{var} \hat{p}(\mathbf{x}) = \frac{1}{N}\left[\frac{1}{(4\pi)^{n/2}|\Sigma|^{1/2}} p(\mathbf{x}) - p^2(\mathbf{x})\right] \tag{9.2.52}$$

For the uniform kernel it is given by

$$\operatorname{var} \hat{p}(\mathbf{x}) = \frac{p^2(\mathbf{x})}{k} \tag{9.2.53}$$

where it is assumed that $\hat{p}(\mathbf{x}) = k/NV$ and that $N \gg k$.

***k* Nearest neighbor (kNN) density estimate.** The kNN estimate of a density function $p(\mathbf{x})$ is defined as

$$\hat{p}(\mathbf{x}) = \frac{k-1}{NV(\mathbf{x})} \tag{9.2.54}$$

where $V(\mathbf{x})$ denotes the volume of the neighborhood $L(\mathbf{x})$, which contains the k samples $\mathbf{X}_i$ which are closest to $\mathbf{x}$ [20].

We recall that in defining the Parzen estimate with a uniform kernel, we assumed V to be fixed and let k be a random variable. The kNN estimate can be interpreted as being obtained from the Parzen estimate when we choose k to be fixed and let V be a random variable. That is, we extend the local neighborhood around $\mathbf{x}$ until the region includes the k nearest vectors. Thus both $L(\mathbf{x})$ and $V(\mathbf{x})$ are random variables and are functions of $\mathbf{x}$.

In a neighborhood where $p(\mathbf{x})$ is large, we might expect that the number, k, of samples $\mathbf{x}_i$ will be large and that correspondingly, $V(\mathbf{x})$ will be small. As can be seen from Eq. (9.2.54), $\hat{p}(\mathbf{x})$ will also be large. Thus the kNN estimate is intuitively appealing.

It can again be shown that the kNN estimate is approximately unbiased with variance given by

$$\operatorname{var} \hat{p}(\mathbf{x}) = \frac{p^2(\mathbf{x})}{k-2} \tag{9.2.55}$$

Nonparametric supervised pattern classification. To find the nonparametric classifiers corresponding to Eq. (9.2.18), given the set of training vectors $\{\mathbf{X}_i, i = 1, 2, \ldots, N_1\}$ belonging to class H_1 and $\{\mathbf{X}_i, i = 1, 2, \ldots, N_2\}$ belonging to class H_2, we replace the density functions $p(\mathbf{x} \mid \mathbf{X}_1^i, \mathbf{X}_2^i, \ldots, \mathbf{X}_{N_i}^i), i = 1, 2$ by the corresponding nonparametric estimates. Thus with the Parzen estimate, the conditional likelihood ratio becomes

$$\Lambda(\mathbf{x} \mid \mathbf{Z}_{N_1}, \mathbf{Z}_{N_2}) = \ln \frac{\dfrac{1}{N_1} \sum_{j=1}^{N_1} \phi_1(\mathbf{x} - \mathbf{X}_j^1)}{\dfrac{1}{N_2} \sum_{j=1}^{N_2} \phi_2(\mathbf{x} - \mathbf{X}_j^2)} \tag{9.2.56}$$

On the other hand, if we use the kNN estimate, the likelihood function becomes

$$\Lambda(\mathbf{x} \mid \mathbf{Z}_{N_1}, \mathbf{Z}_{N_2}) = \ln \frac{(k_1 - 1) N_2 V_2(\mathbf{x})}{(k_2 - 1) N_1 V_1(\mathbf{x})} \tag{9.2.57}$$

A discussion of the error performance of these tests can be found in [11].

We conclude our consideration of pattern recognition with a brief discussion of the important problem of feature selection.

9.2.6 Feature Selection

In our discussion of the pattern classification problem, we have assumed that we can associate with each sample a vector of observables (features). The classification is then determined by this feature vector. If the number of features is large, the computational requirements for correct classification become significant. It becomes necessary to select a fewer number of features from a given set to reduce the computational burden and still be able to classify the patterns satisfactorily. Alternatively, we may seek to transform the feature vector to one of lower dimension. Once we find a reasonable solution to this problem, we need to consider only a feature vector of lower dimension. This problem is therefore referred to as dimensionality reduction. In this section we present an approach to feature selection that is based on a generalized form of the Karhunen–Loéve transformation.

Before considering the problem of dimensionality reduction for pattern classification, let us consider the problem of choosing a vector of reduced dimension, $\mathbf{x}(n)$, to adequately represent a vector $\mathbf{x}(r)$ of dimension $r > n$. If the elements of $\mathbf{x}(r)$ were uncorrelated, one way of choosing the reduced vector would be to retain those components of $\mathbf{x}(r)$ that have the largest variance and discard the others. In general, the elements of $\mathbf{x}(r)$ will be correlated. We may therefore seek to expand $\mathbf{x}(r)$ in a set of orthonormal basis vectors $\mathbf{g}_1, \mathbf{g}_2, \ldots, \mathbf{g}_r$ such that

$$\mathbf{x}(r) = \sum_{i=1}^{r} y_i \mathbf{g}_i \tag{9.2.58}$$

where

$$\mathbf{g}_i^T \mathbf{g}_j = \delta_{ij} \tag{9.2.59}$$

We wish to retain a subset $\{y_1, y_2, \ldots, y_n\}$ of the coefficients of the expansion and still be able to represent $\mathbf{x}(r)$. Let us define $\hat{\mathbf{x}}(r)$ as

$$\hat{\mathbf{x}}(r) = \sum_{i=1}^{n} y_i \, \mathbf{g}_i \tag{9.2.60}$$

We seek the set of orthonormal basis vectors $\{\mathbf{g}_i\}$ to minimize the mean-square error in representing $\mathbf{x}(r)$, $\mathrm{E}\{\|\mathbf{x}(r) - \hat{\mathbf{x}}(r)\|^2\}$. It is clear from Eqs. (9.2.58) and (9.2.59) that

$$\begin{aligned} \mathrm{E}\{\|\mathbf{x}(r) - \hat{\mathbf{x}}(r)\|^2\} &= \mathrm{E}\left\{\left\| \sum_{i=n+1}^{r} y_i \mathbf{g}_i \right\|^2\right\} \\ &= \sum_{i=n+1}^{r} \mathrm{E}\{y_i^2\} \end{aligned} \tag{9.2.61}$$

Since

$$y_i = \mathbf{g}_i^T \, \mathbf{x}(r) \tag{9.2.62}$$

we can rewrite Eq. (9.2.61) as

$$\mathrm{E}\{\|\mathbf{x}(r) - \hat{\mathbf{x}}(r)\|^2\} = \sum_{i=n+1}^{r} \mathrm{E}\{\mathbf{g}_i^T \, V_{\mathbf{x}} \, \mathbf{g}_i\} \tag{9.2.63}$$

where $\mathbf{V}_{\mathbf{x}}$ is the covariance matrix of the vector $\mathbf{x}(r)$.

The problem therefore reduces to one of minimizing the right-hand side of Eq. (9.2.63) subject to the constraint

$$\mathbf{g}_i^T \, \mathbf{g}_i = 1 \tag{9.2.64}$$

We solve the problem by forming the augmented cost function

$$J = \sum_{i=n+1}^{r} [\mathrm{E}\{\mathbf{g}_i^T \, \mathbf{V}_{\mathbf{x}} \, \mathbf{g}_i\} - \lambda_i \, \mathbf{g}_i^T \, \mathbf{g}_i] \tag{9.2.65}$$

where the $\{\lambda_i\}$ are a set of Lagrange multipliers. The optimum $\mathbf{g}_i$ obtained by setting the gradient vector $\nabla_{\mathbf{g}_i} J$ equal to zero is

$$\mathbf{V}_{\mathbf{x}} \, \mathbf{g}_i = \lambda_i \, \mathbf{g}_i \tag{9.2.66}$$

Thus the constants λ_i are the eigenvalues of the covariance matrix of $\mathbf{x}(r)$ and the vectors $\mathbf{g}_i$ are the associated eigenvectors. The expansion in Eq. (9.2.58), therefore, is the discrete version of the Karhunen–Loéve expansion which we have encountered earlier.

The minimum mean-square error is obtained from Eq. (9.2.58) as

$$\mathrm{E}\{\|\mathbf{x}(r) - \hat{\mathbf{x}}(r)\|^2\}_{\min} = \sum_{i=n+1}^{r} \lambda_i \tag{9.2.67}$$

It therefore follows that the optimum technique (in the mean-square sense) to represent a vector $\mathbf{x}(r)$ of dimension r in terms of a fewer number n of components, is to order the eigenvalues of the covariance matrix in descending order of magnitude. We then expand $\mathbf{x}(r)$ in a Karhunen–Loéve expansion and retain only the first n coefficients. If we choose an $n \times r$ matrix $\mathbf{T}$ such that its rows are the first n eigenvectors of $\mathbf{V_x}$, the reduced vector $\mathbf{x}(n)$ is given by

$$\mathbf{x}(n) = \mathbf{T}\mathbf{x}(r)$$

Let us now extend the preceding discussion to the problem of dimensionality reduction in a pattern classification problem. We restrict our discussion to the binary classification problem, in which we are required to choose between two pattern classes. Given the pattern vector $\mathbf{x}(r)$ of dimension r, we seek to obtain a vector $\mathbf{x}(n)$ of dimension n such that we are still able to discriminate between the two pattern classes.

Let $\mathbf{V}_1$ and $\mathbf{V}_2$ be the covariance matrices of the vectors belonging to the classes H_1 and H_2, respectively. Let P_1 and P_2 be the probabilities of occurrence of the two classes. We define a generalized covariance matrix $\mathbf{V}$ as

$$\mathbf{V} \underline{\Delta} P_1\mathbf{V}_1 + P_2\mathbf{V}_2 \tag{9.2.68}$$

Let $\mathbf{D}$ be a matrix of eigenvectors of $\mathbf{V}$ such that

$$\mathbf{D}^T\mathbf{V}\mathbf{D} = \mathbf{I} \tag{9.2.69}$$

where $\mathbf{I}$ is the identity matrix.

If we define the matrices $\mathbf{V}^i, i = 1, 2$ as

$$\mathbf{V}^i = \mathbf{D}^T(P_i\mathbf{V}_i)\mathbf{D} \tag{9.2.70}$$

we can write Eq. (9.2.69) as

$$\mathbf{V}^1 + \mathbf{V}^2 = \mathbf{I} \tag{9.2.71}$$

Let λ_j^i and $\mathbf{g}_j^i$ represent the eigenvalues and eigenvectors of $\mathbf{V}^i$ for $i = 1, 2$:

$$\mathbf{V}^i\mathbf{g}_j^i = \lambda_j^i\mathbf{g}_j^i \qquad j = 1, 2, \ldots, r \tag{9.2.72}$$

It can easily be established that

$$0 \leq \lambda_j^i \leq 1 \qquad \text{for} \quad i = 1, 2; j = 1, 2, \ldots, r \tag{9.2.73}$$

From Eq. (9.2.71) we can therefore write

$$\mathbf{V}^2\mathbf{g}_j^1 = (\mathbf{I} - \mathbf{V}^1)\mathbf{g}_j^1 = (1 - \lambda_j^1)\mathbf{g}_j^1 \tag{9.2.74}$$

Equation (9.2.74) shows that if $\mathbf{g}_j^1$ is an eigenvector of $\mathbf{V}^1$ associated with the eigenvalue λ_j^1, it is also an eigenvector of $\mathbf{V}^2$ associated with the eigenvalue $(1 - \lambda_j^1)$. If we now arrange the eigenvalues of $\mathbf{V}^1$ in descending order, the eigenvalues of $\mathbf{V}^2$ will then be in increasing order. That is, the most important features of class H_1 are the least important ones of class H_2, and vice versa. We therefore choose a $n \times r$ matrix $\mathbf{T}_1$ such that its first n_1 rows are the first n_1 eigenvectors of $\mathbf{V}^1$ and the remaining

$n_2 = (n - n_1)$ rows are the eigenvectors of $\mathbf{V}^2$. We can then choose the vector $\mathbf{x}(n)$ as

$$\mathbf{x}(n) = \mathbf{T}\mathbf{x}(r) \tag{9.2.75}$$

where

$$\mathbf{T} = \mathbf{T}_1 \mathbf{D} \tag{9.2.76}$$

The procedure thus involves expanding the vector $\mathbf{x}(r)$ in the eigenvectors of the generalized covariance matrix $\mathbf{V}$ and retaining only the first n_1 terms (corresponding to the n_1 most significant features of the class H_1) and the last n_2 terms (corresponding to the n_2 most significant features of class H_2). One way of choosing n_1 and n_2 is to choose them as integers closest to $P_1 n$ and $P_2 n$, respectively.

9.2.7 Section Summary

In this section we have attempted to apply the techniques of estimation and detection to the problem of classifying patterns. In Section 9.2.1 we discussed the basic components of the pattern recognition problem. The several steps involved in the design of a pattern recognition system are feature selection, feature ordering and extraction, and pattern classification. Our attention was mostly concentrated on the classification problem. We assume that in every case the functional forms of the underlying density functions are known. The case when the densities are completely known was considered in Section 9.2.2. The techniques of hypothesis testing discussed in Chapter 3 are directly applicable to this problem and the results can be derived in a straightforward manner.

If the class-conditional densities are known only to within certain parameters, these parameters must be learned before we can classify the patterns. The problem is fairly straightforward if we are given sample (test) patterns whose classifications are known. This results in the supervised learning problem discussed in Section 9.2.3. The learning of the parameters is facilitated if, as discussed in Chapter 5, we assume a reproducing form for the prior densities.

When the classification of the test patterns is not known, the problem of learning becomes complicated. The unknown parameters of the class conditional densities must be estimated using these samples of unknown classification, that is, in a nonsupervised manner. This problem is considered in Section 9.2.4, where it is shown that the optimal (Bayes) solution leads to exploding computational requirements. The decision-directed approach to unsupervised classification represents a suboptimal solution that overcomes this problem. In Section 9.2.5 we discussed some nonparametric density estimation techniques as applied to pattern recognition problems. The section concluded with a technique for feature ordering and selection in Section 9.2.6. The technique uses the discrete Karhunan–Loéve transform in selecting the features.

The techniques in this chapter can easily be modified to obtain sequential algorithms for pattern classification. These follow along the lines of our discussion in Chapter 3. Details can be found in [6].

9.3 APPLICATION TO SYSTEM IDENTIFICATION

The design of any system requires a knowledge of its dynamic characteristics. While a simple system may be built with only an intuitive feeling about its behavior, its proper operation depends on a choice of parameters suitable to the characteristics of the particular system. These characteristics must usually be determined by measurements on the system inputs and outputs. In this section we consider the application of the techniques discussed earlier to the identification of systems with random inputs.

A system can be considered to be a transformation of an input into an output. System identification then implies the determination of this transformation based on measurements of the input and the output. The first step is to construct a mathematical model within a specified class of models which describes the physical transformation occurring in the system. To make this meaningful, we select an error criterion and choose the model within the specified class which minimizes the criterion [21]. While several criteria can be used, as in earlier chapters, we will for the most part restrict ourselves to the minimum mean-square error criterion.

Mathematical models can be linear or nonlinear, continuous or discrete-time, deterministic or stochastic, with finite or infinite memory, and so on. The choice of the model used in a particular case depends on the available data, the purpose to which the model is to be put, and the complexity of the algorithms involved. Quite often, the model has little connection with the actual system. For example, a linear time-invariant model is often a sufficiently accurate representation of a nonlinear time-varying system. Once the model is chosen, it can be characterized in terms of suitable descriptions, such as a differential equation, transfer function, impulse response, or power series expansion. The characterization may be parametric or nonparametric. Parametric characterizations assume a functional form (for example, a differential equation) for the input–output relation, whereas nonparametric characterizations (for example, impulse responses) do not. The former are easier to determine since all that is required is the estimation of certain parameter values from measured data, so that the performance criterion is minimized.

The choice of the mathematical model is neither simple nor straightforward in a general situation. Questions such as what models to select or indeed what models are available or whether the solution to the identification problem is unique are not easy to answer. However, for linear and time-invariant systems these problems are fairly easily answered. The characterization is well established and the order of the system determines the maximum number of parameters to be identified. From a practical point of view, the assumption of a linear time-invariant model for the system considerably reduces the complexity of the identification problem. For such an assumption to be valid, the system parameters must vary slowly in comparison to the signals. The linearity assumption is justified if the system is close to normal operating conditions and we are interested in determining changes away from normal.

In this section we are concerned primarily with the identification of linear time-invariant systems on the basis of normal operating records. Since our interest is in discussing algorithms that can be processed in digital hardware, we use discrete-time models exclusively in this section. We therefore assume either a difference equation or a transfer function characterization of the system. Thus the characterization is parametric, so that the problem essentially reduces to one of identifying certain parameters in the system. We could therefore use the parameter estimation techniques discussed in Chapter 5. However, in most applications, the dependence of the output signal on the parameters is only indirect and the parameter estimates cannot be obtained in a straightforward manner.

We can formulate the system identification problem in several ways, depending on which signals are available and how they are characterized. While in some situations we are able to measure both the input and output to the system, in many others we may be able to measure only the output. The observed signals may be characterized as being either deterministic or statistical and may be corrupted by noise. In those cases where we are unable to access the input, we are basically looking for a description of the properties of the output signal. For example, given a stochastic signal, we can describe the statistical properties of the signal by assuming that it is the output of a linear system driven by white noise. The signal is then described in terms of the system parameters and the statistical properties of the input.

While attempts have been made to present the various techniques for system identification in a unified framework, the methods available in the literature represent, for the most part, a collection of individual techniques. We will therefore content ourselves with presenting a few basic algorithms and not attempt to discuss all available methods. We refer the reader to the recent books on identification for a discussion of these techniques [22,23].

A problem closely related to the identification of stochastic models is that of estimating the power spectrum of a stationary, stochastic process. In many of our discussions in previous chapters, we have assumed that the statistics of the underlying distributions are known, at least to second order. In a practical situation, these statistics, if not assumed a priori, will have to be determined on the basis of records of typical data. We have discussed earlier estimation of the mean and variance of a random variable on the basis of observed values. More generally, given sample values of a stationary process, we may be interested in estimating the autocorrelation function or power spectrum of the process. If the underlying process is continuous time, the sample values must be obtained by sampling the signal. If the signal is bandlimited, the sampling frequency must be chosen to be at least twice as large as the bandwidth of the signal (the Nyquist rate). If the signal is not bandlimited, to prevent aliasing, frequently the signal is lowpass filtered before sampling.

There are two approaches that can be used in estimating the correlation function $\phi_x(k)$. In a nonparametric approach, we try to obtain a numerical estimate for each of the M numbers $\phi_x(0), \phi_x(1), \ldots, \phi_x(M-1)$ for some choice of M. In the parametric approach, we might assume a functional form for the autocorrelation function and determine certain associated (unknown) parameters. For example, we may assume that

$$\phi_x(k) = \phi_x(0)\alpha^{|k|} \qquad |\alpha| < 1$$

and estimate the parameters α and $\phi_x(0)$. We note that the signal $x(k)$ can be considered to be the output of a first-order all-pole filter driven by white noise. The problem of spectral estimation then reduces to one of estimating the parameters of the corresponding system model. Thus the approaches discussed in this section can also be used for spectral estimation. The reader is referred to [25–28] for a discussion of power spectrum estimation by various methods.

9.3.1 Identification of ARMA Models

We recall from our discussions in Chapter 2 that a stationary stochastic process can be considered to be the output of a linear time-invariant system driven by white noise. For example, for the process with autocorrelation sequence

$$\phi_x(k) = \phi_x(0)\alpha^{|k|} \tag{9.3.1}$$

the power spectrum in the Z-domain is given by

$$\Phi_x(z) = \frac{\phi_x(0)(1 - \alpha^2)}{(z - \alpha)(z^{-1} - \alpha)} \tag{9.3.2}$$

If we let $\phi_x(0)(1 - \alpha^2) = G^2$, we can write the preceding as

$$\Phi_x(z) = \frac{G}{z - \alpha} \frac{G}{z^{-1} - \alpha} \tag{9.3.3}$$

It then follows that $\Phi_x(z)$ can be considered to be the spectrum of the output of a linear discrete-time system with transfer function

$$H(z) = \frac{G}{z - \alpha} \tag{9.3.4}$$

driven by a zero-mean white sequence $w(k)$ with unity variance. We can therefore write

$$X(z) = \frac{G}{z - \alpha} W(z) \tag{9.3.5}$$

where $X(z)$ and $W(z)$ represent the transforms of $x(k)$ and $w(k)$.

The corresponding difference equation describing the process $x(k)$ is

$$x(k + 1) - \alpha x(k) = Gw(k) \tag{9.3.6}$$

A process defined by an equation such as (9.3.6) is said to be an *autoregressive* (AR) process. The spectrum estimation problem thus reduces to one of identification of the parameters of the autoregressive model of the process. In the sequel, we assume that $x(\cdot)$ is zero mean.

The process described by Eq. (9.3.6) is a first-order process. We also note from the system transfer function Eq. (9.3.4) that $H(z)$ has no finite zeros. We can generalize the model by assuming that $H(z)$ is a general transfer function of the form

$$H(z) = \frac{G(1 + \beta_1' z^{-1} + \cdots + \beta_M' z^{-M})}{1 + \alpha_1 z^{-1} + \cdots + \alpha_N z^{-N}} \tag{9.3.7}$$

so that the system has M zeros and N poles.

Let

$$G\beta_i' = \beta_i \qquad \text{for } i = 1, \ldots, M \tag{9.3.8}$$

Then

$$H(z) = \frac{G + \sum_{i=1}^{M} \beta_i z^{-i}}{1 + \sum_{i=1}^{N} \alpha_i z^{-i}} \tag{9.3.9}$$

The output $X(z)$ is given by

$$X(z) = H(z)W(z) = \frac{G + \sum_{i=1}^{M} \beta_i z^{-i}}{1 + \sum_{i=1}^{N} \alpha_i z^{-i}} W(z) \tag{9.3.10}$$

The corresponding difference equation can be written as

$$x(k) + \sum_{i=1}^{N} \alpha_i x(k-i) = Gw(k) + \sum_{i=1}^{M} \beta_i w(k-i) \tag{9.3.11}$$

This equation represents an *autoregressive-moving average* (*ARMA*) process. We note that if $\beta_i = 0, i = 1, \ldots, M$, we obtain the autoregressive process defined earlier. If $\alpha_i = 0, i = 1, \ldots, N$, then $x(k)$ is a *moving average* (*MA*) process. In the former case $H(z)$ will be an all-pole function with no finite zeros. For the moving average process, $H(z)$ represents an all-zero system.

We now discuss several techniques for the identification of the parameters of the ARMA process defined previously. We first discuss the identification of autoregressive parameters by observing $x(k)$. Later we extend our results to ARMA models.

Autoregressive parameter identification. Let us for the moment assume that $\beta_i = 0$, so that $x(k)$ is an autoregressive process. For this case Eq. (9.3.11) becomes

$$x(k) + \sum_{i=1}^{N} \alpha_i x(k-i) = Gw(k) \tag{9.3.12}$$

Multiplying both sides of Eq. (9.3.12) by $x(k-j)$ and taking expectations yields

$$\mathrm{E}\left\{\left[x(k) + \sum_{i=1}^{N} \alpha_i x(k-i)\right] x(k-j)\right\} = \mathrm{E}\{Gw(k)x(k-j)\} \tag{9.3.13}$$

Now

$$\begin{aligned} \mathrm{E}\{Gw(k)x(k-j)\} &= \mathrm{E}\left\{Gw(k)\left[Gw(k-j) - \sum_{i=1}^{N} \alpha_i x(k-j-i)\right]\right\} \\ &= \begin{cases} G^2 & j = 0 \\ 0 & j \neq 0 \end{cases} \end{aligned} \tag{9.3.14}$$

where the last step follows because $w(k)$ is a white sequence. By noting that

$$E\{x(k-i)x(k-j)\} = E\{x(j)x(i)\} = \phi_x(j-i) \tag{9.3.15}$$

and combining Eqs. (9.3.13) and (9.3.14), we have

$$\phi_x(0) + \sum_{i=1}^{N} \alpha_i \phi_x(i) = G^2 \tag{9.3.16a}$$

$$\phi_x(j) + \sum_{i=1}^{N} \alpha_i \phi_x(j-i) = 0 \qquad j \neq 0 \tag{9.3.16b}$$

The identification problem can now be stated: Given the autocorrelation coefficients $\phi_x(j)$, find the parameter estimates $\hat{\alpha}_i, i = 1, 2, \ldots, N$. These parameters can be determined by using the first N equations in Eq. (9.3.16b). Once the $\hat{\alpha}_i$ have been determined, we can use Eq. (9.3.16a) to find G.

We can rearrange these equations as

$$\sum_{i=1}^{N} \hat{\alpha}_i \phi_x(j-i) = -\phi_x(j) \qquad j = 1, \ldots, N \tag{9.3.17}$$

Quite often an alternative formulation of the identification problem is stated as follows. We define a *residual* or *error* as

$$\epsilon(k) = x(k) + \sum_{i=1}^{N} \hat{\alpha}_i x(k-i) \tag{9.3.18}$$

and seek the values of $\hat{\alpha}_i$ such that the mean-square error $E\{\epsilon^2(k)\}$ is minimized. It can then be easily verified that $\hat{\alpha}_i$ will satisfy Eq. (9.3.17).

We can write the set of equations (9.3.17) in compact form by using vector-matrix notation. We then obtain the Yule–Walker equations

$$\begin{bmatrix} \phi_x(0) & \phi_x(1) & \cdots & \phi_x(N-1) \\ \phi_x(1) & \phi_x(0) & \cdots & \phi_x(N-2) \\ \phi_x(2) & & & \\ \vdots & \vdots & & \vdots \\ \phi_x(N-1) & & \cdots & \phi_x(0) \end{bmatrix} \begin{bmatrix} \hat{\alpha}_1 \\ \hat{\alpha}_2 \\ \\ \vdots \\ \hat{\alpha}_N \end{bmatrix} = - \begin{bmatrix} \phi_x(1) \\ \phi_x(2) \\ \\ \vdots \\ \phi_x(N) \end{bmatrix} \tag{9.3.19}$$

Comparison with Eq. (6.3.69) shows that Eq. (9.3.19) is the same as the forward-predictor equations that we discussed in Chapter 6, with $-\hat{\alpha}_i$ replacing h_i^m. Thus the parameters $\hat{\alpha}_i$ may be considered to be the coefficients of the filter which provides the best linear mean-square one-stage prediction of $x(k)$ based on the previous N values. As such, the model of Eq. (9.3.12) is also referred to as a *linear prediction model* for the signal $x(k)$.

Let us define the matrix $\boldsymbol{\phi}^N$ and the vectors $\hat{\boldsymbol{\alpha}}^N$ and $\boldsymbol{\rho}^N$ as

$$\boldsymbol{\phi}^N = \begin{bmatrix} \phi_x(0) & \cdots & \phi_x(N-1) \\ \vdots & & \vdots \\ \phi_x(N-1) & \cdots & \phi_x(0) \end{bmatrix} \quad \hat{\boldsymbol{\alpha}}^N = \begin{bmatrix} \hat{\alpha}_1 \\ \vdots \\ \hat{\alpha}_N \end{bmatrix} \quad \text{and} \quad \boldsymbol{\rho}^N = \begin{bmatrix} \phi_x(1) \\ \vdots \\ \phi_x(N) \end{bmatrix} \tag{9.3.20}$$

The solution to the set of Eqs. (9.3.19) can then be written as

$$\hat{\alpha}^N = -[\boldsymbol{\phi}^N]^{-1}\boldsymbol{\rho}^N \tag{9.3.21}$$

As noted in Chapter 6, the matrix $\boldsymbol{\phi}^N$ is a Toeplitz matrix and is hence positive definite and invertible. Furthermore, the estimates $\hat{\alpha}_i$ will form a stable set. That is, the polynomial

$$1 + \sum_{i=1}^{N} \hat{\alpha}_i z^{-i} = 0 \tag{9.3.22}$$

has all its zeros inside the unit circle in the Z-plane. This guarantees that the autoregressive model is stable [26].

The minimum value of J obtained by using this choice of $\hat{\alpha}_i$ easily follows as [see Eq. (6.3.75)]

$$J = \mathrm{E}\{\epsilon(k)x(k)\} = \phi_x(0) + \sum_{i=1}^{N} \hat{\alpha}_i \phi_x(i) \tag{9.3.23}$$

It only remains to determine the correlation coefficients $\phi_x(k)$ for $k = 1, 2, \ldots, N-1$. These coefficients may be estimated using any of the techniques discussed in [24–27]. Commonly used estimates are

$$\hat{R}_x(k) = \frac{1}{L - |k|} \sum_{n=0}^{L-|k|-1} x(n)x(n+k) \tag{9.3.24}$$

or

$$\hat{R}_x(k) = \frac{1}{L} \sum_{n=0}^{L-|k|-1} x(n)x(n+k) \tag{9.3.25}$$

where L is the length of the data sequence.

Example 9.3.1

We can relate our results on the estimation of parameters of the AR process to the ML estimates of Chapter 5. Suppose that we are given measurements $x(1), x(2), \ldots, x(L)$. We recall that the maximum likelihood estimate $\hat{\alpha}_{\mathrm{ml}}$ is obtained by solving the log likelihood equation

$$\left.\frac{\partial \ln p(x(1), x(2), \ldots, x(L) \mid \boldsymbol{\alpha})}{\partial \boldsymbol{\alpha}}\right|_{\boldsymbol{\alpha}=\hat{\boldsymbol{\alpha}}_{\mathrm{ml}}} = 0$$

Unfortunately, the determination of this density is not easy in most cases. We can obtain an approximate expression for the density in the case where $w(k)$ is a Gaussian sequence. To do so, let us denote the set $\{x(1), x(2), \ldots, x(R)\} = \mathbf{x}_R$. Then

$$\begin{aligned} p(x(1), x(2), \ldots, x(L) \mid \alpha) &= p(\mathbf{x}_L \mid \alpha, G) \\ &= p(x(N+1) \cdots x(L) \mid \mathbf{x}_N, \alpha) p(\mathbf{x}_N \mid \alpha) \end{aligned}$$

When L is large compared to N, the contribution from the term $p(x(1), x(2), \ldots, x(N) \mid \alpha)$ can be neglected. Since $w(\cdot)$ is zero-mean, Gaussian, white with unity covariance

$$p(w(N+1), w(N+2), \ldots, w(L)) = \frac{1}{(\sqrt{2\pi})^{L-N}} \exp\left(-\frac{1}{2}\sum_{k=N+1}^{L} w_k^2\right)$$

it follows from Eq. (9.3.12) that

$$p(x(N+1)x(N+2),\dots,x(L)\,|\,\boldsymbol{\alpha},G)$$
$$= \frac{1}{(\sqrt{2\pi}G)^{L-N}} \exp\left\{-\frac{1}{2G^2}\sum_{k=N+1}^{L}\left[x(k)+\sum_{i=1}^{N}\alpha_i x(k-i)\right]^2\right\}$$

The log-likelihood function is then approximately given by

$$\ln p(x(1),x(2),\dots,x(L)\,|\,\boldsymbol{\alpha},G) \simeq -(L-N)(\sqrt{2\pi}G)$$
$$-\frac{1}{2G^2}\sum_{k=N+1}^{L}\left[x(k)+\sum_{i=1}^{N}\alpha_i x(k-i)\right]^2$$

The ML estimates $\hat{\alpha}_{ml}$ are obtained by solving the equations

$$\sum_{k=N+1}^{L}\left[x(k)+\sum_{i=1}^{N}\hat{\alpha}_i x(k-i)\right]x(k-j)=0 \qquad j=1,2,\dots,N$$

or, equivalently,

$$\sum_{i=1}^{N}\hat{\alpha}_i\left[\sum_{k=N+1}^{L}x(k-i)x(k-j)\right]=-\sum_{k=N+1}^{L}x(k)x(k-j) \qquad j=1,2,\dots,N$$

It easily follows from Eq. (9.3.25) that

$$\sum_{k=N+1}^{L}x(k-i)x(k-j)=L\hat{R}_x(|j-i|)$$

so that

$$\sum_{i=1}^{N}\hat{\alpha}_i\hat{R}_x(|j-i|)=-\hat{R}_x(j)$$

which is the same as Eq. (9.3.17) with $\hat{R}_x(\cdot)$ replacing $\phi_x(\cdot)$.

Example 9.3.2

We consider the identification of the parameters of a third-order autoregressive model by using Eq. (9.3.21) based on 1024 points of the data sequence. The data were generated by simulating the process on a computer. The model was assumed to be

$$x(k)+0.9x(k-1)+0.36x(k-2)+0.054x(k-3)=w(k)$$

so that the actual parameters are

$$\alpha_1=0.9 \qquad \alpha_2=0.36 \qquad \alpha_3=0.054$$

The input $w(k)$ is a stationary, zero-mean, white sequence with unity variance. The four values of the autocorrelation coefficients needed to obtain the estimates were determined from Eq. (9.3.25) as

$$\hat{R}_x(0)=2.0307 \qquad \hat{R}_x(1)=-1.3893 \qquad \hat{R}_x(2)=0.6778 \qquad \hat{R}_x(3)=-0.3126$$

The estimates as obtained from Eq. (9.3.21) are

$$\hat{\alpha}_1=0.8932 \qquad \hat{\alpha}_2=0.3222 \qquad \hat{\alpha}_0=0.0815$$

There are some disadvantages to determining the estimate $\hat{\boldsymbol{\alpha}}^N$ from Eq. (9.3.21). First, the order N of the model must be known (or at least fixed) a priori. Also, to compute the correlation coefficients, the data sequence must be processed all at once. Hence the technique is a *batch-processing* technique. The computation of the correlation coefficients is time consuming. Finally, Eq. (9.3.21) involves inverting the matrix $\boldsymbol{\phi}^N$. The last can be overcome by using the U-L factorization method discussed in Chapter 6.

We can try to overcome the first one by computing the estimates $\hat{\boldsymbol{\alpha}}^N$ for various values of N and accepting as the true model order the one that yields the least value of J. This, however, requires that Eq. (9.3.21) be solved repeatedly for various N. We can avoid this problem by using the Levinson–Durbin recursion and the lattice structures that we encountered in Chapter 6. We recall from our discussions in that chapter that the Levinson–Durbin algorithm is an efficient method for computing the coefficients of successively higher-order filters starting with zeroth order, in terms of the partial correlation or reflection coefficients. The corresponding algorithm for the identification problem follows by noting the equivalence between the estimated model coefficients $\hat{\alpha}_i$ and the one-stage predictor coefficients $-h_i$.

9.3.2 Lattice Structure for AR Model Identification

Let us denote the model coefficients for the nth-order model by

$$\hat{\boldsymbol{\alpha}}^n = [\hat{\alpha}_1^n \cdots \hat{\alpha}_n^n]^T \tag{9.3.26}$$

and let

$$\hat{\boldsymbol{\alpha}}^{n+1} = [\hat{\alpha}_1^{n+1} \cdots \hat{\alpha}_{n+1}^{n+1}]^T \tag{9.3.27}$$

denote the coefficients for the model of order $n + 1$. Use of Eqs. (6.3.89) then yields the following recursion:

$$\hat{\alpha}_0^{n+1} = \hat{\alpha}_0^n = 1 \tag{9.3.28a}$$

$$\hat{\alpha}_i^{n+1} = \hat{\alpha}_i^n + p_{n+1}\hat{\alpha}_{n+1-i}^n \qquad i = 1, \ldots, n \tag{9.3.28b}$$

where

$$p_{n+1} = \hat{\alpha}_{n+1}^{n+1} \tag{9.3.28c}$$

The coefficients p_{n+1} are the partial correlation or reflection coefficients and can be computed as

$$p_{n+1} = -\frac{1}{J_n}\left(\sum_{j=1}^{n+1} \hat{\alpha}_{n-j+1}^n\right)\phi_x(j) \tag{9.3.28d}$$

where J_n is the minimum cost corresponding to the nth-order model as given by Eq. (9.3.23). J_n can also be computed using Eq. (6.3.89e) as

$$J_{n+1} = J_n(1 - p_{n+1}^2) \tag{9.3.28e}$$

with initial condition $J_0 = \phi_x(0)$.

The lattice structure for the AR identification problem follows from Fig. 6.3.6 directly and is shown in Fig. 9.3.1. The residual $\epsilon(k)$ in Eq. (9.3.18) will now correspond to the forward prediction error. Thus the output of the nth summer along the upper leg of the lattice is the residual corresponding to the nth-order model for the autoregressive process. We refer to this as the forward residual $\epsilon_f^n(k)$. Similarly, we refer to the output of the nth summer along the lower leg as the backward residual $\epsilon_b^n(k)$. From Fig. 9.3.1 it follows that

$$\epsilon_f^{n+1}(k) = \epsilon_f^n(k) + p_{n+1}\epsilon_b^n(k-1) \tag{9.3.29}$$

$$\epsilon_b^{n+1}(k) = \epsilon_b^n(k-1) + p_{n+1}\epsilon_f^n(k) \tag{9.3.30}$$

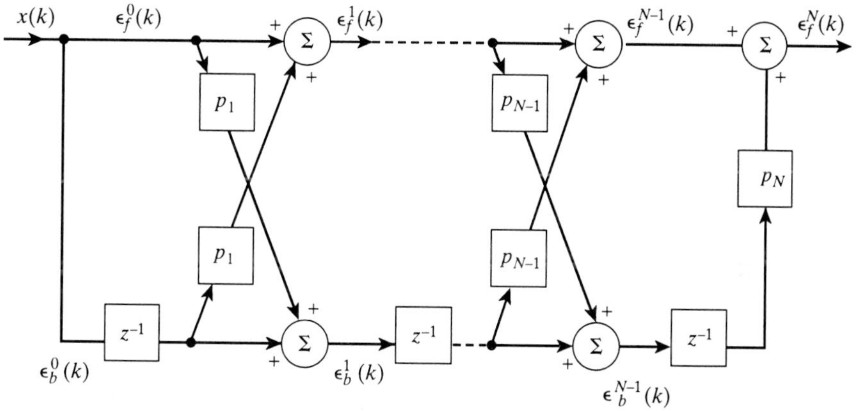

Figure 9.3.1 Lattice structure for identification of AR process.

We can compute p_{n+1} by minimizing either $E\{(\epsilon_f^{n+1}(k))^2\}$ or $E\{(\epsilon_b^{n+1}(k))^2\}$. It can easily be verified that the two estimates for p_{n+1} follow as

$$\hat{p}_{n+1}^f = -\frac{E\{\epsilon_f^n(k)\epsilon_b^n(k-1)\}}{E\{[\epsilon_b^n(k-1)]^2\}} \tag{9.3.31}$$

and

$$\hat{p}_{n+1}^b = -\frac{E\{\epsilon_f^n(k)\epsilon_b^n(k-1)\}}{E\{[\epsilon_b^n(k)]^2\}} \tag{9.3.32}$$

Since $x(k)$ is stationary, it can be shown by direct computation that

$$E\{[\epsilon_b^n(k-1)]^2\} = E\{[\epsilon_f^n(k)]^2\} \tag{9.3.33}$$

We can therefore combine Eqs. (9.3.31) and (9.3.32) to obtain as our estimate

$$\hat{p}_{n+1} = \frac{-2E\{\epsilon_f^n(k)\epsilon_b^n(k-1)\}}{E\{(\epsilon_b^n(k-1))^2\} + E\{(\epsilon_f^n(k))^2\}} \tag{9.3.34}$$

The cross correlations in Eq. (9.3.34) can be computed in a manner similar to Eqs. (9.3.24) or (9.3.25). The expression for $\hat{p}_{n+1}$ obtained by using the correlation estimate of Eq. (9.3.25) is usually referred to as the *Burg algorithm* [30].

Since p_{n+1} depends only on $\epsilon_b^n(k)$, it is clear that p_{n+1} is independent of p_j for $j > n + 1$. This means that in contrast to the computation of the coefficients α_i, we need not recompute p_i when the assumed model order is increased. In fact, the algorithm also gives us a means of determining the model order N. As mentioned earlier, if the parameters are identified correctly, the residual will be white. Let us therefore compute the mean-square value of the residual at each stage of the recursion as

$$J_n = \mathrm{E}\{[\epsilon_f^n(k)]^2\} \tag{9.3.35}$$

and let us form the difference

$$J_{n+1} - J_n \tag{9.3.36}$$

for each n. If this quantity attains a minimum or becomes essentially constant, this implies that the residuals cannot be reduced by further processing. The value of n at which this happens should thus correspond to the model order.

In practice, however, this procedure is not very satisfactory since we have to use estimated correlations. While J_n decreases as n increases, the variances of the correlation estimates increase as the order of the filter and hence the number of parameters increases. It is therefore often difficult to determine when the difference $J_n - J_{n-1}$ becomes a constant. Other criteria that take into account the variance of the estimates have been proposed. Of these, the most widely used seems to be the *Akaike information theoretic criterion* (AIC), defined as [31]

$$\mathrm{AIC}(n) = L \ln J_n + 2n \tag{9.3.37}$$

where L is the number of data values used in computing the correlation estimates. The correct value of n is the one that minimizes the AIC. The AIC has also been used to select the model order in MA and ARMA models [26].

The Levinson algorithm avoids the need for predetermining the model order N. It is, however, a batch processing technique and involves the computation of correlation coefficients. We now discuss several sequential techniques for the identification of AR parameters.

9.3.3 Sequential Identification of Partial Correlation Coefficients

We can obtain a sequential variation of the Levinson algorithm by replacing the expectation operation in previous sections by the time average. Thus given a signal $y(k)$, we replace $\mathrm{E}\{y^2(k)\}$ by

$$\frac{1}{k}\sum_{i=1}^{k} y^2(i) \tag{9.3.38}$$

It follows that we can obtain an equation analogous to Eq. (9.3.34) for the estimate $\hat{p}_{n+1}$ at time k as [29]

$$\hat{p}_{n+1}(k) = -\frac{2\sum_{i=1}^{k} \epsilon_f^n(i)\epsilon_b^n(i-1)}{\sum_{i=1}^{k}\{[\epsilon_b^n(i-1)]^2 + [\epsilon_f^n(i)]^2\}} \tag{9.3.39}$$

We can obtain a sequential algorithm for estimating $\hat{p}_{n+1}$ by writing Eq. (9.3.39) as

$$\hat{p}_{n+1}(k) = -\frac{2\sum_{i=1}^{k-1} \epsilon_f^n(i)\epsilon_b^n(i-1)}{\sum_{i=1}^{k}[\epsilon_b^n(i-1)]^2 + \sum_{i=1}^{k}[\epsilon_j^n(i)]^2} - \frac{2\epsilon_f^n(k)\epsilon_b^n(k-1)}{\sum_{i=1}^{k}[\epsilon_b^n(i-1)]^2 + \sum_{i=1}^{k}[\epsilon_f^n(i)]^2} \tag{9.3.40}$$

or equivalently as

$$\hat{p}_{n+1}(k) = \hat{p}_{n+1}(k-1)\frac{\sum_{i=1}^{k-1}[\epsilon_b^n(i-1)]^2 + \sum_{i=1}^{k-1}[\epsilon_f^n(i)]^2}{\sum_{i=1}^{k}[\epsilon_b^n(i-1)]^2 + \sum_{i=1}^{k}[\epsilon_f^n(i)]^2} - \frac{2\epsilon_f^n(k)\epsilon_b^n(k-1)}{\sum_{i=1}^{k}[\epsilon_b^n(i-1)]^2 + \sum_{i=1}^{k}[\epsilon_f^n(i)]^2} \tag{9.3.41}$$

By adding and subtracting $\{[\epsilon_b^n(k-1)]^2 + [\epsilon_f^n(k)]^2\}$ in the numerator of the first term of the right-hand side of Eq. (9.3.41), it follows that

$$\hat{p}_{n+1}(k) = \hat{p}_{n+1}(k-1) - \frac{1}{\sum_{i=1}^{k}\{[\epsilon_f^n(i)]^2 + [\epsilon_b^n(i-1)]^2\}} \times [\hat{p}_{n+1}(k-1)\{(\epsilon_f^n(k))^2 + (\epsilon_b^n(k-1))^2\} + 2\epsilon_f^n(k)\epsilon_b^n(k-1)] \tag{9.3.42}$$

Equations (9.3.42) together with (9.3.29) and (9.3.30) constitute a complete set of algorithms for sequential identification of the parameters $\hat{p}_i$. The parameters $\hat{\alpha}_i^n$ can again be computed by use of Eq. (9.3.28). The algorithms are summarized in Table 9.3.1.

Example 9.3.3

We use the sequential algorithms of Table 9.3.1 to estimate the parameters of a sixth-order AR model. The data were generated by simulating the system on a computer. The input $w(k)$ to the model was a zero-mean, white, Gaussian sequence, with unity variance.

TABLE 9.3.1 SEQUENTIAL IDENTIFICATION OF PARTIAL CORRELATION COEFFICIENTS

Parameter identification algorithm:

$$\hat{p}_{n+1}(k) = \hat{p}_{n+1}(k-1) - \frac{\hat{p}_{n+1}(k-1)\{(\epsilon_f^n(k))^2 + (\epsilon_b^n(k-1))^2\} + 2\epsilon_f^n(k)\epsilon_b^n(k-1)}{\sum_{i=1}^{k}\{[\epsilon_f^n(i)]^2 + [\epsilon_b^n(k-1)]^2\}} \quad (9.3.42)$$

$$\epsilon_f^{n+1}(k) = \epsilon_f^n(k) + \hat{p}_{n+1}(k)\epsilon_b^n(k-1) \quad (9.3.29)$$

$$\epsilon_b^{n+1}(k) = \epsilon_b^n(k-1) + \hat{p}_{n+1}(k)\epsilon_f^n(k) \quad (9.3.30)$$

Initial conditions:

$$\epsilon_b^0(k) = \epsilon_f^0(k) = x(k) \text{ observations}$$

Conversion of $\hat{p}_n$ to $\hat{\alpha}_i$: For $n = 1, 2, \ldots, N$

$$\hat{\alpha}_0^{n+1} = \hat{\alpha}_0^n = 1 \quad (9.3.28a)$$

$$\hat{\alpha}_i^{n+1} = \hat{\alpha}_i^n + \hat{p}_{n+1}\hat{\alpha}_{n+1}^n \quad (9.3.28b)$$

$$\hat{\alpha}_{n+1}^{n+1} = \hat{p}_{n+1} \quad (9.3.28c)$$

The actual parameter values and the identified values at the end of 250 samples for these two cases is shown in Table 9.3.2. For purposes of determining the accuracy of the estimates the squared error $e_i = \sum_{k=1}^{k_f}[p_i(k) - \hat{p}_i(k)]^2; i = 1, \ldots, N$, was computed for each i for $k_f = 250$ and is also tabulated in the table.

The differences of the squared sum of the residuals J_n calculated for various values of n are tabulated in Table 9.3.3. The quantity J_n attains a minimum for $n = 6$

9.3.4 Sequential Identification of AR Parameters

We now consider the sequential identification of the parameters α_i. The algorithm that we obtain will be similar to the Kalman filter algorithm that we studied in Chapter 6. Let us define the parameter vector $\boldsymbol{\alpha}$ as

$$\boldsymbol{\alpha}^T = [\alpha_1 \cdots \alpha_N] \quad (9.3.43)$$

Since $\boldsymbol{\alpha}$ is a constant vector, we can model it by the equation

$$\boldsymbol{\alpha}(k+1) = \boldsymbol{\alpha}(k) \quad (9.3.44)$$

Our problem is to estimate $\boldsymbol{\alpha}(k)$ on the basis of the sequence of observations $x(1), x(2)$, and so on. The observation model can be obtained from Eq. (9.3.12) as

$$z(k) = x(k) = -\sum_{i=1}^{N}\alpha_i x(k-i) + Gw(k) \quad (9.3.45)$$

TABLE 9.3.2 SEQUENTIAL IDENTIFICATION OF PARTIAL CORRELATION COEFFICIENTS IN EXAMPLE 9.3.2

Index, i	1	2	3	4	5	6
Actual parameter values, p_i	0.885	−0.554	0.754	0.884	−0.638	0.385
Identified parameter value, $\hat{p}_i$	0.9185	−0.6132	0.8475	0.7634	−0.6327	0.4622
Initial mean-square error, e_i	0.7832	0.3069	0.5685	0.7815	0.4070	0.1482
Final mean-square error, e_i	0.1125×10^{-2}	0.3505×10^{-2}	0.8733×10^{-2}	0.1455×10^{-1}	0.2767×10^{-4}	0.5955×10^{-2}

TABLE 9.3.3 COST AS A FUNCTION OF MODEL ORDER

System Order, n	1	2	3	4	5	6	7	8
J_n	0.1142×10^7	0.1114×10^6	0.1482×10^6	0.3270×10^5	0.7112×10^4	0.2280×10^4	0.2782×10^4	0.2644×10^4

Let us define the matrix $\mathbf{C}(k)$ as

$$\mathbf{C}(k) = [-x(k-1) \cdots -x(k-N)]$$

Then the observation equation can be written as

$$z(k) = \mathbf{C}(k)\boldsymbol{\alpha}(k) + v(k) \tag{9.3.46}$$

where we have set

$$Gw(k) = v(k) \tag{9.3.47}$$

so that $v(k)$ is white with covariance kernel G^2. Comparison of Eqs. (9.3.44) and (9.3.46) with the message and observation models of the Kalman filter discussed in Chapter 6 shows the equivalence between the two problems. To complete the equivalence, let us assume that $\boldsymbol{\alpha}(0)$ is a random vector with known mean and covariance. The equations for the estimate $\hat{\alpha}(k)$ follow directly from Table 6.4.1 as

$$\hat{\boldsymbol{\alpha}}(k+1) = \hat{\boldsymbol{\alpha}}(k) + \mathbf{K}(k+1)[x(k) - \mathbf{C}(k)\hat{\boldsymbol{\alpha}}(k)] \tag{9.3.48}$$

The gain vector $\mathbf{K}(k+1)$ is given

$$\mathbf{K}(k+1) = \frac{\mathbf{V}_{\tilde{\alpha}}(k)\mathbf{C}^T(k)}{G^2 + \mathbf{C}(k)\mathbf{V}_{\tilde{\alpha}}(k)\mathbf{C}^T(k)} \tag{9.3.49}$$

and the error variance matrix $\mathbf{V}_{\tilde{\alpha}}(k)$ satisfies the difference equation

$$\mathbf{V}_{\tilde{\alpha}}(k+1) = \mathbf{V}_{\tilde{\alpha}}(k) - \frac{\mathbf{V}_{\tilde{\alpha}}(k)\mathbf{C}^T(k)\mathbf{C}(k)\mathbf{V}_{\tilde{\alpha}}(k)}{G^2 + \mathbf{C}(k)\mathbf{V}_{\tilde{\alpha}}(k)\mathbf{C}^T(k)} \tag{9.3.50}$$

The technique is therefore sequential in nature. The model order N, however, must be fixed a priori. We need to specify the gain G, the initial value for the estimate $\hat{\alpha}(0)$ and the initial error variance $\mathbf{V}_{\tilde{\alpha}}(0)$. The convergence of the algorithm is clearly determined by the rate at which $\mathbf{K}(k)$ tends to zero. From Eq. (9.3.49), it can be seen that this is influenced by the choice of G. To see this, we note that Eq. (9.3.50) can be written as

$$\mathbf{V}_{\tilde{\alpha}}(k+1) = [\mathbf{I} - \mathbf{K}(k+1)\mathbf{C}(k)]\mathbf{V}_{\tilde{\alpha}}(k) \tag{9.3.51}$$

For small G, the product $\mathbf{K}(k+1)\mathbf{C}(k)$ is approximately equal to the identity matrix, so that $\mathbf{V}_{\tilde{\alpha}}(k)$ tends to zero after only a few samples have been processed. Conversely, if G is large, the estimator gain $\mathbf{K}(k)$ tends to go to zero very slowly.

While it may thus appear to be desirable to choose the value of G small so that convergence is accelerated, such a procedure could cause the algorithm to converge to a value that is not the correct value of the parameter. This is related to the phenomenon of divergence in the Kalman filter.

The choice of the initial values of $\hat{\boldsymbol{\alpha}}$ and $\mathbf{V}_{\tilde{\alpha}}$ are intertwined in that the choice of $\mathbf{V}_{\tilde{\alpha}}$ should reflect the degree of confidence placed in the initial choice of $\hat{\boldsymbol{\alpha}}$. If this initial choice is thought to be good, $\mathbf{V}_{\tilde{\alpha}}(0)$ should be small so that the estimates change very little. If the initial choice is thought to be poor (or uncertain), $\mathbf{V}_{\hat{\alpha}}(0)$ should be chosen large.

An alternative way of accounting for the uncertainty in the choice of the initial value for the parameter vector is to model $\boldsymbol{\alpha}(k)$ not as a constant vector but as a Markov process described by the equation

$$\boldsymbol{\alpha}(k+1) = \boldsymbol{\alpha}(k) + \mathbf{u}(k) \tag{9.3.52}$$

where $\mathbf{u}(k)$ is a stationary white sequence. Equation (9.3.51) then becomes

$$\mathbf{V}_{\hat{\alpha}}(k+1) = [\mathbf{I} - \mathbf{K}(k+1)\mathbf{C}(k)]\mathbf{V}_{\hat{\alpha}}(k) + \mathbf{V}_{\mathbf{u}} \tag{9.3.53}$$

where $\mathbf{V}_{\mathbf{u}}$ is the variance of the process $\mathbf{u}(k)$. The choice of $\mathbf{V}_{\mathbf{u}}$ will then determine the rate of convergence of the algorithm. Since $\mathbf{V}_{\hat{\alpha}}(k)$ is now bounded below by $\mathbf{V}_{\mathbf{u}}$, the problem of converging to a wrong value because of the gain vector going to zero too soon is obviated.

Example 9.3.4

We consider the identification of a sixth order model using the sequential algorithms of Eqs. (9.3.48) to (9.3.50). The system was simulated on a computer and the results at the end of 200 iterations are shown in Table 9.3.4. The simulations were carried out using two different values of G^2. The mean-square error obtained in each case, $\sum_{i=1}^{6} (\alpha_i - \hat{\alpha}_i)^2$ is also shown in the table. The initial values for the parameters were chosen to be zero in both cases. Convergence as indicated by the mean-square error seems to be faster for the lower value of G^2. This is in conformity with the earlier discussion on the choice of G^2.

TABLE 9.3.4 SEQUENTIAL IDENTIFICATION OF AUTOREGRESSIVE PARAMETERS

Parameter	α_1	α_2	α_3	α_4	α_5	α_6	Error
True value	−1.052	0.1071	0.5371	−0.53	−1.094	−0.529	—
Estimate for $G^2 = 3$	−1.0602	0.1477	0.4914	−0.4292	−1.069	−0.469	0.0108
Estimate for $G^2 = 100$	−1.008	0.1772	0.4375	−0.5593	−1.026	−0.443	0.0296

As stated earlier, one of the applications of determining the parameters of an autoregressive model of a stationary stochastic process is in the computation of the power spectrum. It is thus instructive to see how well the power spectrum is identified by this technique. Figure 9.3.2 shows the true spectrum and the spectrum obtained from the autoregressive model. As can be seen, the true spectrum has peaks at 500, 1800, and 2200 Hz. The AR model does identify the location of all three spectral peaks fairly accurately. In general, the AR model fits the true spectrum better at spectral peaks than at troughs. This is particularly useful in applications such as speech processing, since it has been determined that the ear is relatively sensitive to the location of the peaks of the speech spectrum.

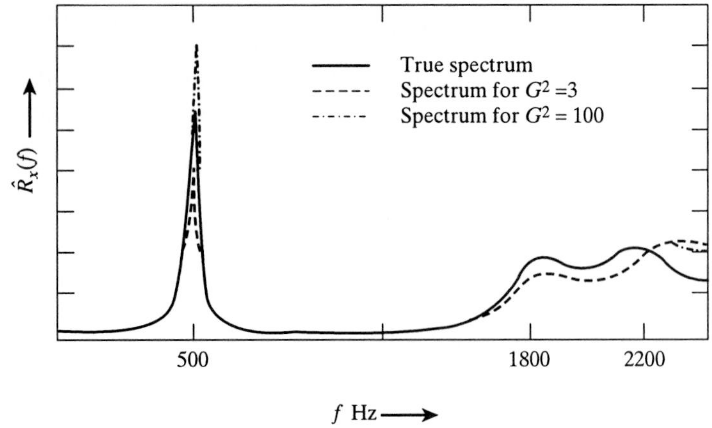

Figure 9.3.2 True and computed spectra for Example 9.3.4.

9.3.5 Estimation of MA Processes

The MA process is modeled as

$$x(k) = Gw(k) + \sum_{i=1}^{M} \beta_i w(k - i) \tag{9.3.54}$$

and is the output of the all-zero system, $H(z) = G + \sum_{i=1}^{M} \beta_i z^{-i}$, driven by the zero-mean, unit variance, white sequence $w(k)$. It can easily be verified that the correlation coefficients of the process are given by

$$\phi_x(k) = G^2 \sum_{i=0}^{M-|k|} \beta_i \beta_{i+|k|} \qquad k = 0, 1, \ldots, M \tag{9.3.55}$$

Thus, given the correlation coefficients $\phi_x(k)$, estimating the MA coefficients involves solving $M + 1$ nonlinear equations. In principle, the equations can be easily solved by dividing all the equations by G^2 and solving for the estimates $\hat{\beta}_i$ in reverse order. That is, we solve for $\hat{\beta}_M$ first, then use this value to solve for $\hat{\beta}_{M-1}$ next, and so on. As may be expected, in general, multiple solutions exist.

An alternative method to solve for the estimates is by noting that the spectrum of $x(k)$ is given by

$$\Phi_x(z) = \sum_{i=0}^{M} \phi_x(k) z^{-k} = H(z)H(z^{-1}) \tag{9.3.56}$$

We can therefore do a spectral factorization of $\Phi_x(z)$ to identify terms corresponding to $H(z)$. Since the roots of $\Phi_x(z)$ occur in conjugate reciprocal pairs with reference to the unit circle in the z-plane, several factorizations are possible. Typically, we associate the roots inside the unit circle with $H(z)$. The order of the model may again be chosen by using the AIC.

We can also use the techniques of Section 9.2 by replacing the MA process by an equivalent higher-order AR process. We explore this approach further in the next section, on ARMA process identification.

9.3.6 Extensions to ARMA Processes

We now extend the algorithms of previous sections to the identification of the parameters of an ARMA process described in terms of either the difference equation, Eq. (9.3.11), or in terms of the z-transform description, Eq. (9.3.10).

$$X(z) = \frac{G(\beta_0 + \beta_1 z^{-1} + \cdots + \beta_M z^{-M})}{1 + \alpha_1 z^{-1} + \cdots + \alpha_N z^{-N}} W(z) = H(z)W(z) \qquad (9.3.57)$$

where $\beta_0 = 1$.

Multiplying both sides of Eq. (9.3.11) by $x(k - j)$ and taking expectations results in

$$\phi_x(j) + \sum_{i=1}^{N} \alpha_i \phi_x(j - i) = G \sum_{i=0}^{M} \beta_i \phi_{wx}(j - i) \qquad (9.3.58)$$

From Eq. (9.3.57), it follows that

$$\Phi_{wx}(z) = H(z^{-1})\Phi_w(z) = H(z^{-1}) \qquad (9.3.59)$$

or, in terms of the impulse response $h(j)$ of the system $H(z)$, we have

$$\phi_{wx}(j) = h(-j) \qquad (9.3.60)$$

where $h(j)$ is the coefficient of the z^{-j} term in the power series expansion of $H(z)$ obtained by dividing the denominator into the numerator,

$$H(z) = G + G[\beta_1 - \alpha_1]z^{-1} + G[(\beta_2 - \alpha_2) - \alpha_1(\beta_1 - \alpha_1)]z^{-2} + \cdots \qquad (9.3.61)$$

Substituting Eq. (9.3.60) in Eq. (9.3.58) and noting that since $H(z)$ is causal, $h(j) = 0$ for $j < 0$, we have

$$\phi_x(j) + \sum_{i=1}^{N} \alpha_i \phi_x(j - i) = G \sum_{i=j}^{M} \beta_i h(i - j) \qquad (9.3.62)$$

We can obtain the $M + N + 1$ equations needed to solve for α_i, β_i, and G by writing out Eq. (9.3.62) for $j = 0, 1, 2, \ldots, M + N$ as

$$\phi_x(j) + \sum_{i=1}^{N} \alpha_i \phi_x(j - i) = G \sum_{i=j}^{M} \beta_i h(i - j) \qquad j = 0, 1, 2, \ldots, M \qquad (9.3.63a)$$

$$\phi_x(j) + \sum_{i=1}^{N} \alpha_i \phi_x(j - i) = 0 \qquad j = M + 1, M + 2, \ldots, M + N \qquad (9.3.63b)$$

These equations are referred to as the *modified Yule–Walker equations*.

We note that Eq. (9.3.63b) can be solved independently of Eq. (9.3.63a) to obtain the AR parameters α_i. The equations are linear in these parameters and can be solved using any of the techniques discussed earlier in parameter identification. Once these parameters have been obtained, we can determine the MA parameters β_i and the gain G from Eq. (9.3.63a) by substituting for $h(j)$ in terms of α_i, β_i. However, the equations are now nonlinear in the parameters and their solution is not straightforward. We can use the spectral factorization approach discussed in Section 9.3.5 to obtain a solution. It is also possible to extend the Levinson algorithm to solve the modified Yule–Walker equations. Details of this algorithm can be found in [26].

As noted earlier, an alternative approach to ARMA identification is to replace the ARMA model by an equivalent higher-order AR model. That is, we replace $H(z)$ in Eq. (6.3.57) with an all-pole transfer function by representing each zero by a sufficient number of poles [32] as

$$1 + \beta z^{-1} = \frac{1}{1 - \beta z^{-1} - \beta z^{-2} - \beta z^{-3} - \cdots} \tag{9.3.64}$$

In most applications it suffices to use just the first few terms in the series. We can therefore approximate the ARMA process $x(k)$ by a purely AR process by writing $H(z)$ in Eq. (9.3.57) as

$$\begin{aligned} H(z) &= \frac{G(1 + \beta_1 z^{-1} + \cdots + \beta_M z^{-M})}{1 + \alpha_1 z^{-1} + \cdots + \alpha_N z^{-N}} \\ &= \frac{G}{1 + r_1 z^{-1} + r_2 z^{-2} + \cdots + r_{N'} z^{-N'}} \end{aligned} \tag{9.3.65}$$

While $N' = \infty$ yields an exact equivalence, the approximation is fairly good for finite N'. The value of N' is generally greater than $N + M$.

To determine the relation between the parameters $\{r_i\}$ of the equivalent AR model and $\{\alpha_i\}$ and $\{\beta_i\}$ corresponding to the ARMA model, we can cross-multiply terms in the two expressions for $H(z)$ in Eq. (9.3.65) to get

$$\begin{aligned} &(1 + \beta_1 z^{-1} + \cdots + \beta_M z^{-M})(1 + r_1 z^{-1} + \cdots + r_{N'} z^{-N'}) \\ &\qquad = 1 + \alpha_1 z^{-1} + \cdots + \alpha_N z^{-N} \end{aligned} \tag{9.3.66}$$

Equating coefficients of like powers of z^{-1} gives us the following set of equations:

$$\begin{aligned} \alpha_1 &= \beta_1 + r_1 \\ \alpha_2 &= \beta_2 + \beta_1 r_1 + r_2 \\ &\vdots \\ \alpha_N &= \beta_N + \beta_{N-1} r_1 + \cdots + r_N \\ 0 &= \beta_N r_1 + \beta_{N-1} r_{i+1} + \cdots + \beta_1 r_{N+i-1} + r_{N+i} \qquad i = 1, 2, \ldots, N' \end{aligned} \tag{9.3.67}$$

where $\beta_j = 0$ for $j = M + 1, M + 2, \ldots, N$. These equations can be put in matrix form as

$$\begin{bmatrix} r_N & r_{N-1} & \cdots & r_{N-M+1} \\ r_{N+1} & r_N & \cdots & r_{N-M+2} \\ \vdots & & & \vdots \\ r_{N'-1} & r_{N'-2} & \cdots & r_{N'-M} \end{bmatrix} \begin{bmatrix} \beta_1 \\ \beta_2 \\ \vdots \\ \beta_M \end{bmatrix} = - \begin{bmatrix} r_{N+1} \\ r_{N+2} \\ \vdots \\ r_{N'} \end{bmatrix} \tag{9.3.68}$$

and

$$\begin{bmatrix} \alpha_1 \\ \alpha_2 \\ \alpha_3 \\ \vdots \\ \alpha_N \end{bmatrix} = \begin{bmatrix} \beta_1 \\ \beta_2 \\ \beta_3 \\ \vdots \\ \beta_N \end{bmatrix} + \begin{bmatrix} 1 & 0 & 0 & \cdots & 0 \\ \beta_1 & 1 & 0 & \cdots & 0 \\ \beta_2 & \beta_1 & 1 & \cdots & 0 \\ \vdots & & & & \\ \beta_{N-1} & \beta_{N-2} & \beta_{N-3} & & 1 \end{bmatrix} \begin{bmatrix} r_1 \\ r_2 \\ r_3 \\ \vdots \\ r_N \end{bmatrix} \tag{9.3.69}$$

While Eq. (9.3.68) represents a set of $(N' - N)$ equations in M unknowns, it must be noted that the coefficients $\{r_i\}$ correspond to an infinite-order AR model. In practice, in order to obtain a solution, we choose $N' = N + M$ and determine the ARMA parameters in terms of the parameters of the approximate equivalent AR model of order N'. The parameters of this approximate model can be found using any of the techniques discussed earlier. The main advantage of this approach is that once the parameters of the approximate AR model are determined, both the AR and MA parameters of the original ARMA model are obtained as the solution of a set of linear equations.

A somewhat different set of equations, which are also linear in the ARMA coefficients is given in [33]. As with the procedure above, the partial correlation coefficients of the equivalent $(M + N)$th-order AR model are first obtained. However, the equations for determining the ARMA parameters use the AR parameters from different orders computed from the partial correlations. Simulation examples presented in [34] seem to indicate that while Eqs. (9.3.68) and (9.3.69) give good estimates of the AR parameters, the method in [33] gives better estimates of the MA parameters.

With either scheme, there still remains the problem of fixing the relative orders N and M. Let $\hat{N}$ and $\hat{M}$ denote assumed values of N and M. Denoting the square matrix on the left-hand side of Eq. (9.3.68) by $A_{\hat{N},\hat{M}}$, we can test the rank of $A_{\hat{N},\hat{M}}$ for values of $\hat{N}, \hat{M}$ such that $\hat{N} + \hat{M} = \hat{N}'$ until for some M and N,

$$\text{rank}\{A_{\hat{N},\hat{M}}\} = M_1 \qquad \text{for} \quad \hat{N} > \hat{N}_1, \hat{M} > \hat{M}_1$$

or, alternatively,

$$\det A_{\hat{N},\hat{M}} = 0 \qquad \text{for} \quad \hat{N} > \hat{N}_1, \hat{M} > \hat{M}_1 \tag{9.3.70}$$

In practice, because of round-off and truncation errors, the determinant in Eq. (9.3.70) is rarely zero. We can therefore use as our criterion

$$\det A_{\hat{N},\hat{M}} \leqslant \epsilon \qquad \text{for some } \epsilon \geqslant 0$$

If M and N are estimated on the low side, there is the possibility of omitting the dynamics essential in signal description. On the other hand, if M and N are overestimated, some negligibly small coefficients are introduced into the model. These correspond to the addition of small-magnitude high-frequency terms and should not cause difficulties in most applications.

Example 9.3.5

We consider the identification of the ARMA model

$$x(k) + 1.5x(k - 1) + 0.625x(k - 2) = w(k) - 0.5w(k - 1)$$

where $w(k)$ is a zero-mean, unity variance, white, Gaussian sequence. The data were again generated by simulating the system on a computer. Table 9.3.5 shows the cost for various values of N'. As can be seen from the table, the cost becomes essentially constant after $N' = 3$. We thus choose $N' = 3$. This is acceptable since $M + N = 3$ for this example. The true values of the parameters and the estimates obtained on the basis of 250 samples are shown in Table 9.3.6.

TABLE 9.3.5 DETERMINATION OF MODEL ORDER FOR EXAMPLE 9.3.5

System Order, n'	1	2	3	4	5
$J_{n'}$	1.976	1.871	0.1362	0.1218	0.1039

TABLE 9.3.6 IDENTIFICATION OF AR AND MA PARAMETERS FOR THE ARMA MODEL OF EXAMPLE 9.3.5

True values:	$\alpha_1 = 1.5$	$\alpha_2 = 0.625$	$\beta_1 = -0.5$
Estimated values:	$\hat{\alpha}_1 = 1.509$	$\hat{\alpha}_2 = 0.703$	$\hat{\beta}_1 = -0.445$

9.3.7 Identification of ARMA Models: Direct Determination

We can set up an algorithm for identifying the $\{\alpha_i\}$ and $\{\beta_i\}$ parameters in the ARMA model simultaneously using a procedure similar to the one for identifying the $\{\alpha_i\}$ parameters in the AR model using the Kalman algorithm [35]. Let $\boldsymbol{\theta}$ denote the size $(N + M)$ column vector

$$\boldsymbol{\theta} = [\alpha_1 \cdots \alpha_N \quad \beta_1 \cdots \beta_M]^T \tag{9.3.71}$$

Since $\boldsymbol{\theta}$ is a constant vector, it can be modeled by the equation

$$\boldsymbol{\theta}(k + 1) = \boldsymbol{\theta}(k) \tag{9.3.72}$$

The observation model is obtained from Eq. (9.3.11) as

$$z_k = x_k = -\sum_{i=1}^{N} \alpha_i x(k - i) + \sum_{i=1}^{M} \beta_i w(k - i) + Gw(k) \tag{9.3.73}$$

If the input $w(k)$ were available for measurement, we can form the row vector $\mathbf{C}(k)$:

$$\mathbf{C}(k-1) = [-x(k-1) - x(k-2) \cdots -x(k-N) \quad w(k-1) \cdots w(k-M)] \tag{9.3.74}$$

and write the observation equation as

$$z(k) = \mathbf{C}(k-1)\boldsymbol{\theta}(k) + v(k) \tag{9.3.75}$$

where again we have defined

$$v(k) = Gw(k) \tag{9.3.76}$$

Equations (9.3.72) and (9.3.75) are similar to Eqs. (9.3.44) and (9.3.46) and hence the Kalman algorithm may be applied to estimate the parameter vector $\boldsymbol{\theta}$. Since, however, in most applications, the input $w(k)$ is not available, we may seek to estimate it at any stage by using Eq. (9.3.73).

We thus write

$$\hat{G}(k)\hat{w}(k) = x_k + \sum_{i=1}^{N} \hat{\alpha}_i(k)x(k-i) - \sum_{i=1}^{M} \hat{\beta}_i \hat{w}(k-i) \tag{9.3.77}$$

We see that the quantity on the right-hand side of Eq. (9.3.77) represents the residual $\epsilon(k)$ so that

$$\hat{w}(k) = \frac{1}{\hat{G}(k)}\epsilon(k) \tag{9.3.78}$$

where

$$\epsilon(k) = x(k) + \sum_{i=1}^{N} \hat{\alpha}_i(k)x(k-i) - \sum_{i=1}^{M} \hat{\beta}_i \hat{w}(k-i) \tag{9.3.79}$$

To evaluate $\hat{G}(k)$ from Eq. (9.3.78) we use the fact that $w(k)$ is white with unity spectral height to obtain

$$\mathrm{E}\{[\hat{G}(k)\hat{w}(k)]^2\} = \hat{G}^2(k) = \mathrm{E}\{\epsilon^2(k)\} \tag{9.3.80}$$

where $\mathrm{E}\{\epsilon^2(k)\}$ can be computed as

$$\mathrm{E}\{\epsilon^2(k)\} = \frac{1}{k}\sum_{i=1}^{k} \epsilon^2(i)$$

If we now replace $w(k-i), i = 1, 2, \ldots, M$ in Eq. (9.3.74) by $\hat{w}(k-i)$ obtained from Eq. (9.3.78) we have

$$\mathbf{C}(k-1) = [-x(k-1) \cdots -x(k-N)\hat{w}(k-1) \cdots \hat{w}(k-M)] \tag{9.3.81}$$

The identification algorithm becomes

$$\hat{\boldsymbol{\theta}}(k+1) = \hat{\boldsymbol{\theta}}(k) + \mathbf{K}(k+1)[x(k+1) - \mathbf{C}(k)\hat{\boldsymbol{\theta}}(k)] \tag{9.3.82}$$

$$\mathbf{K}(k+1) = \mathbf{V}_{\tilde{\theta}}(k)\mathbf{C}^T(k)[G^2(k) + \mathbf{C}(k)\mathbf{V}_{\tilde{\theta}}(k)\mathbf{C}^T(k)] \tag{9.3.83}$$

and

$$\mathbf{V}_{\tilde{\theta}}(k+1) = [\mathbf{I} - \mathbf{K}(k+1)\mathbf{C}(k)]\mathbf{V}_{\tilde{\theta}}(k) \tag{9.3.84}$$

Equations (9.3.77)–(9.3.84) constitute a complete set of algorithms for identifying the parameters of the ARMA model of (9.3.11). It has been shown that under certain circumstances the estimate $\hat{\boldsymbol{\theta}}(k)$ converges to the true value with probability 1 [35].

Example 9.3.6

We consider the identification of the parameters of the ARMA model of Example 9.3.4 using the direct method. Table 9.3.7 shows the true parameter values, and estimates obtained on the basis of 250 data samples and 1000 samples, respectively. The identification of both the AR and MA parameters seems to be slightly better in this example than in Example 9.3.4. However, as pointed out earlier, we need to assume the model order a priori in this technique.

TABLE 9.3.7 DIRECT DETERMINATION OF ARMA PARAMETERS

True values	$\alpha_1 = 1.5$	$\alpha_2 = 0.625$	$\beta_1 = -0.5$
Estimated value, $k_f = 1000$	$\hat{\alpha}_1 = 1.49$	$\hat{\alpha}_2 = 0.604$	$\hat{\beta}_1 = -0.518$
Estimated value, $k_f = 250$	$\hat{\alpha}_1 = 1.463$	$\hat{\alpha}_2 = 0.5825$	$\hat{\beta}_1 = -0.505$

9.3.8 Estimation in Nonlinear Models: Extended Kalman Filter

We have so far considered the identification of the parameters of a linear time-invariant model with stationary white noise input. We now consider a general nonlinear model and present a technique for identifying the parameters of such a model.

Let $\boldsymbol{\theta}$ represent the vector of unknown parameters in the system. Let us consider the system described by

$$\mathbf{x}(k+1) = \mathbf{f}(\mathbf{x}, \boldsymbol{\theta}, k) + \mathbf{g}(\mathbf{x}, \boldsymbol{\theta}, k)\mathbf{w}(k) \tag{9.3.85}$$

where $\mathbf{x}(k)$ is an n-vector describing the state of the system and where $\mathbf{w}(k)$ is a zero-mean, white input sequence with covariance

$$\operatorname{cov}\{\mathbf{w}(k)\mathbf{w}(j)\} = \mathbf{V}_w(k)\delta_{kj} \tag{9.3.86}$$

The observation vector $z(k)$ is

$$\mathbf{z}(k) = \mathbf{h}(\mathbf{x}, \boldsymbol{\theta}, k) + \mathbf{v}(k) \tag{9.3.87}$$

with the observation noise $\mathbf{v}(k)$ being zero mean, white, with covariance

$$\operatorname{cov}\{\mathbf{v}(k)\mathbf{v}(j)\} = \mathbf{V}_v(k))\delta_{kj} \tag{9.3.88}$$

and independent of $\mathbf{w}(k)$.

Assume that $\mathbf{x}(k_0)$ is a random variable with known mean and covariance and independent of $\mathbf{w}(k)$ and $\mathbf{v}(k)$ for all $k > k_0$, so that

$$\begin{gathered} \mathrm{E}\{\mathbf{x}(k_0)\} = \boldsymbol{\mu}_{\mathbf{x}_0}, \ \operatorname{var}\{\mathbf{x}(k_0)\} = \mathbf{V}_{\mathbf{x}_0} \\ \operatorname{cov}\{\mathbf{x}(k_0), \mathbf{w}(k)\} = \operatorname{cov}\{\mathbf{x}(k_0), \mathbf{v}(k)\} = \operatorname{cov}\{\mathbf{w}(k), \mathbf{v}(j)\} = 0 \end{gathered} \tag{9.3.89}$$

If the time evolution of the parameter vector $\boldsymbol{\theta}$ can be described by a difference equation of the form

$$\boldsymbol{\theta}(k+1) = \boldsymbol{\psi}(\boldsymbol{\theta}, k) \tag{9.3.90}$$

we can define an augmented state vector $\mathbf{X}$ as

$$\mathbf{X} = \left[\begin{array}{c} \mathbf{x} \\ \hline \boldsymbol{\theta} \end{array}\right]$$

From Eqs. (9.3.85) and (9.3.90) we therefore have

$$\begin{aligned} \mathbf{X}(k+1) &= \left[\begin{array}{c} \mathbf{x}(k+1) \\ \hline \boldsymbol{\theta}(k+1) \end{array}\right] = \left[\begin{array}{c} \mathbf{f}(\mathbf{x}, \boldsymbol{\theta}, k) + \mathbf{g}(\mathbf{x}, \boldsymbol{\theta}, k)\mathbf{w}(k) \\ \hline \boldsymbol{\psi}(\boldsymbol{\theta}, k) \end{array}\right] \\ &= \boldsymbol{\phi}(\mathbf{X}, k) + \boldsymbol{\Gamma}(\mathbf{X}, k)\mathbf{w}(k) \end{aligned} \tag{9.3.91}$$

Similarly, the observation model can be written in terms of the augmented state vector as

$$\mathbf{Z}(k) = \mathbf{h}(\mathbf{X}, k) + \mathbf{v}(k) \tag{9.3.92}$$

The problem of identification of the parameter vector now reduces to one of estimating the states $\mathbf{X}(k)$ of the augmented system and observation models Eqs. (9.3.91) and (9.3.92). Since the models are nonlinear, we have to use an approximate algorithm for determining the estimates. In fact, we have already seen an application of this technique in Chapter 8 in our discussion of equalizers for intersymbol interference. As in Chapter 8, we may use the extended Kalman filter algorithms of Section 6.6.3 to obtain estimates of the augmented state $\mathbf{X}(k)$ for the message and observation models of Eqs. (9.3.91) and (9.3.92). The algorithms are:

$$\hat{\mathbf{X}}(k+1) = \hat{\mathbf{X}}(k+1|k) + \mathbf{K}(k+1)\{\mathbf{Z}(k+1) - \mathbf{h}[\hat{\mathbf{X}}(k+1|k), k+1]\} \tag{9.3.93a}$$

$$\hat{\mathbf{X}}(k+1|k) = \boldsymbol{\phi}[\hat{\mathbf{X}}(k|k), k] \tag{9.3.93b}$$

$$\mathbf{K}(k+1) = \mathbf{V}_{\tilde{\mathbf{X}}}(k+1|k+1)\frac{\partial \mathbf{h}^T[\hat{\mathbf{X}}(k+1|k), k+1]}{\partial \hat{\mathbf{X}}(k+1|k)}\mathbf{V}_{\mathbf{v}}^{-1}(k+1) \tag{9.3.93c}$$

$$\begin{aligned} \mathbf{V}_{\tilde{\mathbf{X}}}(k+1|k) = {} & \frac{\partial \boldsymbol{\phi}[\hat{\mathbf{X}}(k|k), k]}{\partial \hat{\mathbf{X}}(k|k)} \mathbf{V}_{\tilde{\mathbf{X}}}(k|k) \frac{\partial \boldsymbol{\phi}^T[\hat{\mathbf{X}}(k|k), k]}{\partial \hat{\mathbf{X}}(k|k)} \\ & + \boldsymbol{\Gamma}[\hat{\mathbf{X}}(k|k), k]\mathbf{V}_{\mathbf{w}}(k)\boldsymbol{\Gamma}^T[\hat{\mathbf{X}}(k|k), k] \end{aligned} \tag{9.3.93d}$$

$$
\begin{aligned}
\mathbf{V}_{\tilde{\mathbf{X}}}(k+1|k+1) = \mathbf{V}_{\tilde{\mathbf{X}}}(k+1|k) - \mathbf{V}_{\tilde{\mathbf{X}}}(k+1|k)&\frac{\partial \mathbf{h}^T[\hat{\mathbf{X}}(k+1|k), k+1]}{\partial \hat{\mathbf{X}}(k+1|k)} \\
\times \Bigg\{ &\frac{\partial \mathbf{h}[\hat{\mathbf{X}}(k+1|k), k+1]}{\partial \hat{\mathbf{X}}(k+1|k)} \mathbf{V}_{\tilde{\mathbf{X}}}(k+1|k) \\
\times &\frac{\partial \mathbf{h}^T[\hat{\mathbf{X}}(k+1|k), k+1]}{\partial \hat{\mathbf{X}}(k+1|k)} + \mathbf{V}_v(k+1) \Bigg\}^{-1} \\
\times &\frac{\partial \mathbf{h}[\hat{\mathbf{X}}(k+1|k), k+1]}{\partial \hat{\mathbf{X}}(k+1|k)} \mathbf{V}_{\tilde{\mathbf{X}}}(k+1|k)
\end{aligned}
\tag{9.3.93e}
$$

We note that the algorithms simultaneously estimate the states of the system and the unknown parameters.

Example 9.3.7

We consider the problem of identifying the time constant in a first-order system described by

$$x(k+1) = ax(k) + w(k)$$

The observations are given by

$$z(k) = x(k) + v(k)$$

Since a is a constant parameter, it can be modeled as

$$a(k+1) = a(k)$$

Let us define the states of the augmented model as

$$
\begin{aligned}
X_1(k) &= x(k) \\
X_2(k) &= a(k)
\end{aligned}
$$

Then

$$
\begin{aligned}
X_1(k+1) &= X_1(k)X_2(k) + w(k) \\
X_2(k+1) &= X_2(k)
\end{aligned}
$$

and

$$Z(k) = X_1(k) + v(k)$$

We note that even though the system was linear, the augmented message model is nonlinear.

The results of a computer simulation of the algorithms are shown in Fig. 9.3.3. Three simulations corresponding to different values of the input covariance kernel $V_w(k)$ and the measurement noise covariance kernel $V_v(k)$ are shown. The initial conditions for the simulations were chosen as $\mathbf{X}(0) = \begin{bmatrix} 0 \\ 1 \end{bmatrix}$ and $V_x(0) = I$. The results indicate that the identification of the unknown parameter is better at lower values of V_w and V_v.

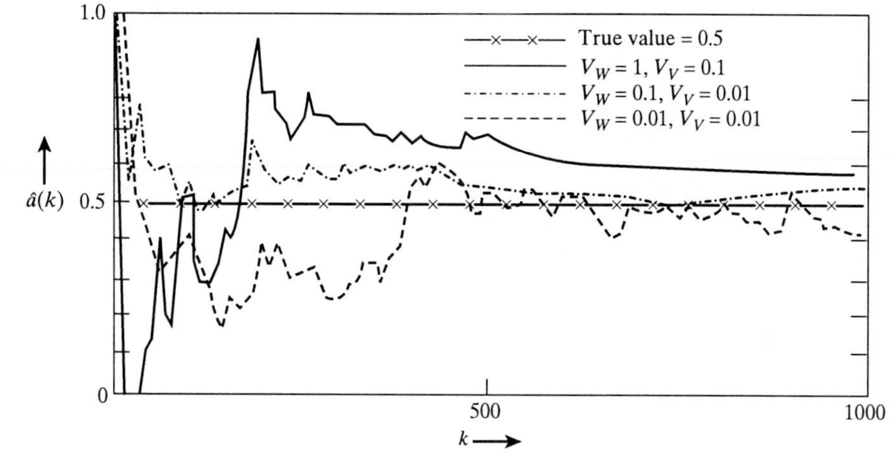

Figure 9.3.3 Identification using the extended Kalman filter.

9.3.9 Maximum Likelihood Identification

In our discussions in Chapter 5 we saw that the maximum likelihood technique plays an important role in the identification of nonrandom parameters. In this section we consider the identification of the parameters of linear time-invariant systems using the maximum likelihood approach. We assume that the system is described by the state-space model

$$\mathbf{x}(k+1) = \boldsymbol{\phi}\mathbf{x}(k) + \boldsymbol{\Gamma}\mathbf{w}(k) + \mathbf{B}\mathbf{u}(k) \tag{9.3.94a}$$

$$\mathbf{z}(k) = \mathbf{H}\mathbf{x}(k) + \mathbf{v}(k) \tag{9.3.94b}$$

where the state vector $\mathbf{x}(k)$ is of dimension n, $\mathbf{u}(k)$ is a p-vector of deterministic inputs, $\mathbf{z}(k)$ is a q-vector of observations. $\mathbf{w}(k)$ and $\mathbf{v}(k)$ are assumed to be zero-mean, stationary, Gaussian, white processes, independent of each other and with covariance kernels $\mathbf{V}_\mathbf{w}$ and $\mathbf{V}_\mathbf{v}$, respectively. $\boldsymbol{\phi}$, $\boldsymbol{\Gamma}$, $\mathbf{B}$, and $\mathbf{H}$ are constant matrices of appropriate order [36].

The most general identification problem arises when some or all of the elements of the matrices $\boldsymbol{\phi}$, $\boldsymbol{\Gamma}$, $\mathbf{B}$, $\mathbf{H}$, $\mathbf{V}_\mathbf{w}$, and $\mathbf{V}_\mathbf{v}$ are unknown. Let us denote by $\boldsymbol{\theta}$ the set of unknown parameters in the system, and let $\mathbf{Z}_N$ denote the set of observations $\mathbf{z}(1), \mathbf{z}(2), \ldots, \mathbf{z}(N)$. The ML estimate of $\boldsymbol{\theta}$ is then obtained by maximizing $p(\mathbf{Z}_N \mid \boldsymbol{\theta})$. We note that

$$\begin{aligned} p(\mathbf{Z}_N \mid \boldsymbol{\theta}) &= p(\mathbf{z}(N) \mid \mathbf{Z}_{N-1}, \boldsymbol{\theta}) p(\mathbf{Z}_{N-1} \mid \boldsymbol{\theta}) \\ &= p(\mathbf{z}(N) \mid \mathbf{Z}_{N-1}, \boldsymbol{\theta}) p(\mathbf{z}(N-1) \mid \mathbf{Z}_{N-2}, \boldsymbol{\theta}) \cdots p(\mathbf{z}(0)) \end{aligned} \tag{9.3.95}$$

so that

$$\ln p(\mathbf{Z}_N \mid \boldsymbol{\theta}) = \ln \sum_{k=1}^{N} p(\mathbf{z}(k) \mid \mathbf{Z}_{k-1}, \boldsymbol{\theta}) \tag{9.3.96}$$

Because of our Gaussian assumptions, it is clear that the random vector $\mathbf{z}(k)$ is Gaussian. Its mean and variance can be obtained from Eq. (9.3.94b) as

$$\begin{aligned} \mathrm{E}\{\mathbf{z}(k)\,|\,\mathbf{Z}_{k-1}, \boldsymbol{\theta}\} &= \mathbf{H}\mathrm{E}\{\mathbf{x}(k)\,|\,\mathbf{Z}_{k-1}, \boldsymbol{\theta}\} \\ \operatorname{var}\{\mathbf{z}(k)\,|\,\mathbf{Z}_{k-1}, \boldsymbol{\theta}\} &= \operatorname{var}\{\mathbf{H}[\mathbf{x}(k) - \mathrm{E}\{\mathbf{x}(k)\,|\,\mathbf{Z}_{k-1}, \boldsymbol{\theta}\}] + \mathbf{v}(k)\} \end{aligned} \tag{9.3.97}$$

By assumption $\mathbf{v}(k)$ is a white process independent of $\mathbf{w}(k)$ for all k. It follows that $\mathbf{v}(k)$ is independent of $\mathbf{x}(k)$ for all k. Also, $\mathbf{Z}_{k-1}$ does not depend on $\mathbf{v}(k)$. If we now make the following definitions:

$$\hat{\mathbf{x}}(k\,|\,k-1) = \mathrm{E}\{\mathbf{x}(k)\,|\,\mathbf{Z}_{k-1}, \boldsymbol{\theta}\}$$

$$\tilde{\mathbf{x}}(k\,|\,k-1) = \mathbf{x}(k) - \hat{\mathbf{x}}(k\,|\,k-1)$$

and

$$\mathbf{V}_{\tilde{\mathbf{x}}}(k\,|\,k-1) = \operatorname{var}\{\tilde{\mathbf{x}}(k\,|\,k-1)\} \tag{9.3.98}$$

it follows that the mean and variance of $\mathbf{z}(k)$ can be written from Eqs. (9.3.97) as

$$\begin{aligned} \mathrm{E}\{\mathbf{z}(k)\,|\,\mathbf{Z}_{k-1}, \boldsymbol{\theta}\} &= \mathbf{H}\hat{\mathbf{x}}(k\,|\,k-1) \\ \operatorname{var}\{\mathbf{z}(k)\,|\,\mathbf{Z}_{k-1}, \boldsymbol{\theta}\} &= \mathbf{H}\mathbf{V}_{\tilde{\mathbf{x}}}(k\,|\,k-1)\mathbf{H}^T + \mathbf{V}_{\mathbf{v}} \end{aligned} \tag{9.3.99}$$

Substitution in Eq. (9.3.96) then yields the log-likelihood function as

$$\begin{aligned} \ln p(\mathbf{Z}_N\,|\,\boldsymbol{\theta}) = &-\frac{1}{2}\sum_{k=1}^{N} \ln\{|\mathbf{H}\mathbf{V}_{\tilde{\mathbf{x}}}(k\,|\,k-1)\mathbf{H}^T + \mathbf{V}_{\mathbf{v}}|\} - \frac{Nq}{2}\ln(2\pi) \\ &-\frac{1}{2}\sum_{k=1}^{N} [\mathbf{z}(k) - \mathbf{H}\hat{\mathbf{x}}(k\,|\,k-1)]^T \\ &\times [\mathbf{H}\mathbf{V}_{\tilde{\mathbf{x}}}(k\,|\,k-1)\mathbf{H}^T + \mathbf{V}_{\mathbf{v}}]^{-1}[\mathbf{z}(k) - \mathbf{H}\hat{\mathbf{x}}(k\,|\,k-1)] \end{aligned} \tag{9.3.100}$$

It only remains to obtain the quantities $\hat{\mathbf{x}}(k\,|\,k-1)$ and $\mathbf{V}_{\tilde{\mathbf{x}}}(k\,|\,k-1)$. Because of the form of the system model Eq. (9.3.94), it is clear that these quantities can be obtained by using the discrete Kalman filter algorithms of Chapter 6. The algorithms are as under:

$$\begin{aligned} &\hat{\mathbf{x}}(k+1\,|\,k) \\ &\quad = \boldsymbol{\phi}\hat{\mathbf{x}}(k\,|\,k-1) + \mathbf{K}(k+1, k)[\mathbf{z}(k) - \mathbf{H}\hat{\mathbf{x}}(k\,|\,k-1)] + \mathbf{B}\mathbf{u}(k) \end{aligned} \tag{9.3.101a}$$

$$\mathbf{K}(k+1\,|\,k) = \boldsymbol{\phi}\mathbf{V}_{\tilde{\mathbf{x}}}(k\,|\,k-1)\mathbf{H}^T[\mathbf{H}\mathbf{V}_{\tilde{\mathbf{x}}}(k\,|\,k-1)\mathbf{H}^T + \mathbf{V}_{\mathbf{v}}]^{-1} \tag{9.3.101b}$$

$$\begin{aligned} \mathbf{V}_{\tilde{\mathbf{x}}}(k+1\,|\,k) &= \boldsymbol{\phi}\mathbf{V}_{\tilde{\mathbf{x}}}(k\,|\,k-1)\boldsymbol{\phi}^T + \boldsymbol{\Gamma}\mathbf{V}_{\mathbf{w}}\boldsymbol{\Gamma}^T \\ &- \boldsymbol{\phi}\mathbf{V}_{\tilde{\mathbf{x}}}(k\,|\,k-1)\mathbf{H}^T[\mathbf{H}\mathbf{V}_{\tilde{\mathbf{x}}}(k\,|\,k-1)\mathbf{H}^T + \mathbf{V}_{\mathbf{v}}]^{-1}\mathbf{H}\mathbf{V}_{\tilde{\mathbf{x}}}(k\,|\,k-1)\boldsymbol{\phi}^T \end{aligned} \tag{9.3.101c}$$

These equations differ from the usual Kalman algorithms in that the quantity $\hat{\mathbf{x}}(k\,|\,k-1)$ is the one-stage predicted estimate of $\mathbf{x}(k)$ conditioned on the parameter vector $\boldsymbol{\theta}$. Also, Eq. (9.3.101a) has been modified to take into account the nonrandom input $\mathbf{u}(k)$.

While equations (9.3.100) and (9.3.101) give us a means of determining the log-likelihood function, there still remains the problem of finding the estimate $\hat{\boldsymbol{\theta}}$ by maximizing $\ln p(\mathbf{Z}_N \mid \boldsymbol{\theta})$. In general, this is a fairly difficult task. We can consider an iterative approach in which the estimate $\boldsymbol{\theta}^j$ at the jth iteration is obtained from the estimate at the previous iteration by using an algorithm of the type

$$\hat{\boldsymbol{\theta}}^j = \hat{\boldsymbol{\theta}}^{j-1} + \Delta\hat{\boldsymbol{\theta}} \tag{9.3.102}$$

where $\Delta\hat{\boldsymbol{\theta}}$ represents a correction term. Usually, $\Delta\hat{\boldsymbol{\theta}}$ is chosen to be of the form

$$\Delta\hat{\boldsymbol{\theta}} = -\mathbf{K}(j)\frac{\partial \ln p(\mathbf{Z}_N \mid \boldsymbol{\theta})}{\partial \boldsymbol{\theta}}\bigg|_{\boldsymbol{\theta}=\hat{\boldsymbol{\theta}}^{j-1}} \tag{9.3.103}$$

where $\mathbf{K}(j)$ represents a suitably chosen matrix. The solution is an off-line solution and the data set $\mathbf{Z}_N$ may have to be used over several iterations to obtain reasonably converging solutions. In general, the computation of $\Delta\hat{\boldsymbol{\theta}}$ can be quite tedious.

If the dimension of the parameter vector $\boldsymbol{\theta}$ is small and if the range of values which $\boldsymbol{\theta}$ takes are known or can be reasonably guessed, we can simply evaluate $\ln p(\mathbf{Z}_N \mid \boldsymbol{\theta})$ for several values of $\boldsymbol{\theta}$ over its range and choose as our estimate, that value which yields the largest value of $\ln p(\mathbf{Z}_N \mid \boldsymbol{\theta})$.

While the solution of the likelihood equations is, in general, difficult, we can obtain simple expressions for the estimates in some special cases. We illustrate this by means of two examples.

Example 9.3.8

Consider the scalar model

$$x(k+1) = \phi x(k) + w(k) + u(k) \tag{9.3.104a}$$

$$z(k) = x(k) \tag{9.3.104b}$$

Let us assume that we want to identify ϕ and V_w so that $\boldsymbol{\theta} = [\phi \quad V_w]^T$. It follows from Eq. (9.3.100) that

$$\ln p(\mathbf{Z}_N \mid \boldsymbol{\theta}) = -\frac{1}{2}\sum_{k=1}^{N} \ln V_{\tilde{x}}(k \mid k-1) - \frac{N}{2}\ln(2\pi) - \frac{1}{2V_{\tilde{x}}(k \mid k-1)}\sum_{k=1}^{N}[z(k) - \hat{x}(k \mid k-1)]^2 \tag{9.3.105}$$

Now

$$\begin{aligned}\hat{x}(k \mid k-1) &= \mathrm{E}\{x(k) \mid \mathbf{Z}_{k-1}, \boldsymbol{\theta}\} \\ &= \mathrm{E}\{\phi x(k-1) + w(k-1) + u(k-1) \mid \mathbf{Z}_{k-1}, \boldsymbol{\theta}\} \\ &= \phi z(k-1) + u(k-1)\end{aligned} \tag{9.3.106}$$

so that

$$\begin{aligned}\tilde{x}(k \mid k-1) &= x(k) - \hat{x}(k \mid k-1) \\ &= z(k) - \phi z(k-1) - u(k-1)\end{aligned}$$

and

$$V_{\tilde{x}}(k \mid k-1) = [z(k) - \phi z(k-1) - u(k-1)]^2 \tag{9.3.107}$$

Furthermore, we can write

$$\begin{aligned} V_w &= \mathrm{E}\{w^2(k)\} = \mathrm{E}\{[x(k+1) - \phi x(k) - u(k)]^2\} \\ &= [z(k+1) - \phi z(k) - u(k)]^2 \qquad (9.3.108) \\ &= V_{\tilde{x}}(k \mid k-1) \end{aligned}$$

Substitution of Eqs. (9.3.106) and (9.3.107) in Eq. (9.3.105) yields the log-likelihood function as

$$\begin{aligned} \ln p(\mathbf{Z}_N \mid \boldsymbol{\theta}) = &-\frac{N}{2}\ln V_w - \frac{N}{2}\ln(2\pi) \\ &-\frac{V_w^{-1}}{2}\sum_{k=1}^{N}[z(k) - \phi z(k-1) - u(k-1)]^2 \end{aligned}$$

The ML estimates obtained by setting the partial derivatives of $\ln p(\mathbf{Z}_N \mid \boldsymbol{\theta})$ with respect to ϕ and V_w equal to zero are

$$\hat{\phi} = \frac{\sum_{k=1}^{N} z(k-1)[z(k) - u(k-1)]}{\sum_{k=1}^{N} z^2(k-1)}$$

and

$$\hat{V}_w = \frac{1}{N}\sum_{k=1}^{N}[z(k) - \hat{\phi} z(k-1) - u(k-1)]^2$$

Example 9.3.9

We consider the scalar autoregressive model

$$x(k) + \sum_{i=1}^{N} \alpha_i x(k-i) = w(k) + u(k)$$

$$z(k) = x(k)$$

where $w(k)$ is zero mean, white, Gaussian with variance V_w.

We want to identify the coefficients α_i and V_w. The parameter vector, therefore, is

$$\boldsymbol{\theta} = [\alpha_1 \quad \alpha_2 \quad \cdots \quad \alpha_K \quad V_w]^T$$

Because the observations are noise-free, it easily follows from our derivation in Example 9.3.8 that

$$\hat{x}(k \mid k-1) = -\sum_{i=1}^{K} \alpha_i z(k-i) + u(k)$$

and

$$V_w = V_{\tilde{x}}(k \mid k-1) = \left[z(k) + \sum_{i=1}^{K} \alpha_i z(k-i) - u(k)\right]^2$$

The log-likelihood function, therefore, is

$$\ln p(\mathbf{Z}_N \mid \boldsymbol{\theta}) = -\frac{N}{2}\ln V_w - \frac{N}{2}\ln(2\pi) - \frac{V_w^{-1}}{2}\sum_{k=1}^{K}\left[z(k) + \sum_{i=1}^{K}\alpha_i z(k-i) - u(k)\right]^2$$

It follows that the estimates can be obtained by solving the set of equations

$$\sum_{i=1}^{K}\hat{\alpha}_i\left[\sum_{k=1}^{K} z(k-i)z(k-j)\right] = \sum_{k=1}^{K}[u(k) - z(k)]z(k-j) \qquad j = 1, 2, \ldots, K$$

$$\hat{V}_w = \frac{1}{N}\sum_{k=1}^{N}\left[z(k) + \sum_{i=1}^{K}\hat{\alpha}_i z(k-i) - u(k)\right]^2$$

We conclude this section by noting that while our results were derived under Gaussian assumptions, as the number of observations become very large, these assumptions are not crucial to our results. A lower bound on the maximum achievable accuracy using the ML technique can be obtained from the Cramér–Rao bound discussed in Chapter 5 [36].

9.3.10 Section Summary

In this section we have considered the application of the concepts of estimation theory to the identification of linear time-invariant systems. We assumed that the system was of known structure so that the problem is essentially one of estimating certain parameters associated with the model. The models that we were concerned with primarily were the autoregressive-moving average model and the state-space models.

In Section 9.3.1 we considered the estimation of parameters in an ARMA model representation of a stationary stochastic process. We started with a discussion of AR models and derived the Yule–Walker equations for estimating the model parameters, given values of the autocorrelation sequence. These equations, which are linear in the AR parameters, are identical to the equations arising in the linear prediction of AR processes considered in Chapter 6. We presented the Levinson algorithm and the corresponding lattice structure for solving the Yule–Walker equations in terms of the partial correlation coefficients in Section 9.3.2. In Section 9.3.3 we derived a sequential version of this algorithm which does not involve computation of the autocorrelation coefficients. In Section 9.3.4 we presented an alternative sequential identification algorithm similar to the Kalman filter algorithms of Chapter 6. Estimation of MA parameters considered in Section 9.3.5 requires solution of a set of nonlinear equations unlike the case of AR models. In Section 9.3.6 we considered the extension of these results to ARMA models. Since the solution of the modified Yule–Walker equations for this case requires solving a set of nonlinear equations for the MA parameters, an alternative technique was presented, which involved replacing the ARMA model with an equivalent higher-order model. Many of these results can also be extended to vector ARMA processes.

Identification of parameters in a state-variable representation of the system was considered in Section 9.3.7. By treating the parameters as states, we can pose the problem as one of estimation in nonlinear state-variable models (Section 9.3.8). Approximate algorithms for the parameter estimates were obtained by using the extended Kalman filter algorithms of Chapter 6.

For nonrandom parameter estimation, ML estimates are very attractive because of their properties. For state-variable models, expressions for the likelihood function were derived in Section 9.3.9 under Gaussian assumptions. The resulting equations for the estimates are usually difficult to solve. In most cases an off-line iterative technique must be resorted to. For some special cases, the algorithms take simple forms.

The modified Yule–Walker equations have been the subject of much research in recent years and several methods for solving these equations have been reported in the literature. The effects of noise in the measurements have also been studied. As noted earlier, the first step in obtaining a solution is to estimate the correlation coefficients of the observed process from observed data. Clearly, the statistical properties of the parameter estimates of the AR or ARMA models will depend on factors such as the length of the data record or the method used for computing the correlation coefficients.

An alternative to using time-averaged estimates to replace the autocorrelation coefficients is to formulate the identification problem directly in terms of the data sequences. That is, instead of defining the cost function as the expectation of the error sequence, we define it as the sum of the squares of the residual over the time interval of interest. Such an approach is referred to as a *least-squares approach* and is essentially equivalent to replacing the expectation of the error by a time average. The interested reader is referred to the literature for discussions of least-squares approaches (see, for example, [22]).

EXERCISES

9.1. In a two-category classification problem, the class-conditional densities satisfy a Cauchy distribution

$$p(x\,|\,H_i) = \frac{1}{\pi\alpha}\,\frac{1}{1+\left(\dfrac{x-\beta_i}{\alpha}\right)} \qquad i = 1,2$$

Assuming equal prior probabilities for the two classes, determine the discriminant function. What is the probability of misclassification?

9.2. The Bhattacharya coefficient is defined as

$$\rho = \int [p(\mathbf{x}\,|\,H_1)p(\mathbf{x}\,|\,H_2)]^{1/2}\,d\mathbf{x}$$

Show that the error probability for a two-class problem satisfies

$$P_e \leqslant \sqrt{P_1 P_2}\,\rho \leqslant \tfrac{1}{2}\rho$$

9.3. In the two-category classification problem, instead of assigning a pattern $\mathbf{x}$ to one of the two classes, we may, under certain circumstances, reject it as being unrecognizable. If we consider rejection as assigning a pattern to class H_3, we can formulate the problem as a ternary hypothesis-testing problem. We make the following assignment of costs:

$$c_{ii} = 0 \qquad c_{ij} = c_s \qquad i,j = 1,2 \qquad c_{i3} = c_r \qquad i = 1,2$$

Show that the optimal classification rule is to assign $\mathbf{x}$ to class i if

$$P(H_i \mid \mathbf{x}) \geqslant P(H_j \mid \mathbf{x}) \quad \text{and} \quad P(H_i \mid \mathbf{x}) \geqslant 1 - \frac{c_r}{c_s} \qquad i,j = 1,2$$

and reject otherwise and that the corresponding discriminant functions are given by

$$f_i(x) = \begin{cases} p(\mathbf{x} \mid H_i)P(H_i) & i = 1,2 \\ \dfrac{c_a - c_s}{c_s} \displaystyle\sum_{j=1}^{2} p(x \mid H_j)P(H_j) & i = 3 \end{cases}$$

9.4. In a binary scalar pattern classification problem, the distributions are Gaussian as follows:

$$p(x \mid H_i) = \frac{1}{\sqrt{2\pi}\,V_i} \exp\left[-\frac{1}{2}\frac{(x - m_i)^2}{V_i^2}\right] \qquad i = 1,2$$

Assume that $P_1 = P_2 = \frac{1}{2}$. Generate a set of samples corresponding to each class and use the decision rules of Section 9.3 to classify them. Obtain an estimate of the probability of error by taking the ratio of the number of correctly classified samples to the total number of samples and compare with the theoretical value. Consider the cases

(a) $V_i = 1$ and $m_i = (-1)^i, \quad i = 1,2$

(b) $V_1 = \frac{1}{2}$, $V_2 = 4$ and $m_i = 0, \quad i = 1,2$

(c) $V_1 = \frac{1}{2}$, $V_2 = 4$ and $m_1 = -1$ and $m_2 = 1$

9.5. In a supervised learning problem, the class conditional densities are given as

$$p(x \mid \theta_i) = \begin{cases} \theta_i e^{-\theta_i x} & x \geqslant 0 \\ 0 & \text{otherwise} \end{cases}$$

where θ_i are nonrandom. Given N_1 samples from class 1 and N_2 samples from class 2, set up the learning algorithm for estimating θ_i. What is the discriminant function?

9.6. In Exercise 9.5, assume that θ_i are random parameters. Find the algorithms for learning θ_i in this case by choosing a suitable reproducing form for the prior density of θ_i. (Refer to Section 5.7.)

9.7. Repeat Exercise 9.6 if the class conditional densities are

$$P(x \mid \theta_i) = \theta_i^{x_i}(1 - \theta_i)^{x_i}$$

where θ_i is the probability that $x_i = 1$.

9.8. Obtain the learning algorithms in Example 9.2.2.

9.9. In a two-category unsupervised learning problem, the class-conditional densities are

$$P(\mathbf{x} \mid H_i, \boldsymbol{\theta}_i) = \prod_{j=1}^{n} \theta_i^{x_j}(1 - \theta_i)^{1-x_j} \qquad i = 1,2$$

so that the mixture density for $\mathbf{x}$ is

$$P(\mathbf{x}\,|\,\boldsymbol{\theta}) = \sum_{i=1}^{2} P(\mathbf{x}\,|\,H_i, \boldsymbol{\theta}_i)P(H_i)$$

Find the maximum likelihood estimate $\hat{\boldsymbol{\theta}}_i$ of $\boldsymbol{\theta}_i$.

9.10. Formulate the nonsupervised learning pattern classification problem for the data in Exercise 9.4, assuming that the mean of class H_1 is not known. Simulate the decision directed scheme for the problem. Plot the curve of estimated mean versus number of learning samples and the graph of an estimate of the probability versus number of total test samples for the scheme.

9.11. Entropy is a statistical measure of uncertainty. Features that reduce the uncertainty within a pattern vector can be considered to be more informative than those that do not. A technique for feature selection therefore is to minimize the entropy of the pattern classes. The entropy of the ith class is defined as

$$H_i = -\int_x p(\mathbf{x}\,|\,H_i) \ln p(\mathbf{x}\,|\,H_i)\, d\mathbf{x}$$

We seek an $r \times n$ matrix $\mathbf{T}$ which, operating on the n-dimensional vector $\mathbf{x}(n)$, will yield a feature vector of reduced dimension $\mathbf{x}(r), r < n$:

$$\mathbf{x}(r) = \mathbf{T}\mathbf{x}(n)$$

Let us consider the two-category problem in which the class conditional densities are Gaussian with equal covariances $\mathbf{V}$ but unequal means. Show that to minimize the entropy, the rows of $\mathbf{T}$ must be chosen as those eigenvectors of $\mathbf{V}$ corresponding to the r smallest eigenvalues.

9.12. The estimate of the autocorrelation sequence of a data sequence of 1000 samples for the first 10 lags is as follows:

$$\hat{R}_x(0) = 1,\ \hat{R}_x(1) = 0.05,\ \hat{R}_x(2) = 0.25,\ \hat{R}_x(3) = 0.025$$

$$\hat{R}_x(4) = 0.075,\ \hat{R}_x(5) = 0.25,\ \hat{R}_x(6) = 0.125,\ \hat{R}_x(7) = 0.025$$

$$\hat{R}_x(8) = 0.125,\ \hat{R}_x(9) = -0.0125,\ \hat{R}_x(10) = 0$$

Assume that we want to fit a 10th order model for the data. Find the estimate of $\boldsymbol{\alpha}$.

9.13. Establish Eq. (9.3.63).

9.14. Consider the system

$$\dot{x}(t) = 0.5x(t) + bw(t)$$

where $w(t)$ is zero mean, white noise with unity variance. The observations are

$$z(t) = x(t) + v(t)$$

where $v(t)$ is zero mean, white with covariance kernel 4, uncorrelated with $w(t)$. We wish to identify the unknown parameter b.

(a) Set up the extended Kalman algorithms for this problem. Can you identify b uniquely?

(b) Define the variable $y(t)$ as $y(t) = (1/b)x(t)$. Set up the problem in terms of the variable $y(t)$ and obtain the equations for the identification of the unknown parameter. Is this scheme better or worse than the scheme in part (a)?

9.15. Consider the system

$$\mathbf{x}(k+1) = \boldsymbol{\phi}\mathbf{x}(k) + \boldsymbol{\Gamma}\mathbf{w}(k)$$
$$\mathbf{z}(k) = \mathbf{H}\mathbf{x}(k) + \mathbf{v}(k)$$

with the usual assumptions on $\mathbf{w}$ and $\mathbf{v}$.

(a) Obtain the ML estimate of $\mathbf{H}$ assuming that $\boldsymbol{\Gamma}$ is known.

(b) Obtain the ML estimate of $\boldsymbol{\Gamma}$ assuming that $\mathbf{H}$ is known.

9.16. Use the state-augmentation method in Exercise 9.15 and obtain algorithms for estimating $\mathbf{H}$ and $\mathbf{B}$.

9.17. Derive the algorithms for estimating $\mathbf{B}$ in the model of Eq. (9.3.94) using the state-augmentation technique. Simplify the algorithms when we have perfect observation of the state so that

$$\mathbf{z}(k) = \mathbf{x}(k)$$

9.18. Consider ML estimation in the model of Eq. (9.3.94) when the observation noise is zero. Assume that the unknown parameters are the elements of $\boldsymbol{\phi}$, $\mathbf{B}$, and $\mathbf{V_w}$.

(a) Derive the equations for obtaining $\hat{\boldsymbol{\theta}}$. Assume $\mathbf{H} = \mathbf{I}$.

(b) Do these algorithms specialize to the algorithms of Example 9.3.8 for the scalar case?

(c) Can you write the algorithms in sequential form?

9.19. Set up the ML estimation algorithms for the model of Eq. (9.3.94) if $\boldsymbol{\Gamma} = \mathbf{0}$. Assume that $\mathbf{V_v}$ and $\mathbf{H}$ are unknown.

9.20. In the problem of Example 9.3.8, assume that only $\boldsymbol{\phi}$ is unknown. Find $\boldsymbol{\phi}_{\text{ML}}$. Is it an unbiased estimate? Find a lower limit on the achievable accuracy in identification. (Use the Cramér–Rao bound.)

9.21. Consider the general ARMA model of the form

$$x(k) + \sum_{i=1}^{M} \alpha_i x(k-i) = \sum_{i=1}^{M} \beta_i u(k-i) + w(k)$$

where $u(\cdot)$ is a known input and $w(k)$ is a zero mean, Gaussian white noise with $\mathrm{E}\{w(k)w^T(j)\} = V_w \delta_{kj}$. Derive the ML estimates $\hat{\alpha}_i$, $\hat{\beta}_j$, $i = 1, \ldots, N, j = 1, \ldots, M$. Assume that $z(k) = x(k)$.

9.22. In Exercise 9.21 assume that $w(k)$ is colored with

$$\mathrm{E}\{w(k)w(k-j)\} = \begin{cases} V_j & j = 0, 1, \ldots, N \\ 0 & \text{otherwise} \end{cases}$$

with observations

$$z(k) = x(k) + v(k)$$

Find the ML estimates of $\hat{\alpha}_i$, $\hat{\beta}_i$, and V_j.

9.23. In Section 9.3.9 we derived the ML estimates of the parameters in a state-variable model of a linear time-invariant system under the assumption that the parameters were nonrandom. Derive MAP algorithms for the model of Eq. (9.3.94) if the unknown parameter vector $\boldsymbol{\theta}$ is Gaussian with mean $\boldsymbol{\mu}_{\boldsymbol{\theta}}$ and variance $\mathbf{V}_{\boldsymbol{\theta}}$.

9.24. [37,38] Let $x(\cdot)$ be a zero-mean, stationary Gaussian process whose correlation coefficients for the first N lags $r(0), r(1), r(2), \ldots, r(N-1)$ are known. Suppose that we now want to determine $r(N)$. One way of doing this is to maximize the entropy of the process defined as

$$H = -\int p(\mathbf{x}) \ln p(\mathbf{x})\, d\mathbf{x}$$

where $p(x)$ denotes the multivariate density function of the process. For the N-variate Gaussian density function with covariance matrix R_N, where

$$R_N = \begin{bmatrix} r_0 & r_1 & r_2 & \cdots & r_N \\ r_1 & r_0 & r_1 & \cdots & r_{N-1} \\ \vdots & & & & \vdots \\ r_N & r_{N-1} & r_{N-2} & \cdots & r_0 \end{bmatrix}$$

show that maximizing the entropy is equivalent to maximizing

$$\ln(2\pi)^{N/2} \det[R_N]^{1/2}$$

Hence determine the value of r_N that maximizes the entropy of the process. Can you interpret your results in terms of the autoregressive models discussed in Section 9.3.1?

REFERENCES

1. R. Schalkoff, *Pattern Recognition: Statistical, Structural and Neural Approaches*, Wiley, New York, 1992.
2. R. O. Duda and P. E. Hart, *Pattern Classification and Scene Analysis*, Wiley, New York, 1973.
3. P. R. Devijver and J. Kittler, *Pattern Recognition: A Statistical Approach*, Prentice Hall, Englewood Cliffs, NJ, 1982.
4. P. R. Krishnaiah and L. N. Kanal, Eds., *Handbook of Statistics, Vol. 2, Classification, Pattern Recognition and Reduction of Dimensionality*, North-Holland, Amsterdam, 1982.
5. T. Y. Young and K. S. Fu, Eds., *Handbook of Pattern Recognition and Image Processing*, Academic Press, New York, 1986.
6. K. S. Fu, *Sequential Methods in Pattern Recognition and Machine Learning*, Academic Press, New York, 1968.
7. T. M. Cover and P. E. Hart, Nearest neighbor pattern classification, *IEEE Trans. Inf. Theory*, **IT-13**, 21–27 (1967).
8. E. A. Patrick, *Fundamentals of Pattern Recognition*, Prentice Hall, Englewood Cliffs, NJ, 1972.
9. C. W. Therrien, *Decision, Estimation and Classification: An Introduction to Pattern Recognition and Related Topics*, Wiley, New York, 1989.
10. J. Tou and R. Gonzalez, *Pattern Recognition Principles*, Addison-Wesley, Reading, MA, 1974.
11. K. Fukunaga, *Introduction to Statistical Pattern Recognition*, 2nd ed., Academic Press, New York, 1990.

12. J. Spragins, Reproducing distributions for machine learning, *Technical Report 6103-7*, Stanford University, Palo Alto, CA, November 1963.
13. N. Abramson and D. Braverman, Learning to recognize patterns in a random environment, *IEEE Trans. Inf. Theory*, **IT-8**, No. 5, September, 58–63 (1962).
14. D. G. Keehn, A note on learning Gaussian properties, *IEEE Trans. Inf. Theory*, **IT-11**, No. 1, January, 126–132 (1965).
15. H. Teicher, Identifiability of mixtures, *Ann. Math. Stat.*, **32**, March, 244–248 (1961).
16. S. J. Yakowitz and J. D. Spragins, On the identifiability of finite mixtures, *Ann. Math. Stat.*, **39**, February, 209–214 (1968).
17. S. C. Fralick, Learning to recognize patterns without a teacher, *IEEE Trans. Inf. Theory*, **IT-13**, No. 1, January, 544–549 (1967).
18. R. F. Daly, The adaptive binary detection problem on the real line, *Technical Report 2003-3*, Stanford University, Palo Alto, CA, February 1962.
19. H. J. Scudder III, Adaptive communication receivers, *IEEE Trans. Inf. Theory*, **IT-4**, No. 5, October, 544–549 (1965).
20. B. V. Dasarathy, Ed., *Nearest Neighbor (NN) Norms: Nearest Neighbor Pattern Classification Techniques*, IEEE Computer Society Press, Los Alamitos, CA, 1991.
21. P. Eykhoff, *System Identification*, Wiley, New York, 1974.
22. L. Ljung, *System Identification: Theory for the User*, Prentice Hall, Englewood Cliffs, NJ, 1987.
23. T. Soderstrom and P. Stoica, *System Identification*, Prentice Hall, Englewood Cliffs, NJ, 1989.
24. G. M. Jenkins and D. G. Watts, *Spectral Analysis and Its Applications*, Holden-Day, San Francisco, 1968.
25. L. H. Koopmans, *The Spectral Analysis of Time Series*, Academic Press, New York, 1974.
26. S. M. Kay, *Modern Spectral Estimation: Theory and Application*, Prentice Hall, Englewood Cliffs, NJ, 1988.
27. S. L. Marple, Jr., *Digital Spectral Analysis with Applications*, Prentice Hall, Englewood Cliffs, NJ, 1987.
28. S. Haykin, Ed., *Advances in Spectrum Analysis and Array Processing*, Vols. 1 and 2, Prentice Hall, Englewood Cliffs, NJ, 1991.
29. M. D. Srinath and M. M. Viswanathan, Sequential algorithm for identification of parameters of an autoregressive process, *IEEE Trans. Autom. Control*, **AC-20**, No. 4, August, 542–546 (1975).
30. J. P. Burg, Maximum entropy spectral analysis, Ph.D. thesis, Stanford University, Stanford, CA, May 1975.
31. H. Akaike, A new look at the statistical model identification, *IEEE Trans. Autom. Control*, **AC-19**, No. 6, December, 716–723 (1974).
32. D. Graupe, D. J. Krause, and J. B. Moore, Identification of autoregressive moving-average parameters of time series, *IEEE Trans. Autom. Control*, **AC-20**, No. 1, February, 104–107 (1975).
33. S. Li and B. W. Dickinson, Application of the lattice filter to robust estimation of AR and ARMA models, *IEEE Trans. Acoust. Speech Signal Process.*, **36**, No. 4, April, 502–512 (1988).

34. S. Li and B. W. Dickinson, A comparison of two linear methods of estimating the parameters of ARMA models, *IEEE Trans. Autom. Control*, **AC-34**, No. 8, August, 915–917 (1989).
35. R. L. Kashyap, Estimation of parameters in a partially whitened representation of stochastic processes, *IEEE Trans. Autom. Control*, **AC-19**, No. 1, February, 13–21 (1974).
36. R. L. Kashyap, Maximum likelihood identifications of stochastic linear systems, *IEEE Trans. Autom. Control*, **AC-15**, No. 1, February, 25–34 (1970).
37. A. van den Bos, Alternative interpretation of maximum entropy spectral analysis, *IEEE Trans. Inf. Theory*, **IT-17**, No. 4, July, 493–494 (1971).
38. J. A. Edward and M. Fitelson, Notes on maximum-entropy processing, *IEEE Trans. Inf. Theory*, **IT-19**, No. 2, March, 232–234 (1973).

APPENDIX A

Bilateral Transforms

We have used the bilateral or two-sided Laplace and z-transforms in the text in our discussions of power spectra of stationary stochastic processes. We recall that the unilateral Laplace transform of a causal function $f(t)$ is defined as

$$\mathcal{L}\{f(t)\} = F(s) = \int_0^\infty f(t)e^{-st}\,dt \tag{A.1}$$

where s is a complex variable. For this integral to exist, there must be a real constant α such that

$$|f(t)| < k_1 e^{\alpha t}$$

The integral is then absolutely convergent for Re $s > \alpha$, so that the region of convergence is to the right of a vertical line in the s-plane through the point Re $s = \alpha$.

For functions defined for negative values of t, that is, for anticausal functions, the one-sided Laplace transform is given by

$$\mathcal{L}\{f(t)\} = F(s) = \int_{-\infty}^0 f(t)e^{-st}\,dt \tag{A.2}$$

If $|f(t)| < k_2 e^{\beta t}$ for some constant $\beta > 0$, the integral exists for values for s such that Re $s = \sigma < \beta$. The region of convergence is therefore to the left of a vertical line through β.

We note that replacing t with $-t$ in Eq. (A.2) yields

$$F(s) = \int_0^\infty f(-t)e^{st}\,dt \tag{A.3}$$

If we therefore define the causal function $f_1(t)$ as

$$f_1(t) = f(-t) \qquad t \geqslant 0$$

we have

$$F(s) = F_1(-s)$$

Thus the properties of the $\mathscr{L}$-transform of the anticausal function $f(t)$ can be determined in terms of the one-sided transform of the causal function $f_1(t)$.

Let $f(t)$ be a function defined over the entire time axis such that

$$|f(t)| < e^{\beta t} \qquad \text{for} \quad t < 0$$

and

$$|f(t)| < e^{\alpha t} \qquad \text{for} \quad t \geqslant 0$$

We define the bilateral transform of $f(t)$ as

$$F(s) = \int_{-\infty}^{\infty} f(t)e^{-st}\,dt \tag{A.4}$$

The region of convergence of $F(s)$ is the vertical strip in the s-plane defined by $\alpha < \text{Re}\, s < \beta$. It follows that for the bilateral Laplace transform of a function to exist, β must be greater than α.

The time function $f(t)$ corresponding to a transform $F(s)$ can be obtained as

$$f(t) = \frac{1}{2\pi j}\int_{c-j\infty}^{c+j\infty} F(s)e^{st}\,ds \tag{A.5}$$

Here the integration is performed along a vertical line in the s-plane corresponding to $\text{Re}\, s = c$ as $\text{Im}\, s$ varies from $-\infty$ to ∞. The line is chosen to lie in the region of convergence of $F(s)$.

In evaluating the integral in Eq. (A.5), we note that if the function $f(t)$ is causal, the region of convergence will be to the right of all the singularities of $F(s)$. If $f(t)$ is anticausal, the region of convergence will be to the left of the singularities of $F(s)$. If $f(t)$ has both a causal and an anticausal part, the singularities of $F(s)$ to the right of the region of convergence will correspond to the anticausal part of $f(t)$ while those to the left will correspond to the causal part. If $F(s)$ vanishes as $s \to \infty$, the integral can be most easily evaluated by closing the contour of integration with a semicircle of infinite radius and using the residue theorem. If the contour is closed in the left half-plane we obtain the causal part of $f(t)$.

For discrete sequences, we can similarly define a bilateral Z-transform. We recall that given the causal sequence $f(k)$ defined for nonnegative integer values of k, we define the Z-transform of the sequence as

$$Z\{f(k)\} = F(z) = \sum_{k=0}^{\infty} f(k)z^{-k} \tag{A.6}$$

The series is absolutely convergent for all z such that $|z| > R_{c^+}$, where

$$R_{c^+} = \overline{\lim}\,|f(k)|^{1/k}$$

Here, $\overline{\lim}$ refers to the largest of the limit points of $|f(k)|^{1/k}$ as $k \to \infty$. For sequences defined for all integer values of k, we define a bilateral or two-sided Z-transform

$$F(z) = \sum_{k=-\infty}^{\infty} f(k)z^{-k} \tag{A.7}$$

Let R_{c^+} and R_{c^-} be the largest and smallest of the limit points to which $|f(k)|^{1/k}$ tends as $k \to \infty$ and $k \to -\infty$, respectively. Then $F(z)$ converges absolutely for all z such that $R_{c^+} < |z| < R_{c^-}$. It follows, therefore, that for the bilateral transform to exist, we must have $R_{c^+} < R_{c^-}$. If $f(k)$ is a causal sequence, its inverse transform is given by

$$f(k) = \frac{1}{2\pi j} \oint_{\Gamma} z^{k-1} F(z)\, dz \tag{A.8}$$

where Γ is any closed contour in the region of convergence of $F(z)$. If $f(k)$ is an anticausal sequence, the inverse transform is given by

$$f(k) = \frac{1}{2\pi j} \oint_{\Gamma} F(z) z^{-(k+1)}\, dz \qquad \text{for} \quad k < 0 \tag{A.9}$$

REFERENCES

1. H. Freeman, *Discrete-Time Systems: An Introduction to the Theory*, Wiley, New York, 1965.
2. G. Doetsch, *Guide to the Applications of the Laplace Transformation*, Van Nostrand, Princeton, NJ, 1961.

APPENDIX B

Calculus of Extrema

We consider the problem of optimizing a function $\phi(x_1, x_2, \ldots, x_n)$ of n variables $x_1, x_2, \ldots, x_n$. A necessary condition that the function have a stationary value is that

$$d\phi = \frac{\partial \phi}{\partial x_1} dx_1 + \frac{\partial \phi}{\partial x_1} dx_2 + \cdots + \frac{\partial \phi}{\partial x_n} dx_n = 0 \tag{B.1}$$

for all permissible values of the differentials $dx_1, \ldots, dx_n$. If the n variables are all independent, this is equivalent to the n conditions

$$\frac{\partial \phi}{\partial x_1} = \frac{\partial \phi}{\partial x_2} = \cdots = \frac{\partial \phi}{\partial x_n} = 0 \tag{B.2}$$

If the n variables are dependent, the method of *Lagrange multipliers* is most useful in obtaining necessary conditions for a stationary point. Let the dependence between the variables be expressed in terms of the *constraint* equations

$$f_i(x_1, x_2, \ldots, x_n) = 0 \qquad i = 1, \ldots, N \tag{B.3}$$

We form the augmented function

$$\phi^1(x_1, x_2, \ldots, x_n; \lambda_1 \cdots \lambda_N) = \phi(x_1, x_2, \ldots, x_n) + \sum_{i=1}^{N} \lambda_i f_i(x_1 \cdots x_n) \tag{B.4}$$

The problem of optimizing the function $\phi(\cdot)$ subject to the constraints, Eq. (B.3) is then equivalent to optimizing the function $\phi^1(\cdot)$ in which the variables are treated as being independent. The constants λ_i are known as Lagrange multipliers.

We can therefore obtain necessary conditions similar to Eq. (B.1) as

$$\frac{\partial \phi^1}{\partial x_i} = \frac{\partial \phi}{\partial x_i} + \sum_{i=1}^{N} \lambda_i \frac{\partial f_i}{\partial x_i} = 0 \qquad i = 1, 2, \ldots, n \tag{B.5}$$

The set of Eqs. (B.3) and (B.5) constitute a set of $n + N$ equations in the $(n + N)$ variables $\{x_1, \ldots, x_n; \lambda_1 \ldots \lambda_N\}$ and can be solved in any suitable manner.

CALCULUS OF VARIATIONS

In many problems the quantities x_i are dependent on another variable (for example, time). The function to be optimized (cost function) is an integral of the form

$$I = \int_{t=a}^{t=b} \phi(x_1, \ldots, x_n, \dot{x}_1, \ldots, \dot{x}_n, t)\, dt \tag{B.6}$$

where

$$\dot{x}_i = \frac{dx_i}{dt}$$

We confine ourselves to the case of a single variable x, so that Eq. (B.6) becomes

$$I = \int_{t=a}^{t=b} \phi(x, \dot{x}, t)\, dt \tag{B.7}$$

To optimize I, we consider a variation δx in x at a fixed value of t of the form

$$\delta x(t) = \epsilon \eta(t) \tag{B.8}$$

where ϵ is a fixed scalar and $\eta(\cdot)$ is an arbitrary function. The corresponding change in the functional ϕ is then given by

$$\begin{aligned} \Delta\phi &= \phi(x + \delta x, \dot{x} + \delta\dot{x}, t) - \phi(x, \dot{x}, t) \\ &= \frac{\partial\phi}{\partial x}\delta x + \frac{\partial\phi}{\partial\dot{x}}\delta\dot{x} + \text{(higher-order terms)} \end{aligned}$$

The first-order variation in ϕ is then defined to be

$$\delta\phi = \frac{\partial\phi}{\partial x}\delta x + \frac{\partial\phi}{\partial\dot{x}}\delta\dot{x} \tag{B.9}$$

and represents a first-order approximation to the change in ϕ as the locus of x changes from one curve to another. The laws of variation of sums, products, ratios, and so on, are analogous to the corresponding laws of differentiation. For example,

$$\delta(\phi_1\phi_2) = (\delta\phi_1)\phi_2 + \phi_1(\delta\phi_2) \tag{B.10a}$$

$$\delta(\phi_1/\phi_2) = \frac{\phi_2\,\delta\phi_1 - \phi_1\,\delta\phi_2}{\phi_2^2} \tag{B.10b}$$

$$\phi\int_a^b \phi\, dt = \int_a^b \delta\phi\, dt \tag{B.10c}$$

In terms of variations, a necessary condition that I should be stationary is that δI should be zero, which yields

$$\delta I = \delta\int_a^b \phi(x, \dot{x}, t)\, dt = \int_a^b \delta\phi(x, \dot{x}, t)\, dt = \int_a^b \left(\frac{\partial\phi}{\partial x}\delta x + \frac{\partial\phi}{\partial\dot{x}}\partial\dot{x}\right) dt \tag{B.11}$$

By integrating the second term by parts, we obtain as a necessary condition for an optimum, the Euler–Lagrange equation

$$\frac{\partial \phi}{\partial x} - \frac{d}{dt}\left(\frac{\partial \phi}{\partial \dot{x}}\right) = 0 \tag{B.12a}$$

and the transversality conditions

$$\frac{\partial \phi}{\partial \dot{x}} \delta x(t) \Big|_a^b = 0 \tag{B.12b}$$

The transversality conditions together with any specified boundary conditions on $x(t)$ or $\dot{x}(t)$ provide necessary boundary conditions for solving Eq. (B.12). The extension to several variables follows easily. The conditions for stationarity of the integral in Eq. (B.6) are

$$\frac{\partial \phi}{\partial x_i} - \frac{d}{dt}\left(\frac{\partial \phi}{\partial \dot{x}_i}\right) = 0 \tag{B.13}$$

with the transversality conditions

$$\sum_{i=1}^{n} \frac{\partial \phi}{\partial \dot{x}_i} \delta x_i(t) \Big|_a^b = 0 \tag{B.14}$$

If there are additional constraints on x or $\dot{x}$ of the form

$$f(x, \dot{x}, t) = 0 \tag{B.15}$$

we form the integral

$$\begin{aligned} I' &= \int_a^b [\phi(x, \dot{x}, t) + \lambda f(x, \dot{x}, t)]\, dt \\ &= \int_a^b \phi'(x, \dot{x}, \lambda, t)\, dt \end{aligned} \tag{B.16}$$

where $\phi'(\cdot)$ has the obvious definition. We can then treat the problem as one of finding the optimum of I' with respect to the functions x and λ. The necessary conditions for stationarity of I' follow from Eq. (B.11) as

$$\frac{\partial \phi'}{\partial x} - \frac{d}{dt}\left(\frac{\partial \phi'}{\partial \dot{x}}\right) = 0 \tag{B.17a}$$

and

$$\frac{\partial \phi'}{\partial \lambda} - \frac{d}{dt}\left(\frac{\partial \phi'}{\partial \dot{\lambda}}\right) = 0 \tag{B.17b}$$

Explicitly, these equations can be written as

$$\frac{d}{dt}\left[\frac{\partial \phi}{\partial \dot{x}} + \lambda f\right] = \frac{\partial \phi}{\partial \dot{x}} \tag{B.18a}$$

$$f(x, \dot{x}, t) = 0 \tag{B.18b}$$

The transversality conditions obtained from Eq. (B.12b) are

$$\left[\frac{\partial \phi}{\partial \dot{x}} + \lambda \frac{\partial f}{\partial \dot{x}}\right] \delta x \bigg|_a^b = 0 \tag{B.19}$$

These equations together form a set of equations to solve for x and λ. The extension to the n-variable case is straightforward.

REFERENCE

1. F. B. Hildebrand, *Methods of Applied Mathematics*, Prentice Hall, Englewood Cliffs, NJ, 1952.

APPENDIX C

Vectors and Matrices

We list here some definitions and properties which are useful. Details may be found in the references at the end of the appendix.

VECTOR SPACES

For a given integer n, the set of all ordered n-tuples is an n-dimensional *vector space*.

A set of k vectors $\mathbf{x}_1, \mathbf{x}_2, \ldots, \mathbf{x}_k$ is said to be *linearly independent* if there exists scalars $c_1, c_2, \ldots, c_k$, not all zero such that

$$c_1\mathbf{x}_1 + c_2\mathbf{x}_2 + \cdots + c_k\mathbf{x}_k = 0 \qquad \text{(C.1)}$$

If no such set of scalars exists, the vectors are independent. A set of vectors $\mathbf{x}_1, \mathbf{x}_2, \ldots, \mathbf{x}_k$ is said to *span* a vector space V if every vector $\mathbf{x}$ in V can be expressed as a linear combination of the vectors $\mathbf{x}_i$; that is,

$$\mathbf{x} = \sum_{i=1}^{k} c_i\mathbf{x}_i \qquad \text{(C.2)}$$

A set of linearly independent vectors which span a space V form a *basis* for V. Given a basis for V, any vector in V is expressible *uniquely* as a linear combination of the members of the basis.

MATRICES

A matrix of order $m \times n$ is a rectangular array of mn quantities with m rows and n columns. The matrix is square if $m = n$.

A matrix with only one column (row) is a column (row) vector. The *transpose* of a matrix $\mathbf{A}$ obtained by interchanging rows and columns is denoted by $\mathbf{A}^T$. The transpose of the product of two matrices is $(\mathbf{AB})^T = \mathbf{B}^T\mathbf{A}^T$. A square matrix $\mathbf{A}$ has an *inverse* $\mathbf{A}^{-1}$

$$\mathbf{AA}^{-1} = \mathbf{A}^{-1}\mathbf{A} = \mathbf{I} \qquad \text{(identity matrix)} \qquad \text{(C.3)}$$

if and only if its determinant is nonzero. **A** is then said to be *nonsingular*. The inverse of the product of two nonsingular matrices **A** and **B** is

$$(\mathbf{AB})^{-1} = \mathbf{B}^{-1}\mathbf{A}^{-1} \tag{C.4}$$

A matrix for which $\mathbf{A}^T = \mathbf{A}^{-1}$ is *orthogonal*. If $\mathbf{A}^T = \mathbf{A}$, it is *symmetric*, and if $\mathbf{A}^T = -\mathbf{A}$, it is *skew-symmetric*. Any arbitrary matrix can be expressed as the sum of a symmetric matrix and a skew-symmetric matrix as

$$\mathbf{A} = \frac{\mathbf{A} + \mathbf{A}^T}{2} + \frac{\mathbf{A} - \mathbf{A}^T}{2} \tag{C.5}$$

If **A** is an $m \times n$ matrix, the expression

$$\mathbf{y} = \mathbf{Ax}$$

represents a linear transformation that maps an element **x** of the vector space V_x into an element **y** of the vector space V_y. If **A** is a square matrix of order n, **A** transforms a vector space into itself. In this case there exist certain vectors whose direction remains unchanged under the transformation. Such vectors are the *eigenvectors* of the transformation (matrix) **A** and satisfy the relation

$$\mathbf{Ax} = \lambda\mathbf{x} \tag{C.6}$$

for some scalar λ. The scalar λ is the *eigenvalue* associated with the eigenvector **x**. Thus any nonzero vector satisfying

$$[\mathbf{A} - \lambda\mathbf{I}]\mathbf{x} = 0 \tag{C.7}$$

is an eigenvector of **A**. The values of λ for which this equation has a nonzero solution are determined by solving the *characteristic equation*

$$\det[\mathbf{A} - \lambda\mathbf{I}] = 0 \tag{C.8}$$

The nth-degree polynomial, $\det(\mathbf{A} - \lambda\mathbf{I})$, is the *characteristic polynomial* of **A**. Thus the matrix **A** has n characteristic or eigenvalues. A square matrix satisfies its own characteristic equation (Cayley–Hamilton theorem). That is, if the characteristic equation of **A** is

$$\sum_{i=0}^{n} a_i \lambda^i = 0 \tag{C.9}$$

then

$$\sum_{i=0}^{n} a_i \mathbf{A}^i = \mathbf{0} \tag{C.10}$$

An $n \times n$ matrix **A** is *similar* to an $n \times n$ matrix **B** if there exists a nonsingular matrix **P** such that

$$\mathbf{B} = \mathbf{P}^{-1}\mathbf{AP} \tag{C.11}$$

P is then said to be a *similarity* transformation.

If there are n independent eigenvectors $\mathbf{x}_1 \cdots \mathbf{x}_n$ of the matrix $\mathbf{A}$, then $\mathbf{A}$ is similar to a diagonal matrix of the form

$$\mathbf{A}_D = \begin{bmatrix} \lambda_1 & & & 0 \\ & \lambda_2 & & \\ & & \ddots & \\ 0 & & & \lambda_n \end{bmatrix} \tag{C.12}$$

where $\lambda_i, i = 1, \ldots, n$ are the eigenvalues of $\mathbf{A}$. The eigenvalues may or may not be distinct. The corresponding similarity transformation $\mathbf{P}$ is

$$\mathbf{P} = [\mathbf{x}_1 \mid \mathbf{x}_2 \mid \cdots \mathbf{x}_n] \tag{C.13}$$

A *quadratic form* $Q(\mathbf{x})$ associated with a square matrix $\mathbf{A}$ is

$$Q(\mathbf{x}) = \mathbf{x}^T\mathbf{A}\mathbf{x} = \sum_{i=1}^{n}\sum_{j=1}^{n} a_{ij}x_ix_j \tag{C.14}$$

The quadratic form associated with a matrix $\mathbf{A}$ is identical to that associated with its symmetric part.

A square matrix $\mathbf{A}$ is *positive definite* if the associated quadratic form $Q(\mathbf{x})$ positive for all $\mathbf{x} \neq 0$. It is positive semidefinite if $Q(\mathbf{x})$ is only nonnegative for all $\mathbf{x}$.

The following results hold for real, symmetric matrices:

1. The eigenvalues of any real symmetric matrix are real.
2. A real symmetric matrix is positive definite if and only if all eigenvalues are positive. Thus any real, symmetric positive definite matrix is nonsingular and its inverse is also positive definite.
3. A real symmetric matrix is positive definite if and only if its principal minors are positive (Sylvester test).
4. A real symmetric matrix $\mathbf{A}$ is positive definite if and only if it can be expressed in the form

$$\mathbf{A} = \mathbf{M}\mathbf{M}^T \tag{C.15}$$

where $\mathbf{M}$ is real and nonsingular.

5. If $\mathbf{A}$ is real, symmetric, and positive definite, there exist matrices $\mathbf{A}^{1/2}$ and $\mathbf{A}^{-1/2}$ such that

$$[\mathbf{A}^{1/2}]^2 = \mathbf{A} \tag{C.16a}$$

and

$$[\mathbf{A}^{-1/2}]^2 = \mathbf{A}^{-1} \tag{C.16b}$$

Thus a real, symmetric positive definite matrix possesses a *square root* which is also positive definite.

A real symmetric matrix $\mathbf{A}$ can always be transformed into a diagonal matrix of its eigenvalues, $\mathbf{\Lambda}$. The transformation matrix $\mathbf{P}$ is also symmetric, and orthogonal ($\mathbf{P}^T = \mathbf{P}^{-1}$):

$$\mathbf{P}^T\mathbf{A}\mathbf{P} = \mathbf{\Lambda} \tag{C.17}$$

SIMULTANEOUS DIAGONALIZATION OF TWO MATRICES

Let $\mathbf{A}_1$ and $\mathbf{A}_2$ be two real symmetric positive definite matrices. Let $\mathbf{\Lambda}_1$ and $\mathbf{\Lambda}_2$ be diagonal matrices of eigenvalues of $\mathbf{A}_1$ and $\mathbf{A}_2$, respectively. We seek a transformation matrix $\mathbf{P}$ such that

$$\mathbf{P}^T\mathbf{A}_1\mathbf{P} = \mathbf{I} \quad \text{and} \quad \mathbf{P}^T\mathbf{A}_2\mathbf{P} = \mathbf{\Lambda}_2 \tag{C.18}$$

We determine $\mathbf{P}$ as follows. Let $\mathbf{P}_i$ be the matrix of eigenvectors of $\mathbf{A}_i, i = 1, 2$. Then, from Eq. (C.17) we have

$$\mathbf{P}_1^T\mathbf{A}_1\mathbf{P}_1 = \mathbf{\Lambda}_1 \tag{C.19}$$

Pre- and postmultiplying Eq. (C.19) by $\mathbf{\Lambda}_1^{-1/2}$, we obtain

$$\mathbf{\Lambda}_1^{-1/2}\mathbf{P}_1^T\mathbf{A}_1\mathbf{P}_1\mathbf{\Lambda}_1^{-1/2} = \mathbf{I} \tag{C.20}$$

Let us define the following matrix:

$$\mathbf{K} = \mathbf{\Lambda}_1^{-1/2}\mathbf{P}_1^T\mathbf{A}_2\mathbf{P}_1\mathbf{\Lambda}_1^{-1/2} \tag{C.21}$$

Let $\mathbf{P}$ denote the eigenvector matrix of $\mathbf{K}$. The required transformation is then given by

$$\mathbf{P} = \mathbf{P}_1\mathbf{\Lambda}_1^{-1/2}\mathbf{P}_2 \tag{C.22}$$

This follows since

$$\begin{aligned}\mathbf{P}^T\mathbf{A}_2\mathbf{P} &= \mathbf{P}_2^T\mathbf{\Lambda}_1^{-1/2}\mathbf{P}_1^T\mathbf{A}_2\mathbf{P}_1\mathbf{\Lambda}_1^{-1/2}\mathbf{P}_2 \\ &= \mathbf{P}_2^T\mathbf{K}\mathbf{P}_2 \qquad \text{[from Eq. (C.21)]} \\ &= \mathbf{\Lambda}_2\end{aligned}$$

The last step follows since the eigenvalues of $\mathbf{A}_2$ and $\mathbf{K}$ are the same. Furthermore, since $\mathbf{K}$ is symmetric, we have $\mathbf{P}_2^T\mathbf{P} = \mathbf{I}$. Premultiplying Eq. (C.20) by $\mathbf{P}_2^T$ and postmultiplying by $\mathbf{P}_2$ then yields

$$\mathbf{P}^T\mathbf{A}_1\mathbf{P} = \mathbf{P}_2^T\mathbf{P} = \mathbf{I}$$

The transformation $\mathbf{P}_1\mathbf{\Lambda}_1^{-1/2}$ is usually referred to as the *whitening transformation* of the matrix $\mathbf{A}_1$.

THE MATRIX EXPONENTIAL

If $\mathbf{A}$ is an $n \times n$ matrix, the exponential $e^{\mathbf{A}}$ is defined by the convergent series

$$e^{\mathbf{A}} = \mathbf{I} + \sum_{n=1}^{\infty} \frac{\mathbf{A}^n}{n!} \tag{C.23}$$

Some of its properties are

$$\begin{aligned} e^{\mathbf{PAP}^{-1}} &= \mathbf{P}e^{\mathbf{A}}\mathbf{P}^{-1} \\ e^{\mathbf{A}+\mathbf{B}} &= e^{\mathbf{A}} \cdot e^{\mathbf{B}} \qquad \text{if } \mathbf{AB} = \mathbf{BA} \\ \det e^{\mathbf{A}} &= e^{\operatorname{tr}\mathbf{A}} \qquad \text{where } \operatorname{tr}\mathbf{A} = \text{trace of } \mathbf{A} = \sum_{i=1}^{n} a_{ii} \\ (e^{\mathbf{A}})^{-1} &= e^{-\mathbf{A}} \\ e^{0} &= \mathbf{I} \end{aligned} \tag{C.24}$$

DIFFERENTIATION OF MATRICES AND VECTORS

If $\mathbf{A} = \mathbf{A}(t)$,

$$\frac{d\mathbf{A}}{dt} = \begin{bmatrix} \frac{da_{11}}{dt} & \cdots & \frac{da_{1n}}{dt} \\ \vdots & & \vdots \\ \frac{da_{n1}}{dt} & \cdots & \frac{da_{nn}}{dt} \end{bmatrix} \tag{C.25}$$

If $\mathbf{x}$ is a vector and $f(\mathbf{x})$ is a scalar function of $\mathbf{x}$, the gradient of $f(\cdot)$ with respect to $\mathbf{x}$ is

$$\frac{df}{d\mathbf{x}} = \nabla_{\mathbf{x}} f = \left[\frac{df}{dx_1} \cdots \frac{df}{dx_n}\right]^T \tag{C.26}$$

If $\mathbf{g}(\mathbf{x})$ is a vector function of the vector $\mathbf{x}$, the Jacobian is

$$\frac{\partial \mathbf{g}}{\partial \mathbf{x}} = \begin{bmatrix} \frac{\partial g_1}{\partial x_1} & \frac{\partial g_1}{\partial x_2} & \cdots & \frac{\partial g_1}{\partial x_n} \\ \vdots & \cdot & & \vdots \\ \frac{\partial g_m}{\partial x_1} & & \cdots & \frac{\partial g_m}{\partial x_n} \end{bmatrix} \tag{C.27}$$

The derivative of a matrix function $\mathbf{F}(\mathbf{x})$ with respect to the vector $\mathbf{x}$ is the partitioned matrix

$$\frac{d\mathbf{F}}{d\mathbf{x}} = \left[\frac{d\mathbf{F}}{dx_1} \cdots \frac{d\mathbf{F}}{dx_n}\right]^T \tag{C.28}$$

NORMS OF VECTORS AND MATRICES

A norm of a vector $\mathbf{x}$ or matrix $\mathbf{A}$ is a generalization of the concept of length.

For a vector $\mathbf{x}$ the norm is a real nonnegative number $|\mathbf{x}|$ satisfying the following relations:

1. $|\mathbf{x}| > 0$ for $|\mathbf{x}| \neq 0$; $|\mathbf{x}| = 0$ if $\mathbf{x} = 0$.
2. For any scalar α $|\alpha\mathbf{x}| = |\alpha||\mathbf{x}|$ where $|\boldsymbol{\alpha}|$ is the absolute value.
3. $|\mathbf{x}_1 + \mathbf{x}_2| \leqslant |\mathbf{x}_1| + |\mathbf{x}_2|$ for all $\mathbf{x}_1$ and $\mathbf{x}_2$ (triangle inequality).

The l_p norm of a vector is defined as

$$|\mathbf{x}|_p = \left(\sum_{i=1}^{n} |x_i|^p\right)^{1/p}$$

The l^2 norm is the Euclidean norm. It can be shown that for $p = \infty$

$$|\mathbf{x}|_\infty = \max_i |x_i|$$

The norm of a matrix $\mathbf{A}$ is a nonnegative number $|\mathbf{A}|$ satisfying

1. $|\mathbf{A}| > 0$ for $\mathbf{A} \neq 0$, $|\mathbf{A}| = 0$ if $\mathbf{A} = 0$.
2. $|\alpha\mathbf{A}| = |\alpha||\mathbf{A}|$ for all scalar α.
3. $|\mathbf{A} + \mathbf{B}| \leqslant |\mathbf{A}| + |\mathbf{B}|$ for all $\mathbf{A}$ and $\mathbf{B}$.
4. $|\mathbf{AB}| < |\mathbf{A}| \cdot |\mathbf{B}|$ for all $\mathbf{A}$ and $\mathbf{B}$.

Typical norms are

$$|\mathbf{A}|_1 = \max_j \sum_{i=1}^{n} |a_{ij}| \qquad \text{the } l_1 \text{ norm}$$

$$|\mathbf{A}|_\infty = \max_i \sum_{j=1}^{n} |a_{ij}| \qquad \text{the } l_\infty \text{ norm}$$

The definitions can be extended to the case where the elements of a vector or matrix are functions of time. For example, for a vector, $\|\mathbf{x}\|_p = \left[\int_{-\infty}^{\infty} (|\mathbf{x}(t)|_p)^p\, dt\right]^{1/p}$ defines the L_p norm of the vector $\mathbf{x}(t)$. For $p = \infty$, this becomes

$$\|\mathbf{x}\|_\infty = \sup_{-\infty < t < \infty} |\mathbf{x}(t)|_\infty$$

A MATRIX LEMMA

We now prove the following useful lemma: *Let* $\mathbf{A}$ *be a nonsingular matrix and let* $\mathbf{B}$ *and* $\mathbf{C}$ *be two matrices such that* $\mathbf{A} + \mathbf{BC}$ *and* $\mathbf{I} + \mathbf{CA}^{-1}\mathbf{B}$ *are nonsingular. Then*

$$(\mathbf{A} + \mathbf{BC})^{-1} = \mathbf{A}^{-1} - \mathbf{A}^{-1}\mathbf{B}(\mathbf{I} + \mathbf{CA}^{-1}\mathbf{B})^{-1}\mathbf{CA}^{-1} \qquad \text{(C.29)}$$

Proof: Let $(\mathbf{A} + \mathbf{BC})^{-1} = \mathbf{M}^{-1}$. Then

$$\mathbf{I} = \mathbf{M}^{-1}\mathbf{M} = \mathbf{M}^{-1}(\mathbf{A} + \mathbf{BC}) = \mathbf{M}^{-1}\mathbf{A} + \mathbf{M}^{-1}\mathbf{BC} \tag{C.30}$$

Postmultiply by $\mathbf{A}^{-1}$ and rearrange to get

$$\mathbf{M}^{-1}\mathbf{BCA}^{-1} = \mathbf{A}^{-1} - \mathbf{M}^{-1} \tag{C.31}$$

Postmultiplying (C.31) by $\mathbf{B}$ and rearranging yields

$$\mathbf{M}^{-1}\mathbf{BCA}^{-1}\mathbf{B} + \mathbf{M}^{-1}\mathbf{B} = \mathbf{M}^{-1}\mathbf{B}(\mathbf{I} + \mathbf{CA}^{-1}\mathbf{B}) = \mathbf{A}^{-1}\mathbf{B} \tag{C.32}$$

Postmultiply by $(\mathbf{I} + \mathbf{CA}^{-1}\mathbf{B})^{-1}\mathbf{CA}^{-1}$ to get

$$\mathbf{M}^{-1}\mathbf{BCA}^{-1} = \mathbf{A}^{-1}\mathbf{B}(\mathbf{I} + \mathbf{CA}^{-1}\mathbf{B})^{-1}\mathbf{CA}^{-1} \tag{C.33}$$

which together with (C.31) yields (C.29).

REFERENCES

1. R. Bellman, *Introduction to Matrix Analysis*, McGraw-Hill, New York, 1960.
2. F. R. Gantmacher, *The Theory of Matrices*, Vols. I and II, Chelsea Publishing Co., New York, 1959.
3. P. R. Halmos, Finite dimensional vector spaces, *Ann. Math.*, *Study* 7, Princeton University Press, Princeton, NJ, 1948.
4. T. M. Apostol, *Mathematical Analysis*, Addison-Wesley, Reading, MA, 1957.
5. K. Hoffman and R. Kunze, *Linear Algebra*, Prentice Hall, Englewood Cliffs, NJ, 1965.

Subject Index

A

B

C

D

E

F

S

T

U

V

W

Author Index

A

B

C

D

E

F

Z